Early wooden railways

Early wooden railways

M. J. T. Lewis

London Routledge & Kegan Paul

First published 1970
by Routledge & Kegan Paul Ltd
Broadway House, 68-74 *Carter Lane*
*London, E.C.*4

Printed in Great Britain
by T. & A. Constable Ltd
*Hopetoun Street, Edinburgh, EH*7 4*NF*

SBN 7100 6674 0

Contents

Plates

Plates 12, 15, 17, 21, 23, 28, 31, 34, 35, 36, 40, 48, 50, 57, 58, 59, 64, 69, 70, 71, 72, 73, 74 are reproduced by courtesy of the Trustees of the British Museum.

Plates 16, 22, 29, 30, 44, 49, 54, 62, 63 are reproduced by courtesy of the Syndics of the Cambridge University Library.

Figures in text

Figs. 4, 5, 9, 12, 14, 18, 24, 48, 49, 52, 55 are reproduced by courtesy of the Trustees of the British Museum.

Fig. 19 is reproduced by courtesy of the Syndics of Cambridge University Library.

Preface

Most people, if asked the date and place of the first railway, would probably suggest the Liverpool and Manchester of 1830, or the Stockton and Darlington of 1825; and might throw in the name of George Stephenson for good measure. In a sense they would be right – in the sense that the Liverpool and Manchester (with its minor predecessor, the Canterbury and Whitstable) was the first public railway to carry passengers and freight with steam traction, the sort of railway (minus the steam) which we naturally think of today. But though the Liverpool and Manchester was a great landmark, 1830 is a recent date in the history of railways in their widest sense. There had been centuries of development before. But whereas the public railway has received concentrated attention from historians, its predecessors have not. Early railways lacked the aesthetic glory of the steam locomotive and the express train, the technical delights of signalling and of great engineering works, the complexities of train-operating, the immense social and economic impact of their successors. Early railways were small, self-effacing and private, being connected almost entirely with mining. They had, on the face of it, little romance and no quality of greatness. But they were the ancestors without which the Railway Age could not have been; and their contribution to industrial and technological development, though not obvious, was very real and very considerable.

This is not to say that early railways have been utterly neglected. A small but devoted band of historians has done much valuable work, led by Charles E. Lee, whose slim but meaty *Evolution of Railways*, published in 1937, first set the subject on a sound footing. This was followed up by a much-enlarged second edition and many subsequent articles from

Mr. Lee's pen; and many others have worked on particular areas or particular railways. Although early Continental railway history has remained untouched, an attempt has twice been made to produce a general, reasonably definitive, history of early British railways. These are C. F. Dendy Marshall's *History of British Railways down to 1830* (1938) and B. Baxter's *Stone Blocks and Iron Rails* (1966), neither of which is satisfactory. The shortcomings of both books lie in their inaccuracies and in their concentration on the immediate forerunners of the Railway Age, the iron railways of the early nineteenth century, to the detriment of earlier wooden ones. Both of these failings stem from the authors' limited use of primary source material.

This book sets out not to cover the whole history of railways in Britain and Europe down to 1830, but a large part of it: that part where the rails were made of wood. This period extends from the Middle Ages down to the 1790s, when the coming of the iron rail spelt the beginning of the end for wood, and it continues considerably later in so far as survivals of the old order lasted well into this century. The title, *Early Wooden Railways*, speaks largely for itself, but a word or two of explanation may be needed on what I have and have not included. The railways of the ancient world – for such there assuredly were – receive short shrift, both because they were not wooden and because they have no bearing at all on what came after. At the other end of the scale, I have dealt with the earliest application of iron to railway tracks because these lines were still basically of wood, merely covered by a layer of armour-plating. Take away the iron, and a wooden railway was left; but take away the wood, and the result was not a viable railway. The wooden railways, and even the wooden railways plated with iron, with which I deal were practically all private lines for carrying mineral traffic only. Their history I have pursued, regardless of date, to the bitter end, either until they died or until they were converted to iron. I have concerned myself only with this first generation of railways. For this reason I have not continued the story to embrace later lines like the G.W.R., whose iron bridge rails were laid on longitudinal wooden baulks, or many an American railway whose wooden rails were covered, if at all, with thin iron plates. They did indeed inherit these features from their parents, but they belong in spirit as well as in date to the younger generation of public passenger railways, and as such fall outside my scope.

If, as I hope, this book presents a fuller and more connected account of early railways than has been done before, it is not the last word. Very far from it. On the British side, I have relied far more on archives than on the printed word. But I cannot claim to have sifted through all the relevant papers there must be; and though I trust I have drawn the outline right, there are still details that need elaboration and maybe amendment. On the Continental side, my work has been much more the discovery and collation of published material than delving into archives. In case I am accused here of ignoring my own principles and relying on secondary sources, I would explain that most of this material is in fact primary – either contemporary descriptions printed at the time or subsequent publications of original manuscripts. But the need remains for somebody who has the necessary linguistic and palaeographical abilities, and above all the time, to make a prolonged tour of Continental libraries and archives, especially in Germany, Austria, Czechoslovakia, Hungary and Rumania. He would be able to advance the story many stages. No one person knows quite what manuscript evidence does survive, but without a doubt it is considerable.

On the Continent, since most mining was carried out, or at least administered, by Germans and German-speakers, the bulk of the evidence is in that language. It has set me some problems. With place-names, I have adhered to those which most commonly appear in mining documents, regardless of present-day forms. Thus the great mining town in Slovakia is always Schemnitz, the German name, not Selymecz Banya, as the ruling Hungarians called it, nor Banská Štiavnica, the native Slovak name which it enjoys now. Again, the rich gold-mine in Transylvania is referred to as Nagyag, the Magyar name most commonly applied, not as the Rumanian Săcărîmb, nor the rare German Szekerembe. Anyone wishing to translate an old name into its modern equivalent or vice versa has only to consult the Index, where cross-references will be found.

Secondly, weights and measures given in the sources belong to a multitude of differing local standards. The Saxon foot, for example, equalled 11·149 modern English inches; the Prussian foot equalled 12·356. But it is by no means always clear which standard is being used, and, since it rarely matters if the figures are misleading by a few per cent, I have simply left them as they stand without any elaborate attempts at conversion. I have made an exception in the few cases where a point of interest or importance does depend on exactitude: when I have done so, I have made it plain. The ell, however, which usually meant 2 ft., but sometimes only 1½, I have always turned into the corresponding feet and inches.

Next, the question of money. Since groschen and guilders and so forth mean nothing to most people, I have ventured to convert all German and Austrian monetary values into English terms appropriate to the period in question. This is a perilous proceeding, and such conversions should be treated with considerable latitude.

Lastly, I have assumed that the reader is not a linguist, and have translated all quotations into English, whatever the original language. A few of the most common German technical terms are mentioned to provide a starting-point for the novice who wishes to consult the sources for himself. There are a few basic words, however, which I have left in German throughout because they have untranslatable connotations: *Hund*, *Truhe* and *Riesen*, all of them types of railway vehicle. A slightly different case is the *Leitnagel*, which steered a Hund. This could perfectly well be translated as 'guide-pin', but I have left it in German to avoid possible confusion with what I term the 'guide-wheel', which belonged to a quite distinct breed of railway.

In speaking of British as well as Continental mines, I have consistently used, for the sake of simplicity, a number of words which some might feel disposed to quarrel with. 'Level', for example, signifies a horizontal, or reasonably horizontal, underground gallery or passage; 'adit' the particular sort of level which debouches into the open air rather than ends at a vertical shaft, be it used for drainage or transport or both. In coal-mines, a more special terminology is needed, which is explained as it arises. The man who pushes underground railway trucks, of whatever sort, always appears as the 'putter'. Since British mining vocabulary differed from region to region, none of these terms, strictly speaking, is universally applicable, but all can be justified for one district or another. It is preferable, it seems to me, to dispense with complete accuracy than invite confusion by using a multiplicity of local terms which all mean the same thing.

In transcribing English quotations I have adhered to the originals, except for expanding the more obscure abbreviations and, in a few instances, improving the punctuation to make them easier to read. All dates given simply as a year are in New Style; Old Style dates with day, month and year have the New Style year added if necessary.

I have at least mentioned all the wooden railways of which I am aware. But the reader who expects a precise account of the route and history of each of them is doomed to disappointment. This book is intended as a general discussion of early railways, illustrated by details culled from the individual lines. To consider each of these individual lines exhaustively and in turn would involve a quite different angle of approach, and a quite unwieldy bulk of material.

Acknowledgements

The compilation of a book of this kind would be quite impossible without the co-operation and active assistance of a large number of individuals and institutions. Such help I have received at every turn.

For permission to quote from manuscripts in their ownership or care, I am deeply in the debt of the following librarians, archivists and others:

The Trustees of the British Museum.

The Controller of H.M. Stationery Office for Crown copyright records in the Public Record Office.

The Curator of Historical Records, Scottish Record Office.

The Archivists of the County Record Offices of Cumberland, Durham, Glamorgan, Northumberland and Shropshire, of the Newcastle and Southampton Record Offices, Leeds City Libraries Archives Department, and British Transport Historical Records.

The Librarians of the National Library of Wales, Gateshead Public Library, John Rylands Library, Manchester, Newcastle City Libraries, Sheffield City Library, Shrewsbury Borough Library, the William Salt Library, Stafford, Newcastle University Library, Nottingham University Library, Swansea University Library, the Library of the Religious Society of Friends, and the Tiroler Landesmuseum Ferdinandeum, Innsbruck.

The Assay Master, Birmingham Assay Office.

Professor E. A. Vincent, Department of Geology and Mineralogy, University of Oxford.

The Council of the North of England Institute of Mining and Mechanical Engineers, Newcastle upon Tyne.

The Council of the Society of Antiquaries of Newcastle upon Tyne for the Cotesworth MSS. in Gateshead Public Library and their papers in the Northumberland Record Office.

The Council of the Yorkshire Archaeological Society.

The Rector of St. Mary's, Gateshead.

His Grace the Duke of Norfolk, E.M., K.G., for the Arundel Castle MSS. in Sheffield City Library.

The Earl of Elgin for his papers at Broomhall, Fife.

Earl Fitzwilliam for the Wentworth Woodhouse muniments in Sheffield City Library.

The Earl of Lonsdale for his papers in Cumberland Record Office.

The Earl of Wharncliffe for his papers in Sheffield City Library.

The Earl of Strathmore for his papers in Durham Record Office

Viscount Ridley for the Ridley MSS. in Northumberland Record Office.

Lord Newton, for the Legh muniments in John Rylands Library.

Sir John Clerk, Bt., for the Clerk of Penicuik muniments in the Scottish Record Office.

Lady Labouchere for her papers in Shropshire Record Office.

Mrs. John R. Greene of Aqualate for the Yarranton papers in the William Salt Library, Stafford.

Miss D. E. M. Wilson for the Wilson MSS. in Northumberland Record Office.

R. Armstrong, Esq., for the Armstrong MSS. in Northumberland Record Office.

C. J. W. Blackett, Esq., for the Blackett (Wylam) MSS. in Northumberland Record Office.

E. S. C. Curwen, Esq., for his papers in Cumberland Record Office.

P. J. Dudding, Esq., for the Carr Ellison and Cookson (Minsteracres) MSS. in Northumberland Record Office.

British Steel Corporation, South Wales Group, for the Dowlais records in Glamorgan Record Office.

Of all these, I cannot help singling out for a special word of thanks the staffs of the Gateshead Public Library, the North of England Institute of Mining and Mechanical Engineers, and the Northumberland Record Office. These three repositories contain a mass of material – the notes will testify to its volume – which was located and fed to me with great efficiency and forbearance during the long hours and days I spent working there. Without such service the labour of searching would have been infinitely harder.

My thanks are also due to the staffs of all other libraries and archives where I have worked; to the persons and bodies who have given leave to reproduce illustrations, and whose kindness is separately acknowledged in the lists of illustrations; and to the following authors for permission to quote from their books: Dr. Arthur Raistrick, *Dynasty of Ironfounders*; Dr. Edwin Welch, *The Bankrupt Canal*; Dr. P. J. Wexler, *La Formation du Vocabulaire des Chemins de Fer en France*; and Mr. A. L. Lloyd, *Come all ye bold miners*.

I am deeply grateful, too, to all those who have given me information, directed me to new sources, joined me in fieldwork, or helped me in one way or another: in this country, Frank Atkinson, D. B. Barton, Ivor J. Brown, H. Clegg, C. R. Clinker, H. F. G. Dalston, J. H. Denton, Sir Arthur Elton, J. Goodchild, Charles Hadfield, Dr. S. B. Hamilton, R. G. Hawkins, J. Hollingworth, R. J. Hunter, A. B. Jeffery, J. Langton, Charles E. Lee, J. Kenneth Major, M. J. Messenger, H. W. Paar, B. C. Roberts, M. Seymour, A. G. Sidebotham, W. N. Slatcher, L. D. W. Smith, Dr. R. S. Smith, Peter S. Stevenson, B. Trinder, C. P. Weaver and M. J. W. Willis-Fear; abroad, Bergassessor aD. H. Conrad of the Bergbau-Museum, Bochum; Oberbergrat H. Dennert of Clausthal-Zellerfeld; Dipl. Ing. G. Weiler of Berlin; Dr. G. Neubert of the Stadt- und Bergbau-museum, Freiberg; M. Badik of the Vlastivedné múzeum, Spišska Nová Ves; Križ Jozef of the Banské múzeum Dionýza Stùra, Banská Štiavnica; Miroslav Ličko of Banská Bystrica; Dr. Peter Ratkoš of the Slovenskej Akadémie Vied, Bratislava; Dr. S. Goldenberg and Dr. A. Neamţu of the Universitatea Babeş-Bolyai, Cluj.

Having been confronted at one time or another with sources in more than a dozen different languages, I have been much assisted in my linguistic difficulties by Christopher Barnes, Joe Lass, and Kurt Paulus, who helped me with Russian, Slovak and German respectively. Above all, without Michael Seymour's guiding hand in French and German I would have gone far astray. And not only in languages. I do not care to think how many hours of his time I have consumed – for me, not one of them wasted – in discussion and argument over the whole field of this book. The same holds true of John Horsley Denton, who has freely imparted his knowledge of the Shropshire railways and helped to correct an incipient bias of mine towards Tyneside. My debt to both of them is deep.

Finally, the typescript has been read with great generosity and with many fruitful suggestions by five people: John Denton, Michael Seymour, W. N. Slatcher, Professor Jack Simmons, and my father, Professor M. A. Lewis, who has been, as always, a tower of strength: I need say no more. Through their advice, many faults have been weeded out. All those that remain, in this comparatively unknown subject, are mine alone.

1

The early railway scene

Die Hundstösser werden von der Jacobiroll zu dem Sturtz
Hund Gänge laufen, der Aufseher zu den Sauberjungen,
Ihr Burschen, komt mit mir in Klementilauf. Die Berge müssen
Dort in rechtem Eilen hinauf gesaubert werden.

From *Die Klänge von der Teufe*, a traditional German-Hungarian play of sixteenth-century origin: József Ernyey and Géza Kurzweil, *A felsömagyarországi banyavárosok német népi színjátékai*, Budapest (1932), 577.

What is a railway? To most people today it means a track of steel rails which carries and guides a train of public passenger or freight vehicles, hauled by some mechanical means. Many of the constituent parts of this definition are in fact comparatively modern: the public railway goes back some 170 years, passenger traffic and mechanical traction about the same distance. The steel rail is a little more than a century old; its predecessor, the iron rail, a bare century older still. The train itself, in the sense of a string of vehicles, has no venerable roots. What then is left? The answer is a great deal. In its most basic sense, a railway is a prepared track which so guides the vehicles running upon it that they cannot leave the track. That, and no more. The vehicle need not even be wheeled, for a sledge will serve. The method by which the vehicle is prevented from leaving the track varies, the flanged wheel being but one of several. For centuries the vehicle was moved by hand, by horse-power or by gravity. For centuries, too, the only material used for the track was wood. This sort of wooden railway can be traced from the high Middle Ages almost down to the present day, although it was gradually ousted from the 1790s onwards as the iron rail, the steam locomotive and other features of the modern railway began to prevail.

Just as there is no doubt that the modern railway was developed from the mine waggonways of England, so too there is no doubt that the earliest wooden railway was evolved in Medieval and Renaissance Europe, again almost exclusively in the mines. Along with the steam engine for drainage and the blast furnace for smelting, it was one of three great inventions that made possible the Industrial Revolution, all of them stemming directly from the exploitation of mineral wealth. There were, however, a number of cases in the ancient world which fall within our definition of a railway.[1]* Many of them are controversial, either as to date or to purpose, and it is quite certain that they had no influence on what came after them. They deserve a detailed study in their own right, and need not here detain us long.

The germ of the railway might be traced back to Neolithic times, when it was sometimes the practice, in making trackways across marshy ground, to lay transverse planks upon the surface, with longitudinal beams of wood pinned along each edge of the trackway thus formed. These beams served primarily to hold the planks together, but incidentally they no doubt prevented sledges from slipping off the track. Examples have been found in Germany, Somerset and Cambridgeshire from the Neolithic to the late Bronze Age (third millennium to roughly 900 B.C.).[2]

Around the Mediterranean early rutways are occasionally discovered – parallel grooves carved into living rock or into artificial road surfaces. Usually, it seems, these ruts were worn down by traffic, not deliberately made, although the authorities disagree about the extensive system of ruts on Malta, which date probably from the second millennium B.C.[3] One case where the grooves undoubtedly were cut on purpose was the *Diolkos*, or portage for carrying ships on wheeled cradles across the Isthmus of Corinth, which was built or at least begun by Periander about 600 B.C., and remained in use until the ninth century A.D.[4] Very occasionally, too, ruts were cut into Roman roads, especially in mountainous country, to prevent traffic from slipping off the road and down a precipice. One feature common to all these early rutways was a remarkably uniform gauge: even in Malta it varied only between 4 ft. 5 in. and 4 ft. 9 in., while in later times it was constant at about 4 ft. 8 in. or 4 ft. 9 in. The obvious resemblance to our modern standard gauge of 4 ft. $8\frac{1}{2}$ in. will be discussed in a later chapter.

go & see it is X by the ship Canal

The only suggestion of wooden railways that we find in classical literature comes from the fertile imagination of Hero of Alexandria, who is well known as the first man to describe a steam engine. His date is very uncertain: the century between A.D. 150 and 250 is perhaps most probable. Like his steam device, Hero's railway was but a plaything; he had no thoughts of applying it to a more useful purpose. He had designed an 'automatic theatre', a mechanical puppet-show which moved on to the stage (and off it) under its own power, the axle being turned by a string and a descending weight inside the body. The theatre's three wheels ran on the floor, which, Hero said, must be flat and free of obstacles; 'but if the floor is not of the kind described, then level planks must be laid upon it,

* Notes appear at the end of the book, pp. 369-411.

The Evolution of Railways by Charles E Lee

along the length of which channels will be formed by means of straight strips nailed on, to keep the wheels rolling in the channels'.[5] He was, in effect, making ruts out of wood.*

If these were the prenatal stirrings of the railway, it was not born until the late Middle Ages. Greek and Roman mining had been on a small scale, relying almost entirely on slave labour. With the collapse of the Empire, mining dwindled almost to nothing, until it was revived in the German-speaking parts of Europe.[7] Iron was being dug in some quantity in the eastern Alps by the ninth century; copper, silver and lead in the Harz Mountains, the Black Forest, the Vosges and the Jura by the tenth. From the twelfth century to the fourteenth German mining techniques were spread far afield by colonists and by traders, notably to the rich silver- and tin-bearing Erzgebirge or Ore Mountains between Upper Saxony and Bohemia; to Silesia, with its lead-mines; to Lower and Upper Hungary (now eastern Slovakia and northern Rumania), well provided with gold, silver and copper; to Transylvania, with every sort of metal; to Austria – Styria, the Tyrol, Carinthia, and Carniola – rich in silver, copper, lead, iron and quicksilver (see fig. 1). The colonists are usually called Saxons, but came in fact from several parts of Germany proper and from the Low Countries. Their mines led Europe in the working of gold, silver, iron and copper – though the copper-mines of Falun in Sweden, developed under German influence, came to rival central Europe. They were not so pre-eminent in tin and lead, fields where England was not behindhand. Coal was hardly known between Roman times and the late twelfth century, but from then on it came gradually into its own, though not on any scale until the sixteenth century. Here the Liège and Mons areas, and especially Britain, were supreme.

This first great outburst of mining was in part, perhaps, responsible for the flowering of the Middle Ages. It released a mass of wealth from which sprang patronage of the arts, leisure for learning, and a curiosity that prompted the first stirrings of scientific enquiry. Silver- and gold-mines belonged to the highest authority, lay or ecclesiastical, which rarely worked them itself, but threw them open to all comers and took its royalty from the produce. Base metals generally belonged to the landlord and again were usually worked under lease by private partnerships. Mining communities sprang up from the twelfth century, some of them so rich that they

* Although so many basic inventions were made by the flourishing technical society of China – such as gunpowder, printing, canal locks – the railway was not one of them. There is, indeed, no evidence for anything in the form of a railway there until 1793, when an English embassy noted a drawbridge which was rolled across a Grand Canal lock chamber on grooved rails.[6]

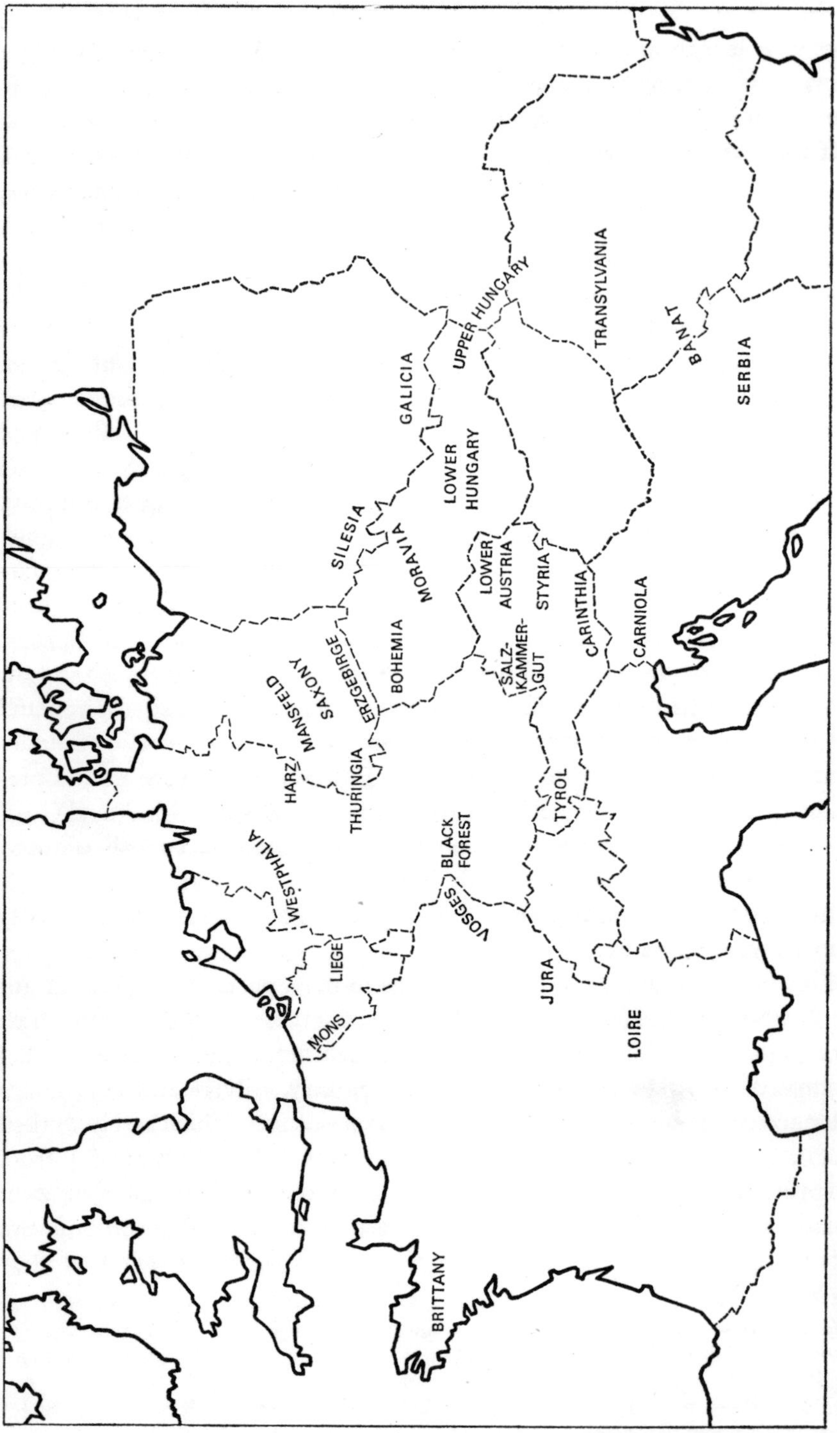

FIG I Mining areas of the Continent.

created new towns – Freiberg in Saxony, Iglau in Moravia, Schemnitz in Lower Hungary all owe their origin to mines. Wherever miners contributed to the wealth of a sovereign prince, they received privileges in return: freedom from taxes and partial self-government. European miners were pioneers and explorers – the German word *Bergmann* means *mountaineer* – and as such, and as bringers of wealth, they were essentially freemen, not slaves; their profession was respectable, not considered degrading.

Life was not, however, all a bed of roses. Mining now prospered, now declined. One such decline came in the fourteenth century, for a mixture of reasons: political troubles, an economic slump, natural disaster in the form of the Black Death, and the working out of the most accessible veins. But from the middle of the fifteenth century the mines recovered and surged ahead in a great revival parallel to the Renaissance and the new spirit for exploration that was inspiring the Continent. Great technical advances were made in Central Europe, and spread out westward as far as Britain. Before, working had been largely by bell-pits – small excavations swelling out at the foot of a shallow pit – with few proper shafts. Now shafts could be sunk deep, drained by tunnels or emptied by water-buckets raised by waterwheels and horse-gins. Occasionally mines might reach 600 ft. below the surface. Underground workings spread out horizontally, demanding a suitable form of transport that was met, as we shall see, by the railway. Power was applied to crushing and smelting the ore; machines came in for working the metal once it was smelted. All this was by no means universal: even some of the largest mines clung to old methods. At Schwaz, for example, the leading mine of the Tyrol and one of the most productive of silver and copper in Europe, the men still drained the mine in 1537 by passing buckets up the shaft by hand. But by and large the period 1460 to 1530 marked the pinnacle of European metal-mining, as it did the pinnacle of the Renaissance. This was the period of Georgius Agricola, whose great *De Re Metallica* was the first technical account of mining, and of Vanuccio Biringuccio, whose *De la Pirotechnia* was likewise the pioneer work on metallurgy. They were the contemporaries of Paracelsus, Dürer, Holbein, and Luther. In their time, the output of silver and of copper increased five-fold. In 1525 in the Holy Roman Empire, Charles V said, 100,000 men were employed in mining and metallurgy. The demand for iron grew, and with it the main iron-producing centres in France, Styria, Carinthia, Westphalia, and Spain. At the same time the reduction in timber resources led to the growing use of coal, although it was now dug on a considerable scale only in England and the Low Countries.

Once again all was not well. The good old days were passing when the mines were run by working partnerships of miners. With the increasing size of the workings and the need for machinery, costs rose and the necessary finance could come from only two sources. The prince or landlord might supply it from his own wealth, or great merchant financiers like the Fuggers and the Haugs might step in. Some mines – in the Harz and Lower Hungary, for instance – became State-owned; some fell into the hands of the merchants, like others in Lower Hungary and in Austria. In either case, the miner became a wage-earner paid by a master, lost many of his privileges, and was reduced to forming guilds and even, occasionally, striking. But he retained much of his old dignity; apart, that is, from the coal-miner, whose dirty occupation was never blessed with the romance that surrounded metal-mining. Red tape and a top-heavy hierarchy of officials grew up, until every trifling scheme needed approval from the top, and rules and regulations proliferated.

Warfare and religious troubles further damaged the mines: much of eastern Europe was lost to the Turks, and the Church became unwilling to invest in mining. In particular, the new discoveries of metal in South and Central America hit the European industry hard. The great Potosi mines in Bolivia, found in the 1540s, and others in Peru provided vast quantities of silver with cheap slave labour. At the same time, some larger Continental mines were becoming worked out. The seventeenth century in general was one of decline, and though there was a resurgence in the eighteenth, especially in parts of eastern Europe recovered from the Turks, the Continent never recovered its former mining glory. Its initiative in technology had been handed over in the seventeenth century to Holland, Sweden, and especially to Britain, all countries protected from the religious struggles and, in the main, from actual warfare. In Britain after the Reformation landlords had leapt at the chance to exploit the mines. These islands are not, on the whole, comparable to the Continent in metalliferous wealth, so that it was on the foundation of coal that the Industrial Revolution was built.

This, then, was the background to the earliest Continental railways. Ore had long been transported underground on wooden trays pulled by rope along the floor; or on sledges; or in wicker corves, or baskets, and leather sacks, carried in the arms or on the back; or, most frequently, in common or garden wheelbarrows. They sometimes ran on the bare floor, but usually the level or adit served for drainage as well as for transport, in which case the drain would be laid down the centre of the level and planked over to form the *Tragwerk*, or wooden floor, for transport. When four-wheeled trucks first appeared, they ran on this *Tragwerk*, no doubt at first without any form of guidance for their wheels.

When the first true railways appeared, wooden rails were laid on the

Tragwerk. They were known as *Gestänge*, a term that was also applied to pump-rods and the like. The trucks were pushed, almost always by hand, from the vein where the ore was hewn, or from a point not far away, along the more or less horizontal galleries to the main shaft, to the winze or subsidiary shaft, or to the adit mouth, depending on the individual mine. If the railway ran straight to the main shaft, the ore was transferred there from truck to kibble or bucket and raised to the surface by the horse-gin; in a winze the kibble was raised by windlass only to the level above, where the ore was again tipped into a railway truck. If the railway ran into the open from an adit, as often happened in hilly country, it usually ended on a raised timber staging or 'bridge' just outside the adit mouth, where the ore was tipped on to a heap below. As the stamps for crushing the ore and the furnaces for smelting it needed power, they had to be on or near a river or stream, which might or might not exist close to the shaft or adit. If the works were nearby, the ore was normally carried there in wheelbarrows; if distant, on carts or packhorses. Rarely did a railway extend any distance from the adit-mouth, and even more rarely was there a separate railway from the shaft-head to the waste tips or the stamps.

The primary purpose of the railway was to carry as much ore as possible as quickly as possible along the narrow levels which were liable, with less efficient transport, to form a bottleneck between the production of ore at the vein and its processing at the works. On the surface it was a different matter. Mountainous terrain often limited the usefulness of a railway. Moreover, if a small truck of underground size was run on the surface, it could not compete in efficiency with an ordinary cart; and the advantages of enlarging the railway truck to a useful capacity for overland transport very rarely occurred to Continental engineers. Though the railway may perhaps have been born just before the great slump of the fourteenth century, it was not until the early sixteenth, with the peak of mining activity, that it came into its own. From the seventeenth century onwards the European railway showed little technical development, although it was widespread geographically from Brittany and England in the west to Russian central Asia in the east. It was the miners' preoccupation with underground transport that prevented it from developing beyond its original purpose; and in the end it died out, leaving no permanent mark on the history of technology. The English coal railway, on the other hand, was essentially for surface use, and was capable of almost infinite expansion.

The generic name for a four-wheeled mining vehicle was *Hund*, a word about whose origin there has been much controversy. In German, it simply means 'dog'. The usual explanation, current ever since 1556, when Agricola wrote, is that 'some people call it a dog because when it moves, it utters a noise which seems to resemble the barking of dogs'. He gives the same etymology for another device that he calls a *Hund*, the travelling cradle on an overhead crane.[8] Others have supposed the truck was so called because, doglike, it accompanied the miner everywhere.[9] In fact, though there can be no certainty, a more prosaic derivation seems likely: from the Czech and Slovak *honit*, meaning to chase or to drive fast, akin to the Old Czech *hyntow* and the Magyar *hintó*, meaning a coach or carriage.[10] Indeed one of the earliest uses of the word in a mining context is in a Czech document of about 1486. A number of early German sources spell it *Hunt* (as do several more modern ones). In those days, before spelling was standardized, this meant little, since *Hunt* and *Hund*, as well as *Hundt*, which is sometimes found, would be pronounced identically. But the deliberate alternative spelling '*Hunte* or *Hunde*' (in the plural) given by a document of 1553[11] does suggest that the truck and the dog were not yet fully equated. The word, along with many other German mining terms, was adopted into a number of European languages: *Hund* in Swedish; *Hunt* in Czech, Slovak and Polish; *Hant* in Magyar; *Hanta* in Serbo-Croat and Turkish. The French, when they did not use *chariot de mine*, translated *Hund* into *chien*, and the Russians likewise turned it into *sobaka*. German influence on English mining, which has left a number of technical terms in our language, was mostly too early for *Hund*. When Englishmen described German examples, they usually used the German word, or simply called them 'waggons'. Many German compounds also appear, such as *Tagehund* (surface Hund) as opposed to *Berghund* or *Grubenhund* (mine Hund), or *Auslaufhund*, the Hund that ran out of the adit. The workman who operated the Hund was usually known as the *Hundstösser* or, especially in the Harz, as the *Hundläufer* (Hund-pusher or runner) – a word that will be translated here as *putter*.

Another name was also in common use, *Truhe* (literally box or chest), with its variant spellings, such as *Truche*, *Trocke*, *Trüge*, *Trugge* and *Druhe*. Its putter was the *Truhenläufer*. Whereas the Hund was always wheeled, the Truhe seems to have begun life, as its name suggests, simply as a rectangular box for ore. If it was carried underground, it rested on a two-wheeled barrow framework called the *Wagen*,[12] and was hoisted bodily up the shaft. The laws of Rammelsberg in the Harz mention underground *Wagenknechte* or waggon-men in 1476,[13] and the accounts of Neusohl copper-mines in Lower Hungary for 1535 record *Truhenläufer* and *Hundstösser* as distinct categories of workmen.[14] It was particularly in use for carting ore around the smelting works themselves – at Neusohl, for example, in 1560,[15] and at Eisenerz in Styria as late as 1788.[16] *Truhe* could also describe the body of a road waggon. But as early as 1533, in many places, *Truhe* had become an exact synonym for *Hund*, and pictures of Truhen show that they were of the same design as Hunds. Austria, and especially the Tyrol, hardly used the word *Hund* at all; in most other areas both were applied, with *Truhe* becoming rarer towards the north, until in Silesia, Saxony, and the Harz it was quite unknown.

When did the Hund and Truhe first appear? It has been said that the railway goes back to the twelfth century,[17] but there is no basis for

FIG 2 Miner pushing truck, Black Forest, *c.* 1350 (Freiburg Cathedral).

this claim. There is in the Minster of Freiburg-im-Breisgau a stained-glass window presented about 1350 by Johann Snewlin der Gresser, owner of the nearby Schauinsland mines. At the foot of two of the lights are small scenes showing miners picking at the rocks, and in the third is a miner pushing what appears to be a box-like vehicle, which cannot be a barrow or a sledge (fig. 2). Only its back is shown, so that we do not know what the wheels were like, nor whether it ran on rails.[18] The laws promulgated by Queen Elisabeth for the Styrian mine of Schladming in 1408 laid down penalties for anyone 'at the mine who takes any rails [*gestenge*] or wooden

B

boxes [*holtzladen*] or tools or anything that is not his own'.[19] But we cannot be sure that the rails actually guided a Hund. The earliest use of the word *Hund* that the writer has found occurs in the laws drawn up in 1427 by Duke Friedrich IV for Gossensass, a lead- and silver-mine opened in 1423 just on the Italian side of the Brenner Pass, but which was then in Austria. The weekly pay for the workmen was to be 'for a foreman, 1 ducat; for a hewer, 1 Rhenish guilder; for a *huntstosser*, 26 kreuzer; for shovellers and sorters, 22 kreuzer each'.[20]

Further east, a tradition was handed down at Königsberg, one of the seven mining towns of Lower Hungary, that 'the mines of Königsberg formerly yielded immense profits, so that the city once made a present to King *Matthias* of a *dog* (as the little carriage used in the mines is call'd) full of ducats'.[21] Matthias Corvinus was King of Hungary from 1458 to 1490. The mine was drowned in the late 1500s, but this tradition was not recorded until 1730. In Bohemia, a code of laws drawn up by King Vladyslav II about 1486 for Kuttenberg silver-mine, some 40 miles east of Prague, ordains that the ore be brought for sale to the works in Truhen and Hunds (*na truckach i na huntjsch* in the Czech original).[22] In all these early examples we are left quite in the dark about the form of the Hund and its means of guidance, if any. It is not until very shortly before 1500 that light begins to dawn.

2

The Leitnagel Hund

Die Truchenlauffer lauffen
Das Ertzt auf Thalben aus,
Dörffens weiter nicht kauffen,
Kommt der Hutmann heraus.

Eisenerz song of 1655, 'Freut euch, ihr Bergleut' alle',
quoted by Matthias Abele, *Continuatio Metamorphosis Telae Judiciariae*, Nürnberg (1684), 35f.

By far the most common type of Hund ran on two parallel planks or *Gestänge*, set close together, but not quite touching. Into the slot thus formed between them fitted the *Leitnagel*, or guide-pin, an iron bar which projected down below rail level from the base of the Hund's body near its front end. How, where and when this exceedingly simple device was first invented we shall probably never know. A Hund of some sort, as we have seen, was in use at Gossensass in 1427; and if the Slavonic etymology for the word is accepted, then Hungary – always an inventive region – may have been first in the field. But early Hunds were not necessarily guided. We cannot be certain that the Leitnagel was in use before the middle of the sixteenth century, although it had very probably appeared before the end of the fifteenth. The first areas to adopt it, as far as we know, were the Saxon Erzgebirge, the Vosges, the Tyrol and Lower Hungary. But, so uncertain are the surviving records, we cannot tell which deserves the credit for its invention.

The mines which eventually invested in the Leitnagel Hund were numerous: too many to discuss in detail or even to list individually here. They may be found on the map in fig. 3. Our knowledge of the Hunds they used comes from a variety of sources: incidental references in early mining laws, inventories, accounts, and ever more detailed descriptions in the countless technical books and periodicals of the eighteenth and nineteenth centuries. As a result, our knowledge is much fuller for recent times than for old. These details will be discussed in due course; meanwhile, let us look at the scattered and fragmentary evidence for Hunds down to 1600. In some cases it is made plain that the Hunds were guided by a Leitnagel. For the rest, it has to be inferred.

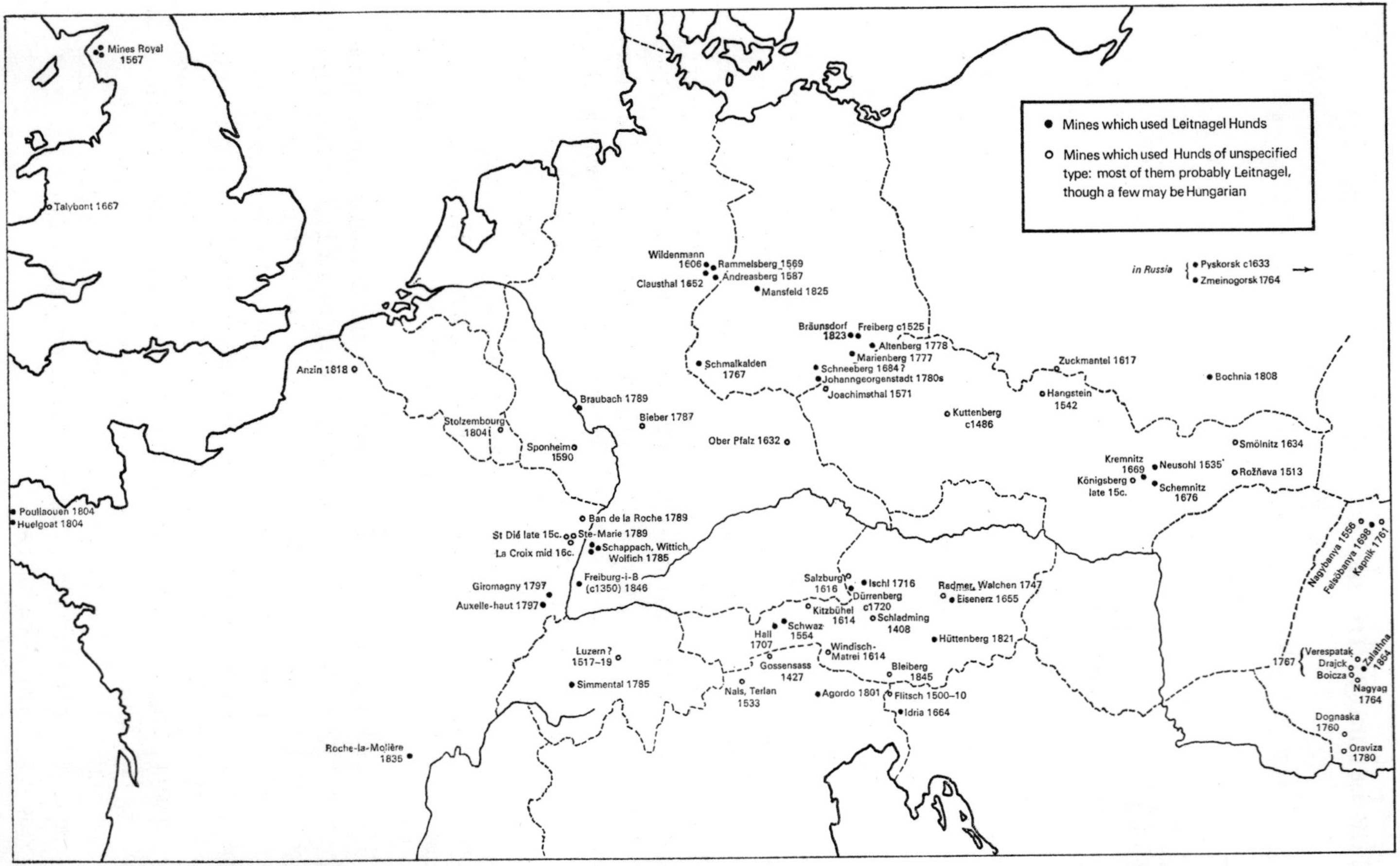

FIG 3 Distribution of Leitnagel Hunds. The dates are those of the first record of Hunds at each place. They are not necessarily the dates when Leitnagel Hunds were introduced.

Starting in the east, and working in an anti-clockwise direction, we come first to Upper Hungary. An inventory, in Latin larded with Magyar and German, of the Nagybanya estates of the Count of the Imperial Treasury was drawn up in 1556. Among the mining and smelting items occur '5 measures or vehicles called running-barrows [*futo par*]', and '6 barrows of ore-bearing mud from the Royal mine'. These may merely be wheelbarrows that run, as opposed to hand-barrows that are carried. But two more items record '8 iron axles belonging to the Hunt carriage [*Was tengel hunt hoz walo*]' and '14 iron discs belonging to the barrows or push-Hunts'.[1] Hereafter no Hunds are recorded at Nagybanya until 1698.[2]

At the other end of Upper Hungary there is in the Cathedral of the mining town of Rosenau (Rožňava) an altar-painting of the Virgin and Child surrounded by mining scenes, dated 1513 (pl. 1). At one point a miner is emerging from an adit, pushing a tall Hund on small wheels, which are clearly running on planks set close together in the manner of Leitnagel Hund rails. Between the rails there flows a drain.[3]

The Hund was early established at the flourishing copper-mines of Herrngrund, near Neusohl in Lower Hungary. As we have noticed, there were Hund-putters and Truhen-putters here by 1535, both finding frequent mention thereafter. An inventory of 1546 lists 94 Hunds and 48 spare rails (*Gestänge*) for repairing and lengthening the track. In 1581 there were 74 Hund-putters.[4] Königsberg, another of the nearby mining towns, was apparently using Hunds in the late fifteenth century (p. 10).

Moving westwards, we find Truhen-putters at Hangstein in Moravia in 1542,[5] and Hunds at Kuttenberg in Bohemia in about 1486 (p. 10), and again in 1570, when 60 Hunds deposited their loads of ore at the main shaft every hour.[6] In the Erzgebirge, both on the Saxon and the Bohemian sides, the Hund was evidently widespread by the mid-sixteenth century. The first hint is supplied by a carved wooden pediment on a house in Freiberg Obermarkt, which, among other mining activities, depicts an underground Hund, though the work is too small to show any details.[7] The great work of Georg Agricola, *De Re Metallica*, gives us our first comprehensive description of a Hund and its Leitnagel (p. 19 and figs. 4 and 5). Agricola (1494-1555) practised as a physician at Joachimsthal, the largest mining town on the Bohemian slopes of the Erzgebirge, and in 1533 moved to Chemnitz on the Saxon side, where he spent the next twenty years collecting material for his *magnum opus*, which was published posthumously.[8] The delightful illustrations come mostly from the pen of the artist Rudolf Manuel Deutsch. Though Agricola was not unacquainted with other mining areas, his book naturally deals in the main with Erzgebirge practice. To him, the Hund was clearly no novelty. The sermons of

Johann Mathesius, pastor at Joachimsthal shortly after Agricola's time, are enlivened by sidelights on the history and life of the mines there. He several times mentions Hunds and their rails, and uses the phrase '*den Hund anhangen*', to hang on to the Hund, meaning to do no work.[9]

In the Harz Mountains the early accounts of Rammelsberg are very vague about the vehicles used for underground transport until 1569, when a Hund-putter is first mentioned, and by 1577 detailed costs are entered for building Hunds.[10] At Andreasberg nearby a Hund-putter first appears in 1587.[11] The mining laws of Sponheim, south-west of Mainz, laid down in 1590 that 'the adit shall be no wider than is necessary for two Truhen running on the rails to be able to pass close to each other'.[12] Double track was distinctly rare in such cases.

When we reach the Vosges, we find the earliest hint of the Leitnagel Hund. Late in the fifteenth century the Chapter of St.-Dié commissioned a fine illuminated gradual or anthem-book, one of whose pages is decorated around the edge with a series of paintings of the local silver-mines at work.[13] Amongst them are two scenes (pls. 2 and 3) depicting Hunds running in and out of the mine upon parallel planks with the suggestive narrow gap between. Both Hunds are seen from an angle that would hide the Leitnagel, if it existed. Not long afterwards – somewhere between 1532 and 1562 – an artist named Heinrich Gross drew a series of admirable sketches of the silver-mines of St.-Nicholas at La Croix-aux-Mines, some five miles south-east of St.-Dié.[14] One drawing (pl. 4) shows the underground workings and the ore being raised up small winzes in buckets. Rails are laid on the lowest level, though evidently not in use, but on the two higher levels Hunds are being pushed, emptied and loaded. In a sketch of the adit mouth (pl. 5), one putter is bringing his Hund out of the mine, and two more have tipped theirs bodily sideways to empty out the contents for the ore-breaker and washer to deal with. In a third sketch of the same place (pl. 6) a shift is just going underground, with one putter wheeling his Hund up to the rails to join his three fellows at the back of the procession. In parenthesis, if the reader is tempted to identify the hooded miners in these early drawings with Snow White's seven dwarfs, he does not err. In German folk-tale they really were miners.

Once again, in Gross's drawings, the rails are identical to those used for Leitnagel Hunds, but no Leitnagels are visible. Did Gross's and the St.-Dié Gradual's Hunds have them? It is tempting to assume they did, if only because of the rails and of the insouciance with which the putters are handling their Hunds, as if they had no steering problems. On the other hand, three Hunds in Gross's drawings have been run several feet off the rails. A Leitnagel Hund could not be so treated, since the pin would rest on the ground rather than the front wheels, unless the front of the Hund was raised slightly in the air by pressing down on the back – a practice not unknown with Leitnagel Hunds elsewhere (pl. 11). The question must be left open. But if rails of this form preceded the Leitnagel itself, it is easy to see how the Leitnagel came to be devised.

Further south again, an ink-and-wash drawing by Hans Holbein the Younger (pl. 7) shows a mining scene set in Alpine country, with workmen driving wedges, climbing ladders and, on the right, pushing a Hund out on to a wooden staging. Holbein (1497-1543) spent almost all his working life in England or in Basel. This drawing is certainly of nowhere in England, and the scenery seems too Alpine for the Basel area. Its most likely date is 1517-19, when Holbein was in the Lucerne area.[15]

In Austria, and especially in the Tyrol, the Hund was in early use. We have met the Huntstösser at Gossensass in 1427, and though there is no more evidence for the fifteenth century, the sixteenth is full of it. In 1517, for example, Maximilian I's mining laws for Austria, Styria, Carinthia and Carniola include the Truhen-putter.[16] An expert from Gossensass, reporting in 1533 on the Nals and Terlan mines (now in the Italian Tyrol), said that the old levels were so badly driven that Truhen could no longer be used for transport, and that the rails were split and neglected and the timbering dangerous.[17] The pride of the Tyrol, however, was the great silver and copper-mine of Falkenstein at Schwaz on the Inn below Innsbruck. Begun in 1409, it had reached its pinnacle of wealth and fame by the mid-sixteenth century: in 1554 it employed 750 Truhen-putters, in 1556 no fewer than 1,132.[18] In this last year the *Schwazer Bergbuch* was compiled, a historical, legal and technical account of the mine, which has survived in eight manuscript copies.[19] It is in some respects the Austrian counterpart of Agricola's Saxon treatise, which was published in the same year as the *Schwazer Bergbuch* was written. But the *Bergbuch* describes and illustrates the Hund itself – or the *Berg Truhe*, as it was always called here – in a great deal more detail, as will appear in due course (pls. 8-11). Lastly in Austria we come to an altar triptych of the first decade of the sixteenth century, painted for Flitsch in Carinthia.[20] In the background, behind St. Christopher and St. Daniel, a Hund is emerging from an adit on to a raised wooden staging, again running on rails which appear to be designed for a Leitnagel.

Having completed the sixteenth-century circuit of the Continent, we must turn to the Leitnagel Hunds (or rather Truhen) imported by Austrian miners to the English Lake District in the 1560s. They deserve a detailed study – more detailed than can be given here – because they were the first railways in Britain, although they had no effect whatever on later developments, and because their existence, though suspected, has never hitherto been proved. The evidence was published many years ago by W. G. Collingwood, but its significance has been missed.[21]

The history of the Lake District mines is briefly this.[22] Continental experts had for long been prospecting and even occasionally working mines in Britain; but by Tudor times we were lagging far behind Europe in technical knowledge and in production of metals. The early Tudors made attempts to rectify this, but with little success. It was only in 1564 that a real start was made. In that year Elizabeth, advised by William Cecil, granted letters patent to Thomas Thurland, a cleric with an interest in mining, and Daniel Hochstetter of Augsburg, a mining and smelting expert whose father had been employed by Henry VIII. They were empowered to mine gold, silver, copper and quicksilver in eight northern and western counties and in Wales, paying royalty to the Queen and employing, if they desired, foreign labour. The finance was to be provided by shareholders, about half the shares being held by or for the merchant firm of Haug, Langnauer and Partners of Augsburg, who had business interests in Schwaz; and in 1568 the Company of Mines Royal was chartered, the first English manufacturing company.

The first batch of miners arrived in 1564 from Gastein in the Tyrol, followed by many more over the next twelve years, until there were at least 150 German-speakers working alongside local men. The home town of many of the Germans cannot be traced, but apart from the first lot of twelve from Gastein, one came from Joachimsthal, one from Freiburg-im-Breisgau, one from Radstadt in Styria and at least twenty-three from Schwaz. They set up a smelting house at Keswick, and worked copper at three principal mines: Goldscope (a corruption of *Gottes Gabe* – God's Gift) in the Newlands Valley just west of Derwentwater, begun in 1566; Caldbeck to the north of Skiddaw, seriously begun in 1568; and Grasmere, begun in 1566, but closed in 1571. Every aspect of the work is redolent of German techniques.

The accounts record several items of expenditure on underground transport – the vehicles were home-made – some of which Collingwood did not understand and therefore mistranslated. At Shrovetide, 1569, we find for Goldscope mine: 'New iron-ware, tallow, sledges [*Schlepp trög*] and other mining implements, carriage from Smelthouses to Newlands, 15/9'. And 'Wolff Prugger, 38 days at 8*d.*, cutting boards for *berg truchen* and sledges, £1 5*s* 4*d*.' Though Collingwood translates *berg truchen* as 'water-courses', it quite certainly means Hunds. The *Schwazer Bergbuch* used the identical term for a Leitnagel Hund, and carpenter Prugger had come to Cumberland from Schwaz in 1567. The Whitsuntide accounts for 1569 record a local builder 'J. Scot making *berg truchen* and laths'; the Hallowmas accounts '51 loads wood for rails [*Gesteng*] 8/6'. English labour was employed for 'dragging ore' underground, at 5*d*. or 6*d*. a shift. Again, at Christmas 1569 the Caldbeck accounts mention 'Taking *berg truchen* and sledges from Smelthouses, 1/-'.[23]

The mines never prospered – most of their life they lost heavily – and German interest waned, especially after Haug and Langnauer went bankrupt. The Germans in the Lakes mostly stayed there, but no new blood came in. From 1580 the Company ceased working the mines themselves, but leased them out, at first to Daniel Hochstetter and Thomas Smythe. An English inventory of the surface buildings, not the mines themselves, taken in 1586, lists the following items, which presumably represent Hunds or *berg truchen*:

> Newlands Mynes furnyture . . . j hand hatchet for bergge trowghes . . . ij small rowle wagons bounde with Iron to bring Ewers [ores] and sparre forthe of the mynes. . . .
> Cawldbeck mynes furnyture . . . Stampinge howse at that place . . . ij small rowle wagons for the mynes. . . .
> Gresmyre Mynes furniture . . . The Black smithie . . . one small rowlle wagon serving for within the mynes.[24]

Things went from bad to worse, as the Germans died off, techniques were forgotten and fuel ran out. The mines struggled on until the Civil War killed them.

The Company, however, remained in existence, exploiting its monopoly of silver and copper in Wales and elsewhere. They farmed out the silver-mines behind Aberystwyth first to Thomas Smythe, who was followed by a succession of lessees. Among them were Sir Hugh Myddleton and Thomas Bushell, who was advised by Joseph Hochstetter, grandson of Daniel, and who employed German labour.[25] By 1667 one of the largest mines was Talybont. An inventory of that year is quoted by Sir John Pettus in his *Fodinae Regales*[26]:

> Materials at the Mines Royal . . . At *Tallabont* . . . one Beam and Scales with Weights for Oare, one great Trewerne with Iron wheels to carry out deads belonging to the Addits, as also Pumps, Timber, Ropes. . . .

What was this great Trewerne? Is it not an English spelling of *Truhen*? *Truhen*, however, is usually plural (though not always), the singular being *Truhe*; but in the corrigenda at the end of Pettus's book Trewerne is amended, not to Trewe, as we might expect, but to Treve. Can this be another mistake?

To return to the Continent. In the seventeenth century, a period of stagnation, the Leitnagel Hund spread little, the most remarkable export being to Russia. In 1633 Russian prospectors began a copper-mine on the upper Kama, at Pyskorsk near Solikamsk on the western slopes of the Urals. They used German labour, but after a few decades the mine was worked out. Then in 1722 the director of mines in the Urals, in opening up the old mine, discovered old lamps and implements and 'half of a mining truck like those used in the mines of Saxony'.[27] Later still the Leitnagel Hund moved on as far as Russian central Asia. Kuz'ma Dimitrievich Frolov, the son of a humble worker in the Urals (where strong ties were maintained with Saxony), moved in 1761 to Zmeinogorsk mines in the Altai, where in 1764 he introduced the Leitnagel Hund, known here as *sobaka*, or dog, to the smelting works. An endless chain, driven by water power, pulled the Hunds up a slight incline from the ore-chute to the stamps, from which they returned empty by gravity.[28] Frolov also used an underground Hund, but that was guided in a different way (see p. 53).

During the eighteenth century European mining looked up, and Leitnagel Hunds became much more thick on the ground, or, rather, under the ground, although they spread little outside the area they had covered in the sixteenth century – from the Vosges to Upper Hungary and from the Harz to Carniola. The only straggler was Zmeinogorsk in Russia. Likewise, in the first half of the nineteenth century there was little expansion. The Leitnagel Hund appeared in the lead-mines of Poullaouen and Huelgoat (Ille-et-Vilaine) in Brittany perhaps as early as 1735, when some of the directors and miners were Germans, but certainly before 1804.[29] It also found its way to Roche-la-Molière Colliery near St.-Etienne (Loire).[30] Hunds were very rarely used for coal – they are known in collieries only here, in Simmental in Switzerland, and at Mansfeld near Eisleben,[31] though degenerate forms were used elsewhere (p. 56) – because coal is so light a material compared with most ore, and is dug in such quantities, that a normal-sized Hund was hardly efficient. The Leitnagel Hund was also introduced into the salt-mines at Bochnia in Galicia[32]: a more suitable use, but again a rare one, being paralleled only at Hall near Innsbruck and Ischl and Dürrenberg near Salzburg[33] (pl. 12).

In some places the Leitnagel Hund had been completely driven out by the Hungarian Hund before 1800; elsewhere, after the 1820s, it fought a losing battle against the English railway. By the 1860s it was still far from unknown, and it seems to have expired in the 1920s or even 1930s.[34] The Hund had evidently found its way into the Balkans, for the word appears in

Serbo-Croat mining vocabulary, as we have seen, and even in Osman mining codes of as early as 1536.[35] But no specific use is known in those parts. There is no record whatever of Leitnagel Hunds in Scandinavia, nor of Hunds of any kind in Spain or in Italy, except in the very north, next to Austria. The Hund did not leave Europe, either, except for Russian central Asia. It is specifically stated that no Hunds were to be found in the Spanish mines of Central and South America.[36] Slaves were cheap enough.

Three early descriptions of the Leitnagel Hund deserve quoting in full. The first is that by Agricola, published in 1556:

> If the lumps of rock or earth are carried out in wheelbarrows, planks joined together are laid upon the sleepers; if in open trucks [Hunds], then two timbers 9 in. thick and wide are laid on the sleepers, and where they join they are usually hollowed out so that the iron pins of the trucks can move as if in a rut. Indeed, these pins prevent the trucks from straying from the groove or rut, either to the right or to the left. . . . The open truck has a capacity half as large again as a wheelbarrow. It is almost 4 ft. long, and about 2½ ft. wide and high. As its shape is rectangular, it is strapped with three rectangular iron bands, and is also protected by iron straps on each side. To its bottom, two iron axles are fixed, on whose ends there revolve wooden wheels which are prevented from falling off the fixed axles by small iron keys; just as the truck is prevented from straying from the groove or rut by the large blunt pin fixed to its base. To push the truck out when loaded with excavated material and to bring it back when empty, the putter holds the back of it with his hands and pushes. Some people call it a dog because when it moves it utters a noise which seems to resemble the barking of dogs. They use this truck when they carry loads out of the longest adits, both because it moves more easily and because it can hold a heavier load.[37]

Fig. 4 shows Agricola's Hund, *C* being a detached axle with linchpins *E*.

The second, fuller, description comes from the *Schwazer Bergbuch* of 1556.

> When the work or adit has been driven many fathoms into the mountain, the floor of the adit is laid as far as the working place with two rows of planks, each about 3 fathoms long, or longer or shorter as necessary. They are always set close to each other, separated by about 1½ in. On them one walks into and out of the mine, and transports the rock and whatever else is necessary. They are called the rails [*Gestenng*].

A *Pergtruchen* [*Bergtruhen*] consists of a chest, about 3 ft. long, 2 ft. high and 8 in. wide. Underneath is fastened a rectangular beam running the length of the Truhe and projecting a hand's breadth at each end. On the front end the lamp is set. This Truhe is strapped with iron bands, so that it weighs, including the iron and wood, about 50 Viennese pounds. The Truhe has two iron axles nailed to the rectangular beam underneath, and on these axles are four wheels called rollers. These rollers are made entirely out of wood cut across

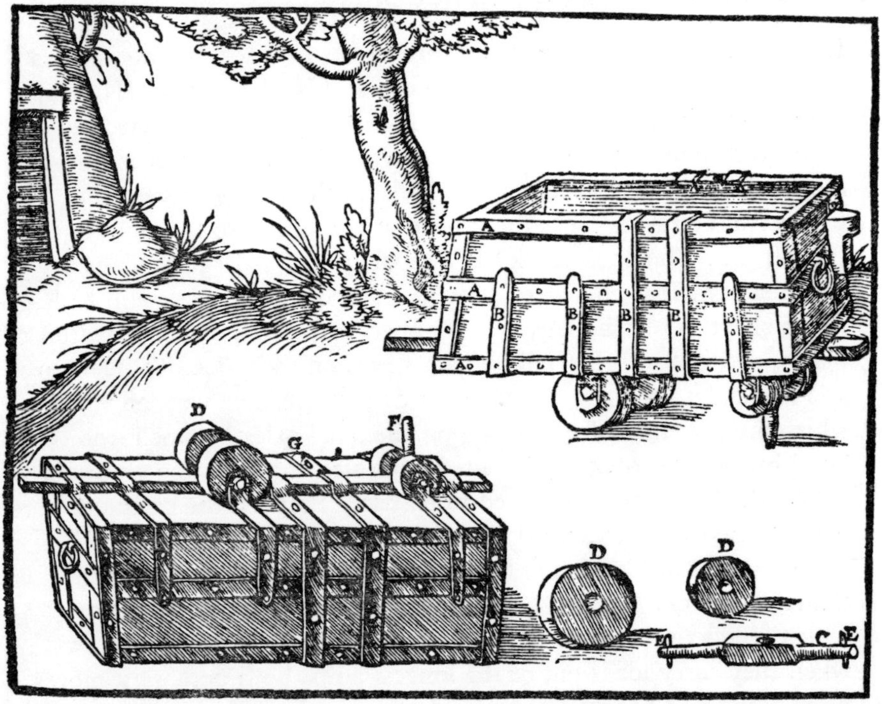

FIG 4 Leitnagel Hund, Erzgebirge, 1556 (Agricola).

the grain, tyred with iron hoops and pierced by iron bearings to rotate on the axles. The front two rollers are smaller, and the back two are larger, nearly 9 in. high. These two larger back wheels have to carry most of the weight of the Truhe. With these Truhen the rock and other material is run and carried out of the mine to the surface. The cost is 12 pounds Bernese or 2 Gulden 24 Kreuzer.

Leitnagel. This belongs to the front of the Truhe, is entirely of iron, a good 9 in. long and rectangular, and is wrought into a point at the end where it is driven into the Truhe. At the other end is a revolving iron ring about 1½ in. high. This pin is driven into the

FIG 5 Leitnagel Hund on 'bridge', Erzgebirge, 1556 (Agricola).

> Truhe about 1½ in. in front of the rollers. When the Truhe is in use, the Leitnagel reaches down between the two rails and holds the Truhe on them so that it cannot fall off the rails. If the Truhe has no Leitnagel it cannot be used or kept on the rails.
>
> A *Laner* [literally wire] is a small flat iron nail with a plain head that is stuck in the Truhe axle outside the rollers, to prevent the rollers falling off the Truhe. A washer is made of flat iron, about 1½ in. across with a hole through it about ¾ in. across. It is used on both axles of the Bergtruhe, outside the rollers, so that their friction is reduced and they revolve more easily.

Pl. 8 shows a *Pergtruhen*, according to the *Schwazer Bergbuch*, standing off the rails.

The third quotation is the earliest description of a Hund in English, written by Edward Browne, son of Sir Thomas, who visited the gold-mines of Kremnitz in Lower Hungary in 1669.

> They carry their Ore under ground from one place to another; or to the bottom of the Pit whence it is drawn out, in a Box or Chest which they call a Hundt or Dog; this runs upon four wheels, is higher behind than before, and hath a tongue of Iron at the bottom, which being fitted into a channel of wood framed in the middle of the bottom of each passage, it can no ways deviate, but keeps allwayes in the middle; and by this means a little Boy will run full speed with three or four hundred pound weight of Ore or Earth before him, wherever you command him, without any light, through those dismal dark passages of the Mine: and it was very new to me to hear the rattling they make in the Mine, and the alteration of the sound as they are nearer or further from us; and to see them come with that swiftness out of the Rocks, overturn their little Charriot, where they are to leave their Ore, then turn again and enter those dark Caves with such a force and swiftness.

He also visited Herrngrund copper-mines near Neusohl, where much timbering was necessary to support the roof, and some passages were inclined.

> In one place where we descended obliquely to go to a remarkable part of the Mine, we found that the Earth had fallen in and had stopped up the passage, but one of our Guides unwilling to go about, and the Earth being yet loose, he made a Burrough into it and digged his way through, although the Earth continually fell upon him and covered him; and got at length through and tumbled down amongst the Work-men below, with a great quantity of Earth after him; with

which they so speedily laded their barrows and Hundts, of which I spake before, and the Boys ran away with them with such swiftness, that in a short time he made the passage clear again up to the place where he had left us.[38]

More often than not, the body of a Hund was not rectangular, but the sides converged towards the top and sometimes towards the front. The reason was to keep the centre of gravity down – the higher it was the more unstable the Hund would be – and to keep it back, over the main wheels. None the less, plenty of Hunds were completely rectangular, especially in the Harz (pl. 13). The only drawing which shows a Hund with outward-flaring sides, in Lower Hungary in 1703, is untrustworthy.[39] Table 1 gives some of the recorded dimensions of Leitnagel Hunds.

Table 1

Inside dimensions of Leitnagel Hunds

				Width				
				Back		Front		
		Length in.	Height in.	Top in.	Bottom in.	Top in.	Bottom in.	Capacity cu. ft.
Underground								
1556	Erzgebirge (Agricola)	*c.* 48	*c.* 30			*c.* 30		
1556	Schwaz	*c.* 36	*c.* 24			*c.* 8		
1773	Schemnitz[40]	31	12	9·5	11	9·5	11	2·7
1807	Poullaouen[41]	32	11	11	11	10	10	2·12
1846	Freiburg iB.[42]	28·5	13·2	11·7	12·8	10	11·1	2·5
1802	Himmelsfürst (Freiberg)[43]	36	16	10	10	10	10	3·33
1763	Harz[44]	44	16	17·5	17·5	17·5	17·5	7·15
1819	Andreasberg[45]	*c.* 44	23	16	21	16	21	10·8
1790	Braubach[46]	50	22	18	18	18	18	11·5
Surface								
1773	Schemnitz[40]	48	16	14	14	14	14	5
1788	Freiberg[47]	34	17	13	13	13	13	4·35
1788	Freiberg[47]	36	21·9	13	13	13	13	5·8
1788	Freiberg[47]	42	25·6	14	14	14	14	8·7

Agricola's figures are very suspicious. His Hund would hold 25 cu. ft., which is impossibly large; and whereas he says that his Hund contains half as much again as a barrow, an ordinary Hund of 4 or 5 cu. ft. was reckoned to hold two and a half barrow loads. Other figures, like the

Schwazer Bergbuch's, do not take into account the sloping top of the Hund or, sometimes, the tapering of the sides. An average Hund had a capacity of about 4 cu. ft., which could be increased considerably by piling the ore. The weight of the contents depended largely on the type of ore, but 1 cwt. to 1 cu. ft. is a rough approximation. The few Hunds designed solely for surface use naturally held a lot, but equally large underground Hunds were restricted to a few areas. In the Harz, which went in for them, a Hund could only be moved by two men. The Saxon Erzgebirge also experimented widely with Hunds of up to 14-cu.-ft. capacity, but again two men were often necessary.[48] The large Hunds at Braubach[49] on the Rhine and at Roche-la-Molière (Loire)[50] were singularly successful in that they could be managed by one man.

Hunds were almost always home-made by the mine carpenter and smith, or even by a special Hund-maker. To avoid waste of energy, they had to be as light as was compatible with durability. The planks forming the sides were normally $\frac{3}{4}$ in. thick, although at Poullaouen in Brittany, where the Hunds were heavy and of small capacity, the sides were $1\frac{1}{2}$ in. The floor was normally made somewhat thicker. To protect the body against the rough and tumble of underground work, iron bands were strapped round it, a necessary protection, but a considerable addition to the weight and cost. Occasionally a Hund had a small handle fixed to the back for the putter to hold (pl. 12); and often it had an iron ring on the front, either for the rope if the Hund was worked by two men, or for letting it down the shaft when new. The single beam under the body which held the axles was called the *Rücken* or backbone, and was some 6 in. wide by 4 in. high.

The axles were of wrought iron, as light as possible, and were almost always nailed or bolted to the *Rücken* – very rarely did they revolve with the wheels. The wheels were held on by linchpins, and the bearings were protected from grit by washers inside and out. The material was wood – usually elm – with iron tyres and iron naves until the late eighteenth century, when cast iron began to come in for the purpose.[51] The iron discs recorded at Nagybanya in 1556 (p. 13) were probably washers, not wheels. Wooden wheels, however, were never completely driven out. On a few very early Hunds, on most of the massive Harz models, and on a very few others[52] the wheels were all of equal diameter. But by far the most usual practice was to make the rear wheels larger – from $6\frac{1}{2}$ to 15 in. in diameter – and the front ones smaller – 4 to $8\frac{1}{2}$ in. The tread might be 1 or $1\frac{1}{2}$ in. wide on the small wheels, 1 to 2 in. on the large. One Hund, however, has rollers rather than wheels, their width being quite equal to their diameter. This is apparently the only Leitnagel Hund to survive, being preserved in the Verkehrshaus der Schweiz in Lucerne. It came from a coal-mine in Simmental in the Bernese Oberland and dates from 1785 (pl. 14). The positioning of the wheels also varied. They were outside the body again on a few early Hunds, on most of the Harz ones, but on few others[53] (fig. 7 – but compare pl. 13, a Harz Hund with wheels underneath). Normally they were placed completely underneath the body, to save on the overall width of the vehicle and on the width of the track. Very often, too, the gauge of the front pair was less than that of the rear pair, so that the wear on the rails could be reduced by making front and rear wheels follow different paths. Such gauges might be 6 in. for the front and 7 in. for the back. The front wheels were generally set fairly close to the front of the Hund, but the back ones might be anywhere from the back end to almost under the centre of gravity. The further forward they were, and the more weight they carried, the easier it was to turn corners and the less the Leitnagel wore away the rails. The wheelbase was sometimes therefore barely 1 ft.

Leitnagel was the original word for the guide pin, from Agricola onwards; but after the late eighteenth century it gradually fell out of use. Its place was taken by *Spurnagel* (track- or groove-pin), which is first recorded in 1767[54] and remained to the end. To avoid confusion, Leitnagel has been adhered to throughout this book. It projected 2 in. or so below rail level, was always of wrought iron, up to 1 in. thick, and usually had a revolving iron roller on its lower end to reduce the friction when it made contact with the side of the groove. It might be fixed into the *Rücken* between the front wheels – sometimes to the axle itself – or a short distance in front. When a Hund came to a bend, the Leitnagel inevitably hit the side of the rail with some force, and in a short time wore the wood away. Several attempts were made to overcome this problem by modifying the Hund itself. One such device was a single wheel on a pivoting axle instead of the Leitnagel and the two ordinary front wheels. It ran in the groove between the rails, and was supposed to guide the Hund round bends with greater ease and less wear. It was tried at Schneeberg about 1790 on a large Hund whose main wheels were 16 in. in diameter and 4 in. on the tread; but it proved unstable and liable to derail, so the single front wheel was replaced by two ordinary ones of 8 in. diameter and a Leitnagel.[55] Another experiment of the same kind at Freiberg in 1841 likewise failed.[56]

Another device was to pivot the front axle in the centre, with a curved bar projecting forwards from the axle to embrace a plank laid between the running rails (pl. 15). When this bar came to a bend, it set the front axle to the correct angle for steering the Hund round. Such Hunds were introduced into two Freiberg mines before 1791, at one of which they survived at least ten years.[57] But the most satisfactory variation on the theme was to attach an ordinary Leitnagel to a bar projecting forward from the pivoting axle. This is first heard of at Schmalkalden in Thuringia in 1767,[58] and was still to be found in 1823 at Bräunsdorf near Freiberg.[59] Very full details survive of its application in 1790 at Braubach silver-mines on the Rhine near Coblenz, where the levels were so tortuous that hitherto only barrows had been feasible. Herr Stockicht, the mine's agent, introduced a pivoting-axle Hund, which, when modified by experience, held the large quantity of 12½ cwt. (pl. 16).[60] Because there was no room to turn the Hund at the unloading point, and the Leitnagel afforded no guidance when the Hund was being pushed backwards, a secondary Leitnagel could be inserted between the large wheels for the return journey. A final and very large example of this sort was replaced at Roche-la-Molière Colliery near St.-Etienne in 1835 – it is not known when it came in. Here the Leitnagel was a long bar ending in a hook, supported at the right height by a chain from the Hund's front[61] (fig. 6). At both places where the performance of these Hunds is known – Braubach and Roche-la-Molière – it was exceedingly good (p. 31), so much did the pivoting axle reduce the friction of Leitnagel and wheels on curves.

An ordinary Leitnagel Hund weighed up to 1 cwt. 40 lb., depending on the ironwork it carried; unusually large and complicated versions, like those at Braubach and Roche-la-Molière, weighed 2 cwt. 91 lb. and 2 cwt. 40 lb. respectively. A Hund at Schwaz in 1556 cost 2 guilders 24 kreuzer complete, or something in the area of 12*s*.; in the Harz in 1577 the woodwork cost a mere 8 groschen, the ironwork 5 florins (at 20 groschen to the florin),[62] the total being much the same as at Schwaz. During the late eighteenth and early nineteenth centuries the outside limits seem to be about 5 reichsthaler for a small underground Hund at Freiberg in 1789, and 41 reichsthaler 16 groschen for a very large surface Hund, again at Freiberg, in 1826.[63] A thaler was worth, very roughly, 5*s*. (24 groschen to a thaler).

The life of a Hund was generally reckoned to be at least two years, depending on the intensity of its use. It is more closely stated as 2,000 to 3,000 shifts[64]: one, two or three shifts were worked in a day, depending on the state of the mine. But running repairs might cost as much as, or even more than, its original price.

In the track, the sleepers (usually called *Stege* or *Unterlage*) were laid directly on the floor, or on the wooden covering to the drain, at intervals of anything from 2 to 6 ft. A typical size was 27 in. long by 5 in. wide by 3 in. deep, but these dimensions were capable of great variation. On them were nailed the rails, for which *Gestänge* was by far the most common name, though other words, such as *Laufbretter* (running planks) or *Strassbäume* (way-beams), were known. Each rail might be from 9 to 24 ft. long, 4 to 14 in. wide, and 1½ to 5 in. thick (Agricola's figure of 9 in. thick seems unlikely). The most common size was about 6 by 3 in. Very occasionally

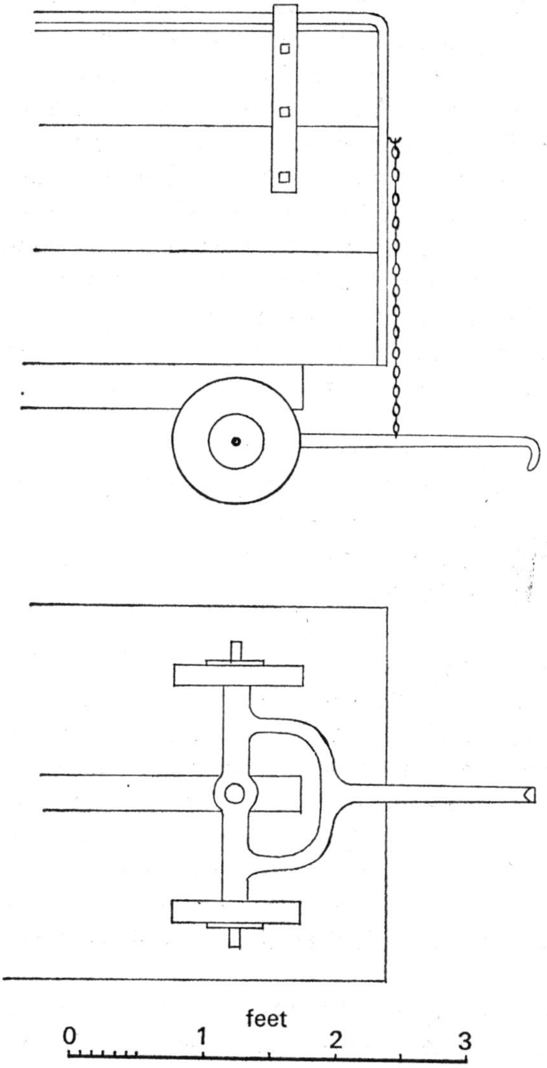

FIG 6 Hund with pivoting axle and projecting Leitnagel, Roche-la-Molière, 1836.

we find the rails recessed into the sleepers. The groove or gap between the rails, nearly always called the *Spur*, could be from 1 to 5 in. wide, depending on the diameter of the Leitnagel roller. The norm was 2½ or 3 in. Points were rare: the only known description of them is at Braubach in 1790 (pl. 16), where an iron stop was inserted into one groove or the other to direct the Leitnagel as required.

Mention has already been made of the wear suffered by the rails, both from the wheels and from the Leitnagel. A hardwood was always preferred – oak, hornbeam and especially beech – and 'when they are pretty worn out on one side, they are turned over, and used until they are quite worn through'.[65] How long it took to wear the rail out depended on the traffic and the wood. At Himmelsfürst mine at Freiberg in 1802, it was a mere two months before a rail had to be turned; at Marienberg in Saxony in 1777, nine years.[66] If by chance a softwood had to be used, a 'wearing strip' about 1 in. thick could be laid on top and turned over when worn, to save the rail itself.[67] Cavities worn by the Leitnagel on the inside of the rail at bends had to be filled with hardwood plugs to prevent the Hund sticking. Occasionally, more drastic steps were taken. The *Schwazer Bergbuch* of 1556 describes and illustrates (pl. 9) an iron plate:

> The *Reibeisen* [rubbing-iron] is about a fathom long, some being longer than others; wrought very thin and rather more than an inch wide. The *Reibeisen* is fastened or nailed to the rails at the rubbing-points or bends of the adit, so that the rails may last longer and not be worn out thereafter by the Truhe.

The *Reibeisen*, it seems, did not protect the rail from the wheels – it was too narrow for that – but from the Leitnagel. But whether it was nailed to the top of the rail along the edge of the groove, or to the side of the rail actually in the groove, is not made clear. Combes, writing in 1845, speaks of it in a general way, saying that it can be laid on the outside rail on bends, which can even be plated entirely with sheet iron or cast iron.[68] But the Reibeisen cannot be traced to any particular mine except for Schwaz, although it must have been used elsewhere. It features in an early phrase, *Reibeisen abreissen* (to lift the rails), a symbolic term for abandoning a mine.[69]

Mines of any size employed special workmen to make and maintain the track. Schwaz had seven of these rail-layers (*Gstenglöger*) in 1638,[70] and the *Bergbuch* of 1556 illustrates their storage hut for rails (pl. 10). In smaller concerns, the Hund-putters had to repair their own track.

The Hund was filled by the shoveller, or *Sauberer* (called 'sweeper' in the Mines Royal), generally a young boy, who either transferred the rock or ore dug by the hewer directly into the Hund, or carried it to the nearest railhead on a sledge. This might take from three to ten minutes per Hund. Occasionally the Hund was filled from a chute leading down from a slightly higher level. The Hund-putter then took over, and pushed it with all speed to the shaft or adit. In early days he sometimes worked entirely in the dark,[71] but it became usual practice to carry an oil lamp on every Hund. If the destination was a shaft, the putter would lift the back of his Hund bodily (pl. 4) – with assistance if it was really large – and empty the ore on to a heap or occasionally down a chute; it was then shovelled into buckets. In the Mansfeld coal-mines, the Hund body was drawn up the shaft, leaving the frame and wheels on the track.[72] If the level ran into the open air out of an adit, the ore might be tipped over the end of a wooden staging (fig. 5, pl. 7), or in a pile beside the track (pls. 1 and 5). Occasionally rails were laid to the smelting works if they were nearby, or to the waste tips. Only two exceptions are known to the general rule that Hunds were emptied by tipping them bodily over: the three-wheeled experiment at Schneeberg in 1791 (p. 25), which had a flap in its base, and the Hund at Roche-la-Molière (fig. 6), whose side was hinged. Once emptied, the Hund was turned round by pivoting it either on the Leitnagel or on the back wheels. Turntables were extremely rare.[73] Emptying and turning occupied only one or two minutes. The putter then began the return journey. As the track was practically always single – double track is only once recorded (p. 14) – and a loaded Hund always had right-of-way, passing points were provided. Whenever he reached one, the returning putter shouted, and if an answering call told him that a loaded Hund was coming, he moved his empty one into the passing place, manhandling it off the track if no rails were laid. On a 500-fathom run, the time spent in the passing places varied from four to twenty-three minutes, depending on the amount of traffic. The putter might take advantage of these enforced halts by trimming his lamp or greasing the axles.

The same putter always worked his own Hund the whole distance, except in Agordo copper-mine near Belluno in northern Italy, where the Hunds were worked by five relays of putters. Skilled German putters might laugh, said the reporter, but it was an excellent system for the Italian workmen.[74]

In the Harz Mountains, the Hund was always pushed by one man behind and pulled by another in front[75] (fig. 7). The same not infrequently happened in Saxony in later days,[76] but hardly anywhere else: one instance was at Hüttenberg in Styria in 1821, where Hunds were pushed and pulled along temporary levels on the bare floor, presumably with the Leitnagel removed.[77]

Non-human traction was limited to inclined levels and the like, and was very uncommon. We have seen how Frolov's Hunds in the Zmeinogorsk works were hauled by an endless rope driven by water-power (p. 18).

Herr Mende introduced a similar system in the main adit of Frisch Glück mine at Johanngeorgenstadt in Saxony in the 1780s,[78] and Rammelsberg mine in the Harz used a balanced incline, whereby the down-going loaded Hunds pulled up the empties on a parallel track by means of a rope round a pulley at the top.[79]

In the early days the putter was a man of mature years, but as time went by he tended to be a youth in his late teens, putting being the intermediate stage in the hierarchy between the shovellers and the hewers. In Lower Hungary women often put, but elsewhere they were hardly to be found underground at all.[80] The putter came under the immediate jurisdiction of the Hund overseer, who recorded the number of Hunds run and ensured that they did not carry short measure, nor yet so much that the ore spilt on to the track.[81] There were always fewer putters than hewers, often

FIG 7 Two-man Hund in the Harz, 1617 (after Loehneyss).

many fewer. At the small mine of Windisch-Matrei in the Tyrol in 1614, two putters served nine hewers[82] (it was noted in Silesia in 1653 that a hewer could supply 288 Hund-loads a week).[83] At the other end of the scale, the great Falkenstein mine at Schwaz employed in 1554 (out of a total of 7,460 men) 3,560 hewers, 836 shovellers and 750 putters. By 1556 the putters had grown to 1,132, but thereafter fell off as prosperity declined: 515 in 1589, 286 in 1638.[84] At Herrngrund, near Neusohl, 256 hewers kept 99 shovellers and 74 putters busy in 1581; but by the end of the century, with a big programme of extending the workings, nearly half the total labour force was occupied in transport (including the shovellers): in 1600 there were 326 hewers to 253 shovellers and 258 putters. When the programme was completed and things settled down again, the number of hewers returned to the previous level, but the number of putters remained high to deal with the longer hauls. In 1629 there were 257 hewers to 139 putters; in 1639 248 to 131.[85]

Putters were practically always paid by the shift. Their wages gradually increased over the centuries in Austro-Hungarian parts from 5·2 kreuzer a shift at Gossensass in 1427 to a maximum of 18 kreuzer in Lower Hungary and Transylvania in the late eighteenth century.[86] It would not be wildly wrong to reckon a kreuzer as a penny. In Saxony in the early nineteenth century the putters were rather worse off, earning 4 groschen a shift, or barely 1*s*.[87] The overseer kept the tally of Hunds each putter had run. In the Harz, however, putters were paid by distance – so much per Hund over a given run – and from 1660 most mines there struck copper tokens (*Bergwerksmarken*), one of which was handed to the putter for every Hund he tipped, to be cashed at the pay office. A typical example is illustrated in fig. 8; the obverse simply reads 'OM', for Obere Massen mine.[88]

For his pay (except in the Harz) the putter had to fulfil a carefully regulated stint of work – so many runs per shift. How many it was depended on the length of the run, the weight of the loaded Hund, and other factors,

FIG 8 Harz Bergwerksmarke – token for one Hund.

such as gradient and quality of track. A number of these operating figures are listed in the Appendix, all of them translated in the last column into a common unit of work, so that their efficiency can be compared. The great variety of performance will be seen: how an extremely small Hund like that at Idria did extremely badly, and how coal, an unsuitable substance, could lower the efficiency, as at Mansfeld. The two top places are both occupied by Hunds with pivoting front axles (p. 26), a measure of their success; still more so in that the Roche-la-Molière colliery heads the table, where Mansfeld comes pathetically last. The longer the run, the better the figures, since less time was consumed in loading and emptying, and more in putting; and likewise several Hunds' performance was boosted by favourable downhill gradients – uphill runs against the load were avoided if humanly possible. But the figures do show that a Leitnagel Hund was on average more than twice as efficient as a wheelbarrow, and maybe more, since the Marienberg barrow seems to have been unusually good. Given a run of reasonable length, and a not too cumbrous build, a Leitnagel Hund was a very worthwhile investment. Its original design remained virtually unaltered to the end of its days (fig. 9),[89] for, apart from the pivoting axle, it offered no real scope for improvement. Greater efficiency was possible only with other means of guidance.

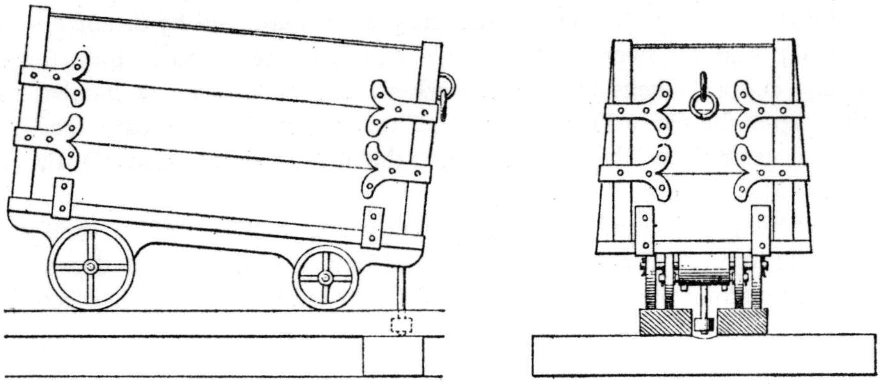

FIG 9 Leitnagel Hund, 1870s.

3

The Hungarian Hund

Sie sind vor Ort, die Lämpchen rund
Wie Irrwischflämmchen aufgestellt.
Die Winde keucht, es rollt der Hund,
Der Hammer pickt, die Stufe fällt.

'Ein Sommertagstraum', Annette Freiin von Droste-Hülshoff, *Gedichte*, Stuttgart (1844), 178.

The Hungarian Hund, as its name implies, was originally a product of Hungary, or rather, in modern terms, of Slovakia. It was the direct descendant of the Leitnagel Hund, and in so far as it was steered only by the putter, with no means of mechanical guidance, it was not strictly a railway within our definition. It would be foolish, however, to ignore it, for in the Hungarian Hund European mine transport reached its highest level of efficiency until the English railway swept the board. The Hungarian Hund was constructed in just the same way as the Leitnagel Hund, except that it had no Leitnagel, normally ran on a single plank instead of on two, and had its larger wheels set almost under the centre of gravity. The putter pressed down on the back, raising the front wheels off the rail, and ran the Hund on the rear wheels alone, steering it as he went. It was no mean skill. The front wheels were used only when the Hund was stationary and on very steep slopes and sharp corners. The wheel gauge was small to allow the plank to be as narrow as possible. The great advantage lay in the reduced friction – only two wheels as against four wheels and a Leitnagel. Its disadvantages included the difficulty of training good putters, and the fact that the Hungarian was no better than the Leitnagel Hund if the levels were low, narrow or particularly crooked. It was beyond even the best putter's power to maintain a respectable speed and to keep the Hund on the track if he had to bend down to avoid hitting his head on the roof, or squeeze through narrow passages, or pick his way along a serpentine plank.

How did the Hungarian Hund originate? There is some evidence that Leitnagel Hund-putters, if their vehicle was suitably built, used to raise the front wheels off the rails when they were in the open air and had light to see and room to manœuvre, simply to reduce the friction and so to move faster. The *Schwazer Bergbuch's* Truhen-putter is doing just this (pl. 11). A seventeenth-century plan of Herrngrund mine near Neusohl includes similar small sketches of putters running their Hunds on the back wheels only, even though the rails are double, as if for the Leitnagel.[1] It was evidently in these mines of Lower Hungary that it was decided, some time in the early or middle seventeenth century, that the Leitnagel was more trouble than it was worth, and should be done away with. The first definite evidence we find is in a mining dictionary of 1667 which deals with Lower Hungary practice[2]:

> The rails are of two sorts, single and double. It is single when the putter runs on only one plank or rail, and the Hund or *Bergtruhen* has no Leitnagel. There are double rails when two small planks are laid close to each other, so that the Leitnagel can run between these two rails. With a Hund or Truhe of this sort the putter can run from the place where he fills his Hund to the adit mouth in the dark, which cannot be done so easily on the other sort of rail without a Leitnagel.

The Hungarian Hund appears in other drawings of 1695 at Schemnitz, in which the artist does not even bother to draw the front wheels.[3] Thereafter it became more and more the norm in Lower Hungary, although the Leitnagel Hund was not finally driven out until the early nineteenth century. The putters at Schemnitz in 1759 were paid the princely wage of 20 to 24 kreuzer per shift (about 1*s.* 8*d.* to 2*s.*), considerably more than their brother Leitnagel putters ever received, simply because of their extra skill.[4]

To the end, Schemnitz remained the true home of the Hungarian Hund. Our best account of the underground transport there comes from Wilhelm Becker, a Freiberg mine captain who paid a visit in 1805.[5] The reasons for the Hund's efficiency, he said, lay mainly in its construction, which was the same in all the mines of the area (pl. 17). The design was basically the same as elsewhere, but special attention was paid to it by taking the best features of all Hunds and moulding them together. Two special Hund-makers were employed full-time at Schemnitz, turning out vehicles to this one pattern. Elsewhere, Hunds were made by non-specialist workmen who had other preoccupations. The excellent construction was backed up by admirable maintenance, and by each putter having his own particular Hund, which he treated with loving care. The Hund was built, like many of the Leitnagel versions, with the sides converging upwards and forwards to bring the centre of gravity down and back. In operation, the putter held the handle on the back with his right hand, and laid his left on the side or on the load itself, this being found the best position for steering. The front wheels were kept small, so that the body did not have to be tilted so far back as to obscure the putter's vision. A method of double-loading was often practised, whereby an inverted saddle acted as an extra container on top of the ordinary load in the Hund. Two curved boards (*ab*, *bc* in figs. 4 and 5 on pl. 17) were strapped and hinged together, with further planks on top (*af*, *ch*). They were held at the right distance apart by a semicircular board and a cable (*fh*) at one end, which also prevented the ore from overflowing. When the Hund itself had been loaded, this contraption was placed on top, its base resting on the Hund sides, and was filled up. It was folded up for the return trip. The simple Hund held $4\frac{1}{4}$ cwt. at least, or 6 cwt. on average; with the saddle it could carry 9 cwt. Over any distance it was far superior to a Leitnagel Hund because of its lightness, speed and, in skilled hands, its ease of control. A good Schemnitz putter scorned the Leitnagel variety, which he called the *Bauernhund*, the farmer's Hund. It was a complete waste of energy, Becker said, to supply artificial guidance, since a putter necessarily accompanied the Hund and could perfectly well steer it himself. Likewise, doors for emptying it, and excessive strapping for durability led to an increase in weight, whereas on underground vehicles lightness was all-important.

Admittedly the Schemnitz levels were high and broad – the main adits 12 ft. by 6 ft., the smaller levels $7\frac{1}{2}$ ft. by at least 5 ft., which meant that the putter could go flat out with no danger of hitting the sides. The putters had to work Hunds for eight years before being promoted to hewers. They were taken on at the age of fourteen to sixteen, and thus put until they were twenty-two or twenty-four, the best years for dexterity and balance. They formed societies so that, should any of them fall short of his quota, his loss was made good by a general contribution. They ran barefoot even in winter, but kept stockings at the adit mouth for outside work. The putters did their own loading to get the balance right, and paid special attention to the rails, which were maintained by the youngest of them, since wetness reduced their speed and the smallest inequality led to oscillation. If the Hund did rock, the putter held it by the back corners and was usually strong enough to prevent it from overturning. Bends were often taken on one wheel. But their skill must have been phenomenal to manage such heavy loads in murky passages at an average of 240 ft. per minute: the outside gauge of the wheels was $8\frac{1}{2}$ in., the width of the rail only 9 in. A durable Hund, with little friction, capable of high speed and heavy loads, together with a skilled and strong putter, made a harmonious unit unsurpassed elsewhere.

Becker well knew what he was talking about:

> I experienced the speed attained by a skilful Hund-putter of this sort in a way that astonished me, when on 17 September 1805 I toured the Finsterort Mine Co. at Hodritsch [near Schemnitz]. I squeezed myself into a Hund [fig. 7 on pl. 17] of the sort always used for tours by officials, and was pushed the whole way, there and back, from the mouth of the Francisci drainage adit to the horse gin, a distance of 1,200 fathoms or so, half an hour's walk. This level was so winding that seldom were four rails laid in a straight line together; most of it was wet or very muddy, and one can say that for 200 fathoms it was under water, sometimes 2 in. deep, so that the position and direction of the track could only be guessed. What was more, the level was unusually low and narrow, which caused me, sitting in my Hund, immense difficulties in bending low enough. During the journey in, we had to pass seven loaded Hunds, which several times involved going back to find the proper passing-places, and then, when I had been run off the rail, waiting until the loaded Hund had passed before we could go on our way again. The delay at such a passing-place can amount to 1½ minutes, and we stopped for four periods of ¾ minute to trim the lamp, so that the total time of our stops amounted to 13½ minutes. Nevertheless, I covered the whole distance to the horse gin in 36 minutes – that is 22½ minutes on the move – and although they had been kind enough to give me the most experienced putter, my two companions arrived at the same time. I admit that I had expected a still greater speed; so my Slavonic coachman, who was waiting with his cabriolet until we had seen the machine, was told in his unintelligible language that I would take note of the time, so that he could gain credit and earn a good tip. This so spurred him on that I covered the return journey of 1,200 fathoms in 16 minutes. The narrowness and constriction of the low tunnels oppressed me less, as my confidence had grown. I was told that if I had not huddled myself up I could easily have broken an arm. The props flashed by me so fast that I could hardly notice what sort of timbering they used, and all my attention had to be concentrated on avoiding their dangerous projections. The lamp was trimmed only twice, and we did not have to pull off the road at all; as a result, our speed was so great that water spurted into my eyes, and my companions arrived 5 and 7 minutes after me, even though their putters had sweated and panted as hard as mine.

From the mid-eighteenth, if not from the seventeenth century, the

FIG 10 Distribution of Hungarian Hund and Mansfeld Schlepphund.

Hungarian Hund was likewise favoured at other mines in Lower Hungary, such as Kremnitz and Herrngrund near Neusohl – one of whose Hunds is preserved in the museum there (pl. 18) – and at Upper Hungary mines, such as Smölnitz and Felsöbanya.[6] It found its way into Transylvania before 1839, and into Silesia at Tarnowitz lead-mines by 1810.[7] Further west it met with a mixed reception. Hungarian Hunds were first introduced from Hungary to the Saxon Erzgebirge by Friedrich von Trebra, one of the leading mining engineers of his day, who installed them in 1775 at Marienberg and Johanngeorgenstadt, and shortly afterwards at Ehrenfriedersdorf.[8] At these three places, and at Schneeberg and Joachimsthal, where they had appeared by 1791,[9] the Hunds immediately showed their worth. At Marienberg they only replaced wheelbarrows, proving themselves two and a half times cheaper; but elsewhere they began to oust the Leitnagel Hund by virtue of their low initial cost, ease of maintenance and greater efficiency. In these mines there was no difficulty in training putters, although they never approached the Schemnitz putters in skill. At Freiberg, however, it was a different story. Here the Hungarian Hund was in use fairly widely by 1785,[10] but it proved unpopular and ineffective because, for some reason, the putters never really mastered the knack of steering at speed. By the 1790s some mines here were already reverting to the Leitnagel Hund,[11] although some Hungarian Hunds stayed in use for most of the nineteenth century.

Elsewhere in western Europe (fig. 10) the Hungarian Hund was in use at Clausthal in the Harz Mountains by 1785, but remained a rarity[12] (pl. 19). It was to be found at Schirmeck iron mines in the Vosges by 1789,[13] in a very inefficient form at Freiburg-im-Breisgau from 1840,[14] and at Kitzbühel in the Tyrol and Schäferötz near Werfen in the Salzkammergut by 1854.[15]

Table 2

Internal dimensions of Hungarian Hunds

				Width				
				Back		Front		
		Length in.	Height in.	Top in.	Bottom in.	Top in.	Bottom in.	Capacity cu. ft.
1805	Schemnitz[16]	35¼	14⅛	10¾	12¾	9⅜	11⅜	3·2
1802	Freiberg[17]	36	18–20	12	13	12	13	5
1826	Freiberg[18]	39	17	11¾	13¾	10¾	12¾	4·7
1878	Freiberg[19]	43¼	17⅛	13⅛	14⅛	13⅛	14⅛	5·3
1826	Clausthal[18]	44	14½	14½	15¼	13½	14½	5·35
1878	Freiburg-i-B.[19]	29⅜	13⅝	12	13¼	10¼	11½	2·5

The cubic capacity of the Hungarian Hund tended to be small – the bodies were never really large, like the two-man Leitnagel Hunds – although as we have seen the Schemnitz loads were often increased by adding a second layer. The dimensions which survive (Table 2) tend to be untypically large, a little over 3-cu.-ft. capacity being the norm.

At Freiberg in 1878 the birch *Rücken* or longitudinal beam underneath was 5½ in. wide and tapered in height from 5½ in. over the rear axle to 4 in. at the front. The rear axle was set 17½ in. from the back, the front axle 3 in. from the front of the body. The wheels were usually of iron-rimmed wood, although cast iron was sometimes used at a late date. At Schemnitz their diameters increased from 6 in. and 3 in. in 1773[20] to 7½ in. and 4½ in. in 1805; elsewhere the larger ones were nearly always 8 in., the smaller from 4 to 6½ in. The treads were generally 2 in. wide, and the wheel-gauges about 6 in. in front and 8 in. behind, though a 5-in. gauge was known.[21] Freiburg-im-Breisgau started off in 1840 with an experimental Hund with only one wheel in front, which proved unstable even when stationary. Next they tried one with the axles equidistant from the centre of gravity, and running on all four wheels, which produced too much friction. Only then did they adopt the traditional type.

A Hund might carry 40 lb. of ironwork (Freiberg, 1835).[22] The Upper Hungarian mine of Felsöbanya was so entangled in red tape as early as 1785 that the smiths had to complete daily forms recording their output of 'Straps for Hund-bodies or wheelbarrows; Hund axles; Hund bearings; Hund wheel tyres; rail drills'.[23] The cost of a Hund varied from 3 reichsthaler 3 groschen (about 15*s.*) at Marienberg in 1785 to 6 reichsthaler 12 groschen (some 32*s.*) at Freiberg in 1826, plus more than half as much again for maintenance throughout its life, which was estimated at one year and at 850 shifts respectively.[24]

The rails were often nailed directly to the floor, though they sometimes rested on sleepers. Their thickness was 1¼ in. at Schemnitz, but 2 in. elsewhere; their width ranged from 9 in. at Schemnitz (with an 8½-in. wheel gauge) to 14 in. at Freiburg-im-Breisgau (with an 8-in. gauge) – a measure of the comparative skills of the putters. Once again a rail's life depended on the traffic and on the conditions of the mine, but it usually lasted considerably longer than a Leitnagel rail, since the Hund wheels wandered from side to side and did not wear grooves by always following the same path. At Marienberg its life was six years, at Freiburg nine and a half.

The Appendix gives some figures of the performance of Hungarian Hunds, which at Schemnitz was always better than any Leitnagel Hunds, and sometimes was elsewhere. The trouble was, as Becker himself admitted, that if the putters were good the performance was good; if they were not good they were worse than useless. Whereas most Hungarian Hunds had succumbed to the English railway, where the putters could hardly go wrong, well before the end of last century, those in Hungary proved very resistant to change. They were still being happily trundled up and down the levels at Smölnitz in 1955,[25] and possibly still survive in a few remote mines.

*

While on the subject of non-guided underground transport, we cannot pass by another minor variety. The copper-schist mines of Mansfeld, twenty miles west of Halle on the Saale, were distinguished by the lowness of their underground levels. The main ones were large enough to accommodate wheelbarrows, but to solve the problem of transport in the secondary roads, which were rarely more than 2 ft. 6 in. high, a special vehicle was evolved, usually called the *Mansfeld Schlepphund* (drag-Hund), although it received many other names – *Schlaghund*, *Treckhund*, *Flötzhund*, and so on. Its origin doubtless lay in the ordinary sledge, but by the time it was first described in 1747 it consisted of a floor of two planks somc 6 ft. long, with two small side ledges to retain the load, all mounted on two long rollers running on the bare floor.[26] The putter crawled along the level, pulling the Schlepphund behind him by means of a leather strap tied to his ankle. This early version with rollers seems never to have acquired any form of rail. By 1825 it was becoming slightly more sophisticated: the two wide wooden rollers had given way to four narrower ones, 7 in. in diameter by 4 in. wide, running (sometimes) on a simple track of two parallel planks, with no form of guidance. The friction of the rollers on the planks and on the axles was considerable and the running costs at this date were more than three times those of English waggons carrying a similar traffic. A Hund cost nearly 12 reichsthaler, or around £3, plus 16*s*. worth of repairs a year. Its body measured 4 ft. 8 in. long by 1 ft. 6 in. wide by 6 in. high (about 3½ cu. ft.), and its life-span was only a year.[27] By 1878 the Schlepphund had reached the peak of its perfection, if such it can be called, with four iron wheels 11½ in. in diameter. The other dimensions were similar to those of fifty years before, and the tops of the wheels were usually the highest point on the truck. Though the putter still often pulled the Schlepphund by a rope, he sometimes pushed it by foot instead (fig. 11), wearing shoulder and leg pads for the purpose.[28] In 1891 the Schlepphund was still going strong.[29]

Similar and even identical Schlepphunds on rollers or wheels and with or without planks were in use in a number of other restricted copper-schist- and coal-mines in the Mansfeld area, and in various parts of Hessen (fig. 10).[30] Further afield, they had appeared in low levels at Schemnitz by 1773, at Tarnowitz in Silesia by 1805, and at St. George and Lavencas (Aveyron) collieries in the south of France by 1818.[31]

FIG 11 Mansfeld Schlepphund and putter, 1878.

4

Channel- and flanged-rail railways

Durch lange Stollen, düstre Gänge
  Wallt schon der Bergmann heiter fort;
Der Wagen klirrt auf dem Gestänge,
  Glückauf! er lagert sich vor Ort.

G. Schneider, 'Der ernste Stand', in Moritz Döring, *Sächsische Bergreyhen*, Grimma, i (1839), 103.

The wooden rail with an upward-projecting strip attached to one or both sides, forming an L- or U-shaped channel in which the wheel could run, was merely a translation into timber of the natural or artificial rut in an ordinary road. We have seen how such rutways in the ancient world occasionally served as railways, and how Hero of Alexandria proposed a wooden version, albeit for a slightly frivolous purpose (see p. 2). So too the men of the late Middle Ages were doubtless influenced by the ruts which bedevilled their own roads when they applied the channel rail to three different activities where a special form of transport was involved. Mining was one of these activities, though here the use of the channel rail was limited, compared with the ubiquitous Hunds already described. The other two, military and canal engineering, represent the only Continental application of railways for a non-mining purpose between ancient times and the nineteenth century. Since they were to this extent untypical, let us deal with them first.

The military use of rails was confined to fortification-building – for lifting earth from ditch to rampart by means of inclined planes. The earliest example of this is also the earliest medieval railway that we know, in the sense that here for the first time the wheels were clearly guided by the track. It occurs in a volume of drawings in the Munich Staatsbibliothek, dated about 1430, by an unknown artist (generally referred to as the Anonymus der Hussitenkriege) who was active in Bohemia and Bavaria at the time of the Hussite Wars.[1] The drawings are mechanical sketches, largely concerned with military subjects – artillery, assault machines, prefabricated bridges and so forth – with a number of civilian subjects, such as cranes, diving suits, windmills, and paddle-boats. His unfinished drawing of a railway truck (pl. 20) is in the military category (witness the crenellated wall behind), and, since there is no descriptive text, its interpretation is enigmatic. At the top left is a large crane, running with six wheels on baulks of wood. Near the top of its vertical member is pivoted a long cross-beam, from one end of which is suspended a water-tank, filled by a wooden leat and provided with a rope, apparently to empty it. One assumes that the weight of the tank, when full, raises the other end of the jib, from which hangs an angled framework. Near the top of this frame rests a four-wheeled truck, prevented, it seems, from sliding down the slope by a long, stout timber, and being emptied (or filled?) by two men. Whether the framework pivots at the angle or on the two pins which pierce it at the bottom we are left to guess, but the principle seems to be that the truck is raised to the working level by the weight of water. What concerns us most, however, is that the two rails on which the truck wheels run can clearly be seen, especially at the bottom, to be of channel section. The two pins here, indeed, look more like stops to prevent the truck running off the end of the track. The hen-coop at the bottom is a mobile cannon-shelter, and probably has no connection with the rest of the picture.

Two later proposals for military railways, both Italian, are fortunately more comprehensible. Ramelli's drawing of 1588 (pl. 21) shows a horse-gin raising and lowering two trucks on a ramp up from the bottom of the ditch.[2] The ramp is equipped with raised ledges at the sides and in the middle, and the trucks have unequal wheels to keep their bodies level on the slope, although Ramelli does not condescend to explain how the same trucks manage to remain horizontal when on the flat, as his picture shows. Lorini's plan of 1596 (fig. 12) is similar, except that the track is single, the truck has but two wheels, the rail is of channel section, and a treadmill provides the power.[3] The rail is 1 ft. wide and 6 in. deep, and has its two flanges only on the incline itself; on the level it simply supports the wheels without guiding them. The truck, with a capacity of $\frac{1}{2}$ ton, is manhandled on the level by the bar *IK*; its front *TQ*, normally held by the bar *VR*, can be removed for emptying the body. Both these Italian examples were proposals only, and we do not know if they were ever built.

The slipway for portaging boats from one level of a canal or river to another was a well-known medieval device, and usually consisted of a gentle slope equipped with transverse rollers and, with luck, a winch, after the manner of punt-rollers on the Thames and Cam today. It was a novelty when the slipway at Fusina, built in 1437 to link the River Brenta with the lagoon five miles from Venice, was fitted out with a horse-gin and wheeled carriage on which the boat rested for the journey. Hans Georg

Ernstinger, travelling along the canal in 1593, came to 'Lucefusina' where, he found,

> one cannot pass straight through in a boat, since a strong wooden weir here prevents the canal, which lies several feet higher than the sea, from running into the sea. Over it the boat is taken on a great sledge with rollers, made for the purpose, by means of a clever traction with horses, as far as the sea.[4]

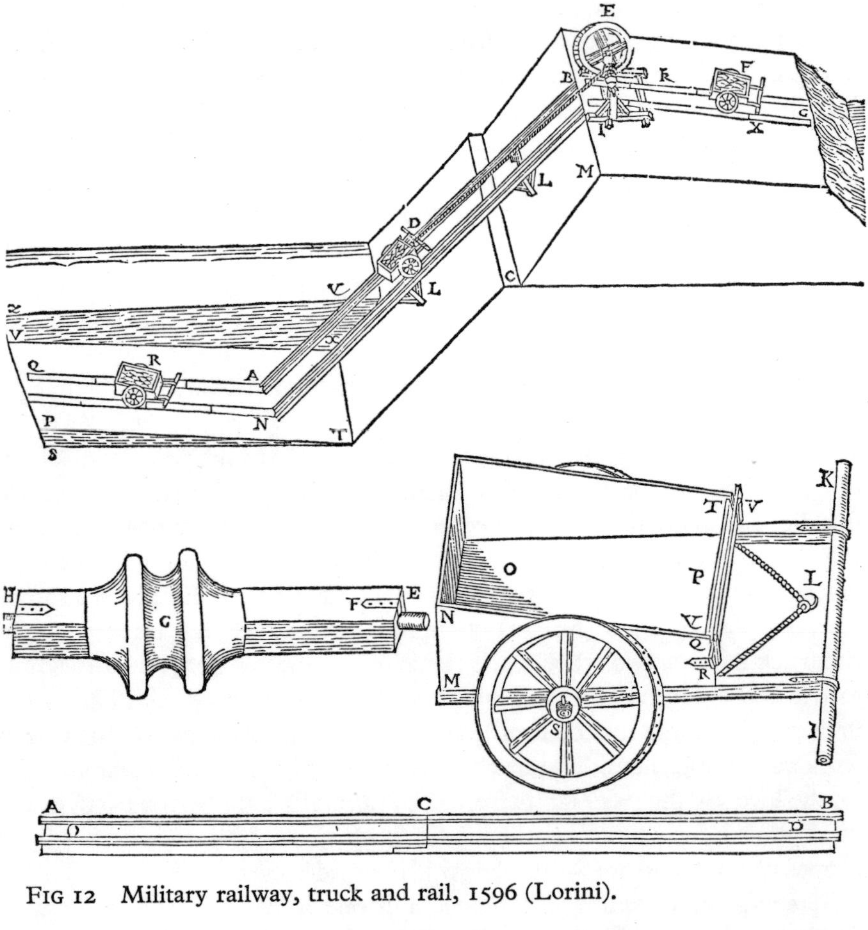

FIG 12 Military railway, truck and rail, 1596 (Lorini).

Vittorio Zonca's illustration (pl. 22), published in 1607, shows how the carriage wheels, of iron-strapped nut-wood, ran between the restraining kerbs *BB*.[5] The double track on the incline is his own suggestion for speeding the passage of traffic, for there was in fact but the one. This type of canal railway found occasional use thereafter, either with stone-built kerbs to guide the carriages,[6] or, later and especially in Britain, with iron rails (see pp. 289-90).

The channel rail was little favoured in mines in contrast to the Leitnagel

FIG 13 Distribution of wooden flanged rails (add Chagirsk and Zmeinogorsk in Russian Central Asia).

and Hungarian Hund. Most of the examples on record are either early or, if late, are in coal-mines, and there is little geographical cohesion about them (fig. 13). Many, however, are inclined planes, for which engineers seem to have thought this type of rail especially suitable. One feature of the Leitnagel and Hungarian Hund, it will be remembered, was that the wheels were almost always (except in the Harz Mountains) mounted underneath the body. Trucks that ran on flanged rails, on the other hand, usually had their wheels set outside the body and close to the corners. There is a whole series of representations of vehicles of this sort dating from the half-century or so before 1550. Some show no rails at all, some do show rails, but apparently unflanged ones, and in a few the rail is certainly flanged. The pictures vary so much in quality that it is dangerous to draw too many conclusions from what the artist did or did not include. If he drew no rails, it may be because he did not notice them, or did not understand their purpose, or because there really were none. Again, if he

FIG 14 Mining truck (in Saxony?), 1503 (Reisch).

drew unflanged rails, either the purpose of the flanges escaped him, or again there were none. But for all the divergence in the track that they show, these pictures seem to form a group and deserve consideration together, for they may represent the development of the principle of the plain wheel guided by a flanged rail. It is the sort of improvement we might expect – plain rails first replacing the bare ground, followed by guiding flanges added to the rails to relieve the putter of the burden of steering the truck.

The illustrations which show no rails are easily disposed of. A simple woodcut in an alchemical treatise of 1503 (fig. 14) depicts an elfish miner pushing a wide truck out of an adit; the place is probably Saxony, for the book was first published in Freiberg, though the same cut appeared in later Strasbourg and Basel editions.[7] Of much higher quality is the work of an anonymous artist known as the Petrarch Master, who illustrated the first German translation of Petrarch's *De Remediis utriusque Fortunae*, published at Augsburg in 1532, although the woodcuts for it had been finished by 1520.[8] The subject of one (pl. 23) is a busy scene in a large underground gallery, with fire-setters, hewers, and one miner pushing a

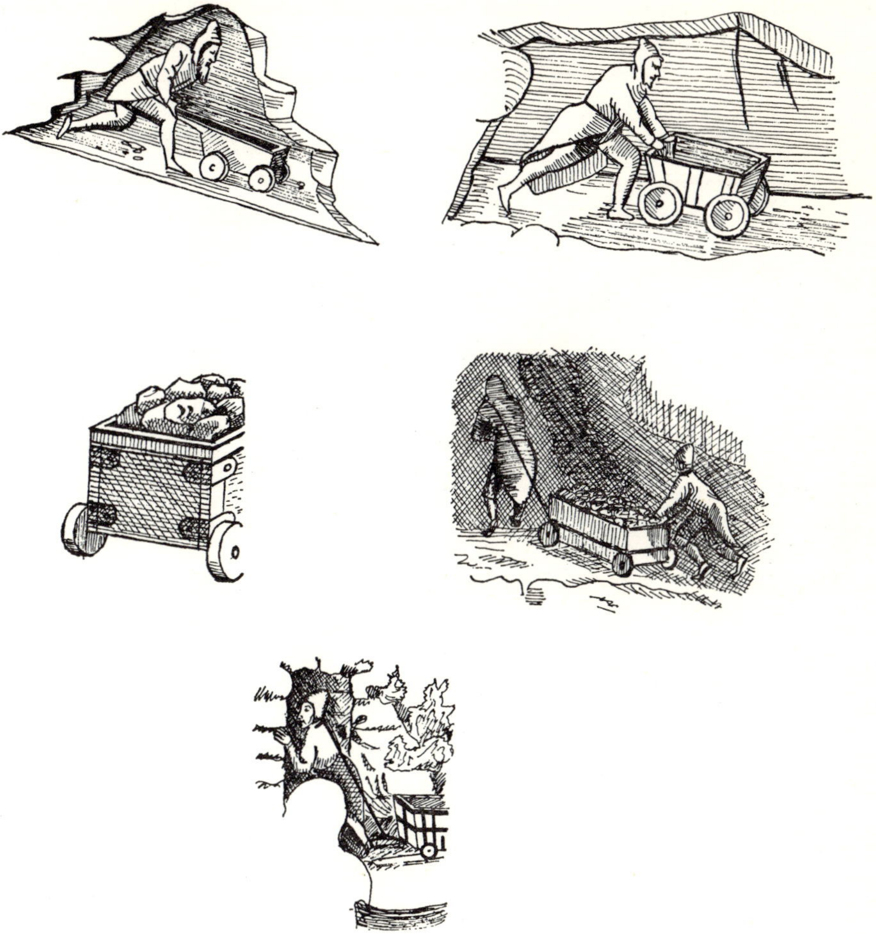

FIG 15 Miscellaneous barrows and trucks in the Erzgebirge, 1521-90 (after Agricola, Albinus, and Annaberg triptych).

large and heavily strapped truck. Again no rails. To these two examples can be added the railless vehicles visible in some of Agricola's and Albinus' woodcuts, which find no mention in their texts. Both relate to Saxony, in 1556 and 1590 respectively.[9] One sort has handles – a form of overgrown wheelbarrow, rare but not unknown elsewhere – the other has not (fig. 15). It is possible that these specimens, along with one or two still to be discussed, are more relevant to the history of the Leitnagel Hund (if they belong to railway history at all); but they are mentioned here merely by virtue of their outside wheels, more typical of the truck for flanged rails than of the Leitnagel Hund, which preferred its wheels to be set under the body.

The silver chain of office worn by the chief goldsmith of Gand, said to be German work of the late fifteenth century, consists of sixteen links, each bearing a scene in relief of one of the processes of mining, smelting and smithing.[10] On the fifth link (pl. 24), a miner is pushing a truck out of an adit. There are rails, but no details are shown. In 1521 Hans Hesse

FIG 16 Mining trucks, Saxony, *c.* 1525 (Schön).

painted a triptych for the miners' altar in the new church at Annaberg in the Saxon Erzgebirge.[11] On the edge of one panel, appropriately adorned with mining scenes, a small, well-strapped truck, running apparently on wooden rails, is being pulled into a level by a rope over a miner's shoulder. Last in this section, and most interesting, is a single-sheet publication entitled *Das sächsische Bergwerk* (fig. 16), and consisting of a woodcut by the Nürnberg painter and designer Erhard Schön, and some rather trite verses below.[12] The date is about 1525. In the picture miners are busy everywhere. Two of them are on transport duty, one bringing a full truck out into the light of day, the other returning with his empty. The rails end a few yards outside the adit-mouths, but for the first time their shape is clearly drawn. Here they are plain baulks, with no flanges, set at a fair distance apart, and quite unlike the Leitnagel Hund rails, which were both wider and set much closer together. If this is an accurate picture, the vehicles must have been infernally difficult to keep on the track underground, with no light to steer by except the glimmer of the oil-lamp fixed to the front. Working a Hungarian Hund must have been child's play compared with a truck like this, for the secret of its manœuvrability lay in the fact that it ran on two wheels only, which were also set at a very much smaller gauge. It is easy to see, from a picture like Schön's, why flanges were added to the rails.

A woodcut in a rare tract on mining law, *Der Ursprung gemeynner*

FIG 17 Mining truck, 1515-38 (*Ursprung*).

Berckrecht wie die lange Zeit von dem Alten erhalten worde . . ., printed for Johan Haselberger of Reichenau in Lower Austria at an unspecified date, but between 1515 and 1538 (fig. 17), gives us our first hint of the flanged rail in mining. In the usual adit scene, the only wheel visible seems to be running in a rail of channel section, which hides the putter's feet. The text helps not at all, for it goes no further than describing the *Tragwerk*, the wooden flooring over the drainage channel in a level. Almost every mine had a *Tragwerk*, whether pedestrians, wheelbarrows, Hunds or other vehicles passed along it. Next comes a painting, 'Les Travaux de la Mine', made in 1544 by Lucas Gassel (*c.* 1500-*c.* 1570), an artist of the Flemish school (pl. 25).[13] Again a truck is emerging from an adit, with an oil lamp fixed to the front. The wheels are apparently spoked and therefore quite possibly of iron, and run upon narrow wooden rails, which look as if they have inside flanges, though one cannot be sure. But too much emphasis should not be put upon these details, for the painting is clearly composite and not done from life. Gassel could have found neither mining nor

FIG 18 Mining truck, Ste.-Marie-aux-mines, 1550 (Munster).

scenery of this kind in his native Brabant, and he never, it seems, strayed far from home. One could surmise, from the similarity, that his miner and truck were based on Erhard Schön's woodcut.

We reach complete certainty with one of the mining scenes which decorate the base of an outsize and exuberantly ornate drinking cup, the Rappoltsteiner Pokal.[14] This was made for the family of Rappoltstein, the owners of a remarkably rich mine at Ste.-Marie-aux-mines (Markirch) in the Liepvre Valley (Leberthal) of Alsace. Tradition says (and the other evidence agrees) that it was first used at a Rappoltstein wedding in 1543. The maker was Georg Kobenhaupt, the Strasbourg master silver-smith, and the material was silver from Ste.-Marie itself. The relief work on the base, portraying the gamut of mining operations, includes yet another truck-in-adit scene (pl. 26). The truck is strapped with iron, the wheels, though below the level of the body, project slightly outside it, and the rails are indisputably flanged on the outside. This is not all, for the cup has another valuable contribution to make in shedding light on the woodcut in Sebastian Munster's *Cosmographia Universalis*,[15] which is perhaps the best-known, after Agricola's Hund drawing, of all illustrations of early mining railways (fig. 18). Munster's picture is also of the Ste.-Marie mines, and the close similarity between his figures and those on the cup (not only the truck-putter, but the hewers, windlass-men, and diviner as well) makes it quite clear that both shared a common source. The history of the woodcut is that, on the publication in 1544 of Munster's first (German) edition, Johann Hubinsack, the Rappoltsteins' mining judge in Leberthal, wrote to Munster pointing out the insufficiency of his account of the Ste.-Marie mines. As a result, Munster visited the mines in 1545, and incorporated in his second German edition of the same year a brief history and description of the works. To his first Latin edition of 1550 he added extra drawings by the hand of Hubinsack himself, as well as a much-extended description from the same source.[16] We may thus guess that Hubinsack either designed the base of the cup or had access to it. It becomes clear, too, that Munster's truck, like that on the cup, is running on flanged rails, though it is far from obvious in the woodcut. One can perhaps see the flange, with an effort of the imagination, in the second length of track in the picture, the bridge or tipping-point directly over the putter's head. Munster's description of 1550 is as follows (that of 1545 is less detailed):

> However much rock or earth is excavated in the levels, there are workmen standing by with containers or trucks running on four small iron wheels, to take it away to the nearest shaft. Here other workmen are stationed with windlasses, to raise whatever is brought to them, and they pass it on to other workmen to take away in trucks. Thus the excavated earth is brought from the deepest parts of the earth, by many levels and shafts, and through many hands, and in a short time a great heap rises up in the open air by the shaft-head, until finally it grows into a mountain.

The Latin words used for the trucks are *alvei ceu capsae*: the German version has *Trügen*, the French *tombereaux*. This is also the earliest reference to iron wheels on a railway.

After 1550 the silence of 150 years falls upon the flanged rail in mines. This is probably due to the chance survival, or loss, of evidence, and not because the rail was forgotten: it was still known to canal and military engineers. It reappears in 1698 on an inclined plane at Humboberg copper-mine near Smedjebacken in Dalarna, Sweden, where the main shaft was not vertical, but sloping at an angle of 48 degrees. It was the work of the great engineer Christopher Polhem (1661-1751) and was an adaptation of his system for winding in vertical shafts, where the tub was raised not by rope, but by two reciprocating rods provided with hooks. The incline was destroyed by a roof-fall in 1717, but was later rebuilt, for it was the subject of a drawing by C. J. Cronstedt in 1729[17] (pl. 27) and a description by A. Hülphers in 1757.[18] The motive power was a water wheel 42 ft. in diameter, which drove overland rodding 1,288 yards long. At the incline head, a bell-crank attached to the rodding moved a rack backwards and forwards, which engaged four pinion wheels mounted in the same plane as the incline. Between these wheels, six more racks were driven in pairs, and were attached to the heads of three pairs of reciprocating rods. One pair (the nearest in the drawing) worked pumps. Each of the other two pairs was fitted with upward-facing hooks at intervals equal to the stroke of the rods, with the result that the loop on the end of the two-wheeled truck, which ran in wooden channel rails astride the reciprocating rods, was automatically passed from a hook on one rod at the end of its upward stroke to a hook on the other rod, which was then at the end of its downward stroke. The incline was 115 fathoms long. No means are visible for transferring the waggons from one track to the other, so presumably each stayed on its own line. But how they were returned underground when empty neither the drawing nor description reveals.

Jacob Leupold in his *Theatrum Machinarum* of 1724 deals with a rope-worked incline serving an alum mine (he does not say where), constructed to replace the old hard system of carting the ore down the mountain in wheelbarrows.[19] Two raised trestles carried the loaded and empty trucks (fig. 19), which were linked by thin ropes to the endless haulage cable. This passed round large pulley wheels at top and bottom (the top one being powerdriven), and was supported at intervals by small vertical pulleys. The trestles 'were made with raised ledges fastened to each side, so that the trucks, especially those going down, cannot deviate sideways'.

We have seen how the Saxon Leitnagel Hund had found its way to the Urals by 1633, and how Kuz'ma Dimitrievich Frolov had introduced it to Russian central Asia in 1764 (p. 18). This was not, however, the first railway in the bleak foothills of the Altai. In the early eighteenth century, Peter the Great's policy of exploiting the resources of the further-flung parts of his empire was being pursued in southern Siberia, where Akinphi Demidov, also with the assistance of Saxon miners, opened up a large number of mines near the borders of Kazakhstan and Mongolia.[20] One of these mines, at Chagirsk (some 120 miles south-east of Novosibirsk), boasted a self-acting incline, illustrated (presumably in an abbreviated form) in a drawing of 1752 in the Altai Local State Archives[21] (fig. 20). It shows two raised tracks of massive wooden rails with outside flanges, for

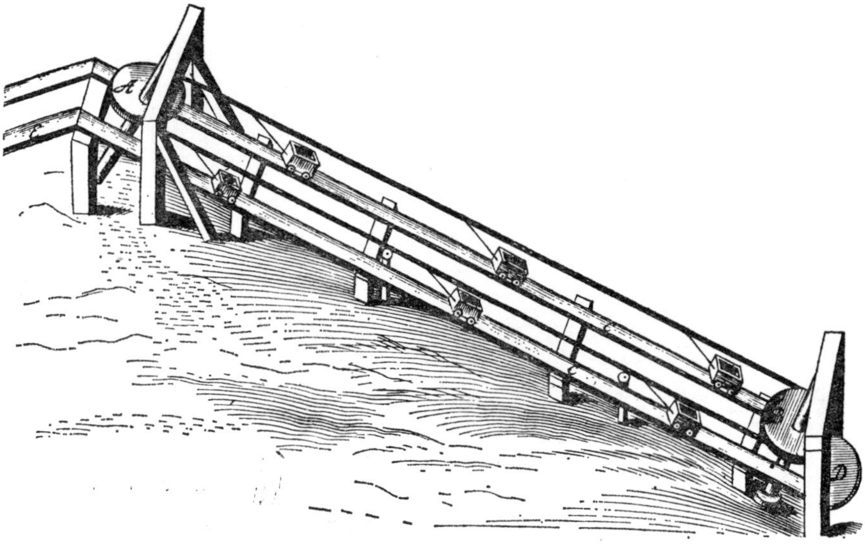

FIG 19 Rope-worked incline, 1724 (Leupold).

elongated trucks with thick wheels. Both trucks are linked by rope to a drum at the incline head, the principle being that the weight of the descending loaded truck pulled up the empty one. A brake on the drum controlled the speed of the operation. At Zmeinogorsk Frolov's Leitnagel Hund had operated inside the smelting works. He also built another railway there in the mine itself, again on a slight incline, and powered like all his machinery by a water wheel. This railway, first recorded on a drawing of 1783, was single track with a passing loop half-way, and it seems to have had flanged rails like those at Chagirsk, though clinching evidence is lacking.[22] The waggons were loaded automatically and were emptied by tipping at the adit mouth. We know much less than we would like about the lines in this remote outpost of western civilization, but it can still be said that Frolov's system of integrating all the processes of mining and smelting, including railways, into a single mechanized whole reached a stage of efficiency matched hardly anywhere in Europe. His son, Pyotr Kuz'mich Frolov, later carried on the same tradition (p. 347).

With three systems in Austria, the tale of the flanged rail outside the coal-mines is almost completed. The first, limited originally to Dürrenberg and Aussee salt-mines in the Salzkammergut, was unusual in that the vehicles were two-wheeled and were pushed by means of two shafts. These carts were probably rail-guided as early as 1750.[23] A description of

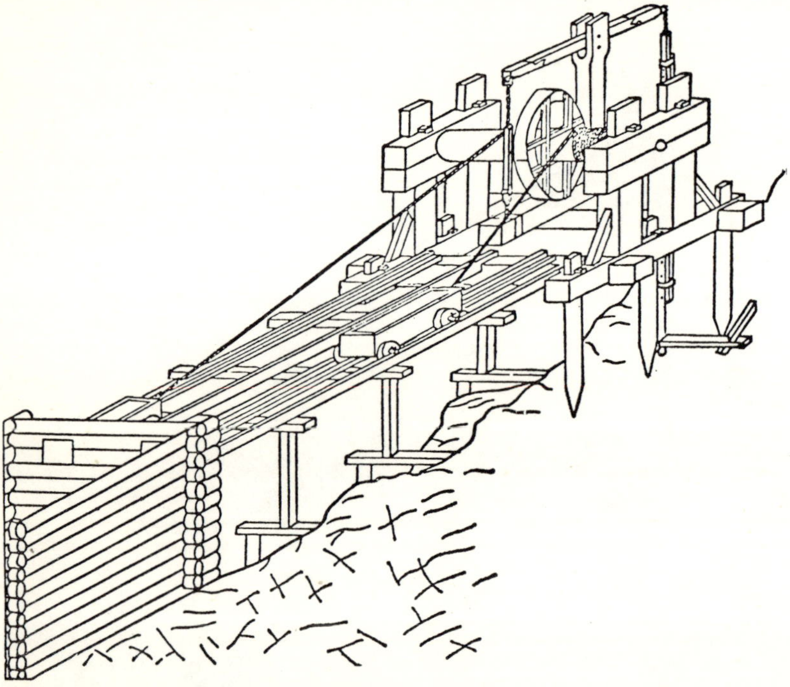

FIG 20 Balanced incline, Chagirsk, 1752.

1775[24] states that the long, level adits were floored with transverse planks, with longitudinal baulks some 4 in. high nailed to the sides, to guide the 2-ft. diameter wheels. On reaching the tips, the putter tilted the cart forward and emptied it through a door in the front. He was expected to make three or four return journeys, each of a mile and a half, in three hours. The writer was inclined to doubt the efficiency of these carts, compared with the Leitnagel Hunds or Truhen that the mine also used, since the large wheels took up so much room. The adits were 6 ft. high, but only 1 ft. 8 in. wide at the top and 3 ft. 4 in. at the bottom. By 1855, however, the carts held 6 cu. ft., compared with the 4 cu. ft. of a Hund, for their bodies measured 4 ft. 5 in. long by 11 in. wide by 1 ft. 6 in. deep:

their wheels too had increased in diameter to 2 ft. 10 in.[25] They were still very much in use in 1884, when they had also been copied at some open cast coal-mines in Saxony.[26]

A long transport system, for carrying logs from the Graf Hoyos' forests of Neuwald down to the Neustädter Canal a little south of Vienna, was built from 1805 to 1827 by Georg Hübmer.[27] Most of the system consisted of dry slides for the timber (see p. 70), but at one point a cable-worked incline was installed, with outside-flanged wooden rails and powered by a

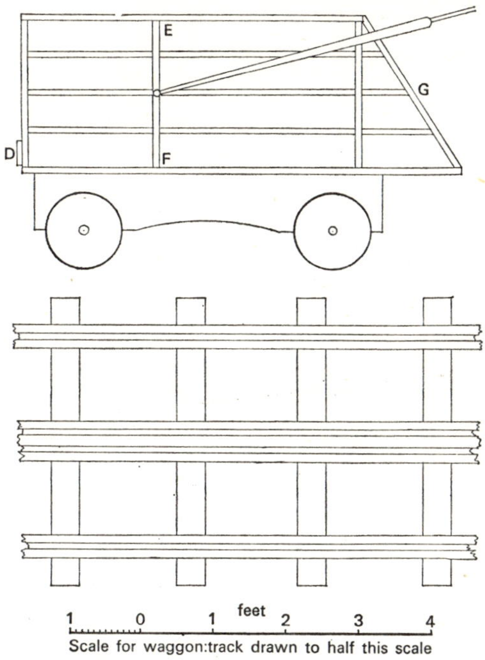

FIG 21 Truck and double track channel rail, Bleiberg, 1845.

water wheel, to surmount a 100-metre rise. Another Austrian incline, this time underground, was built before 1845 in Bleiberg lead-mine near Villach.[28] The track was double, with double-flanged or channel-section rails of about 2-ft. 5½-in. gauge (fig. 21), and, unusually, the bottom of the channel was lined with thin iron strips to reduce friction and wear. The incline was cable-worked and self-balancing, but, as it was to raise the load up to the main gallery, the descending empty trucks were filled with water to outweigh the ascending trucks carrying ore. A truckful of ore, however, might well weigh more than a truckful of water, and so the iron-bound wooden body (measuring 2 ft. 3½ in. high, 2 ft. wide and 6 ft. 2 in. long at the bottom) was divided by a vertical partition (*EF*, fig. 21).

Ore was put into the portion *EFG* only, but both compartments were filled with water for the downward trip. At the bottom, they were emptied by drain plugs at *D* and *F*. The wheels were of iron, 1 ft. 2 in. in diameter and 2¾ in. wide on the tread.

In the later years of the eighteenth century, the term 'German Hund' had meant any form of vehicle that was self-steering – usually by Leitnagel, occasionally by guide-wheel (see pp. 72-3) – while the Hungarian Hund was the truck steered only by the putter. In the early nineteenth century, however, with the modification of old types and the widespread introduction

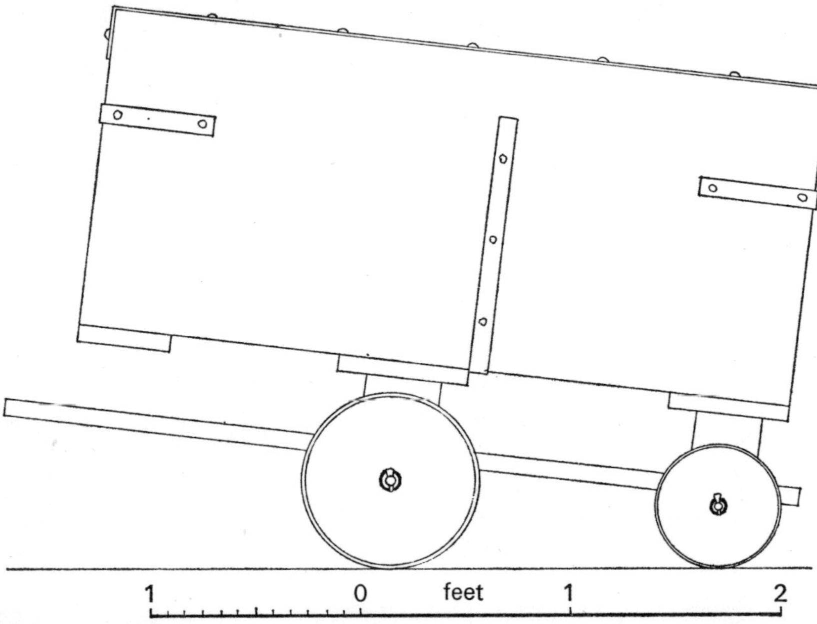

FIG 22 Ruhr 'Hungarian' Hund, 1823.

of new, miners tended to base their terminology on a different principle. The word 'Hund' was now often reserved for trucks whose fore and rear wheels were of different diameters. In the German Hund, still usually but by no means always guided by the Leitnagel, the centre of gravity came between the two axles; in the Hungarian Hund, which was in general still steered by the putter, the centre of gravity fell over, or nearly over, the rear axle. The word 'waggon' (*Wagen*) denoted a vehicle with four wheels of equal size, either the German waggon with plain wheels running on flanged rails, or the English waggon with flanged wheels running on plain rails.[29] This was the usual classification, but for some writers the word 'Hund' retained its original meaning of any mining truck, and was occasionally applied to English waggons.

The collieries of the Ruhr in 1823 employed both types of Hund and both types of waggon.[30] We will have occasion to discuss the English waggons in a later chapter; the other three vehicles all concern us here, even the German and Hungarian Hunds. A Schemnitz putter would have laughed to scorn this decadent descendant of his own well-bred Hund, for though it aped the traditional Hungarian variety in the arrangement of its wheels, the ham-handed colliers had destroyed all its advantages by adding guide-flanges to the rails and by working it with all four wheels on the track. This Hund (fig. 22) was used at St. Mathias and Herzkampf mines, where it held 4 bushels of coal and, unusually, was unloaded through a door in its back wall. The underframe was unusual as well. Three heavy cross-beams below the floor were linked underneath by planks. The two front cross-beams carried the axles, on which turned two 7-in. and two 10-in. wooden wheels, all with iron rims and naves. The wheels were underneath the body, and the gauge was therefore small – 11 in. at the most – though still greater than usual for a Hungarian Hund. Thin strips to act as flanges were nailed to the outside of the two planks which formed the track. The rear wheels took the brunt of the load, although the front wheels, which did most of the steering, also received some extra load through the putter pushing behind. In St. Mathias mine the Hunds were pulled up a final incline to the surface by a winch and rope, and were let down when empty by a brake on the winch. This particular version weighed 1½ cwt. empty and 8 cwt. full.

One mine in the Ruhr used a German Hund with unequal wheels (though mounted outside the inclined body), and many had German waggons with equal wheels and detachable bodies (fig. 23). Both types ran on the same track. Oak sleepers about 5 in. by 2 in., set at 6-ft. intervals, supported rails of a similar section, to which were nailed outside flanges 1½ or 2 in. square. The Hund held 3 bushels on 7-in. and 11½-in. diameter wheels; the waggons only 1½ bushels on 7-in. wooden wheels. The Ruhr used a number of different gauges, from about 1 ft. 6 in. to 2 ft. measured inside the flanges. This allowed ½ to 1 in. play for the wheels. Iron flanged rails first appeared here on a surface railway in about 1826,[31] but wooden track was still in service underground in 1853.[32] Although plank-ways were not infrequently laid, both here and elsewhere, as tracks for ordinary wheelbarrows, the Ruhr barrower, alone of his kind, was often blessed with a railway of his own: two rows of wooden strips, 1 or 2 in. square, were nailed to the single line of planks, leaving a 1½-in. gap between them for his wheel to run in.[33]

In general, the Hund proper was not favoured for coal-mines. In the first place the putters were not sufficiently skilled to handle them properly.

Secondly, both Leitnagel and Hungarian Hund, in their best form, had their wheels underneath their bodies. This did not, however, make them particularly unstable, for a heavy load like metalliferous ore kept the centre of gravity well down; but when they were used for coal, so much lighter a material, the centre of gravity rose dangerously high if a worthwhile load was carried. Once the wheels were set further apart to overcome

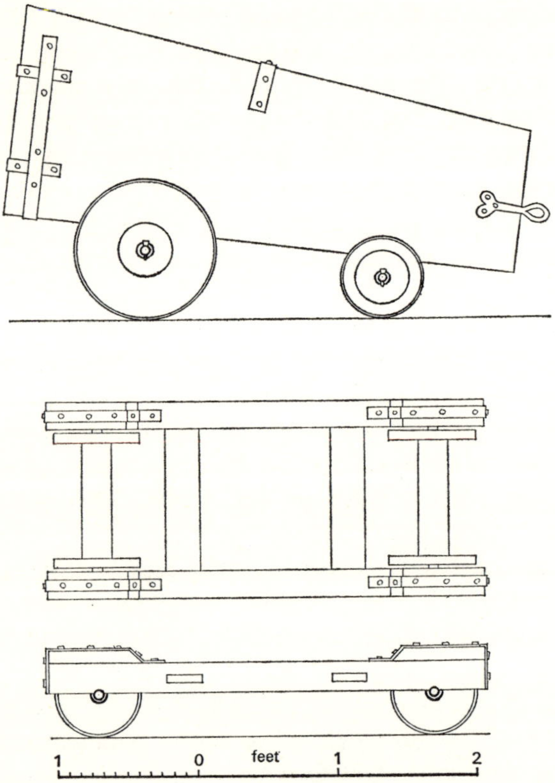

FIG 23 Ruhr 'German' Hund and German waggon, 1823.

this top-heaviness, the main virtue of the Hund was lost. It could no longer take sharp bends with ease, whether steered by Leitnagel or by the putter, and flanges were sometimes, as in the Ruhr, added to the rail to keep the Hund under control. Once this stage was reached, the few remaining Hund features, the sloping body and unequal wheels, had become pointless. It is significant that another area where Hunds were given the extra flanged rails was the Harz, where the older form of transport seems to have been decidedly inefficient. It will be remembered that the

Leitnagel Hunds here, unlike almost every other mining centre, had outside wheels at the corners, and were so heavy both in weight and friction that two men were regularly needed to move them, one pushing and one pulling. They did not use the Hungarian Hund much, and when they did the efficiency was often uncommonly low because the wheels were set far apart.[34] With this heritage, it is not surprising that the Harz engineers of the late eighteenth and early nineteenth centuries not only pioneered the introduction of the English flanged wheel to the Continent, but attempted to remedy the existing state of things by adding flanges to the wooden rails of both the Leitnagel Hund (also removing its pin) and the Hungarian. Before 1838 iron rails were also used for the same purpose.[35] Similar conversions of Leitnagel Hunds are found in 1846 at Bräunsdorf near Freiberg and Siegen in Nassau.[36]

It is difficult to say how justifiable it was to call the flanged rail 'German'. Certainly the wooden version was first applied in Germany, but equally certainly the iron version was an English invention (see p. 316), even though this too was called the 'German rail' on the Continent. Whether or not the flanged rail was in continuous use in metalliferous mines, the development of underground railways in some Continental collieries, such as those of the Mons area, seems to have been parallel to, but independent of, that of English mines. Other coal areas, however, such as Silesia and the Ruhr, were under distinct English influence, and the flanged rails in these collieries may well have been copied from English prototypes. Whether the wooden rail was native or imported, it is clear that when reports came in soon after 1800 of the English iron rail, many Continental collieries adopted it. They still called it the German rail, in succession to their own wooden version and to avoid confusion with the English rail for flanged wheels, whose origin they did acknowledge.

Under the influence of Graf von Reden and William Wilkinson, the coal and iron industries of Silesia adopted many English techniques (see pp. 339-41). One such was the underground canal in the Fuchsgrube at Waldenburg, copied from the Duke of Bridgewater's Worsley mines. Like its prototype, this canal was also fed by rail, originally of the English type for flanged wheels (see pp. 339). But this was replaced in 1803 by a system of container transport on flanged rails, whereby the coal was loaded at the face into wooden boxes and carried by rail to the canal side, where it was craned into barges for the second leg of its journey out of the mine. The barges held eight 4-bushel containers, each 2 ft. 6 in. high, 2 ft. wide and 3 ft. 4 in. long, plus two 3-bushel ones. The railway trucks, pushed by two men, had cast-iron wheels, which ran on wooden rails with outside flanges, set at a gauge of 2 ft. 9 in.[37] (pl. 28). The rails themselves measured 7 in.

by 3 in., the flanges 3 in. square. Smaller-scale wooden lines were still at work in Silesia in the 1850s.[38] This type of track also found favour at ironworks: the blast furnaces at Kreuzberg, for example, were supplied in 1802 by waggons that ran on outside-flanged rails of about 2-ft. 8-in. gauge, and were raised to the furnace mouth by a vertical lift, later replaced by a transporter incline.[39]

The mines of the Saar remained very primitive until they came under Prussian control in 1815, when they were given iron surface railways – with flanged wheels by 1819 and with flanged rails by 1821.[40] There was no doubt a corresponding improvement underground, but there are no details until 1856, when they were using both sledges and waggons.[41] This railed sledge-way was almost unique on the Continent. Planks 6 in. wide were nailed to sleepers set at 3-ft. intervals, and were equipped with outside flanges, 2 in. high and at a gauge of 2 ft., to guide the sledge runners. In one mine the putter held the sledge by two handles like a wheelbarrow, and, when he had a heavy load, could tilt it to bring a pair of wheels at the front into contact with the rails, instead of using the runners. The only other mention known to the present author of a Continental sledge-way is as late as 1891, but it names no places.[42] The track here consisted of two rows of planks laid side by side, with a third row laid centrally along the top. The sledge runners slid along the lower planks and were guided by the upper planks, which they straddled. For ordinary waggons in the Saar, sleepers measuring 6 in. by 1 in. by 5 ft. were laid at 3-ft. 4-in. intervals. They supported oak or beech rails 5 in. by 2½ in., on which were pegged flanges 1½ in. wide and 2 in. high: unusually for the Continent, the flanges were on the inside of the rails. The gauges varied, but were usually within the limits of 1 ft. 11½ in. and 2 ft. 3¾ in. Cast- and wrought-iron plates were also in use by this time.

The development of underground transport in the Mons coalfield followed a course closely similar to that of England.[43] The *sclon*, corresponding to our tram, was in medieval times a plain sledge carrying a corf, or wicker basket, of coal and pulled, by means of a rope attached to the belt, by relays of putters. During the early eighteenth century, if not before, the *sclon* acquired wheels and a planked floor to run on. Before long, wooden strips were nailed to the floor to help guide the *sclon*, and some collieries even protected these strips against excessive wear with iron bands – they are first mentioned at Gade et Hanat pits in the Bois de Boussu in 1766. Two-wheeled barrows, apparently on the same track, had appeared by 1801. Asses and sometimes dogs provided motive power, but the dogs at least proved unmanageable and were banished. Horses were not in general use underground here until 1819. Flanged rails of cast iron drove out the wooden ones from 1820 onwards, and were in turn superseded by English edge rails.

The rest of the colliery tale is soon told. The Liège coalfield may have had a similar history in underground transport, although details are very scanty. By 1773 some of the age-old sledges had been replaced by the *sployon* on small wooden wheels and by the *galhiot* or *gaillelot* on comparatively large iron ones. To begin with at least, the wheels of these *gaillelots* were unflanged, one pair 1 ft. 7½ in. in diameter and the other 1 ft. 2 in., set at about 2 ft. 5 in. wheelbase and 2 ft. gauge. They were quite possibly guided by lateral strips on the planked track, which began to be replaced by iron flanged rails about 1810.[44] The isolated colliery of Faymoreau (Vendée) had typical wooden flanged rails in 1830.[45] The coalfields of St. Etienne (Loire) and Alais (Gard) were late in developing, and started off with iron or (in two cases) stone flanged rails.[46]

The history of the English version of the flanged rail will be pursued in ch. 17.

5

Guide-wheel railways

Cette espece de traineau qui est tiré par un cheval, est composé d'une caisse . . . supportée par quatres petites roues verticales en forme de cylindres, qui tiennent de chaque côté à une piece de bois; au-dessous sont fixées quatre autres petites roues horisontales, qui servent à guider le traîneau & l'empêcher de s'écarter.

Gabriel Jars, *Voyages Métallurgiques*, Lyon, ii (1780), 147, on Lower Hungary Riesen waggons.

All three types of waggon so far discussed were propelled by man-power in the main, and were designed basically for underground use. The only exceptions were where mining trucks emerged from adits into the light of day for a brief journey to the tips, or where rope-worked inclines on the surface served a special military or mining need, for which the flanged wheel was especially favoured. Until the English waggonway was introduced in the late eighteenth century, the only mining centre on the Continent where horse-operated railways of any length could be found was that of Lower Hungary (now Slovakia), where the three great mining towns of Schemnitz, Kremnitz, and Neusohl, with four lesser satellites, were clustered together in the steep, heavily wooded valleys of the Carpathians.

The mines around these towns, producing much copper and lead, with some silver and gold, were for long the richest jewels in the mineral crown of Hungary, for most of them were imperially owned, although working was also allowed to the *Waldbürger*, a class who owned mineral rights not unlike those of the free miners of the Forest of Dean. Partly because of its isolation from Western Europe, but rather more because of the geography of the country, Lower Hungary tended to be inventive in mining practice. We have already seen how it evolved the Hungarian Hund. The mines themselves were often set in the hills, well away from the valleys, where alone was sufficient water-power available for operating the sizeable stamps and bellows necessary for the furnaces. The large quantities of undressed ore thus had to be carried, often for a considerable distance, from the shaft-heads to the stamps, a need which demanded, almost more than anywhere else on the Continent, an efficient system of overland transport.

Plain four-wheeled carts carried the traffic in the late Middle Ages, and although they continued to do so for many centuries, they suffered from all the disadvantages inherent in heavy vehicles constantly traversing steep and unmade tracks. In the 1560s we first hear of a special railed vehicle, which gradually achieved a position of great importance in the transport economy of the mines around Kremnitz and at Herrngrund to the north of Neusohl. The waggon was known as the *Riesen*, *Riesenkasten* or *Riesenwagen*; the track itself as the *Riesen*, *Risswerk*, *Riesengesteng*, *Riesenweg* or other variations on the same theme. The only description of the Riesen to have appeared in English is that by Edward Browne (see p. 22), who visited Kremnitz in 1669:

> Not much unlike this [the Leitnagel Hund] is another instrument they have to bring the Ore from the mouth of the Mine, or from the Hills down to the *Buchworke* [stamps] where they pound it and wash it; but instead of a tongue it hath eight wheels, or four rowlers and four wheels, and the way is made with firre in such manner and at such a distance that the rowlers rowle upon the wood of the Firre-trees. And these rowlers and wheels are so contrived, that these Chests can never overturn nor go out of the way, and a child draweth them, and sometimes a dog serves the turn. To one *Buchworke* alone, they carry every week three or four hundred of these Chests full, and each Chest holdeth four hundred pounds weight.[1]

Browne's description of the eight wheels has been taken to indicate flanged wheels – four rollers, each with a wheel of larger diameter behind it on the same axle.[2] That this is not so is proved by all the other descriptions that exist. Four of his eight were running wheels that rolled along the top of the rails, and four were guide-wheels, mounted horizontally under the body. They bore against the inside of the rails, and so prevented the Riesen from running off the track. The fullest description occurs in a dictionary of mining, compiled in Vienna in 1667, but relating to Lower Hungary,[3] which reads:

> A Riessen is that on which the ore or waste is transported on the surface by horse from the shafts to the stamps or the tips.
>
> Riessen-beams: these are long timbers laid at a constant distance apart and fastened together lengthwise by strong oak nails; on them the Riessen-chest runs. When the Riessen-beam becomes badly worn and impedes the Riessen-chest, long, wide and strong strips of timber called lining-wood are nailed evenly upon these beams.
>
> Riessen-chest is the truck [Truhe] for transporting the ore or waste. Underneath, at the ends, it has four long rollers which run upon the

> Riessen-beams or lining-wood; and under the floor, inside each long roller, it has a small one called the guide-roller. These four small rollers run inside, between and against the Riessen-beams and, by preventing the truck from leaving the Riessen, guide it forward. The horse can be harnessed to the Riessen-chest at either end, to avoid having to turn the truck round.

Similar but less elaborate descriptions are given by Gabriel Jars[4] and Christoph Delius.[5] Jars, who visited Schemnitz in 1758, adds that the body is 4 ft. 6 in. long, 2 ft. wide and 1 ft. 2 in. deep. He confirms that the vehicle is used only for long-distance transport. Delius's volume of 1773 provides the only known detailed illustration of a Riesen (fig. 24), which shows that the rails, some 4 in. high by 7 in. wide, are laid without benefit of sleepers upon the bare ground, at a gauge of about 2 ft. 1 in. The iron-strapped body of the waggon, measuring roughly 4 ft. 6 in. long by 1 ft.

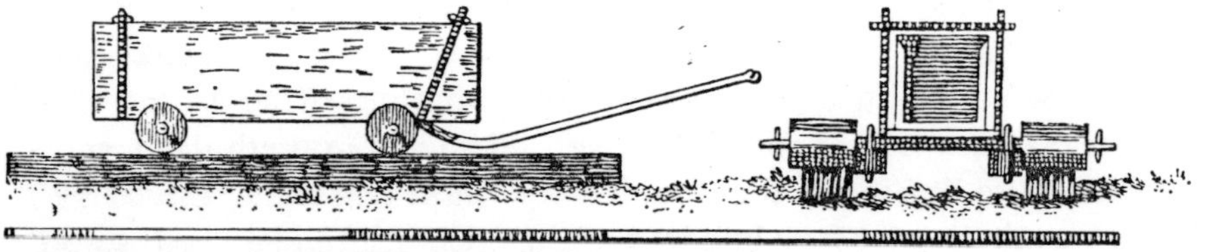

FIG 24 Lower Hungary Riesen, 1773 (after Delius).

4 in. wide by 1 ft. 3 in. deep, runs on rollers about 7 in. in diameter by 10 in. wide. One end of the body may be detachable for unloading, and, instead of a hook at front and back, Delius gives the Riesen what looks like a long shaft or handle, perhaps for hand-pushing. The shape of the wheels must have created a considerable resistance, while the end view, with its projecting wheels so resembling a child's toy truck, seems to modern eyes to be pitifully wasteful of carrying capacity. None the less, the Riesen seems to have marked a great improvement over the plain cart. Browne is, incidentally, alone in referring to boys or dogs as the motive power. Delius also says that the road waggon still used for the same purpose in some parts of Lower Hungary was likewise known as the *Riesenkasten*. In fact, the distinction between a road *Wagen* and a railway *Riesen* or *Riesenkasten* is generally quite clear in the sources.

Two other engravings reveal little. One of Kremnitz in 1703 (pl. 29) depicts three Riesen *en route* from the mines to the stamps, their horses being led by the drivers;[6] on the other, of Herrngrund near Neusohl in 1726 (pl. 30), two drivers are riding and whipping their horses as they haul the Riesen along the top of the tips.[7] Although in neither drawing are the rails or the means of guidance visible, the very small wheels suggest that guide-wheel Riesen, and not road-carts, are depicted.

This, then, was the form of the track and the waggons. To fill out the picture of their development and of their operation, we are vouchsafed a few glimpses from the voluminous reports, instructions and decrees issued and received by the Royal Office of Mines at Budapest (later at Vienna). Although these documents never tell us how the *Riesenwagen* were kept on the *Risswerk* or *Riesengestänge* that they mention, we may be reasonably sure that they did not differ in principle from the vehicles and track already described, which belonged to the same places at about the same times, and bore the same names.

The first indication of their use appears in a list of proposals for improvements at Kremnitz,[8] dated 10 September 1564: 'at Alte Richterin stamps to instal a new roof, and the long rails (*Langgestenng*) or *Risswerkh* with their appurtenances, for transporting the gold ore'. The revision in 1573 of the old mining laws of Kremnitz and Königsberg, among many provisions for individual *Waldbürger* to mine on their own alongside the royal establishments, contains a most interesting section which authorized compulsory purchase, or at any rate compulsory leases, of land for railways.[9] Shafts could be sunk in any field, and the requisite machinery installed, including the *Risswerk* for carrying ore away.

> Concerning the Risswerk and the regulations by which it is to be laid out and built, so that it is not hindered by heavy gradients, the said Risswerk, where it appears necessary, shall be laid along a route [literally a scar] through meadows, wheatfields, gardens or ploughed fields without hindrance; but it is to be so arranged that an agreement can be reached with the owner of the said fields, and where such an agreement cannot be reached amicably, permission and consent is to be obtained by application to a suitably competent judge or magistrate. So long as the Risswerk is built and maintained, the oldest Risswerk shall always have the right of way, provided that essential paths are not thereby obstructed but an outlet from these paths to the highways shall be maintained. If it should happen, as has often happened in the past, that one Risswerk is built across another, the newer one shall always yield the embankment to the older one [sic: *sol . . das Jüngste dem eltesten den aufwurff halten*], and neither shall deliberately hinder the other with its traffic. Furthermore it is also to be known that when one person constructs a Risswerk through the field of another, as mentioned above, the said *Waldbürger* [i.e. the owner of the field], if he wishes to transport waste, rock and ore to

> and fro, shall not be hindered by the Risswerk, but the owner of the Risen shall accommodate him; and if any *Waldbürger* should wish to sink a shaft, he [the owner of the Riesen] shall give way to him and cause no obstruction with his traffic.

The sections of this code dealing with the royal mines lay down that surface workers – on the tips, sorting floors, and *Risswerk* – shall work a nine-hour shift, since their labour is lighter than that of the men underground, whose shift is eight hours. Drivers, 'whether with [road] carts or on the Riesen', must drive with due care and attention.[10]

By 1603 a commission of enquiry at Kremnitz was complaining about the expense of raising the ore and of transporting it from the older shafts to the stamps.[11] Too many gin and Riesen horses were employed, they said, and too many workmen; the number must be reduced and their efficiency improved. The waggon-steward's duties are listed in 1607 as supervising the waggoners and Riesen-men, buying and maintaining waggons and horses, and issuing the grease for lubrication.[12] Hereafter the Kremnitz Riesen escape specific mention in the official documents, except that the *Risen fuhrleuthe* (leaders or drivers) got a pay rise in 1646, and from 1750 received an annual bonus of 3 guilders 20 kreuzer (about 16*s.* 8*d.*) for shovelling snow off the track in winter.[13]

The Herrngrund copper-mines near Neusohl, which stood at such a height above the works that wear and tear on carts and horses was exceedingly heavy, were the next to adopt the Riesen. In 1583 the Office of Mines, being reminded that nine years before it had authorized various measures to save transport expenses here, wrote to the Director of Herrngrund:[14]

> To transport the ore from the mine to Hermanz Hillock, a channel or Rinnwerk [wooden chute] was immediately built, in which the ore could be brought down from the mine and the tips; also a stamping mill with 150 stamps was built beside Wissten Water; and from here to the new works a Risswerk was made, all of which cost 4,257 guilders [about £1,050], against which can be set a saving of 2,000 guilders [about £500] a year on transport. Now that we have received report that the chute or Rinnwerk, together with the boxes for carrying the ore, is already complete,

would he provide further details of the cost, of its usefulness, of its saving on transport costs, and on the state of the *Riss* and stamps. The answer has not survived, but it was two years before the Office had completely digested it. Then at last, its mills grinding with bureaucratic slowness, it came out in 1585 with an order to extend the *Risengesteng* (rails) from the ore-store at the new works down to the furnaces. Two years later again, the Office produced another order to the director of the mines[15]:

> Instead of the existing Rinnwerk, now partly falling into disuse, at the Herrngrund copper-mine, in order to save expense in transporting ore and to increase the volume of traffic, a Risengesteng is to be built and erected like that used at Kremnitz, which has however been abandoned for various reasons and never put to work. We have been reliably informed that a similar Risengesteng from Herrngrund mine smithy to Hermanz Hillock can be completed for 500 guilders [about £125], whereby several draught horses can be laid off, and 2½ pfennig saved on the cost of every ore-carrying journey.

What was the matter with the Kremnitz railway, and wherein (if at all) the proposed line differed from the two sections of Riesengesteng that already existed at Herrngrund, we do not know. Nor, regrettably, are we told more of the Riesen which was to replace the old chute; its route must have been either very steep or very circuitous. The weekly saving on transport by means of Riesen was 19 guilders 50 pfennigs in 1607;[16] if the 1587 estimate of 2½ pfennigs per trip was accurate (at 60 pfennigs or kreuzer to the guilder), an average of 476 Riesen-journeys were made per week, or 79 per day.

The Riesen never seems to have achieved popularity at Schemnitz, where the need was less – most of the mines are within easy reach of their accompanying works. In the late seventeenth century the ore was taken from the shafts to the stamps, in some mines at least, in small Hund-like trucks pushed along a rough road.[17] Riesen had arrived, however, by 1749, when a report says that the rock was tipped at the shaft-head, and either removed by Riesen or run to the waste-tip in small Hunds.[18] These were sometimes Leitnagel Hunds, larger than the underground breed, and according to Delius were called *Riesenhunde*, either because they were of giant (*Riese*) size, or because they served the same purpose on the tips as the *Riesenkasten*. By his time (1773) they had been abolished, except in one washery.[19] The syllabus of the Mining Academy at Schemnitz, on its foundation in 1770, included the art of transport by Riesen,[20] and a building account of 1778 includes the item: 'To an arched bridge of masonry leading from the Magdalene shaft to the tips, and to installing a Riesen Lauf [run] thereon, 1,032 florins 18¼ kreuzer – a tidy sum of about £250.[21] From the paucity of the references, the Riesen seems to have been a rarity at Schemnitz, compared with Kremnitz and Herrngrund.

During the nineteenth century the mines of Lower Hungary lost much of their former glory – Kremnitz indeed had done so some time before – partly because the veins were becoming exhausted, partly because of rising competition from other parts of the world. The Riesen seem to have faded away during the first half of the century, probably driven out by the advent of the all-pervading English flanged-wheel waggons.[22] As far as is known, no Riesen has survived anywhere.

One other mining area of Eastern Europe may have favoured the guide-wheel Riesen for a time: the Nagybanya district on the borders of Upper Hungary and Transylvania, some 220 miles east of Schemnitz, with three large mines at Nagybanya, Felsöbanya and Kapnik. Riesen here are first mentioned in the Office of Mines papers in 1761, when the inefficient double handling of ore at the Felsöbanya shaftheads was abolished by tipping straight into the Riesen instead of tipping on to the ground and reloading on to the *Riesenweg*. It was to be the responsibility of the mine overman to make sure that the Riesen were well loaded, and to record the number driven away. The duties of the supervisory staff were reallocated in 1780: the clerk at the stamps now kept tally of the number of *Riesenkasten* run; the surface overseer ensured that the Riesen were properly driven, without spilling ore on the track; and the stable and Riesen overseer looked after horses and drivers, saw that the *Riesenkasten* were well filled (but not overfilled), measured the Riesen once a week, paying special attention to the adjustable cross-straps which regulated the width of the body (and in the hands of unscrupulous drivers reduced the capacity), and inspected the straps, axles, and rollers. At Kapnik, Riesen were introduced in 1761 to take ore from the St. Barbara adit to the stamps. A *Riessen-Führer-Aufseher* (overseer of Riesen drivers) is mentioned at Nagybanya in 1766.[23]

The form of guidance of the Riesen of this particular area is not on record. It is possible that they were of the flanged-wheel type, as used in Transylvania, to be discussed in the next chapter – Nagybanya is nearer to Transylvania than to Lower Hungary. On the other hand, Nagybanya always adhered to the orthodox mining methods of Lower Hungary rather than to the individualistic practices of Transylvania, and it remained, like the Schemnitz area, a part of the kingdom of Hungary throughout the time that Transylvania, to the detriment of her mining, was under Turkish suzerainty. The Felsöbanya Riesen sounds much more akin to the (comparatively) sophisticated Lower Hungary one than to the exquisitely primitive Transylvanian version.

There now arises the question of the origin of the guide-wheel. An obvious answer is the Leitnagel Hund, with the plank-rails set much further apart, and the Leitnagel itself multiplied by four and equipped with larger rollers to guide the large vehicle. It may be so. But another, more obscure, possibility exists, to appreciate which it is necessary to consider the origin of the word *Riesen*. The use of the term, meaning a truck with guide- or flanged-wheels used for overland transport, or the rails it ran on, is confined to what are now Slovakia and western Rumania. It appears either as *der* or *die Riesen*, often in compounds and with its variant spellings of *Risen*, *Rissen* and *Riessen*, or more rarely (and earlier) as the compound *Risswerk*, meaning the track. *Der Riese* is the German for 'giant', which might be applicable to the vehicle, but hardly to the rails. In fact, the only two exceptions to this geographical distribution probably do mean 'giant': a *Riesenhund* used in the Harz in 1785, which was merely the largest of three different sizes – Bear-, Boar- and Giant-Hund – and a *Riesenkasten* noted in 1856 in Austria, a two-wheeled (and therefore unusually large) wheelbarrow.[24]

Die Riese, however, means an artificial channel, usually in the form of a continuous timber framework, used by foresters for sliding logs by gravity down a steep hillside to the valley below. These Riesen first appeared, it seems, in Styria, Tyrol and the Salzkammergut, where countless references to them are found from as early as 1303. Many villages in the mountain valleys had their own slides to transport their timber, both for firewood and for building, quickly and cheaply down from the forests. Many are the names given to them, but always a variation on the same theme: *Risswerk*, Risen, wood-Risen, log-Risen, tree-Risen, monastery-Risen, local-authority-Risen. Individual houses even had their private branches from the slides.[25] As time went by, they spread to other parts of Europe – the Black Forest, Galicia, the Carpathians, the Banat of Temesvar – and they spread especially to mining areas.[26]

A forestry section was an indispensable adjunct of any mining venture, to supply timbering for the mines themselves, and particularly the vast quantities of charcoal consumed by the roasting and smelting furnaces. As the woods immediately around the larger mines were used up, so the need for a cheap form of timber transport from more distant sources became imperative; and the answer, especially in Lower Hungary, lay in floating logs down the mountain streams, or, if there were no handy streams, in the Riesen. In 1547 a Styrian specialist engineered some streams at Neusohl for log-floating;[27] but this was not enough, and in 1561 the chief foresters of all the Lower Hungary mines were instructed to send to Eisenerz in Styria and Aussee near Salzburg for experts in Riesen-building.[28] Thereafter the forestry orders of the Office of Mines make frequent reference to Risen, *Risswerk* and *Rissgefährten* (Riss-tracks) – decisions to lay them, instructions to maintain them, and so forth.[29] The copper-mines around Neusohl especially favoured Riesen as their timber supply-lines, and at their greatest extent they boasted 106 miles of them.[30]

The similar, and often identical, terminology used for both the forestry slides and the guide-wheel railways of Lower Hungary (and, to a lesser extent, for the flanged-wheel railways of the Banat: see ch. 6) leads one irresistibly to the conclusion that the railways of that name were somehow derived from the much older-established forestry slides. Regrettably, there is a complete lack of descriptive works on forestry before the nineteenth century, and what the Riesen of the sixteenth-century forests looked like can only be deduced by analogy from later times.[31] From a very early date there was a sub-species known as *Wasserriesen* or *Clausen* ('flumes' in English), a wooden channel filled with water for floating logs down; and,

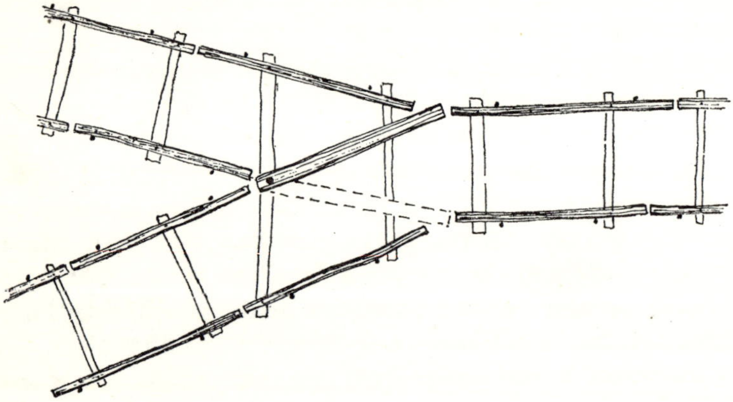

FIG 25 Forestry Wegriesen points.

in winter, slides of ice or packed snow. The dry slides, which are more relevant to our enquiry, generally needed a gradient of 1 in 4 to 1 in 6, and were subdivided, at least by the early nineteenth century, into two main types. The *Holzriesen*, or U-shaped channel built up entirely of wood, was favoured especially in the Black Forest and Tyrol. In the *Wegriesen*, *Riesweg* (Ries-way) or *Holzbahn* (wood-road), used all over Europe but especially in the Carpathians, the timber slid down a road-like formation with occasional cross- 'sleepers' and edged with logs where necessary to prevent it leaving the track. Where branches were needed, there were pivoted 'points' (pl. 31 and fig. 25). This is the nearest approach of a forestry Riesen to what the track of a guide-wheel Riesen must have looked like – this is the only form of point suitable for guide-wheels which project below rail level (see fig. 27). The traffic on these *Rieswege*, however, was not wheeled vehicles, but plain logs or trunks with the boughs lopped off, sliding endways. Though the word *Rissgeferten*, used several times in the 1560s at Neusohl, could be translated either as 'Riss-carriages' or as 'Riss-tracks', the latter must be meant, since the *Rissgeferten* were 'laid'. In the late nineteenth century sledges were sometimes run down these *Rieswege*, but the only case of wheeled vehicles using them seems to have been in Lower Austria in 1607, when it was laid down that 'A Riesen should be wide enough for a waggon to pass along it'.[32] It is indeed possible that the Lower Hungary Riesen for the transport of ore were evolved from running ordinary carts down between the retaining timbers of a forestry *Riesweg*, although one would have thought that, on a gradient sufficient to keep a rough tree-trunk in motion, a horse would have excessive difficulty in pulling an empty waggon up, let alone in holding back a descending waggon full of ore. Alternatively, the ore Riesen may be a slightly more sophisticated version of the forestry Riesen, provided with continuous smooth side timbers for rollers to run on. The word 'Riesen' was restricted to timber-sliding and to ore-carrying in waggons. When a wooden slide was used, for example, for shooting ore down steep gradients, whether the ore was in boxes, as at Herrngrund (see p. 66), or was loose, as illustrated by Agricola,[33] it was called a *Rinnwerk*, not a *Risswerk*; and railways, whether with wood or iron rails, were not used for purely forestry purposes until the nineteenth century (see p. 55).

Although the guide-wheel was the distinctive feature of the Riesen of Lower Hungary, it was not unknown in western Europe from the early years of the eighteenth century. In the great Stora Kopparberg Mine at Falun in Dalarna, Sweden, copper ore was carried in the smaller levels by sledge or barrow, but the main levels to the shaft bottoms were large enough to accommodate a *hästväg* (horse-way), on which ran waggons (*hästkärran*) guided by a varying number of horizontal wheels running between the rails. The records first mention them in 1711, when they were introduced into the Smårumstollen, whence they quickly spread to other parts of the huge workings. The earliest specimen certainly had two guide wheels, one attached to each of the running-wheel axles, and, as in Hungary, the horse could be harnessed to either end.[34] A late eighteenth-century drawing of a somewhat clumsy specimen (pl. 32) ascribes its invention to Christopher Polhem;[35] it may be so, since he was engineer to Falun at the time, but yet, though rarely reticent about his own achievements, he never claimed that he invented it. He had travelled widely in western, but never apparently in central, Europe; none the less, the similarity is such that the inspiration, whether Polhem's or another's, probably came from Lower Hungary, if only indirectly. The waggon in the drawing measures 5 ft. 7 in. by 1 ft. 4 in. inside at the top, by 1 ft. 2 in. deep, supported on four heavy iron-rimmed wooden wheels 1 ft. 3 in. in diameter and 6 in. thick.

The two horizontal wheels, carried by brackets and hooks, ran between two massive rails 1 ft. by 10 in., set 1 ft. 3 in. apart. These dimensions are not completely certain, since the scale on the drawing bears no figures, but they are confirmed by a report of 1751, according to which a waggon was 1 fathom long.[36] This writer, as well as two German travellers in the 1760s, also noted the two guide-wheels.[37]

The Falun stock, however, was not limited to the one type. A waggon in a drawing of about 1750 boasted four horizontal wheels; but, unlike those under the Hungarian Riesen, with one horizontal wheel set inside each vertical one, here they are all strung out in line astern[38] (pl. 33). This waggon – again the scale lacks figures – seems to measure 5 ft. 6 in. long by 1 ft. high, with all the wheels about 1 ft. in diameter. In contrast to this well-endowed example, many waggons had to be content with a single guide-wheel. The great Linnaeus of botanical fame, visiting Falun in 1734, wrote of

> a waggon with four small wheels, under whose floor is a revolving horizontal wheel. There are set on the floor of the mine two beams, on which the four wheels run, but the fifth horizontal wheel runs between the two beams, so that the four wheels may run on the broad surfaces at the sides.[39]

The same arrangement was noted by two Frenchmen: Outhier in 1737[40] and the indefatigable Gabriel Jars in 1767, who also found waggons of a similar kind at Sala silver-mines in Västmanland, where the levels were wide enough for a double track.[41] These guide-wheel waggons survived until they were ousted by English flanged-wheel trucks in the first decade of the nineteenth century.

In Germany, the guide-wheel principle made its first appearance in 1724 at Schulenberg in the Harz Mountains, where the mine engineer Kristian Schwarzkopf built an inclined plane on the surface to carry ore from the shafthead down to the valley bottom.[42] It is significant that Schwarzkopf spent two and a half years in his youth as pupil to Christopher Polhem at Falun, at the time when guide-wheel waggons first appeared there.[43] Pl. 34, though its perspective is haywire, shows the Schulenberg incline, on which ran two small waggons or Hunds, 6 ft. 2 in. long, 2 ft. 6 in. wide, and rising in height from 7 in. at the top to 3 ft. 2 in. at the bottom. The sides were linked by an iron strap (*b*). The iron-rimmed wheels, 3½ in. wide and 9 in. in diameter (top wheels) and 11 in. (bottom) were held on the iron axles by split pins. Just below each wheel a strong iron pin projected 4 in. below rail level, with a tall iron ring revolving on it to reduce friction. In this respect they resembled the pin of the Leitnagel

1. *Above* Hund in adit, Upper Hungary, 1513 (Rožňava triptych).

2/3. *Right and Below* Hund in adit, Vosges, late fifteenth century (St-Dié Gradual).

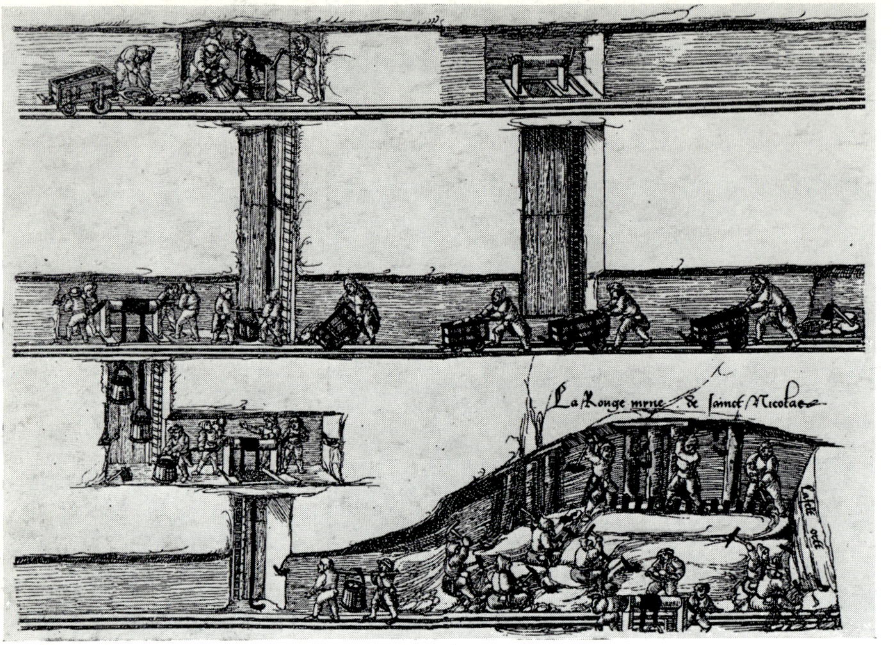

4. *Above* Hunds underground, La Croix-aux-mines, 1532–62 (Gross).
5. *Below* Tipping Hunds, La Croix-aux-mines, 1532–62 (Gross).

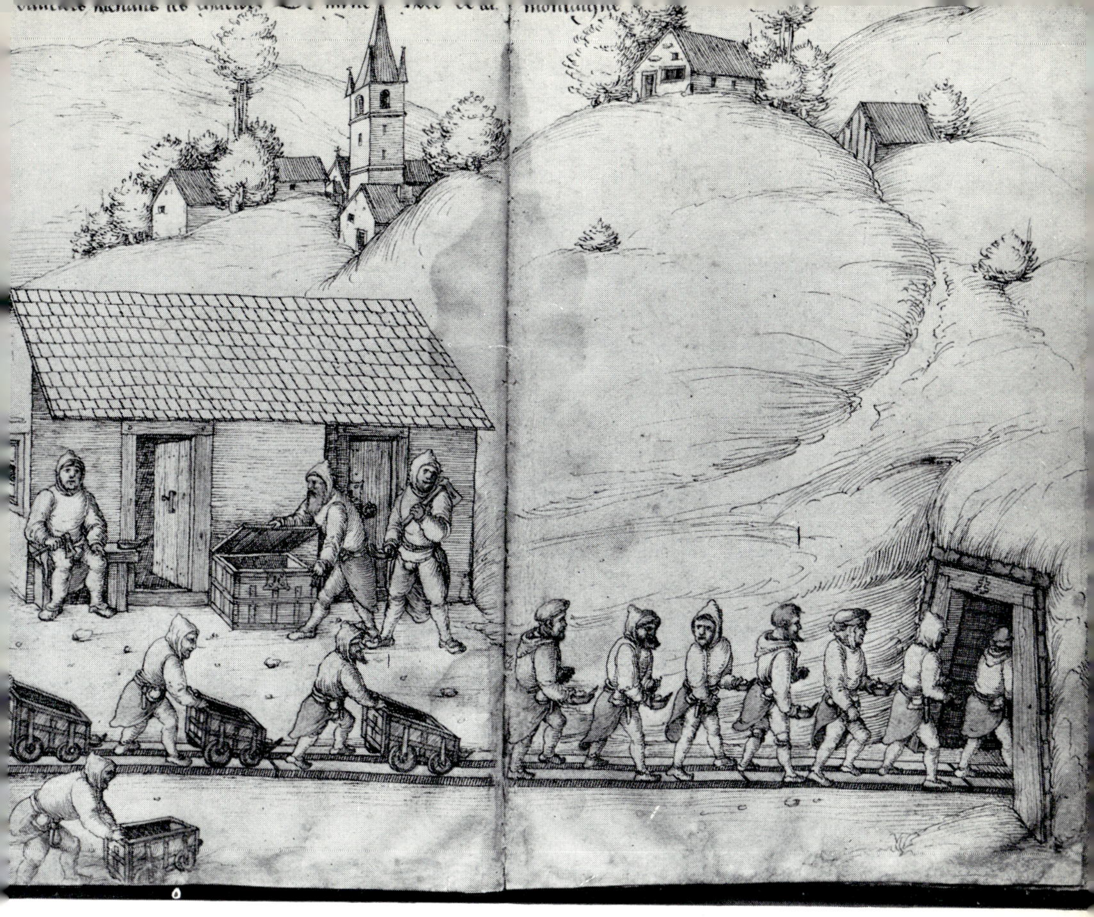

6. *Above* Hund-putters going on shift, La Croix-aux-mines, 1532–62 (Gross)

7. *Left* Miners at work, Switzerland, 1517–19? (Holbein).

8/10. *Above and Below* Leitnagel Hund (Bergtruhen) and storehouse for rails, Schwaz, 1556 (Schwazer Bergbuch).

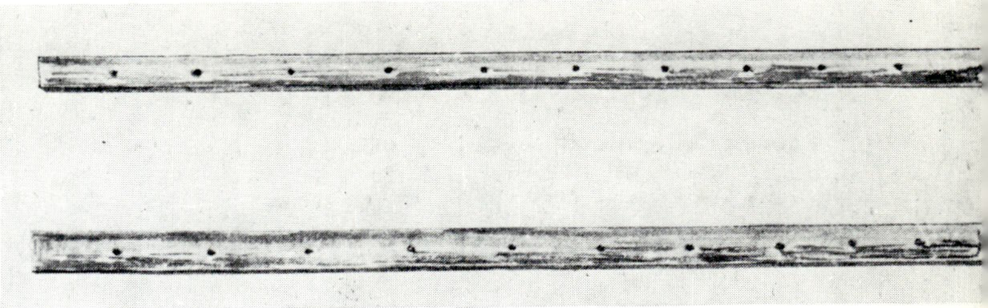

9/11. *Above and Below* Iron rails (Reibeisen) and Hund-(Truhen-) putter, Schwaz, 1556 (Schwazer Bergbuch).

12. *Above* Hund-putters, Hall, 1707 (Gaun).

13. *Below* Model of Harz Leitnagel Hund.

14. *Left* Leitnagel Hund from Simmental coal mine, Switzerland, in Verkehrshaus der Schweiz, Luzern.

15. *Below* Hund with pivoting axle and guide-bar (figs. 1 to 3), and outside-flanged English waggon (figs. 4 to 6): Freiberg, 1791.

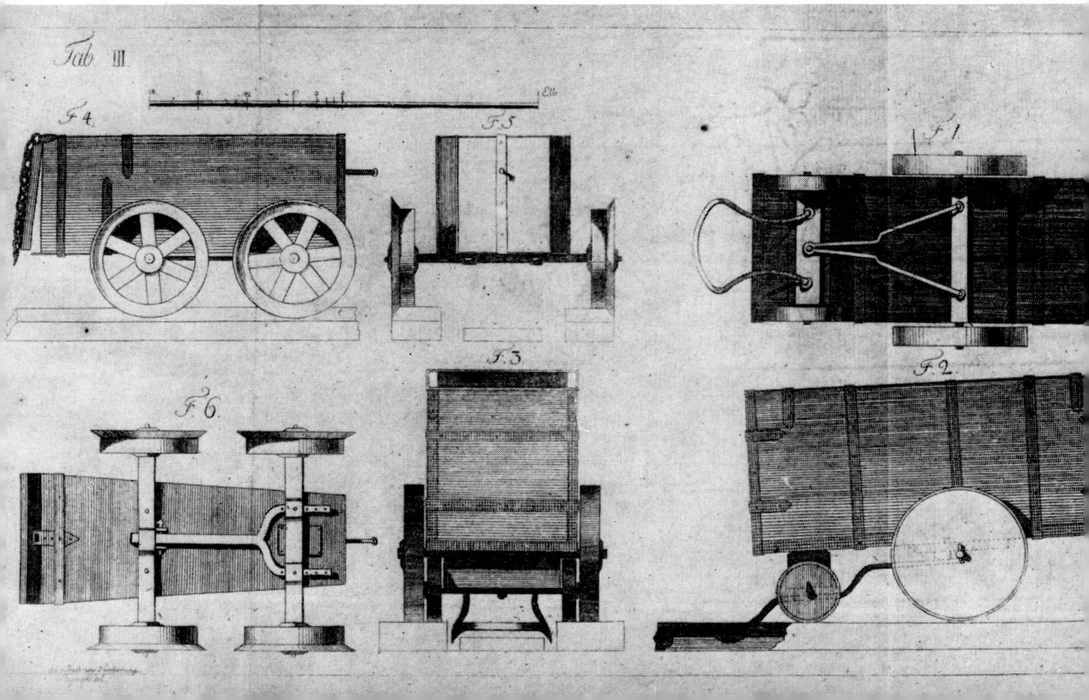

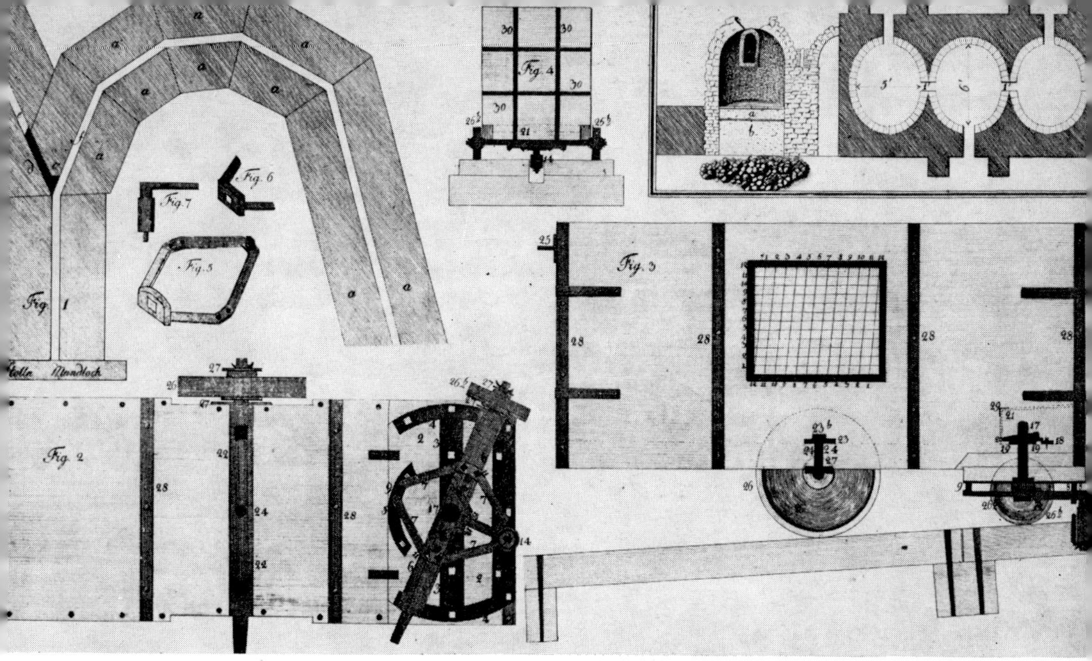

16. *Above* Hund with pivoting axle, and pointwork, Braubach, 1790.
17. *Below* Hungarian Hund, Schemnitz, 1805 (Becker).

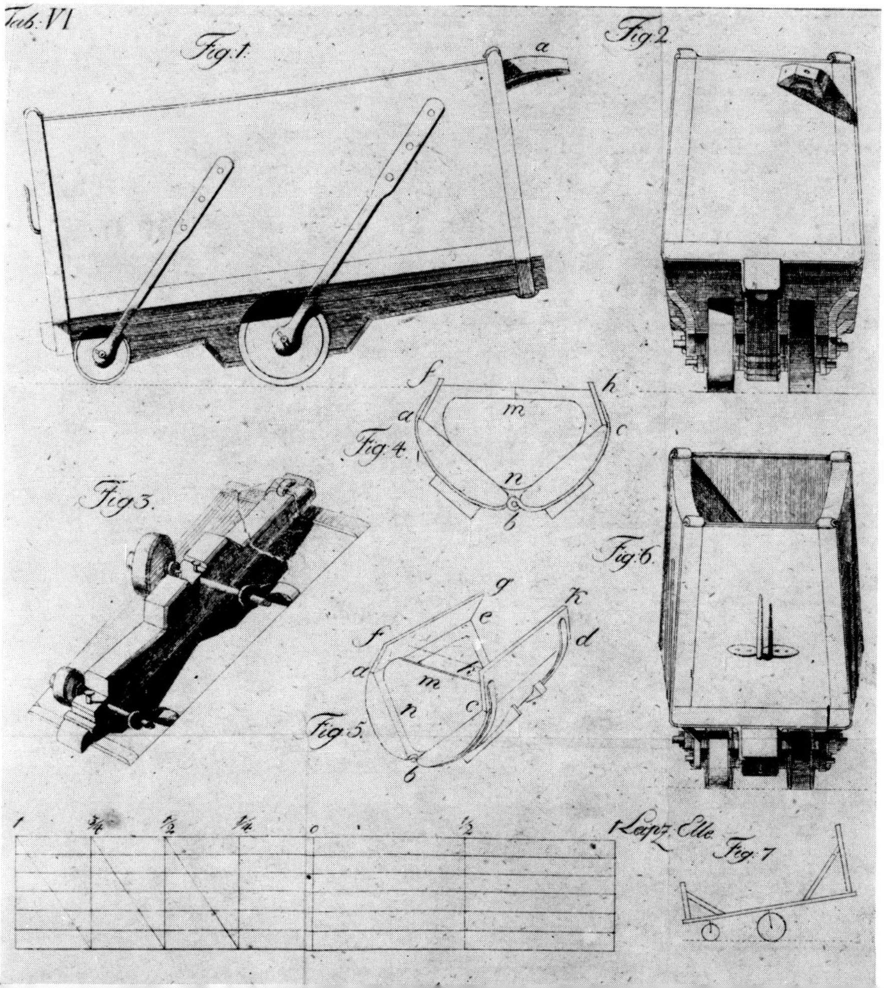

18. *Above* Hungarian Hund in Vlastivedné Múzeum, Banská Bystrica (Neusohl).

19. *Below* Hungarian Hund from the Harz, in Zellerfeld Heimatmuseum.

20. Military railway, *c.* 1430 (MS. of Anonymus der Hussitenkriege).

21. Military railway, 1588 (Ramelli).

22. Fusina canal incline and railway, 1607 (Zonca).

23. Mining truck, *c.* 1520 (Petrarch).

24. *Right* Mining truck, late fifteenth century (Gand collar).

25. *Centre* Mining truck, 1544 (Gassell).

26. *Bottom* Mining truck on flanged rails, Alsace, 1543 (Rappoltsteiner Pokal).

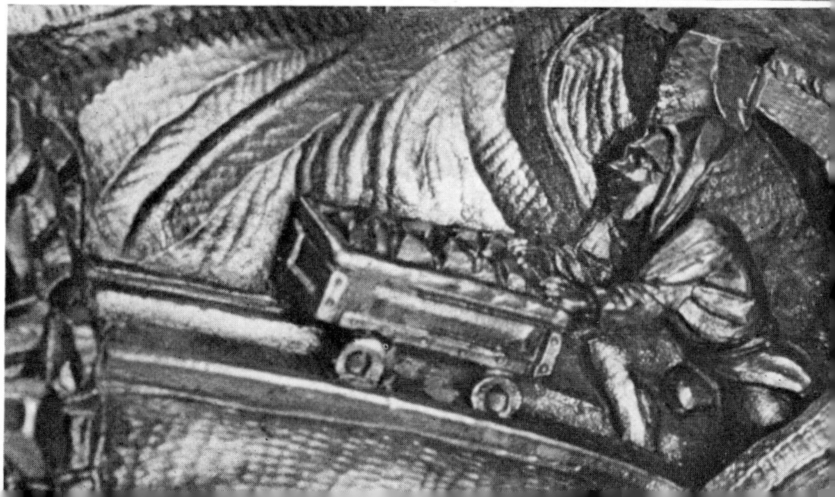

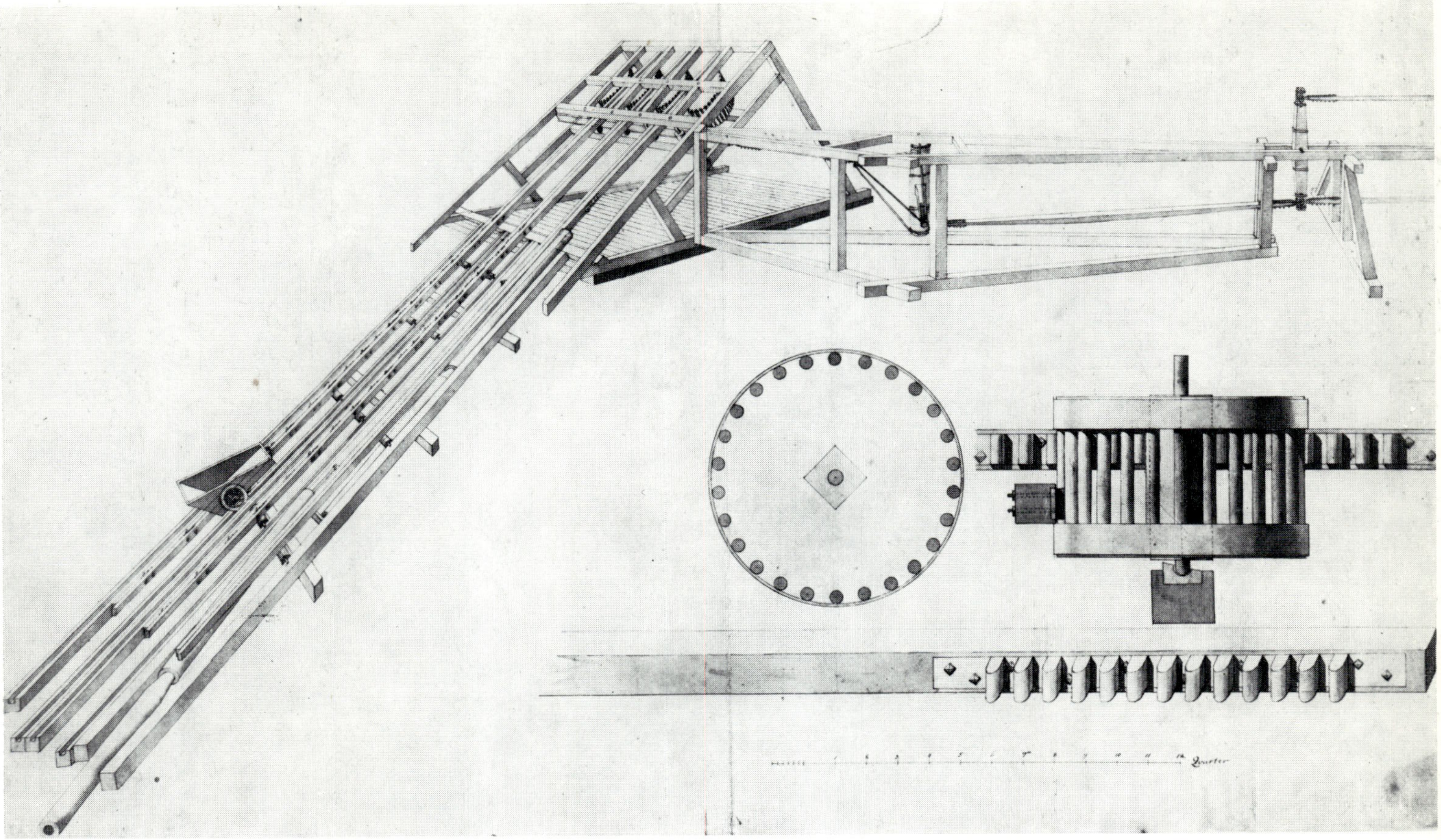

27. Flanged-rail incline and truck, Humboberg, 1729 (Cronstedt).

29. *Above* Kremnitz mine, 1703, with four Riesen visible above and below the waterwheel at left centre.

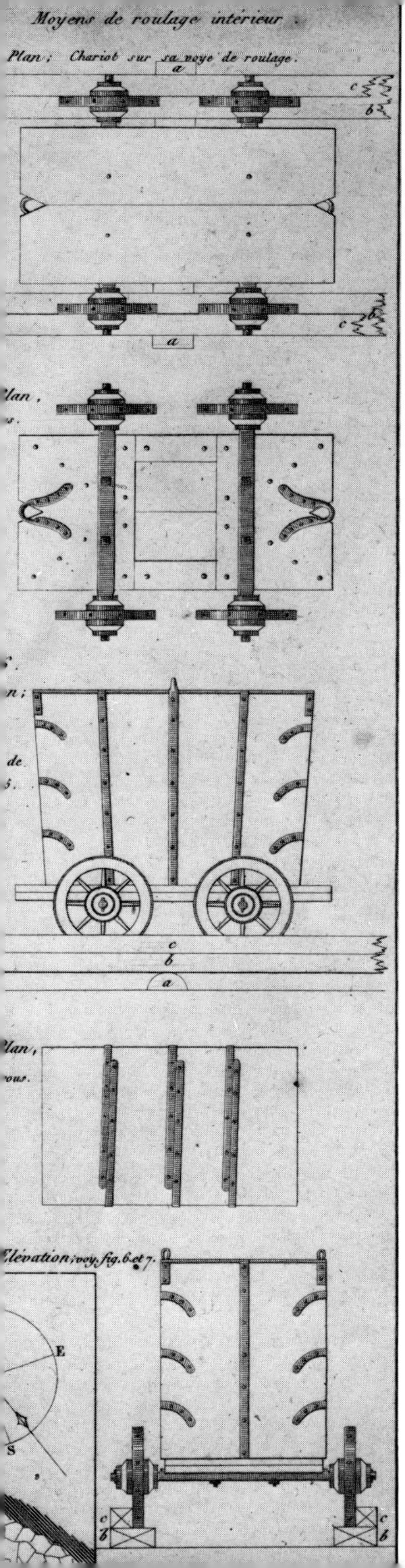

28. *Left* Flanged rails and truck, Waldenburg, 1819.

30. *Below* Herrngrund mine tips, with Riesen, 1726.

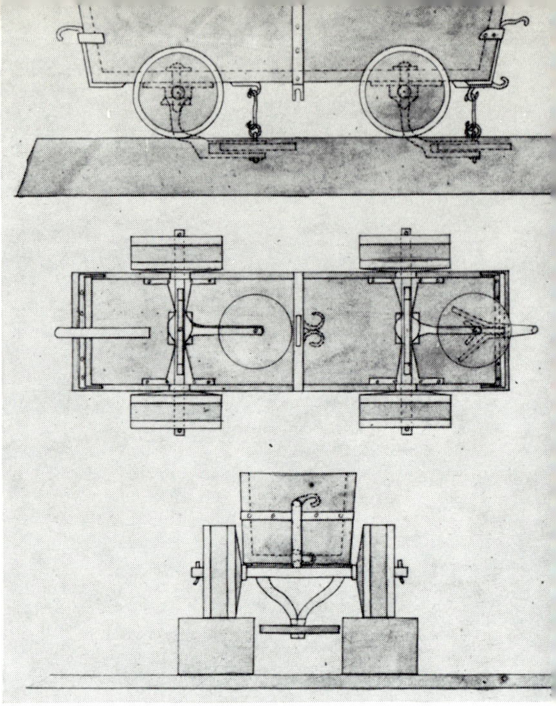

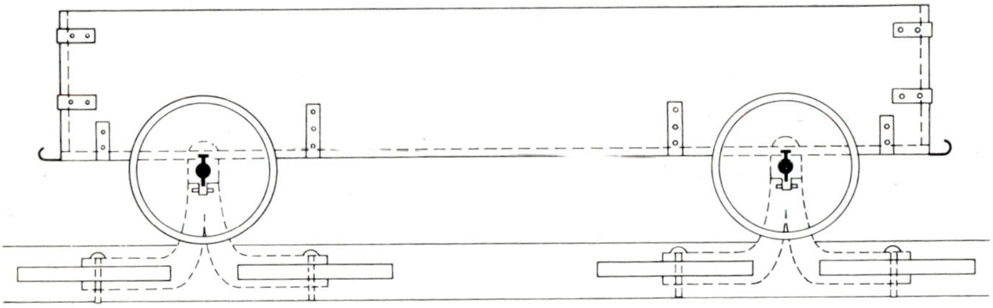

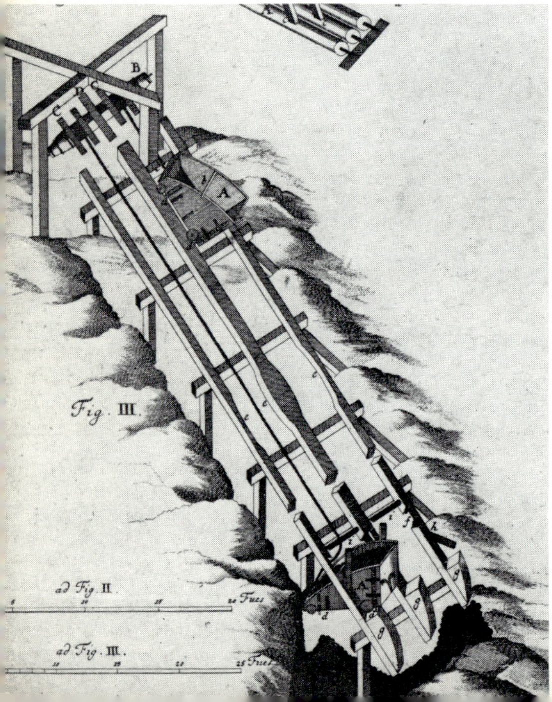

31. *Top Left* Forestry Wegriesen (road slide), 1863.

32. *Top Right* Falun guide-wheel waggon, late eighteenth century.

33. *Middle* Falun guide-wheel waggon, c. 1750.

34. *Left* Balanced incline, Schulenberg, 1763.

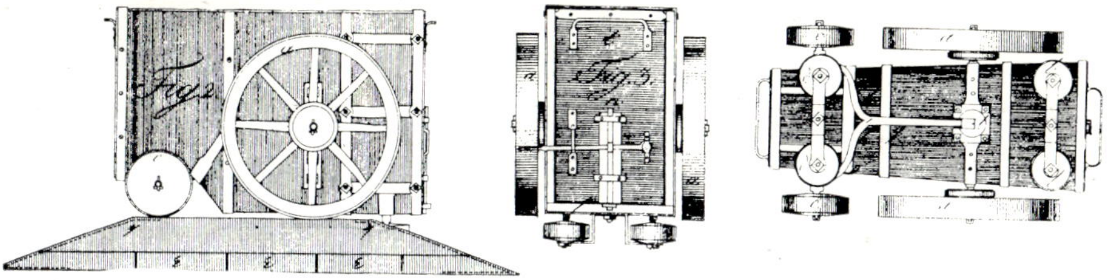

35. *Above* Guide-wheel waggon, Schneeberg, 1794.

36. *Below* Erbisdorf elevated guide-wheel railway, 1792.

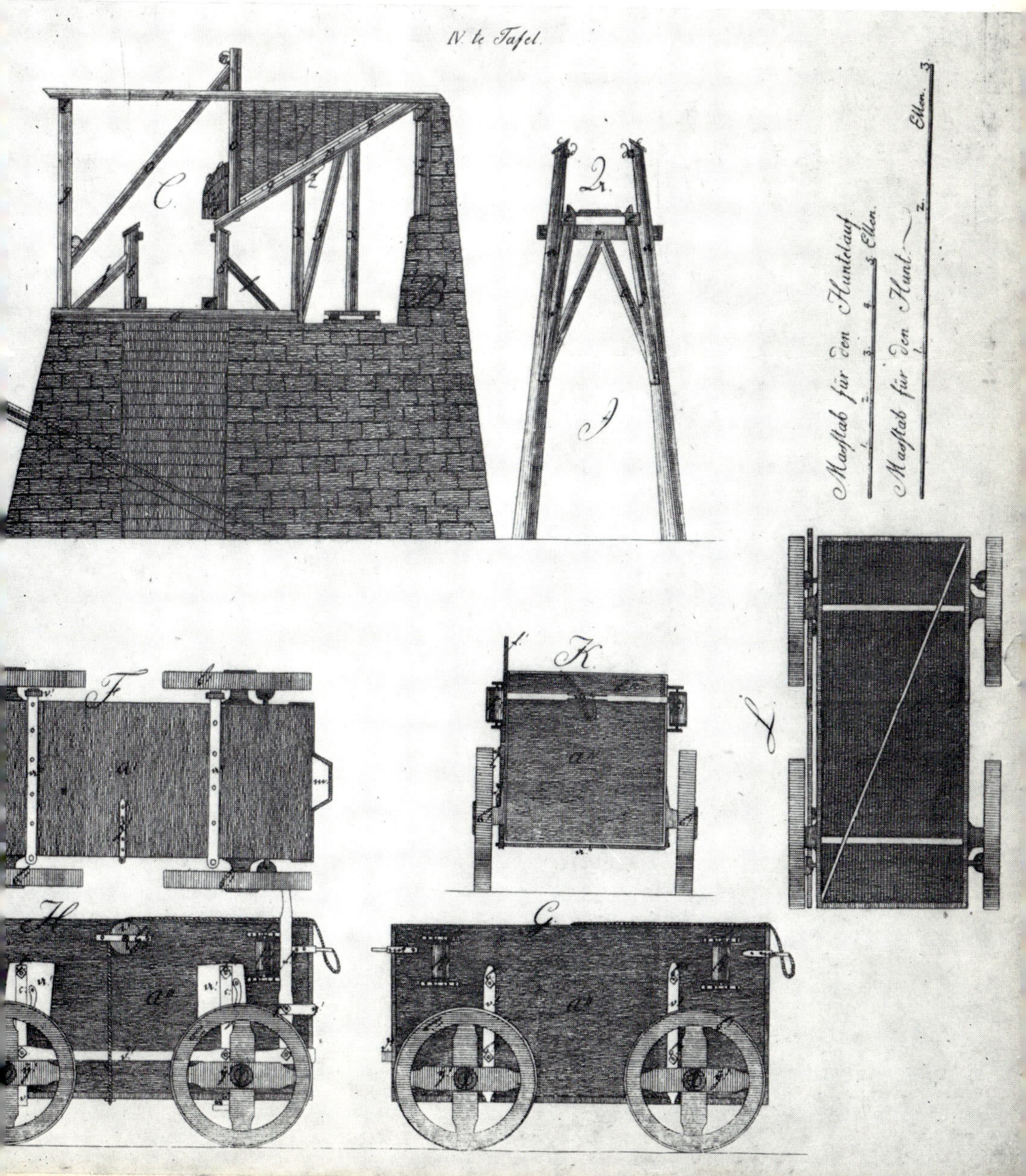

37. *Above* Transylvanian Riesen and track in Berlin.

38. *Below* Transylvanian Riesen in Bochum.

Hund rather than the wooden rollers of Hungary and Sweden, and were indeed called *Leitnägel*. The outer rails were 7 in. wide, the common inner rail 10 in., except at the halfway point, where it bulged out to 14 in. (with corresponding cavities in the outer rails) to allow the waggons to pass. At the incline head, two ropes were wound round drums (*CC*) so that when one waggon was up, the other was down; and the weight of the loaded one descending pulled the empty one back. A brake bore on the wheel (*D*) to regulate the speed. At the incline foot, a section of each rail was cut away (*f*), just below which a piece of angled iron (*h*) was pivoted. The lower wheels of the waggon, having crossed the gap in the rails on the upper arm of the angled iron, then knocked the lower arm down and the upper arm up, thus leaving the gap open for the upper wheels to fall through. The whole waggon swung downwards, held only by the lower wheels, now bearing against the stop-blocks (*g*), and emptied its contents on the tip below. This Schulenberg railway is the earliest known example of the self-acting incline, a principle much used in later years, especially in England.

Three mines in the Saxon Erzgebirge adopted the guide-wheel in the 1790s, though it never proved satisfactory enough to survive for long. The Tiefer Marx Semler level at Schneeberg was equipped in 1791 with large hand-pushed 'German' Hunds running on four iron-rimmed wooden wheels and guided by four small horizontal wheels between the rails, whose gauge was 1 ft. 6 in. But ore emerging through the door in the bottom clogged the guide-wheels; it was expensive and heavy; it held 8·4 cu. ft. and only five runs could be managed, totalling 42 cu. ft., in an eight-hour shift over 390 fathoms.[44] So three years later a smaller and cheaper waggon appeared (pl. 35) to hold 6 cu. ft. and to run seven trips per shift, again a total of 42 cu. ft. But instead of needing a man to work it as before, it was now just light enough to be moved by a boy, whose wages were less. The axles of the eight-spoked wooden front wheels, 2 ft. in diameter, were fixed to the sides, and the front guide-wheels were likewise fastened to the bottom. The small (9-in.) rear wheels, however, with their attendant guide-wheels, were all assembled on an iron arm which ran forward to end in a ball-and-socket joint under the front wheels, thus giving the vehicle great flexibility on curves and, in particular, allowing the body to be tipped forward to empty the contents through the top-hung door in the front, while all four wheels remained on the rails. The cost of the truck was 16 reichsthaler, or around £4. The engineer, Herr Baldauf, was a man of ingenious bent; perhaps too ingenious, for so great was the friction in a loaded Hund that the wretched boy putters were barely able to push it up a regular gradient of 1 in 178. Nothing daunted, Baldauf relaid the track in a series of steps, each about 12 yards long, of which the first 8 yards were level. The next 3 yards then dropped $\frac{1}{2}$ in. to allow the Hund to get up speed to surmount the main step or 'leap', a wedge-shaped rail 20 in. long which rose $3\frac{1}{8}$ in. to the next level (fig. 26). To ease the passage of this switchback, the rails immediately above and below the

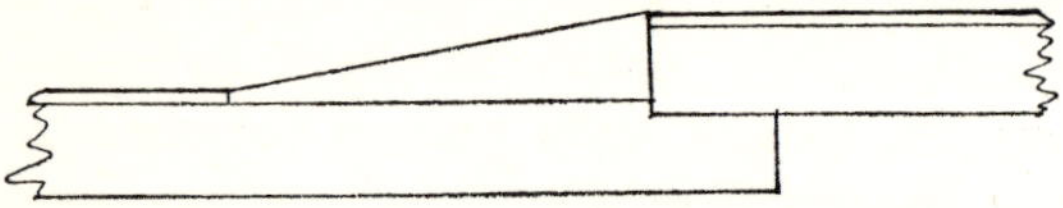

FIG 26 'Step' in track at Schneeberg, 1794.

main step were covered with thin iron plates, and the rails elsewhere had replaceable beech strips, 3 in. by 1 in., nailed to their surfaces. Baldauf also introduced turntables for sharp corners, and to enable Hunds to pass each other he used three devices: a double-track turntable, one track accommodating the empty Hund while the full one ran through on the 'main line'; a short dead-end siding leading off a single-track turntable, used in a similar way; and an elaborate railed platform suspended from a vertical pivot, which swung the empty Hund bodily off the track.[45]

At Freiberg the guide-wheel Hunds were more orthodox. A specimen

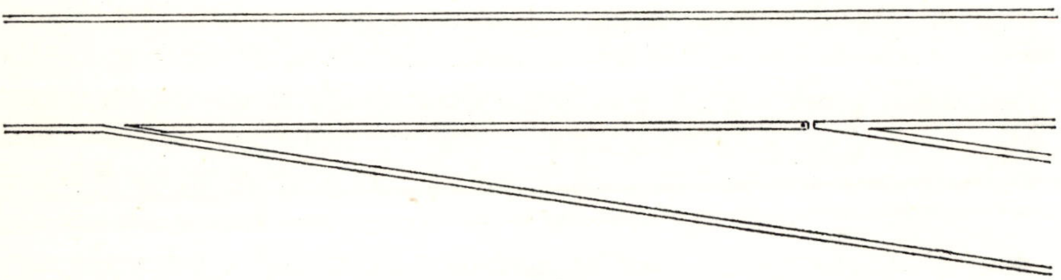

FIG 27 Guide-wheel railway point, Freiberg, 1792.

used in 1792 at Neu Morgenstein Mine looked very like Herr Baldauf's masterpiece: the wheels were similar, and the large ones were so near the centre of gravity that the back could easily be lifted to empty the contents out of the front. But there was no pivoting of axles here, the body was large (14 cu. ft.) and provided with flaring ledges on top to prevent spillage on to the track, and it was a surface Hund, carrying waste to the tips. Hardwood rails, 6 in. square, were set at 2-ft. 4-in. gauge on cross-sleepers at 6-ft. intervals, and we are here vouchsafed our only glimpse of the type of point used on a guide-wheel railway (fig. 27), with one rail pivoted to swing right across the track.[46] D'Aubuisson records that this line was still in use in 1802, when he found similar vehicles underground in the same mine, so heavy both in weight and friction that two men were needed to move them. The rails they used, at 1-ft. 6-in. gauge, were 3 in.

high by 4 in. wide, and were capped English-wise by a strip of wood 4 in. by 1 in., easily replaced when worn.[47]

The last German example to notice, built at Erbisdorf, near Freiberg,

FIG 28 Distribution of guide-wheel railways.

in 1792, was in two respects a peculiarity. A surface line, only 400 ft. long from the roasting ovens to the stamps, it was elevated all the way upon a wooden viaduct about 10 ft. high; and though it ran in the usual way upon four iron-bound wooden wheels, its guide-wheels did not bear against the inside of the wooden rails, but consisted of tall vertical rollers mounted high up the sides of the body and running against baulks attached like hand-rails to the viaduct (pl. 36). The reason for the viaduct was that the ore had to be delivered to the stamps from a height; the reason for this form of guidance was probably to keep it out of reach of the ore that was emptied through a door in the bottom. The Hund – a large one, 4 ft. 11 in. by 1 ft. 10 in. by 2 ft. 3 in. – was filled by chute, pushed by one man, and on reaching the stamps an upward-projecting handle was automatically knocked back to release the catches of the bottom door.[48]

Except in a special case like Falun, with extremely large levels, the guide-wheel Hund was far too elaborate for the rough-and-tumble of underground work, where the levels were usually restricted in size, and where speed, and therefore lightness, was essential. As Becker, the apostle of the Hungarian Hund, was always preaching, the best results were obtained by reducing weight and friction to an absolute minimum, which involved, if the workmen were skilful enough, doing away with mechanical guidance completely. On the surface, where speed mattered less, where horses could be used and where there was more room to fit in the extra wheels, the system was less objectionable. In the event, as on the surface, the underground guide-wheel succumbed to the all-conquering English flanged wheel, except where the Hungarian Hund had already been substituted. But the guide-wheel has never been forgotten. Its use, or reuse, has been proposed from time to time, the best-known cases being Baader's plans of 1822[49] and Prosser's diagonal guide-wheel of 1844,[50] and is now being actively revived, notably on some lines of the Paris Métro and on the new Montreal Métro.

6

Flanged-wheel railways in Eastern Europe

Man ist überzeugt, dass man durch Abwerfung der Riesen und durch Einführung der Schemnitzer Hundeförderung viel ersparen würde; aber man hat die Wallachischen Arbeiter noch nicht dahin bringen können, diese kunstlichere Arbeit zu erlernen.

W. H. G. Becker, *Journal einer Bergmännischen Reise durch Ungarn und Siebenbürgen*, ii, Freiberg (1816), 192, on transport at Nagyag.

The development of each type of railway so far described has proved to be a cul-de-sac: each of them has died, to all intents and purposes, a natural death. The flanged-wheel railway, on the other hand, grew from the most humble of beginnings to blossom out during the nineteenth century into the railway we know today, which socially, economically and even physically changed the face of Britain and of much of the world. There is no shadow of doubt that this plant, slow and insignificant in its early growth, but all-embracing in its maturity, was trained and tended in England up to and indeed far beyond that famous day when it suddenly burst into flower in 1830. But where the seed was first sown has been in dispute, though there cannot be much doubt.

There are but two contenders for the honour: what is now Rumania and England. There is not the faintest hint that the flanged wheel was used anywhere else until about 1775, when the first genuinely English railway was exported to the Continent (see p. 333). The principle of flanged wheels guiding a vehicle along rails may seem supremely simple to us today, but although the grooved pulley was well known to the ancient world, and was applied during the Middle Ages to the aerial ropeway,[1] it went no further until, in some dark mine, the first crude flanged wheel was fashioned. Did this happen in the vigorous and expanding England of the first Elizabeth or in a primitive corner of eastern Europe then subservient to the yoke of Turkey? We cannot be categorical, but the evidence is enough to suggest very strongly that the honour belongs to England.

The basic position is this: the first recorded railway in Britain (ignoring the Leitnagel Hunds at the Mines Royal in Cumberland, a very different kettle of fish) was built at Wollaton near Nottingham in 1603-4. Although proof that English railways used flanged wheels does not appear till later in the century, the likelihood is strong that the first English railway wheels were flanged, as will be explained in ch. 7. The case for the priority of Transylvania rests solely on a truck and a length of wooden track preserved in Berlin (pl. 37), to which a sixteenth-century date is ascribed.[2] There is no doubt about their provenance: they were acquired in 1889 by A. Haarman, director of an Osnabrück steel company, from the Zwölf-Apostel level of Ruda goldmine near Brad in Transylvania, for his unique collection of permanent-way materials. From him they passed, with his collection, to the Verkehrs- und Baumuseum in Berlin, where they survived the Second World War, but are now stored inaccessibly in the Russian sector. There appears to be no evidence for the date ascribed to them, bar their primitive appearance, which, as will be seen later, has nothing to do with the case. Before examining the actual records of such vehicles in Transylvania and the Banat, a brief digression on the early mining history of these countries is necessary, although information is woefully meagre.

The rugged mountains of Transylvania, now the north-western province of Rumania, had been extensively mined by the Romans, especially for gold. After the Gothic invasion, there was little exploitation until bands of the ubiquitous Saxons were attracted here in the eleventh century to form commercial and mining colonies: Ruda, Zalathna, Boicza and Verespatak were among their early mines, Zalathna in particular earning a reputation for wealth.[3] On the defeat of Hungary by the Turks at Mohacs in 1526, Transylvania, which had been more or less Hungarian for centuries, moved into the orbit of the Ottoman Empire. Though nominally independent, an island in a sea of Islam, it was in fact subjected to Turkish suzerainty, and in 1661 became a mere province of Turkey. The Turkish defeat at Vienna in 1683 put an end to this, and Transylvania soon became Hungarian once more in effect, although the union was not ratified until 1699. In the Turkish period the economic climate had been far from favourable to mining development: at best it was a period of stagnation, and we are told that most mines were abandoned to avoid the greed of the Turks. A traveller in 1687, who found any number of mines in Hungary proper, could discover only one (Zalathna) in Transylvania.[4] After the union with Hungary, the Imperial Government took a certain interest in the Transylvanian salt-mines, but only after the loss of Silesia, with all its mineral wealth, to Prussia in 1742 did they turn their attention to the metalliferous mines, and even then the big reorganization was deferred until the 1760s.[5]

The Banat of Temesvar is the district north-east of Belgrade and south-west of Transylvania (now partly in Yugoslavia and modern Hungary,

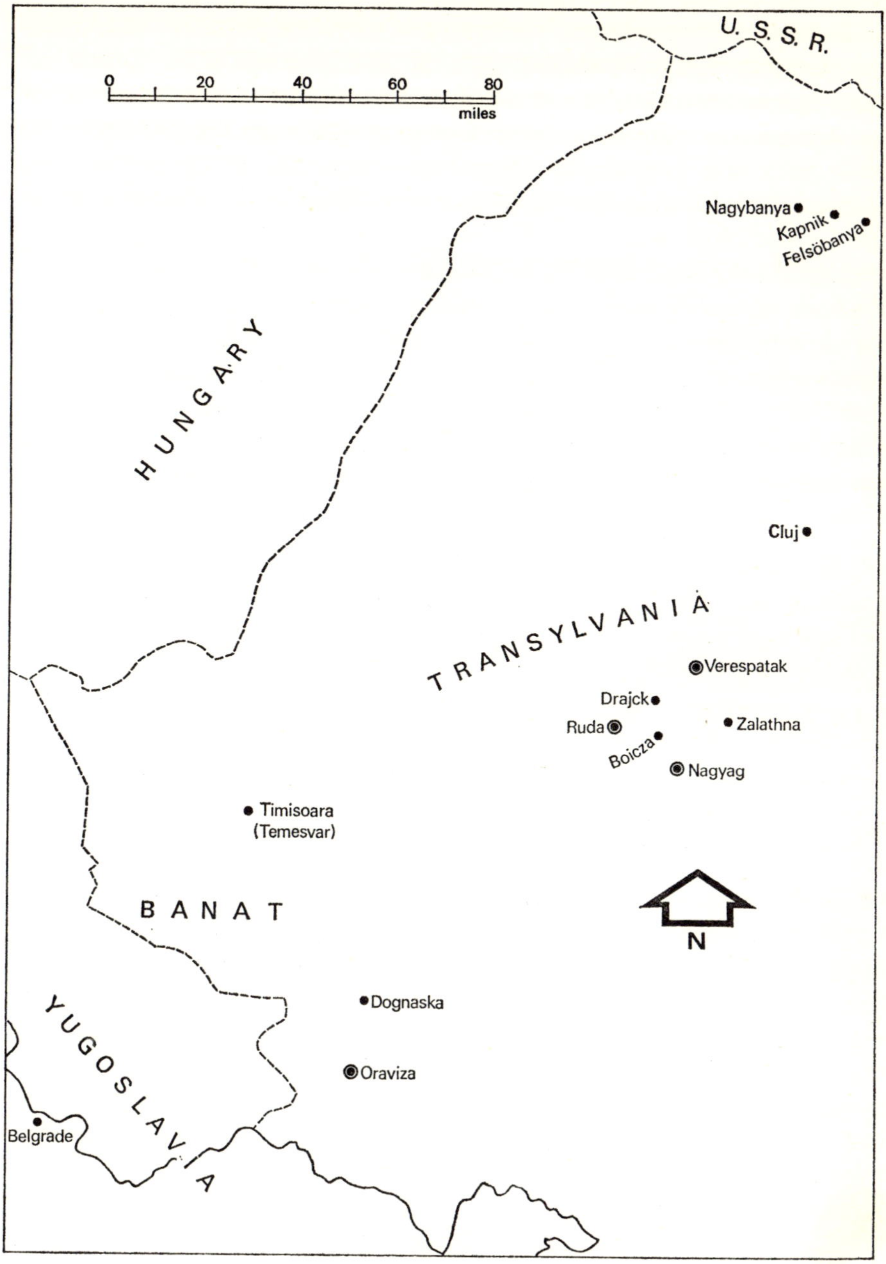

FIG 29 Distribution of the flanged wheel in Eastern Europe.

but mostly in Rumania). It fell under Turkish control in 1552, and suffered far more than Transylvania, for it was largely depopulated, and mining was almost entirely abandoned. In 1716 Hungary recovered it, and set about opening and restoring mines with the utmost vigour.[6] Orders about the Banat mines appear suddenly and in profusion in the papers of the Office of Mines from 1717; a brand-new organization was set up, and squads of miners were drafted in from the Tyrol to the main centres of Oraviza and Dognaska.

The first mining trucks in the Banat were no doubt imported by the Tyrolese, probably their usual Leitnagel Hunds. A list of miners whom the imperial agent at Schwaz was instructed to send to the Banat in 1722 includes 70 windlass-men and Hund-putters,[7] and although this particular draft was called off, many parties found their way there. The Hund certainly came into regular use underground at Oraviza and Dognaska.[8] But it was not until 1794 that we first find definite evidence of the flanged wheel. In that year a Dane, Jens Esmark, visited the Oraviza mines:

> The transport here, as in almost the whole of Banat, is carried out with so-called Riesen, a sort of Hund, which are drawn by horses. They have no wheels, but instead four rollers of 8 or 9 in. diameter, on which are fixed, on the side nearest the body, discs or ledges of 14 in. diameter. These rollers run on rails (*Strassbäume*) with convex surfaces, and the friction is therefore inconsiderable. At the front and rear ends of the frame is fixed an iron hook, to which the end of the horse-harness is attached. The Hund is not turned, but the horse is led round and yoked to the rear when it has to take the Hund back.[9]

In Transylvania the Riesen were likewise preceded by ordinary Hunds. Of the four mines which certainly used the flanged wheel, the best documented is Nagyag, an old Roman gold-and-silver-mine deep in the mountains, which was first reopened in 1746 by the father of the great mineralogist Ignaz Born. The Office of Mines exhorted the manager there in 1764 to continue with 'the improved Hund-system', which might mean the Riesen, though it probably refers to the ordinary Hund, very likely of the Hungarian type.[10] A table of pay for workers at Nagyag, Verespatak, Boicza and Drajck, issued by the Office of Mines in 1767, lists both Hund-putters and overseers of Riesen-drivers. The table, however, is not good evidence for Riesen at these places, for it was copied verbatim from one of the previous year designed for Nagybanya, where Riesen were certainly at work, although probably, as we have seen, of the guide-wheel type.[11] Indeed, no Riesen of any sort had appeared at Verespatak by 1767, and probably none at Nagyag. Born himself, in a description of Nagyag in 1770, speaks of the ore being drawn from the mine, apparently by hand, in wooden *Bergtrügen*,[12] a word elsewhere applied only to the Leitnagel or Hungarian Hund. Then at last the Riesen made its first undoubted appearance here. It is noticed by Andreas Stütz, a visitor of 1795, who is quite definite about the *terminus post quem* of its introduction.

> The transport to the surface of the good ore and the waste rock is one of the new arrangements for the mine made since 1774. It is performed in the Philip and Bernhard adits by Riesen, or large Hunds, which run on four wheels and are drawn by a horse. With these Hunds the Riesen-man goes deep into the mine, and quickly brings out a mass of rock to the tips, or to the crushers and stamps. The company pays for the hire of such a horse 36 kreuzer per shift, and the Riesen-man must lead five or six loads per shift, depending on whether he has to bring his load from far inside the adit or from near the mouth.

He adds a story about the presence of mind of a Riesen-man who, on seeing the roof-timbers in the Josef adit beginning to sag, whipped his horse forward and leapt back himself. The roof collapsed between them; both were missed, but the Riesen was crushed.[13]

Becker inspected Nagyag in 1805, and as usual, except where he found his beloved Hungarian Hund, he was not polite about the transport:

> The Riesen are large rectangular chests holding 10 cwt. of ore and running on four wooden rollers. These run on a sort of Hund-rail of hardwood nailed to the track-bed. The rollers also have, on the side which faces inwards, a projecting flange of larger diameter than the rest, which slides against the vertical face of the rails, and thus prevents the Riesen leaving the road.

A horse can only haul a Riesen for 800 fathoms three times in eight hours; better, he says, to abolish the Riesen and use the Hungarian Hund, although he is not sure if the Wallachian workmen are up to the specialist task of putting.[14] The driver was paid 54 kreuzer a shift, with lamp-oil thrown in, but had to supply his own lubricating grease. Of this large sum of money (about 4*s.* 6*d.*) Stütz tells us that some 3*s.* was for the use of the horse, and only 1*s.* 6*d.* was actual wages. Although the performance of the Riesen, when compared directly with other types of Hund (see Appendix), was not unsatisfactory, its running costs were huge.

One remote mine, deep in the mountains, had been connected with the outside world only by a steep, dangerous, roundabout and expensive pack-horse trail, which Felix Franzenau, the Nagyag director, replaced by an inclined plane shortly before 1806. It was self-acting, with the loaded waggon pulling up the empty; it greatly increased the output of the mine, and reduced the time, expense and danger of transport. The waggons ran on flanged rollers, the lower ones being larger than the upper to compensate for the slope. The rails were wooden and circular in section.[15]

The most illuminating comments on the Nagyag Riesen come from the pen of Otto Freiherr von Hingenau, reporting on the mine in 1857.[16] He wrote that the mines were not being worked to their full capacity because of the unsatisfactory and expensive transport facilities. There were indeed iron railways in the Franz and Josef adits (and had been since 1835[17]), where a horse could deal with a far greater traffic, and the carriage was let out by contract to proper Riesen-waggonmen. But in the other adits the construction and loading of the Riesen was shameful:

> As I was staying beside the Bernhard adit, I had the Banat Riesenway which leads out of it before my eyes every day. This merely consists, after the fashion of *Holzriesen* for the transport of timber in the mountains, of unhewn and often knobbly and crooked tree-trunks, laid at approximately constant distance apart; but the space between is often much widened by the tramping of horses and is full of deep, unfilled potholes. It serves to carry the Hunds (Riesen) which are not much bigger than the underground Hunds of Lower Hungary, and have wide cylindrical wooden wheels, which create so much friction that a horse can drag out only a single loaded waggon, without a saddle [supplementary container on top: see p. 35]. As the waggon is used on the long tip, and is simply overturned bodily and emptied without any regular tipping equipment, it follows naturally that it cannot have a life of any length. This crude system of transport, so wasteful of time and effort, is no longer to be found in many adits, even at Verespatak, where the mines are not famous for their technical progress, and where I have found wooden ways (Riesenways) which were at least made of squared and sawn oak, much preferable to those at Nagyag. What a man and a horse perform here is accomplished at Schemnitz and elsewhere by a single skilled Hund-putter on regular rails with a saddled Hund.

At Nagyag, he goes on, the class of Hund-putter as such is non-existent. The motive power underground is sometimes afforded by hired horses, sometimes by the hewer himself. One trouble is that the difference in temperature between the long underground level and the long tip, especially in winter, is very deleterious to the driver's health, which is one reason why the long-proposed plan to lay an iron railway in the Bernhard adit has not been fulfilled – the increased speed of working would merely increase the driver's discomforts.

Johann Grimm, the late director at Nagyag, replied to von Hingenau's criticisms by pointing out that the Hungarian Hund had long been used on the lower levels, pushed by the hewers themselves. The ore was then transferred either to the iron railways or to the Riesen-ways. As to the Riesen track, sawn beech rails had to be imported from two days' journey away, since there was no saw-mill near, and were therefore costly. Plain beech trunks, however, were cheap, as they were available nearby. The capacity of a Riesen was four and a quarter times that of a Hungarian Hund, so that if the use of the Hund were extended, the putters would have more than four times further to travel. He agreed that in some respects skilful Hund-putters gave better and cheaper performance than the present crude system with horses; but if special Hund-putters were introduced, most of the young and efficient hewers would be lost as such. Hunds were once tried experimentally on a small adit, and the Rumanian yokels who normally drove the Riesen were roped in as Hund-putters; but all of them hated the work and left, and he had to revert to the Riesen-way. He would have liked to lay an iron railway in Bernhard adit when he was at Nagyag – he did build the other two iron lines – but here the adit was too narrow and winding.[18]

This illuminating exchange raises a number of points of interest. The Riesen-way at Verespatak, mentioned by von Hingenau, was built later than 1805, when Becker found that pack-horses were the sole form of transport here, although at Zalathna in 1794 Esmark had found Riesen identical to those he saw at Oraviza (p. 80).[19] Von Hingenau's description of the extreme crudity of the Nagyag Riesen and track brings us back to the truck, now in Berlin, from the nearby mine of Ruda (pl. 37), a vehicle which von Hingenau might equally well have been describing, with its irregular wheels, crooked and knobbly rails set at a gauge of roughly 1 ft. 7 in., its simple pointwork and diminutive, primitive and battered body. It is crude, and therefore seems ancient, but appearances can be deceptive. From what von Hingenau says of Nagyag, the Ruda truck, so far from being a product of the sixteenth century, could equally well come from the mid-nineteenth; and from what another writer says, it seems clear that Riesen were not introduced to Ruda until the late eighteenth century at the earliest. This author, a Jesuit named Fridvaldszky, inspected the Ruda mines in 1767. About the underground workings, the tips, the stamps and the reservoir, he has some cynical remarks to make in his execrable Latin; but of the ore transport he can find nothing more interesting to describe than the pack-horse and its panniers – he mentions no wheeled vehicles at all.[20] After a chequered career through the nineteenth century, the mine was bought by a new company in 1887 (two years before Haarman acquired his Riesen from them), who modernized the equipment – iron rails, steel trucks (still called Riesen), and, after a year or two, electric locomotives.[21] But in one corner of the mine the old order clung on, and a few of the primitive wooden Riesen remained in use until as recently as 1930. In 1933 the Bergbau-Museum at Bochum acquired one of these waggons (pl. 38), the brother of that at Berlin. It is a little less battered, perhaps, and the wheels are not much worn, but otherwise there is nothing to choose between them. It is inconceivable that such a vehicle could remain in use (and in very rough use) for the best part of four centuries, and the Bochum Museum, with more realism than Berlin, ascribe their specimen to the first few decades of the nineteenth century.[22] Even this might be considerably too early a date. Thus all the evidence combines to suggest a nineteenth-century origin for the Berlin truck, and to reject the idea that the flanged-wheel Riesen date back to the sixteenth century.

When von Hingenau talked of the Banat Riesen-way at Nagyag, he presumably implied that the Banat was the home of the Riesen-way, just as the term 'Newcastle waggonway' in eighteenth-century England meant a railway of the kind developed at Newcastle. If the flanged-wheel Riesen is a native of the Banat – and it certainly was not imported from the Tyrol, where there is never a hint of it – it need not surprise us, for the decrepit mines of the Banat were restored earlier and more thoroughly than those of Transylvania. Just as the guide-wheel Riesen of Lower Hungary apparently owed something to the forestry Riesen (p. 71), so too it seems significant that von Hingenau likened the crude Banat Riesen-way at Nagyag to the *Holzriesen* of the mountain forests. The forestry Riesen was an early arrival in the Banat after its liberation from Turkey – the workers ordered there from Schwaz in 1722 included 'experts in flumes and Ries-tracks',[23] while the *Risswerk* for logs was well-established at Oraviza by 1727.[24]

To sum up the results of the written evidence: the flanged-wheel Riesen may have been derived from forestry Riesen, probably began life in the Banat, and first appeared at mines here and in Transylvania as follows:

Oraviza	possibly before 1775, certainly before 1794.
Nagyag	just possibly before 1764, almost certainly after 1774 and certainly before 1795.
Verespatak	after 1805, before 1857.
Zalathna	before 1794.
Ruda	after 1767.

Taken in conjunction with the history of mining in these parts, this pattern strongly suggests a date of introduction in the middle or second half of the eighteenth century. The faint possibility still remains that the flanged wheel could have been invented here before the Turkish period, or during its earlier years, to lie dormant until resuscitated two centuries later. But is this likely? And if it was so, was the idea somehow exported to England, out of a land embattled by Islam and across the wide expanse of Europe, without dropping off any seeds by the wayside? It seems improbable. The two ends of Europe were poles apart, and neither attracted travellers from the other. More probable far, if the flanged wheel really were an early invention of the Banat, that it was also an early and independent invention of England.

In fact, we are almost forced by the evidence to assume that the Banat flanged wheel first appeared in the eighteenth century. This poses the same question the other way round: Was it influenced from England? The chances had become much greater now, with the Ottoman power pushed back, and Europe more open to the exchange of ideas and to travellers, among whom Englishmen could always be numbered. Descriptions of railways could have been carried from Tyne to Danube by visiting engineers, whether English or German, or by some treatise of international circulation like Jars' *Voyages Métallurgiques*. But there still remains the possibility of independent invention: either a late modification of the guide-wheel Riesen, with which the Banat Riesen shares its rollers and its name; or else another, slightly different, imitation of the forestry Riesen.

English railways before 1660

And the said Huntingdon Beaumont hath used new and extraordinary invencions and practises for the spedy and easy conveyance away of the said coales, and especially by breaking the soyle for layinge of rayles to carry the same upon with great ease and expedicion . . . and by drawinge of certen carrylaggs laden with coales uppon the same rayles.

Francis Strelley on the Wollaton Railway, 1615
(*Chancery Proceedings*, Jas. I, S. 27/4).

Up to the accession of Elizabeth, British coal was mined in a small-scale, medieval, way, to supply markets that were largely local. But between then and the end of the Stuart period there was a startling change. Production increased roughly fourteen-fold, and during the eighteenth century it continued to grow steadily, but more slowly. Professor Nef's estimates for the annual output in tons from the various mining districts are as follows[1]:

	1551-60	1681-90	1781-90
Durham and Northumberland	65,000	1,225,000	3,000,000
Scotland	40,000	475,000	1,600,000
Wales	20,000	200,000	800,000
Midlands	65,000	850,000	4,000,000
Cumberland	6,000	100,000	500,000
Others	14,000	132,000	395,000
Total	210,000	2,982,000	10,295,000

The reason for this first great boom in the coal trade lay partly in the changed conditions of supply and partly in those of demand. The dissolution of the monasteries had transferred many rich coal-bearing lands from ecclesiastical control, via the State, into the hands of capitalist adventurers, who were able and indeed eager to work their coal with more vigour and on a larger scale than had been possible before. At the same time, since national supplies of timber were dwindling alarmingly, the use of coal was becoming ever more popular, both for domestic purposes and as a fuel for manufactures, notably in the expanding brick, copper and glass industries.

Once the coal had been dug, it had to be taken to its destination. If it were merely for local consumption, it might be sold at the pit-head itself or very nearby. But if the market were more distant, then the coal had to be taken from pit to river or port: the obvious example is the Tyneside coalfield, which supplied most of London's needs. A seam near the surface and near the river was naturally the most profitable; and when the shallower seams were worked out, then deeper seams would be mined as long as traditional pumping methods could cope with the drainage. But once this drainage limit was reached, those pits near the river remained out of action until the arrival of the steam engine in the early eighteenth century allowed them to be worked yet lower down. Thus, throughout the seventeenth century the general movement of mining on Tyneside was away from the river, an inevitable movement with an inevitable corollary: the problem of overland transport.

Some areas had to rely on the tedious and expensive use of pack-horses, each carrying a load of 2 cwt. or slightly more: the Cumberland coalfield and Coalbrookdale in Shropshire are examples.[2] In Northumberland and Durham, however, wheeled vehicles were the norm. The smallest sort, the two-wheeled cart, was pulled by one horse and held some 3¾ bolls of coal, or about 8¾ cwt. The larger wain was more common, with a usual capacity of 1 fother (7½ bolls or 17½ cwt.), and was hauled by two oxen and two horses.[3] Throughout the seventeenth century, and well into the eighteenth, these wains thronged the fields and lanes as thick as locusts, beating them into dust-bowls in summer, and in winter churning them into quagmires until traffic became impossible. The wains belonged either to the colliery proprietors or to transport contractors: in 1595 the Privy Council was driven to complain to the Mayor of Newcastle that coal prices were soaring because of abuses by three or four richer coalowners, who 'outbid and hire from the rest all the colewaynes that bringe coles to the waterside'.[4] Whickham, the largest colliery, or, rather, group of pits, of the period, had at least 700 wains in daily use. Even as late as 1726 Jesmond colliery is said to have needed 700 wains to carry its coals to Ouseburn, and 600 to 700 wains from Elswick, Benwell and Fenham converged on Scotswood staiths until these collieries acquired waggonways. It was estimated that in 1696 there were something like 20,000 horses at work in the Newcastle coal trade.[5]

Quite apart from the sheer physical disadvantages of a large number of small vehicles using utterly inadequate roads, there was the problem of wayleaves, a millstone that was to incommode the coal-owners' necks for centuries. A lucky few happened to own the land which intervened between their pits and the staiths or shipping-points; but if the land belonged to somebody else, a wayleave or permission to carry traffic over his ground had to be obtained. Such a landowner had the coal proprietor at his mercy, and knew it. Often, in fact, the rents charged for wayleaves were not unreasonable, but abuses were frequent and astronomical demands were not unknown, especially when the landowner in question was a rival in the business. Wayleaves were a very old-established practice: there is mention in 1245 in the Forest of Dean of 'rents for making coal-pits and for way-leave' (*mercedem . . . pro fossatis faciendis de carbone illo et pro cheminagio*).[6] An early Tyneside wayleave was granted by the Bishop of Durham in 1530 for

> carriages by wayne, cowpe [a kind of cart], or horses, from the cole mynes and pitts now opened, or which shall be opened, in Ravensall and Eighton (the inheritance of Richard Gascoigne, Esquier) through all the grounds, waistes and mores of the said Rev. Father.

The term was twenty-one years, the rent 5*s*. a year.[7] The Ravensworth coal-mines, which will frequently reappear in subsequent pages, give us another good example of the problems inherent in the wayleave system and transport by wains. The estate was sold by the Gascoignes in 1607 to Thomas Liddell, who about 1620 obtained a wayleave from his pits to the staiths on the River Team (fig. 30) from a number of copyholders of Whickham, who held their lands from the Bishop. All went reasonably well until 1633, when the copyholders 'did take and uniustlie detayne' six of Liddell's coal wains, along with twelve oxen and twelve horses that were pulling them. When Liddell sued them in Durham Chancery Court, the copyholders, in good legal tradition, denied that the wayleave existed; but if it did exist, they said, it covered only two pits, and was limited in width to 7 yards, with a rent of £10 per year plus 2*s*. for every wain that passed along it. But now, they claimed, Liddell used it to carry the coal from many more than two pits, 'and hath beaten out the same wayes by his said carryages in some places to Twenty fower yeardes or thereaboutes and that by reason thereof such great quantityes of dust be raised that the pasture arrable adioyninge to those wayes are much spoiled'. Besides this, other coalowners were using the way as well. The judgment was that Liddell could use the wayleave provided he 'make watergates, trenches and ditches . . . for avoydinge of water of the said Waie' and kept it at a maximum of 16 yards broad. If he led coal down it from Blackburn pits as well as from Ravensworth, he was to pay more.[8] Liddell seems to have got off lightly: but then he was an ex-Mayor of Newcastle, and a power in the land.

If a wayleave could not be obtained in comparative amity, the frustrated coal-owner often petitioned the Privy Council, which was in the habit of granting a compulsory wayleave, on suitable terms, on the grounds that it would benefit the common weal, the coal trade, and so forth. But sometimes it was sterner. Two owners asked leave in 1624 'to carry and passe theire coales to the said river [Wear] by the wayne and carte wayes now used by the owners and farmers of lyke coale mynes thereaboutes which . . . they are denyed and barred of'. Their petition was refused, because revenue due to the Exchequer was at stake. The Council could not allow them to boost the Sunderland coal trade, which paid no duty, at the expense of Newcastle, which paid 1*s*. per chaldron.[9]

The wayleave problem was capable of no solution short of drastic legislation that no government dared to pass, but the other part of the transport question was. The crowds of wains surging up and down the narrow lanes and spilling over into the adjoining fields when the roads became incapable of holding them could not go on increasing for ever. If the condition of the roads was hideous in summer, in winter it was unspeakable, and traffic was normally suspended until spring. True, shipping was also reduced during the winter gales, but if coal had to be stock-piled at all, the colliery-owners much preferred to do it at the staiths than in the confined surroundings of the pit-head. What was wanted was a means of transport at once more economical and more efficient than the wains, that would carry more coal more quickly and with fewer horses, could operate if necessary all the year round with equal ease, and would not reduce its route to a desert of dust or to a bottomless sea of mud. The answer was the railway, or, as it was usually called, the waggonway.

The birth of the English waggonway was one direct result of the first great boom in the coal trade. In the same way, the growing output of coal in the eighteenth century, and in the early years of the yet greater nineteenth-century expansion, led to the waggonway's maturity and its transfiguration into the public railway that went out to conquer the world. Tyneside became the home of the waggonway, and remained so for two centuries; but the waggonway was not born there. It was born in the Midlands, and was baptized, so to speak, in Nottinghamshire in the first year of the first Stuart, although, as we shall see, it may have first seen the light of day further to the west, in the last years of the Tudors.

The evidence for the earliest waggonways in England has recently been re-examined in a most valuable article by Dr. Richard Smith.[10] The coalfield immediately west of Nottingham, in the parishes of Strelley, Bilborough and Wollaton, was worked throughout the sixteenth century, and well into the seventeenth, on a large scale compared with most other parts of the Midlands at the time. The coal was carted in wains down to the end of Wollaton Lane, where the local dealers collected it either for distribution to Nottingham or else for transhipment on to Trent barges, which carried it down-river, especially to Newark. In 1601 there appeared on the scene Huntingdon Beaumont, the younger son (now aged forty) of a coal-owning family of Coleorton in Leicestershire. His temperament was energetic and imaginative, if impetuous, and he had gained considerable experience of mining at Bedworth near Coventry.[11] He leased and worked a number of pits at Wollaton from Sir Percival Willoughby, and in October 1603 obtained from Sir Philip Strelley the right to work coal for fifteen years in the manors of Strelley and Bilborough to the north of Wollaton. This lease makes no mention of rails, but in October 1604 he made another agreement with Sir Percival and his heir that he should pay them half of his Strelley profits in return for confirmation of their permission for Beaumont to drain off Strelley water through Wollaton lands, and to carry Strelley coals through Wollaton 'alonge the passage now laide with Railes, and with suche or the lyke Carriages as are now in use for that purpose'. The date of this first waggonway is thus fixed to between October 1603 and October 1604.* The line was two miles long, and did not run to the Trent, as might be expected (there may have been wayleave difficulties outside Sir Percival's lands), but to the old wain terminus. A document of 1609 says that the coal was to be stacked 'at Wollerton lane end at the new rayles end'. Wollaton Lane was comparatively accessible for Nottingham and the Trent; Strelley was not. The cost of these two miles is revealed in an account submitted by Beaumont to Sir Percival for the year ending 30 September 1605: 'Item the halfe of the charge of the rayles is 83li. 0*s.* 0*d.*'[12]

The superiority of the waggonway over what had gone before is emphasized by a letter to Sir Percival from his coal factor in May 1610:

> I beseeche you take order with Sir Thomas [Huntingdon's brother] that we maie have libertie to bring coales down the rayles by wagen, for our caridages onely, and we will bring them downe by raile ourselves, for Strelley cartway is so fowle as few cariadges can passe.

The Strelley coal was sold at the pits, or at the waggonway terminus in Wollaton Lane, or at Trent Bridge, or at Newark and elsewhere down the river. The surviving figures for the year 1613-14 show that, out of a total of at least 4,608 rooks carried down the waggonway (excluding, that is, the Strelley sales), 'the sale at the railes for the yeare precedent is 1,600 roo and better'.[13] A rook weighed perhaps $1\frac{1}{2}$ tons. In January 1614/15

* The date of 1598 has sometimes been given to the Wollaton line. One of the documents is dated 1698, in obvious error; but the correct emendation is certainly 1608, not 1598.

Francis Strelley complained in Chancery that Beaumont was overworking the pits, and described the waggonway in the terms quoted at the head of this chapter. By these means, he said, Beaumont was selling at least 120 rooks of coal per week more than he could have done with ordinary means of transport. Beaumont replied that in 1608 Nicholas Strelley had confirmed his original lease, and in return for 'great somes of money' gave him 'free lybertie to laye rayles in or upon any parte or partes of ye premises in Strelley and Bilburgh . . . and for usinge of Rayles from pitt to pitt, for carryinge and conveyinge away of coals upon the said Rayles' with 'Rayles horses wagons or other caryages'.[14] We do not know the outcome of the case, but it hardly mattered. Beaumont's pits in Strelley were proving obstinately unprofitable. He had thereby incurred Sir Percival's hostility, and in 1618 the Strelley lease ran out. Apart from this one example, whose life was fifteen years at the outside, there is no evidence for waggonways in the Nottinghamshire coalfield until late in the eighteenth century.

Wollaton and Strelley, however, were not Beaumont's sole coal-mining concern. He had other irons in the fire far to the north, and it seems clear that the honour of introducing the waggonway to Northumberland was due to him. A very great honour it was to prove, although he himself reaped no benefit at all from his venture. The credit was indeed given to him, not many years after his death, by Gray, the first historian of Newcastle:

> Some south Gentlemen, hath upon great hope of benefit, come into this Country to hazard their monies in Coale-Pits, Master *Beamont* a Gentleman of great ingenuity, and rare parts, adventured into our Mines with his thirty thousand pounds; who brought with him many rare Engines, not then known in these parts; As the Art to Boore with Iron Rodds to try the deepnesse and thicknesse of the Coals: rare Engines to draw Water out of the Pits: Waggons with one Horse to carry down Coales from the Pits, to the Stathes, to the River, &c. Within few years, he consumed all his money, and Rode home upon his light Horse.[15]

Gray's own copy has the £30,000 altered in manuscript to £20,000,[16] but even this sum is probably exaggerated.

Beaumont seems to have built three lines in Northumberland, all of which apparently ran down to the River Blyth, although we know deplorably little about them. The places they served, but not the lines themselves, are marked on fig. 31. Probably the earliest was at Bedlington, where he leased pits from the Bishop of Durham in September 1605. The document has not survived, but a new lease from the Bishop, in September 1608, specifically grants 'sufficient way leave to digge sincke wynne gett and drawe coales and the same to carrye and recarrye either upon rayles or other carryages'.[17] It may have been also in 1605 that he began work at various pits and salt-pans at Cowpen and Bebside, just east of Bedlington. Meeting with no success, he gave up in 1614 and handed over his interests to two partners, Paramore and Sanders, who in turn abandoned the work in 1616. It transpired in a lawsuit in 1636 that they had left behind at Bebside three boats

> & many and diverse Rayles Utensells and implements. . . . And that after theire Desertion and leaveinge as aforesaid one Edward Delavale gentl. tooke and carryed away all the said Rayles sett upon the land and Ground of Bebside for five hundred paces in the Waggon way on both sides of the way.

Likewise at Cowpen, the new farmers of the pits and pans removed another three boats and both waggonway rails for 500 paces, along with their accessories. The Latin version of the same report expands a little on the 'Rayles Utensells and implements' of the English translation: '*multos et diversos cancellos sive Sigilla Anglice Rayles . . . in via vehicularia Anglice wagon way . . . cum Vehiculo Anglice Waggono*'.[18] Although the Court assumed that Paramore and Sanders had laid these lines, Beaumont's previous presence makes it almost certain that he was responsible. He had hoped to build up the Cowpen and Bebside pits until they rivalled the great collieries around Newcastle. His efforts, and those of his successors, only succeeded in raising the shipments from Blyth from 644 tons in 1609 to 3,053 in 1616. So he failed here as he failed at Strelley, and he died embittered and deep in debt in Nottingham Gaol in 1624.

Immediately before the passages just quoted about Beaumont's activities in the north, Gray has this to say about the coal trade:

> Many thousand People are imployed in this Trade of Coales; many live by working of them in the Pits; many live by conveying them in Waggons and Waines to the River Tine; many men are imployed in conveying the Coales in Keels from the Stathes aboard the Ships.

His distinction between waggons and wains is interesting. In the Tyneside coal trade, waggon always means a railway vehicle, never a road one; and wain always means a road vehicle, never (with but one exception) a railway one. Since Beaumont's Bedlington lines had come to a dismal end long before Gray wrote, the inference to be drawn from his remark about the waggons is that the waggonway was by no means a rarity on

Tyneside in 1649. This is not surprising, since wains and impossible roads in combination, as we have seen, were strangling the expansion of the coal trade. It is frustrating that we are unable to pin down more than two of these post-Beaumont but pre-Restoration waggonways, and one apparent proposal for them.

This proposal comes in a scheme put forward in 1628 for placing the control of the coal trade in the hands of the Crown, both to prevent abuse of their powers by the wealthy coal-owners and to direct the profits of what could be a highly profitable business into the royal coffers. The King was to take over all the mines in the Bishopric of Durham, and the working and transport of the coal was to be eased

> by makinge compitent passages for the waters, that anoye the mynors, or pyonors by makinge one Portable Ryver, from those parts, where the Coales are to be acquired; by makinge Highwaies passable, and reformed Cariages, to convey the Coales to that Ryver; by providinge all necessaryes for speedye conveyance to, and from the Shipps.[19]

It is hard to resist the conclusion that the proposed 'reformed Cariages' were waggons on railways. The plan, however, was eventually abandoned, as being too dangerous for the King to handle.

The date of the building of the Ravensworth waggonway, which is usually said to have been the first to the River Tyne, is sometimes quoted as 1632.[20] This is due to a misunderstanding of Taylor's remarks in 1852, when he was trying to establish the year when waggonways first appeared here.[21] He found no reference to them in a division deed of Winlaton lordship of 1632, and argued that they must have been introduced between then and 1649, when Gray mentions them. In fact, the Ravensworth line, as we shall see, was by no means the first, for it was not opened until 1669. We have already looked at a case of 1633 (p. 88), which shows that the Liddells' Ravensworth coal was then being carried to the staiths near the mouth of the River Team by wain. Very close to this wainway, however, where it passed through the Manor of Whickham, there was an early waggonway, the earliest we do know of on the Tyne. The only evidence we have for its existence comes from the burial registers of Whickham parish.[22] Some entries record accidents with road wains, such as:

> 1633 Willm Harrison sonne to James Harrison slaine by a fall fooorth of a Colle waine Aug 28.
>
> 1646 N or M [son] to Michaell ffenwick, slaine at Swalwell steathes wth a waine the 4th of June.

But other entries refer to waggons, and it is clear that road wains and

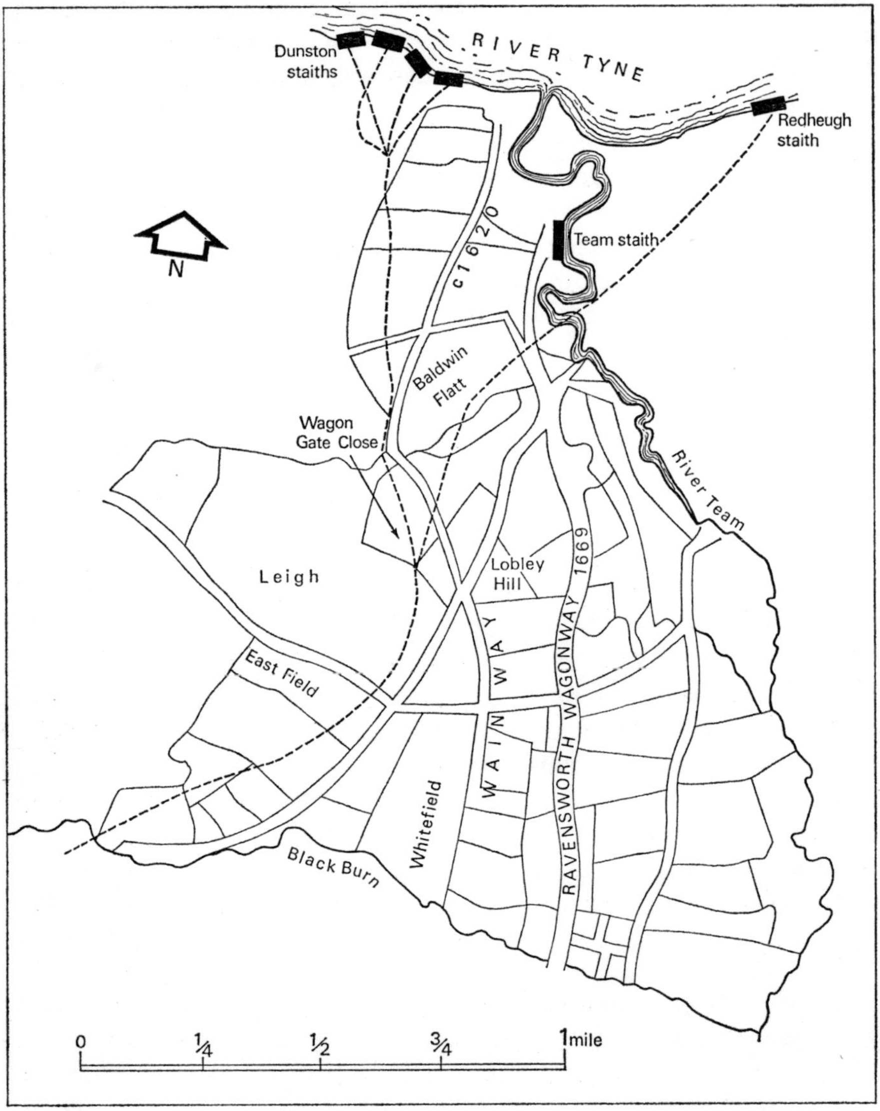

FIG 30 Map of east end of parish of Whickham, showing Ravensworth wainway of *c.* 1620, waggonway of 1669, and (dotted) Tanfield waggonway of 1725. The Whickham waggonway in existence by 1645 passed through East Field, Leigh, Waggon Gate Close, and Baldwin Flatt.

railway waggons were already differentiated in Tyneside terminology. Gray makes the distinction, which was always maintained thereafter. These entries read:

> 1645 Two Childre to wedow Howborne at waggon yate in the Leige [Leigh] the 22th of Octoberr.
>
> 1650 John Craggs slaine in the Leigh in a sinking pitt near the waggon way 9th app.
>
> 1650 Henry [son] to Henry Bird slaine by a waggon in Bawden fflatt the 9th of Aprile.
>
> 1650 N or M [son] to Robt Middlemast slayine with a waggon in East ffeilds the 24th of Juliy.

The sites of these accidents allow us to locate the line, which is in just the right place to carry the coal from the eastern pits of Whickham colliery, worked by Thomas Surtees and Sir Peter Riddell[23] and then the largest in the area, to Dunston staiths (fig. 30). It ran almost if not quite on the line of the Tanfield waggonway (of 1725), some way to the west of the Ravensworth waggonway of 1669, and it began a short distance to the west of the Ravensworth wainway described in 1633, which, significantly enough, it crossed at the 'Waggon gate', the first glimpse we have of a practice common in later years (p. 224), a level crossing with cottage attached, manned by a widow.

On 15 September 1656 the Common Council of Newcastle granted Edward Greene a twenty-one-year lease to build a staith in the Salt-meadows, on the south bank of the Tyne just east of Gateshead, where the town owned a number of coal-pits. The payment was to be

> £17 10*s*. fine and 12*d*. yearly rent for 40 yards of ground upon the east of Backworth [not to be confused with the Backworth north-east of Newcastle] and from a full sea mark 10 yards southward with liberty for a way for waggons.

We do not know if the line was actually built[24] (fig. 31).

These are the only traces we have of waggonways in Northumberland and Durham before the Restoration – few, yet enough to suggest that there may have been a number more, could we but find them.

While Huntingdon Beaumont was busy working his waggonway at Strelley and carrying the good news to Northumberland, two similar lines at Broseley in the Shropshire coalfield first come to our notice. Both were the subject of lawsuits heard before the Star Chamber in 1606 and 1608. In the first case[25] the facts were these. High up above the Severn, where it runs through the Ironbridge Gorge, Richard Willcox and William Wells had an interest in a field named Birchleasow, alongside the village of Broseley and a mile distant from the river. They also owned land at Calcutts, near the river bank (fig. 43):

> And your saide subjectes mindeinge and intendinge not to reape to themselves onelie a privatt commoditie and benefitt by the said delphe and coalmyne, but rather purposeinge the genarall good of the whole contrey thereaboutes did about October now last paste [1605] cause and procuer to be made and sett up a very artificiall Engine or Instruments of Timber, which stood your subiectes in above twoe hundreth markes, therewith to convey coles from the said ground called Birchleasowes by and through other certain growndes parcell of the said demeasnes of the said Mannor of Broseley now in the possession of your said subiectes into and through the said Calcottes and soe unto the banck of the River of Severn, from thence to be conveyed by water . . . to the great ease and common good of all the countrie thereabouts which is verie much distressed with want of fewel.

Evidently Willcox and Wells were not certain of their rights, for they had taken the precaution of applying for authority before they built their 'engin' (the word then meant nothing more than 'device'),

> which your Subiectes were licenced and authoriside to sett up and make by youer highness moste honorable privie counsell by speciall direction from them sente unto your highnes Councell established in the Marches of Wales, unto whom the purpose and use of the said Engin beinge made knowne was exceedinglie well liked and allowed of long before it was finished, in regard it was for the common and generall goode of the whole countrey thereaboutes.

Once again, the advantages of the waggonway over road transport were emphasized by a witness.

> The complaynants [Willcox and Wells] were before the making of the said Engin or Instrument enforced to convey their coale from thence unto the said places [the markets up and down the river] on horse-backe, or by drawghte & waynes, which was a great Trowble & very chargable unto them, and many tymes their oxen horses or drawghte came to be spoyled in bad wayes & ill passage. And the Devise of the said Engin was such that one man as this deponent beleiveth might therby have conveyed more coales in one day, by the said devise, unto the said River, for the servise of the Countrey as aforesaid, then a Wayne with six Oxen was able to doe and that also

> without any danger at all. . . . The said engin would by all likelihode have conveyed unto the said River with little help, more Coales in a day than a wayne & 6 oxen were able to convey to any of the said places in 3 or 4 days.

James Clifford, Lord of the Manor, owner of the land across which the railway ran, and rival to Willcox and Wells in the coal trade, had, it was alleged, brought into the district 'a nomber of lewd persons being the scummes and dregges of many countries from where they have bine banished and driven for their lewdness of life'. At 10 o'clock on the night of 14 November 1605, he collected 'sundrie rioutous routers and unrulelie persons' with axes and 'did cutt hack hewe & displace the said Engin, and the Rayles or sawed Timber there layd & placed for the passage of the said Engine', 'as that xx li [£20] did not suffice to reduce and bringe the same into such repair and serviceable use as it was before'. No sooner was it repaired than Clifford again assaulted it, causing £30-worth of damage.

Clifford's minions, when questioned, had very different, and differing, stories to tell. Jeffrey Hall was asked whether the country would not have benefited 'if the said Engin, fframe or Rayles soe laide had been suffered to be putt to the right and perfect use for which it was first devised and made'? No, he replied, it was not in the public interest, for now there were 'yearelie gotten more coales than can be transported by water'. John Davies and Richard Basset said that when Clifford, as Lord of the Manor, heard that Willcox meant to carry coal 'in & over that way & passage' for the first time, he sent them, armed with nothing more than a bill and a pikestaff, 'to the gates of the said passage', where they, 'seing these Complainants carriage cominge that way with colls, did in peacable manner make stay thereof'. Davies deposed that the rails were little damaged, and were made perfectly serviceable by turning them over, at a total cost of about 10*s*. He thought it was no loss to Willcox and Wells that they did not have an undisputed right to use their engine, for they had managed quite satisfactorily without it before. Basset went further, reporting a rumour in Broseley that Willcox and Wells had realized their folly in building so costly a contraption which would put them to even greater expense than their previous means of transport, and had done the damage themselves in the hope of blaming it on to Clifford and of being awarded compensation:

> And this defendant is the rather perswaded to beleeve the same to be true, for that the railes of the same Engine were not so cutt or hackt but that the same served againe by turninge the said railes and layinge the upper syde downewarde. And that x s. or thereabouts (as

> this defendant hath heard) did suffice to repaire the same Engine, and the railes thereof againe, which if they had bine cutt or hewed by any Adversaries they would by all likelyhoode have so spoiled the same railes as that they should not have served againe.

We do not know the outcome of the case, although it seems that Clifford and his merry men were in the wrong. But the proceedings do give some idea of the uncertain nature of early wayleaves, even though expressly approved by the Council in the Marches, when they passed across the lands of a hostile rival. At the same time, a little emerges about the cost of building such a line. The 200 marks (£133 6*s.* 8*d.*) for a mile of track at Broseley compares with £166 for two miles at Wollaton (see p. 90), although, as Dr. Smith observes, it was in Willcox's interest to exaggerate the sum that the line cost him.

The records of the Council in the Marches at this date are unfortunately lost, but a glimmer of light is cast on the affair by a memorandum written by Sir John Weld in 1619. In the previous year he had bought the Manor of Willey, near Broseley, from Sir Francis Lacon, who had inherited it in 1608 from his father, Rowland Lacon. On taking Willey over, Weld made a list of the legal documents pertaining to the estate: 'In two black buckram baggs are contained these writings following. . . . Proceedings in the Marches of Wales before the counsell there concerning wayes to Severn between Lacon & others & Clifford.'[26] This doubtless refers to the case under discussion, and reveals that Rowland Lacon, although not mentioned in the Star Chamber proceedings, was involved in the original granting of wayleaves. Birchleasow, the site of Willcox and Wells' mine, belonged to the Willey estate. We shall meet both Sir Francis Lacon and Sir John Weld again.

In the second Star Chamber proceedings[27] which concern a Broseley railway, the roles are reversed. Clifford had worked coal in Calcutts under lease from Willcox, but in 1605 or 1606 (no doubt as a result of the first imbroglio) Willcox had leased part of Calcutts to Wells, in order to stop Clifford mining. On 26 October 1607 Willcox's men seized the cottages that Clifford had built in Calcutts, and removed the pit-head gear and 'yo subjects tylting rails in Calcotts aforesaid being yor said subjects moveable & proper goods'. At the same time Thomas Prescott, who was peaceably 'carying Colls from a pitt in Calcotts aforesaid about a quarter of a mile from the said Cottages down from thence to the ryver of Severne', was arrested by Willcox. Clifford 'prepared another paire of Tilting railes', which were promptly removed by Wells' servants on 3 November, and another set of replacements was lifted on 6 November. Wells admitted all the charges, but pleaded that the tilting rails were really his. He agreed that he did

> comaunde Ralfe Wheelwrighte & some others of the persons mentioned in this Interrogatory to take downe certen Tilte Railes which had been sett upp in the said grounde by the now Complainants meanes, & to bringe them into the Cottage . . . and to cutt the ropes wherewith Prescott and his companie did lett the coles downe to Seaverne . . . and he this defendant himself did pull owte the pynnes of the Tiltinge Railes.

Wells' servants likewise admitted that they were responsible on 3 November for 'pulling downe and taking away the tilting Rayles placed upon the foresaid Railes' and that on 6 November they went 'with divers others unto the Railes aforesaid, and there tooke away other tilting Rayles that were newly placed', and 'did help to carry certen Rayles (which were made fast to certen of the grownde in Calcotts aforesaid) into one of the said cottages'. Once again the verdict is not known. Wells and Willcox died in 1610 and 1614 respectively,[28] and it was not long before Clifford gave up mining in Broseley.

Before Clifford finally left the scene, however, he had built another railway somewhat west of Calcutts, serving horizontal drift mines in Yates Coppice, which apparently lay a little inland, somewhere between Ladywood and Calcutts. On 22 September 1608 he leased to Jesse Whittingam of Gloucester the right to all coals and ironstone

> digged wthin ffower Insetts or pitts allreadie begunne by the saied James Clifford in one parcell of ground . . . called . . . Yeates coppies, And all the waies water courses and Railes from the saied ffower Insetts

for a term of five years. Clifford was to

> finde and appointe from time to time sufficiente timber trees . . . for the laying of Railes from the Wharf places by the River of Severne unto the face of the wall of the Coles for and duringe all the said terme, All which Timber and wood is to be falne and cutte and the Railes to be laid at the Costes and charges of the saied Jesse.

He was also to satisfy the tenants for 'Trespass to be done in cominge upon the said ground to digge the said coles to laie Railes ther for the Carridge thereof'. If at the end of the term Whittingam had made a profit, he was to repay Clifford the expenditure laid out on the insets before the lease began; if he had made a loss, Clifford was to pay him the cost incurred in

> settinge forth the said Insetts and in layinge the said Railes wch Railes together wth all and singular the saied Insetts the said Jesse covenanteth wth the saied James Clifford to leave at the end of the saied terme in good reparacon.

In 1612, however, Clifford disposed of his manor of Broseley. William Porter, the next owner but one, split it up in 1618 by selling it to a number of different proprietors, including Lord Bromley, John Weld, Francis Old, Francis Adams and the Huxleys. The ownership of the various parts now becomes complicated and obscure, since these gentlemen and their offspring immediately engaged in bitter dispute and endless litigation about their respective rights, both among themselves and with other claimants, such as the Langley and Young families. We do not know if the Calcutts and Birchleasow lines survived for any length of time, but railways continued at Ladywood and Yates Coppice. In 1615 William Porter mentioned in a survey of his Priory land in Broseley:

> There is one Insett of Cole wch yeldeth me 40^{l} p ann in the occupacon of Edward Pacie wth ye value put at 100^{li} 0^{s} 0^{d}. There may be more Insetts made wch will yealde a greate profitt.

The inset of Pacie's in the Priory land is apparently marked on a map of Broseley of about this date as 'Priorie Inset' between Ladywood and the Severn, with what appears to be a 100-yard railway running from it down to the river. Alongside is another nameless inset, and a third marked 'Mr. Cages Insett'. John Cage had married Mary, daughter of James Clifford. The Pacie family was still working its inset in 1632, and before long extended its interests. In 1650 John Huxley, who had bought this part of the Broseley estate from William Porter, leased Yates Coppice and Beard's Coppice to Henry Langley, with leave to mine 'by Insett or any other wayes or meanes' and with 'libertie to lay Rayles and make waggon wayes for the carrige and transportinge of the said coles downe to Severne'; but Langley immediately sublet to William Pacie.[29]

Meanwhile, railways were appearing a little further inland. Sir Francis Lacon, as we have seen, sold the Manor of Willey in 1618 to Sir John Weld, who at the same time bought up the Manor of Marsh from John Slany and a portion of the Manor of Broseley from William Porter. At once he began to exploit the resources of his new estates, building an iron furnace at Willey and energetically continuing the mining which the Lacons had begun. In 1631 he compiled a memorandum book which has much to say about coal and ironstone mining on the Willey and neighbouring estates, and especially about wayleaves.[30]

I estimate my coles and Ironstone in Willey Parke if I can procure a constant way to Severne will be worth at least in time	5000–00–00
If I can gett a way to Severne by m^r^ Thomas Lacon [Sir Francis' brother] and Mrs Jukes for my coles and ironstone in Rudgwood Birch Lessow Horsley more the Brickfeilds and Darley grounds and my grounds in the Deane, it may be worth to me in time	2000–00–00
Sinking of Pitts and making of Waggon wayes and boreing in Willey Park and Rudgwood &c hath cost	[blank]

His Birchleasow coal might be taken to the Severn along the route of the 1605 waggonway, which may or may not have been lifted. The obstacle here was the inheritors of other parts of the fragmented Broseley estate, one of whom was Francis Adams, who owed Weld money. Weld felt he could 'gett a lease from Mr Young of Calcots inregard I may carry my coles in the Worfe that way, & may hinder Mr Adams to carry his coles to Severne unless he compound with me or give me way for my coles through his grounds'. Alternatively, he could claim the debts owed him by Lady Bromley and the Old family, 'or else in leiu thereof a waggon way through there lands to Severne', a line running eastwards down Tarbatch Dingle. The route he proposed for his coal in the Willey, Dean and Rudgwood area was down Dean Dingle, through Thomas Lacon's and Mrs. Jukes' land (see fig. 43). Inspired by the current claims that iron could be smelted with coal instead of charcoal, Weld pronounced: 'It may fall out, Iron may hereafter be made with pitt cole, then my cole will stand me instedd for my furnis, which coles may be brought to my furnis by wagons.' John Benthall, Weld's neighbour to the west, was also deeply involved. In 1629 he offered (in vain) to get Weld's coal in the north-west corner of Willey estate and pay him 1*s*. a ton, if Weld sank two pits and made 'the way to Severne at my [Weld's] charges'. On both estates good coal seams outcropped, so the early workings were shallow, but as they went deeper Benthall and Weld tried to come to an agreement to share the cost of drainage. Benthall badly needed an outlet for his coal, and 'did offer me a fine or any other assurance to settle a waggon way if we fell to an agreement'.

In September 1634, a year after John Benthall's death, his son Lawrence finally reached this agreement with Weld. Benthall received both a consideration of £50 and leave to get coal and ironstone under Weld's

Woodland Green or Broseley Wood 'in an Insett or Coleworke' driven from his own land. He would

> at his owne coste and charges before the twentieth daie of October next cominge make and drive a sufficient waggon way (by way of Insett) in by and through the land of the said Lawrence Benthall as levell will permitt by layinge Railes and other necessaries up and into the said coles and Ironstone if any there be of the said John Weld. And shall gett take and carrie away all those Coles and Ironstone wch may or shalbe gott in and by the same Insett so to be made and driven, and shall lay Railes for the conveyinge of them to the River of Seaverne.[31]

It is not clear whereabouts on the river this railway was to end, but the nearest point is at the west end of the Priory land below Ladywood, where a short line already existed. Indeed, there existed another railway in the neighbourhood of Woodlands itself, for in 1631 James Lacon (another of the family who retained interests in the area) renewed the lease of two insets at the Woodlands to a syndicate of three proprietors. They were to 'keepe the lower of the sayd two Insetts open and in repayer to give Ayer to a lower Insett now in the occupation of the said James Lacon . . . soe longe as the said lowest Insett lasteth', and could 'gett coles in the Middle Insett Twentie yrds belowe the wagon waye'.[32]

The details and the exact sites of these lines may be in doubt, but a picture emerges of intense mining activity in which waggonways already played an established part throughout most of the first half of the century. This was appreciated over 100 years ago by Professor Pepper when he described Coalbrookdale – a name applied in early days to both sides of the Ironbridge Gorge – as 'celebrated as the locality where railroads formed of wood were first used in the years 1620 and 1650'.[33] The area, however, is still not as celebrated as it might be for its early use of the railway.

The railway also spread before the Restoration to at least one other place in the West Midlands in which, significantly, coal-proprietors from both Wollaton and the Broseley area were concerned. In June 1613 Sir Francis Lacon, of Willey and of Kinlet at the north of Wyre Forest in Shropshire, leased some coal-pits in Lime Pit Fields, Shropshire, and at Bower Hill, Staffordshire, to Sir Percival Willoughby and to John Slany, merchant of London and Lord of the Manor of Marsh, between Broseley and Much Wenlock. Lime Pit Fields are in Lacon's Manor of Kinlet, on the west bank of the Severn five miles above Bewdley; Bower Hill is a mile to the south in the parish of Upper Arley, then in Staffordshire, but now in Worcestershire. In October 1613 Burton Goodwin, a Strelley man who was serving as Willoughby's agent in the area, complained to his master: 'Also the rails which he [Lacon] did promise to have laid before this time in Bower Hill are yet unlaid, neither do I know when they will be laid, the lack of which hath hindered us the taking of 100 marks.' Two years later, however, the boot was on the other foot, for in November 1615 Lacon entered a bill of complaint against Willoughby and Slany, saying that he had been put to great expense 'as well in preparing of pits as in the making and setting up of rails and in providing of tools necessary for the getting of the said coals', but that Willoughby and Slany had not worked the pits nor paid the sums they promised.[34]

There remains one possible pre-1660 waggonway. In November 1606 Thomas Tulloch of Inveresk (just east of Edinburgh) petitioned the Privy Council of Scotland for a patent. Having, he said,

> imployit the maist part of his youth in uncuth nationis in searcheing and learning the knawledge for making and practizeing of ingynis and workis for the commodious and aisie transporting of coillis betuix the colpotis [coal-pits], sey and salt panes of this realme, and haveing now attanit a suir knawledge thairof . . . quhilk interpryse, wark, and travill will benefit the leigeis in sa fer that the samen will ease thame baith of travill and expenssis in the fourt pairt quhilk they bestowit and wairit thairuponn of before,

he humbly desires that, since 'the samyn work and ingyne' will cost him a great deal before he perfects it, he may have an Act of the Council securing his property in his design. He was granted sole power to use the 'ingyne', provided it was 'ane work and ingyne nocht knawin in this kingdome at na tyme before'.[35] This sounds as if it were a railway of some sort, but nothing more is known about it, except that Tulloch's 'ingyne' may have found a use at Stacks on the Forth to the west of Edinburgh, where James Allan was granted the right in 1609 to make coal-pits 'and vther necessary Ingynes for wyning and transporting of the saidis coales to sea and land and to mak frie gaittis and passages to that effect . . . with frie ische [exit] and entrie of wayis and passages for . . . transporting of the saidis coales'.[36] The only nation in the world at that date which used 'ingynes' for transporting coal was England, which to a Scot at the time of the union of the crowns certainly was uncouth or unfamiliar, as the word then meant. So Mr. Tulloch may have served his time at Broseley or at Wollaton.

With this we reach the sum total of our certain information about pre-Restoration waggonways in Britain: one in Nottinghamshire, perhaps five in the North-East, up to eight on the Severn, and possibly one in Scotland. Before going on to discuss the questions which the information raises, it will be well to clear out of the way a few early references to rails which might confuse the issue. The word 'rail' did not always mean a beam for waggon wheels to run on, not did a sleeper necessarily mean a cross-tie under the rails. 'Rail' was a common word for a bar of a fence, and 'sleeper' could mean a foundation-beam for a building. When these words appear in early mining documents it is tempting, but dangerous, to assume a railway connotation unless there is supporting evidence. A simple example, first, from a seventeenth-century document referring to Tudhoe in County Durham: 'Everie pitt ought to be filled and rayled or else well covered.'[37] There are several early usages of this sort from Wollaton itself, usually referring to the gate and fence around the coalfield:

12 June 1551 Paid for mendyng the rayle and the yate at the hey field iv d.
11 April 1579 Paid to Moke and his fellow for vi daies a pece sawing sleepers in Shortwood ix s.[38]

An instance at Prescot in Lancashire in 1594 has been interpreted as a waggonway without real reason.[39] The Court Leet in that year fined Philip Layton, a colliery-owner, for felling trees in Prescot Wood and with them making 'rayles'. In fact, the records show that one of the more common misdemeanours in Prescot at the time was making 'stakes for hedges' with ill-gotten timber, and Layton was almost certainly doing nothing more significant than making fencing-bars.[40] As a last example of this trap for the unwary, the plaintiff in a case of 1632, the lessee of a mine at Softley in County Durham, said that the 'raleing and paveing of the waie with other charges about the repairing of the said waye did cost the lessee . . . seaven or eight pounds'. It sounds just like a railway, but read further on, and one finds the way was railed 'that noe trespass or hert shold be done in the other groundes' belonging to the landowner (the defendant), and that the way did not even carry wheeled vehicles, but only packhorses, some of whose panniers were slit by the landowner in question.[41]

What did the early waggonway look like? The evidence at Wollaton and in the North tells us little: no more, in fact, than that waggons were pulled by horses along the rails, had the advantage of being able to work when the ordinary lanes were impassable, and, at Whickham at least, were heavy enough to cause fatal accidents. The Broseley documents tell us a little more, and reveal a considerable difference. If the mines served were insets or adits, the rails were laid without a break 'from the Wharf places by the River of Severne unto the face of the wall of the Coles'. This presupposes a small waggon for underground use and manpower to push it. There is no evidence at all for horse or oxen haulage in Shropshire until the very end of the century. If the mines were worked from shafts, as at Birchleasow, the railway began at the pithead, but still, it seems, employed small waggons. If the line had to negotiate the steep side of the Severn Gorge, then the waggons were let down by rope. It is specifically stated that Clifford's waggons were so dealt with, and Willcox and Wells' may have been similar, for their line dropped 375 ft. in a mile, with stretches that must have been very steep. The term 'tilting rails' is curious. One use of it, in the phrase 'the tilting Railes placed upon the foresaid Railes', prompts the thought that there may have been two layers of wood, the lower one to carry the weight, the top 'tilting' rails somehow to steer the waggon. But on the whole the usual explanation seems the most probable, that they were so called simply because the line was of necessity on a very steep grade down the valley side. The rails here were held down by pins, and, when hacked about, could still be used by turning them over.

How were the waggons guided? Guided they must have been, or else the phrase 'new and extraordinary invencions and practises' would hardly be applicable to the Wollaton line. Did they use one of the means of guidance employed on the Continent – the flanged rail, the Leitnagel, or the guide-wheel? The flanged rail seems to be ruled out by the fact that the Broseley rails were reversible. The Leitnagel Hund was certainly in use at the three copper-mines in the Lake District, but there were essential differences. Whereas the Hund was designed, with small wheels and small body, for pushing by manpower in restricted galleries underground in metalliferous mines, the whole point of the English waggonway as it developed in the North (even if not originally in Shropshire) was that it should carry large quantities of coal overland by horse-power. For this task the Leitnagel Hund was quite unsuited. As we have seen, the Hund was very rarely used on the surface, and, as far as is known, graced no colliery until the eighteenth century. In its purpose, though not in its design, the guide-wheel Riesen of Lower Hungary differed from the English waggon only in the material it carried; and of all the Continental systems the guide-wheel waggon was indeed the most suitable for transporting coal. But there is never a hint of it in England. Nor, apart from the Mines Royal, is there a tittle of evidence for any import, at any time, of any Continental railway ideas.

It is difficult to assess how much British coal-mining owed in other respects to German metal-mining techniques. A certain amount, we know, but probably not very much. Some improvements in the methods of pumping and winding undoubtedly were derived from Agricola's *De Re Metallica*, or from the Company of Mines Royal, or perhaps from the several foreigners who were granted nebulously worded patents covering such devices. Wollaton was certainly not ignorant of Continental practice, for the Willoughbys knew their Agricola, and left among their papers an engineer's advertisement for a water-pump which concluded: 'at Wollaton . . . ther are models to be seene of all the water-workes that are of any worth or valewe in Italye, Garmanye, or the Low Contryes'.[42] On the whole, however, the coal industry seems to have solved its own peculiar problems itself.[43] One such solution was the boring-rod for locating coal seams without the expense of sinking shafts, an invention which has strange parallels with the waggonway, although it spread much more quickly. It is first mentioned, in a general way, in 1613, and by 1615 it was in use at Newcastle, introduced, we are told, by Huntingdon Beaumont. It is not known if Wollaton had boring rods, but Sir John Weld employed them at Willey between 1618 and 1631.[44] Did they emanate originally from Shropshire or from Nottinghamshire?

Another of the English coal industry's inventions, it seems clear enough, was the waggonway, and the hallmark of the English waggonway was the flanged wheel. True, we have no evidence of it before the Restoration, but thereafter it was the *sine qua non* of every surface railway in this country for a century. Since the flanged wheel is consistent with what pre-Restoration evidence there is, and there is no hint of a change-over at any time, it seems legitimate to suppose that the English waggonway used it from the beginning.

One can hardly hope to discover surviving remains of seventeenth-century vehicles on the surface, but their preservation underground is possible. In the Broseley area, the ground being so hilly, the most accessible seams were mined in the early days by horizontal adits; and in several such workings in and around Broseley old flanged wheels have been found at various dates from the mid-nineteenth century. Wooden rails and wicker baskets mounted on wheels were discovered at the Tuckies and in the Yew Tree Pit above the Calcutts, both collieries which James Clifford had worked. Another pair of wheels came from the Deer Leap at Benthall, carved from solid blocks of wood and fixed on an iron axle which revolved in brass bearings cast with the letters PB.[45] It is tempting to connect these initials with Philip Benthall, Lawrence's son and owner of the estate from 1653 to 1713. Unfortunately, none of these relics can now be traced, but one wheel, found at the turn of this century at Caughley, is still in existence (pl. 39).[46] It is also carved from a single block of elm, with a hole, originally about $1\frac{1}{2}$ in. square, through its centre. The diameter over the flange is $9\frac{1}{2}$ in., over the tread 8 in., and the tread is $3\frac{3}{4}$ in. wide. It has been little used, for the turned groove where flange meets tread has not been worn away. It was apparently discarded because the wood split; it was mended with a pair of S-dogs, but the split widened and they fell out. The date of these wheels cannot be fixed. Such a rabbit-warren of workings was dug under the parish of Broseley that few can be tied with any certainty to documentary evidence, and carbon-14 tests are not yet able to give anything like precise enough an answer. One must resist the temptation to say that, because the wheel looks old, it must be old – the example of the Transylvanian trucks at Berlin and Bochum is sufficient warning (p. 83). But in view of the fact that Coalbrookdale ironworks just across the Severn began turning out quantities of small cast-iron wheels for waggons just such as these in 1729, we may tentatively ascribe these wooden wheels to a date before then. They may indeed belong to the seventeenth century.

The question of priority now arises. There is no doubt that the Wollaton line, built between October 1603 and October 1604, was the first in the Nottingham coalfield, nor that Beaumont took the idea of the waggonway to Tyneside between 1605 and 1608. The gap, however, between the Wollaton railway and the first Broseley one, completed in October 1605, is small. Willcox and Wells' statement that they obtained the approval of the Council in the Marches of Wales long before their line was finished must close the gap still more, but, as the records both of the Privy Council and the Council in the Marches are lost for this period, it is impossible to fix the date of their application. Clifford's railway at Calcutts was built at some unspecified date before October 1607, and his line at Yates Coppice at some date, also unspecified, before September 1608. So the question of the origin of the English railway boils down to this: did Broseley copy Wollaton or vice versa, or did both places perhaps invent the railway independently, or even copy it from some other, unknown, source? The last two alternatives must remain as possibilities, however unlikely. The usual view is that Broseley followed Wollaton's example, and it may well be so. The mines in and around Wollaton had been worked vigorously throughout the sixteenth century, and had reached a point where technical improvements were almost essential to allow their further development. Huntingdon Beaumont was just the inventive, progressive and adventurous type of man who would be likely to introduce innovations such as the waggonway. The Willoughbys had business connections with the West Midlands, though none with Broseley men, as far as is known, until their lease of Kinlet and Bower Hill from Lacon in 1613. They had particular links with collieries and ironworks in the Penkridge, Walsall,

Wolverhampton and Halesowen area,[47] which could well have been the means by which news of the Wollaton waggonway was transmitted to Shropshire.

While it is thus quite possible that Wollaton did precede Broseley in priority, it is worth while looking at Broseley's claims as well. The difficulty here is the almost complete lack of records of the Shropshire coalfield in Tudor times. The output in the sixteenth century is generally said to have been small,[48] but had the papers of, for example, James Clifford's Broseley estate survived, we might well find that the coal industry was much more flourishing than is supposed; and there do gleam through the darkness a few hints that it was so. Most of the incumbents in the area, regrettably, were not of a chatty enough disposition to enter the causes of deaths in their burial registers, but the parson of Much Wenlock, on the border of the mining district, did record a fair number of deaths in coal-mines in the second half of the century, one at least in a pit of Clifford's.[49] Likewise, Clifford was found by a Royal Commission in 1575 to have cast all the rubbish from his mine at the Tuckies into the deepest part of the Severn, to the impediment of navigation.[50] If we assume, for the sake of argument, that the industry was healthier in late Tudor times than has been supposed, were the circumstances right for the invention of the railway? They could well have been. Although vertical shafts were certainly met with, insets or horizontal adits were the norm for the earliest mines near the river banks, and they would seem to provide a likely breeding-ground for the idea of the railway – a continuous line from coal-face to river which would avoid the need to transfer the coal at the adit mouth from barrow or basket to wain. It is no accident that the term 'footrid', which with variants became a popular Shropshire word for a railway, originally meant a drift or adit (p. 233).

If these conditions were favourable for the invention of the railway, is there any suggestion that one or more examples did precede Willcox and Wells' line of 1605? We move here into the realms of guesswork alone. In 1601 the Privy Council received a petition from one Richard Wightman, who stated that John Brewster, who had worked some pits at Broseley, had had a wayleave from the landowners between his mines and the Severn 'that a passage should be suffered thorough their groundes for the carriage of the said coales to the river side'. Brewster had died in 1597,[51] but now that Wightman had taken over his pits, on the same conditions as Brewster, he was denied this passage. He, and the country, was liable to incur a great loss thereby. Would the Privy Council please help him? They passed his request on to the Council in the Marches, where we lose track of it.[52] Now, although there is no mention of rails, and the petition may simply involve an ordinary wayleave for wains, the process is identical to that followed by Willcox and Wells in 1605, when they wanted to build a railway across the land of the openly hostile Clifford. This may be significant, or it may not. Then whereas our sources on the Wollaton waggonway emphasize its novelty, those for the Broseley railways give no indication that they were anything wonderful. Again, the waggonway at Wollaton soon withered away, whereas at Broseley it took immediate and deep root.

None of this, of course, proves that Broseley invented the English railway. It is set out merely to show that Wollaton was not necessarily the first in the field simply because its waggonway, which certainly was the first in that area, appeared before the earliest one at Broseley of which record has been found. Had our main source for Wollaton, the Middleton Collection, not survived, the priority would ostensibly have belonged to Willcox and to Wells, not to Beaumont. By the same token, evidence may yet appear that there were railways in Shropshire before 1603.

The Newcastle waggonway: its spread on Tyneside and elsewhere

There's wood for to cut, and coals for to hew,
  And the bright star of Heaton will carry us through:
Ridley for ever, and Brandling for aye,
  There's plenty of coals for our waggon way.

'Brandling and Ridley', a song of Heaton colliers quoted by John Bell, Jun., *Rhymes of the Northern Bards*, Newcastle (1812), 300.

Under the Commonwealth, the political and economic climate was not favourable to the growth of the coal trade. Around Newcastle in particular many of the richer coal-owners had sided with the Royalists, and had had their estates confiscated for their pains. The progress of the waggonway was thus arrested for a time, until the Restoration of the Monarchy in 1660 gave the signal for a massive expansion in the trade, which in turn stimulated the waggonway into steady growth. The early seedling at Wollaton had long since withered away, leaving Tyneside and Shropshire as the two nurseries which were to improve the breed. They remained very largely independent of each other – Shropshire practices had no influence on those of Tyneside, and Tyneside had only a limited effect on Shropshire – until late in the eighteenth century, when the spread of canals began to break down regional barriers. Tyneside used its own particular terminology, and made large waggons running on track of a comparatively broad gauge; Shropshire had its own vocabulary, too, and favoured smaller waggons on a narrower gauge. Both areas colonized other coalfields, and on the whole kept within their own spheres of influence, so that, broadly speaking, the country was split into two empires. Tyneside-type waggonways were to be found in Cumberland, in Northern Ireland, and in Yorkshire; Shropshire-type railways spread over the West Midlands, north and south Wales, and probably Lancashire. Scotland was debatable country – after Shropshire had gained an early foothold, Tyneside eventually prevailed.

The railway's roots in Shropshire were deep and venerable, and Shropshire has all the credit for an invention that was to prove of epoch-making importance, the cast-iron rail. But for all that, it was Tyneside that boasted a far larger system of waggonways than circumstances permitted in Shropshire or in any of Shropshire's colonies; and it was Tyneside that stimulated the evolution of the waggonway into the modern concept of the railway. At the same time, the surviving evidence for Tyneside waggonways is far greater in bulk than for any other area. So with Tyneside, 'the native land of railways', as it was already being called in 1838,[1] let us begin.

Before discussing the engineering and operation of waggonways, we must first see where, when and by whom they were built. It is not the purpose of this book to describe their routes in minute topographical detail. Even if more of this sort of information survived, such a treatment would be disproportionately bulky in a general survey of this kind. At the same time, space forbids any detailed account of the politics of the coal trade, a hideously intricate subject involving intrigue and counter-intrigue, alliances and open warfare, families divided against themselves: a running battle, in fact, with ever-changing alignments and fortunes, between rival coal-lords and the tyrant wayleave merchants. 'Every man for himself', wrote one coal-owner to another in 1712 in the middle of a particularly virulent dispute over a waggonway, '& God for us all'.[2]

A general outline is necessary, however, of waggonway routes (fig. 31) and their begetters, both to give an historical picture of their growth and to help the reader to identify the lines and the people whom he will meet in later chapters. As to terminology, 'Newcastle waggonway' is a useful generic description for the whole breed of railway, but is misleading when a distinction is needed between the lines of the homeland – the Tyne and Wear – and those of the colonies elsewhere in the country. 'Northumberland and Durham' or 'the Great Northern Coalfield' are over-cumbrous; so the term 'Tyneside' has been adopted, at the expense of strict accuracy, to describe the waggonways which ran to the Northumberland coast and to the River Wear, as well as those on Tyneside proper.

Working geographically from north to south, we come first to the River Wansbeck, where the Black Close Colliery had a staith and waggonway in 1755.[3] On the next river to the south, the Blyth, we do not know the exact site of Beaumont's ill-fated ventures. A line which is shown on Gibson's map of 1787 as running from Barrington or Netherton Colliery to the north bank of the river a little below Bedlington is certainly distinct from, and later than, Beaumont's.[4] On the south bank, a five-mile waggonway from Plessey mine to Blyth Harbour was built before 1699, perhaps by Ralph Brandling, who bought the estate in about 1690: he certainly had waggonways in his Felling collieries on the Tyne by this date. Plessey changed hands with remarkable frequency. The fourth owner after

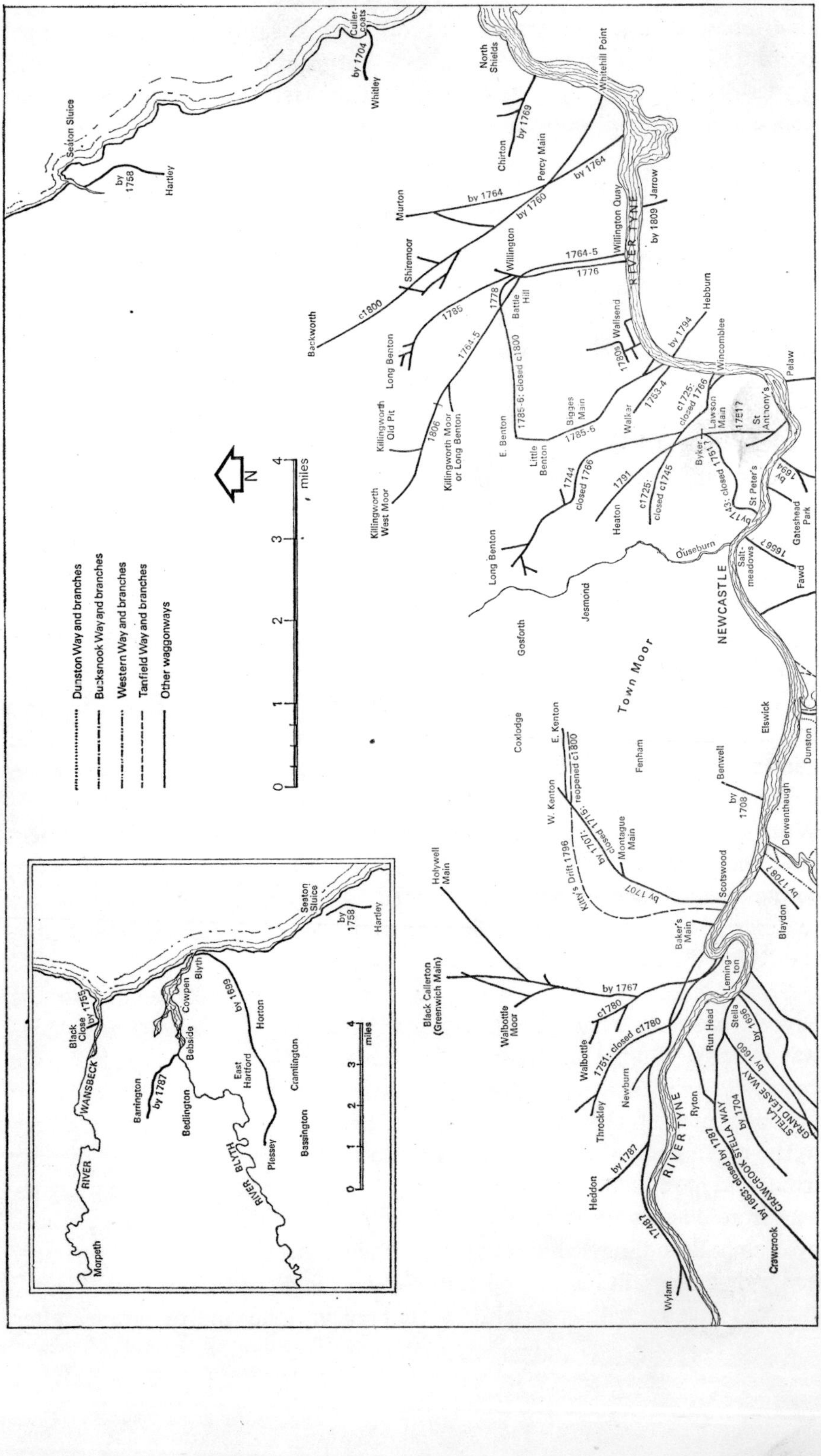
Dunston Way and branches
Bucksnook Way and branches
Western Way and branches
Tanfield Way and branches
Other waggonways
0 1 2 3 4 miles
N
Seaton Sluice
by 1758
Hartley
Culler-coats
by 1704
Whitley
North Shields
Whitehill Point
Chirton
by 1769
Percy Main
Murton
by 1764
by 1760
by 1764
Shiremoor
Willington
Willington Quay
RIVER TYNE
Jarrow
by 1809
Backworth
c1800
Long Benton
1785
1778
Battle Hill
1764-5
1776
Wallsend
Hebburn
by 1794
Killingworth Old Pit
1764-5
1785-6; closed c1800
1760s
Wincomblee
Pelaw
Killingworth West Moor
1806
Killingworth Moor or Long Benton
E. Benton
Bigges Main
Walker
1753-4
c1725; closed 1766
Lawson Main
St Anthony's
Little Benton
1785-6
1744
closed 1766
1791
c1725; closed c1745
Byker
Heaton
Long Benton
Ouseburn
St Peter's
Gateshead Park
Salt-meadows
Fawd
Gosforth
Jesmond
NEWCASTLE
Town Moor
Coxlodge
E. Kenton
Fenham
Elswick
W. Kenton
reopened c1800
Kitty's Drift 1796
by 1707
Montague Main
by 1708
Benwell
Dunston
Derwenthaugh
Holywell Main
Scotswood
Baker's Main
Blaydon
Black Callerton (Greenwich Main)
by 1767
Walbottle Moor
c1780
Leming-ton
Walbottle
1751; closed c1780
Throckley
Newburn
Run Head
Stella
by 1660
GRAND LEASE WAY
STELLA WAY
by 1704
Ryton
Heddon
by 1787
RIVER TYNE
CRAWCROOK STELLA WAY
Crawcrook
Wylam
Morpeth
RIVER WANSBECK
Black Close
by 1755
Barrington
by 1787
Bedlington
Bebside
Cowpen
Blyth
by 1689
Horton
East Hartford
Cramlington
Bassington
Plessey
RIVER BLYTH
Seaton Sluice
by 1758
Hartley
0 1 2 3 4 miles

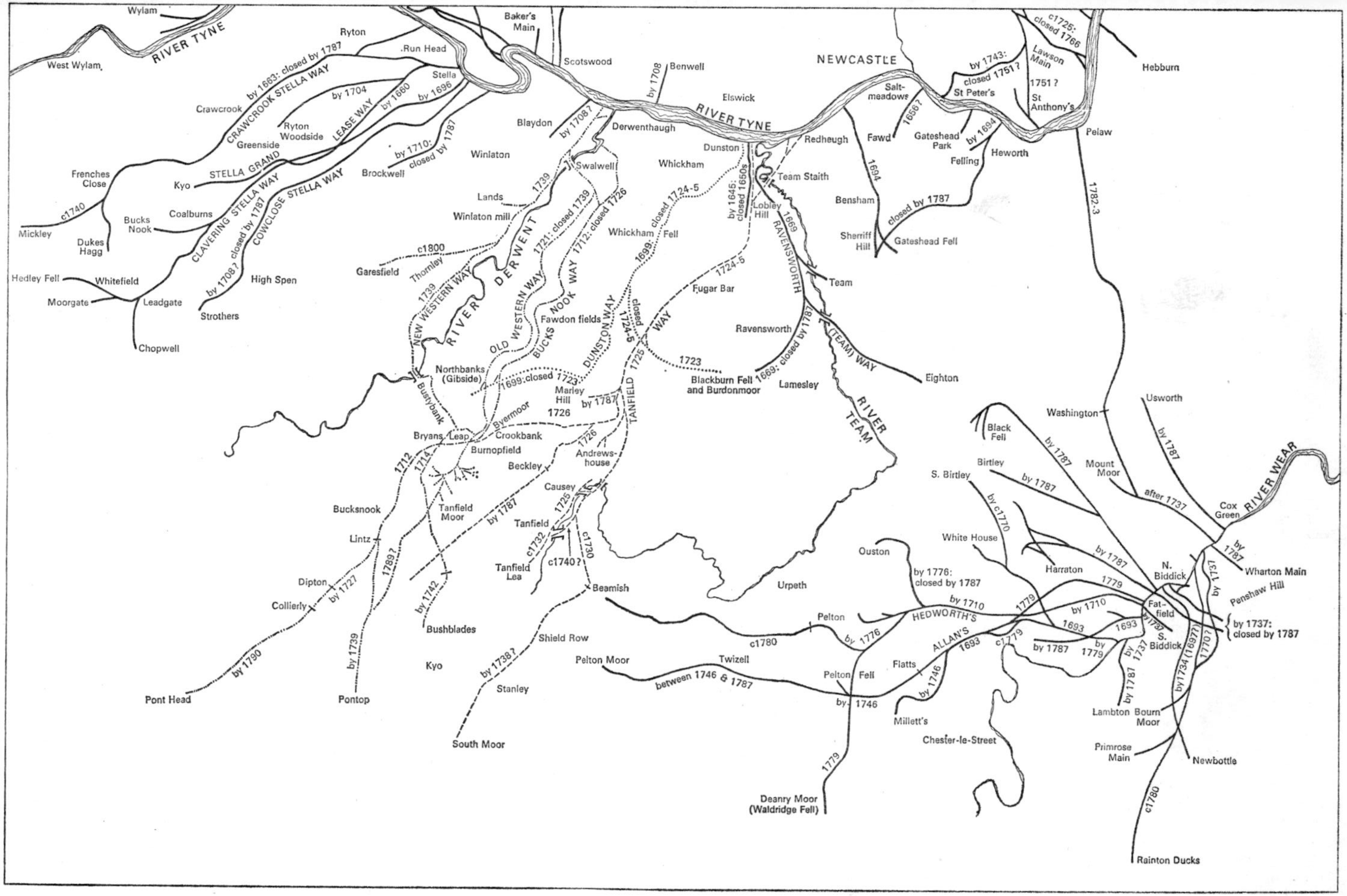

FIG 31 Waggonways of the Tyne, Wear and Northumberland coast.

Brandling was Lord Widdrington, who was attainted for his part in the 1715 rebellion and forfeited his estates. In 1722 the colliery was bought from the Crown by Richard Ridley, whose family worked it until its closure in 1813.[5] Further south, a short but important line to Seaton Sluice through Hartley, the property of the Delaval family, is first heard of in 1758.[6] South again, another short waggonway from Whitley pits to Cullercoats Harbour has been ascribed to the period 1621-62, but it can hardly antedate 1677, when the Harbour was built. It certainly existed in 1704.[7]

Turning the corner at Tynemouth and moving westwards up the Tyne, we reach North Shields, where a one-mile waggonway came down from Chirton. It was in existence by 1769.[8] At Whitehill, near Percy Main, was the staith of a three-mile waggonway from Shiremoor and Flatworth, built by William Gibson, Matthew Bell and William Brown between 1755 and 1760. Late in the century it was extended north to Backworth.[9]

The next group to the west, based on Willington, is of considerable importance, for it formed the main connecting link between the old type of colliery waggonway and the modern public railway. It was at Killingworth Colliery, which was a part of the Willington system, that George Stephenson carried out his early railway experiments. When, as an engineer of more than local repute, he went out to build railways elsewhere, he drew on the experience he had gained at Killingworth, and in particular he gave them the same track gauge as he was accustomed to. Willington is the home of the standard gauge.

The history of the Willington and Killingworth waggonways has been much confused by the existence of three distinct collieries called Long Benton. The earliest one, in the extreme west of Long Benton parish, was worked by the Grand Allies, the powerful partnership of coal-masters, from 1744. It was connected to the Heaton waggonway, and will be discussed shortly. When this first Long Benton Colliery was on the verge of exhaustion, the Grand Allies transferred their attention to the east of the parish, two and a half miles from their previous pits. The pits they sank here in 1761 were known as Long Benton New Winning, or occasionally as Killingworth Moor, a name worth using to avoid confusion. In 1762-3 the Allies obtained wayleaves to build a waggonway three miles long, via Willington Square to Willington Quay. It was completed in 1764 or 1765. When Killingworth Moor was becoming worked out, they sank a new pit at Killingworth West Moor, rather over a mile to the north-west, and extended their waggonway to it in 1806.[10] Meanwhile, the first pit of a new colliery at Willington Square was completed by William Gibson, Matthew Bell and William Brown, and was joined to the staiths at

Willington Quay by a new waggonway, close and parallel to the Killingworth way, but independent of it, which was opened on 6 March 1776. They obtained wayleaves in 1778 for a short branch to Battle Hill from Willington Colliery, and next year leased a new mine, the third to bear the name of Long Benton. It lay two miles north-west of Willington Colliery and half a mile north-east of the Grand Allies' Killingworth Moor or Long Benton mine. Gibson, Bell and Brown were given wayleave to lay an extension of their waggonway to the new Long Benton in 1782, and completed it apparently in 1785. At the same time they were sinking several scattered pits which formed Bigges Main Colliery, two miles due west of Willington Colliery. These were linked in 1785 and 1786 by new waggonways both to Battle Hill (and so to the Willington waggonway proper) and to new staiths at Wallsend on the Tyne.[11]

The three sides of a square thus formed were called 'the Willington waggonway' indiscriminately, even after the cross-link from Bigges Main to Willington was lifted about 1800. The first mile of the Tyne to Bigges Main line was used in 1808 by the new Kenton and Coxlodge Railway, and caused some trouble when the latter was converted to rack rails for the use of a Blenkinsop locomotive. Since this section still belonged to the Willington owners, and was still wooden (it was not converted to iron till 1819), the engines could not travel all the way to the staiths.[12]

In between the two legs of the Willington system lay Wallsend Colliery, with a short waggonway that dated from the 1780s;[13] and to the west of the Bigges Main line was another short, independent waggonway serving Walker Colliery, granted its wayleave in 1751, but probably not built till two or three years later.[14]

The waggonway to Heaton and the first Long Benton collieries had a disturbed history. George Liddell and William Cotesworth bought Heaton in 1717, but the Corporation of Newcastle steadfastly refused to grant them a wayleave because Alderman Richard Ridley's Byker Colliery would suffer from the competition. At last, in desperation, Liddell and Cotesworth petitioned the King in 1724, who graciously informed the Corporation that because of

> the hardshipp and oppressions the p^{r}sons now or hereafter Interested in Heaton Colliery may suffer from a partiall or Totall obstruction of the said Wayleave, Its therefore his Majesties will and Pleasure that you grant wayleaves to both partyes desireing the same upon Just Reasonable and Equall Terms.

The waggonway was soon built, down to staiths at Wincomblee, a little upstream of the Walker Colliery staiths, and in 1726, on the formation of the Grand Alliance (see p. 122), came under their wing.[15] As the lower pits suffered more and more from water and the accessible coal was worked out, the line was extended inland to the Long Benton estate, leased by the Allies from the Earl of Carlisle. This extension was first envisaged in 1735, and wayleave was obtained the next year; but the Long Benton pits were not seriously worked and the waggonway was not laid until 1744, the year when Heaton Colliery was closed. The wayleave also permitted a line to Killingworth, which was not made. In 1766 Long Benton too was worked out, and the waggonway was closed completely.[16]

Upstream again, Richard Ridley's pits at Byker were linked by a waggonway before 1743 to staiths at St. Peter's, replaced probably in 1751 by new ones at St. Anthony's.[17] After one or two false starts, Heaton Colliery was again worked from 1791, and was joined by a new waggonway to the head of the old Byker system at Walker Hill or Lawson Main pit. Part of the line was relaid in 1797 with the first all-iron rails on Tyneside, but it remained largely of wood until 1821.[18]

Matthew White's collieries at Coxlodge, Gosforth and Jesmond to the north of Newcastle had no waggonways for many years. A counsel's opinion given in 1763 on a conveyance dated 1672 which concerned South Gosforth and Coxlodge collieries said:

> Some short time before this conveyance a new method was invented for carrying Coals to the River in large machines called Waggons made to run on Frames of Timber fixt in the Ground for that purpose and since called a Waggon Way which frames must of necessity lye very near, if not altogether upon a level from the Colliery to the River and therefore wherever there are any hills or vales between the colliery and the river and the same cannot be avoided, it is necessary in order to the laying of such waggon ways, then to make cutts through the Hill or level the same, and to raise or fill up the Vales so that such Waggon Way may lye upon a level as near as possible.[19]

Although this has been thought to indicate a Coxlodge waggonway of 1672, it is not good evidence. In fact, all the signs are to the contrary. In 1708 White obtained a waggonway wayleave in Jesmond, but in 1720 he was still leading his coals by wain when he got another wayleave from the Corporation for a waggonway from Gosforth, Coxlodge and Jesmond along the edge of the Town Moor to the mouth of the Ouseburn just east of Newcastle. As far as can be seen, it was not built, and another proposal of 1763 for a line on the same route met the same fate. This was the occasion of the counsel's opinion, the question being whether an old wayleave for carts and wains could be deemed to authorize waggonways as well – a moot point which cropped up elsewhere. Two further plans in 1790 and 1800, which envisaged a waggonway extending westwards to include East Kenton Colliery, likewise failed. It was not until the iron Kenton and Coxlodge Railway appeared in 1808 that these collieries obtained surface rail access.[20]

West of Newcastle, Tempest's and Carr's pits at Elswick may have had a short waggonway by 1698, and the Montagues' at Benwell by 1708.[21] Kenton Colliery had a waggonway running down to Scotswood by 1707, and perhaps as early as 1688 when a wayleave along the north-western edge of Newcastle Town Moor was granted to Sir William Blackett, who was then leasing the colliery. Although Kenton was drowned in 1715, the track was not lifted immediately. A viewer (colliery engineer) reported in 1717:

> An Estimate or Computation of Fenham Colliery lying near Kenton Waggon Way belonging to Sr William Blackett Barr^t^ & now by reason the Colliery of Kenton is drowned and so the way Waggons and Trunk [staith] are become useless and therefore it is believed this Colliery of Fenham may be of Pticular advantage to Sr William and yeild him considerable yearly proffitt.[22]

The suggestion was not apparently taken up, and the Kenton way was cut back to Montague Main, leaving a bare two miles of track down to Scotswood. Kenton took on a new lease of life in 1796, when Christopher Bedlington drove a three-mile tunnel, known as Kitty's Drift (see p. 324), from Scotswood to East Kenton Colliery, both to drain the pit and subsequently to transport its coal to the Tyne. With the mines thus dewatered, the surface waggonway was extended north again, probably very much along the original route, to serve West Kenton pit.[23] Kitty's Drift was abandoned for transport (though not for drainage) when the iron-railed Kenton and Coxlodge Railway was opened in 1808, curving round above Newcastle and joining the Bigges Main waggonway to terminate at Wallsend.

The most westerly staiths on the north bank of the Tyne were at Lemington, where three waggonways ended. The first, built before 1767, came down from Holywell Main and Greenwich Moor collieries, three and three-quarter miles away. The second line also served two collieries: Matthew Bell and William Brown's Throckley mine, whose waggonway was completed in 1751, and Walbottle, owned by the Duke of Northumberland. When Throckley was closed about 1780, Walbottle was linked instead by a new branch to the Holywell Main waggonway.[24] The last line to Lemington had come down five miles from Wylam, running close to the Tyne all the way. It belonged to the Blackett family, was built (it is said) in 1748, and was early in the locomotive field. An engine was built for Blackett in 1805 at Gateshead by Richard Trevithick's assistant, John Steel, and it was satisfactorily tested in the works. Blackett, however, had second thoughts, realizing perhaps that it would splinter his rails to matchwood, and did not take delivery. But once the line was relaid in 1808 with angled plate rails, the way was clear for Hedley's *Puffing Billy* and its brethren to haul Wylam coals.[25]

Crossing over the Tyne a little upstream of Lemington, and working back along the south bank, we come first to a group of lines which ran down to Stella freehold staiths near Ryton. The output of the mines they served had, by the Restoration, overtaken that of Whickham, which explains the early dates of these Stella waggonways. The most northerly, the Crawcrook way, was in existence by 1663 – a lawsuit of 1665 refers to a bond of two years earlier which promised delivery from Crawcrook Colliery of 20 tens 'of good merchantable ship coles accompting twenty waggons to a tenn, every waggon to contain fifteen bowles usual cole measure'.[26] If wains had still been in use, the ten would certainly have been defined in terms of fothers or wain-loads. Crawcrook is barely three miles from Stella, but as the nearer pits were worked out the line was extended to Frenches Close, and about 1740 to Mickley, nearly seven miles from the Tyne.[27]

Next to the east came the Stella Grand Lease waggonway, running up to Greenside, with a branch, apparently, to Ryton Woodside. It shared the first half-mile from the staith with the Crawcrook way.[28] In 1660 Sir Richard Tempest sold ten keels (the lighters plying between staiths and ships) and 'a quarter part of the wood or timber laid upon trenches, bridges, and waggon ways, or unlaid upon the same'. Again, in 1666 Sir Thomas Tempest leased out 'fower other keelroomes lying in the Western Waggon Way boundringe Eastward on Sr John Marley's keelroomes, westward on a Keelroome of M^r^ Li[onel] Maddyson'.[29] The purchasers and lessees involved were a partnership which included members of the Carr, Vane and Blackett families until about 1708, when Sir Henry Liddell and Edward Wortley bought the Grand Lease pits.[30]

The third line was the Clavering Stella waggonway, which is first heard of in 1696, and long remained the property of the Clavering family. A steep and serpentine line, it passed through Greenside (where it paralleled the Grand Lease way for some distance) to Coalburns, where it threw off a branch to Bucksnook Colliery (not to be confused with the Bucksnook near Burnopfield, of which more later). It proceeded southwards to end in an array of branches of various dates to Moorgate, Whitefield, Hedley Fell and Chopwell pits, some six miles from the Tyne. The main line of the Clavering Stella way ran alongside the last few miles of the Lead Way or Lead Road, a packhorse track for the transport of lead from the inland mines of Allendale and Blanchland to Stella staith. At least from 1710, some lead was transferred from horse to rail at Leadgate and Chopwell, and the Clavering Stella became the only Tyneside waggonway to carry a regular traffic in any commodity apart from coal.[31]

About the fourth Stella line, the Cowclose waggonway which served the Strothers near High Spen,[32] little is known. The last, the Winlaton or Brockwell waggonway, may have been built soon after 1677, when a lease permitted the use of waggons, and it was certainly in existence by 1710, when a man was killed on it. It belonged to Lord Widdrington until he forfeited his estates in 1715, when it was bought by Sir Edward Blackett, Francis Brandling, Richard Ridley and John Simpson.[33]

After a short and unimportant line from Blaydon Colliery, we come to an intricate complex of waggonways which in the first place carried the coals from the upper part of Whickham parish down to Dunston staiths at the mouth of the Team, and to Derwenthaugh staiths at the mouth of the Derwent. They are best considered chronologically rather than geographically. As the lower collieries were worked out, these lines were extended inland to tap the rich areas of Tanfield, Pontop and Stanley. One of the waggonways which ended at Dunston was the Tanfield, perhaps the most famous of all the Tyneside lines. It was not, however, of such venerable antiquity as has often been claimed, for it has been almost inextricably confused in historians' minds with the older Ravensworth or Team waggonway, which also ran down to staiths at the mouth of the Team, where it crossed the Tanfield way. Charles E. Lee, for example, in a paper on the Tanfield entitled 'The World's Oldest Railway', suggested that it was built perhaps in 1632, and certainly by 1671.[34] It may be recalled (p. 93) that the date 1632 is due to a misunderstanding, and 1671, although another misleading date, belongs to the Ravensworth way. In fact, the Tanfield, as it existed until 1962, was built only in 1725, although it had a predecessor on a quite different route to Dunston which dated from 1699.

The history of the waggonways to Dunston and Derwenthaugh, as far as it can be disentangled, was this. We have already considered (p. 93) a short line from some of the Whickham pits to Dunston, which was in existence by 1645 and probably disappeared during the Commonwealth. The next waggonway (often known as 'The Old Way') was that from the Liddell family colliery of Ravensworth, with later branches to their Team, Eighton and Blackburn pits. It ran down the River Team to Dunston, where its Team staith was distinct from the Dunston staith proper (fig. 30). In 1810 John Bailey, enquiring into the origins of waggonways on Tyneside, wrote:

> Upon examining the books of Ravensworth Castle, Mr Robson, (the present agent), informs me, that the first staith bills are in 1671, in the time of Sir Thomas Liddell . . . Joshua French was the staithman, and from his bills, beginning with that year, it appears that coals were then led by waggons to Team Staith.[35]

1671, he makes it clear, was merely when the surviving records began, not when the line was built. This date can, however, be pinned down with accuracy. In 1669 a marriage was arranged between Sir Thomas Liddell's eldest son and the daughter of Sir John Bright of Badsworth in Yorkshire. Bright, wishing for information on the state of Liddell's Ravensworth Colliery in connection with the marriage settlement, surreptitiously engaged a Tyneside viewer to report on it. This he did in August 1669, saying that it was in flourishing condition, with revolutionary drainage engines and enough coal for generations, and that if 2,000 tens* a year were worked (as was quite feasible on present showing), it would yield a profit of £2,500 a year: 'They may work what quantyty they please, and in regard they have begun a waggon way, they may lead what quantyty they please.' In October Liddell suggested that Bright should send an observer to inspect the mines, not knowing that he had already done so. The colliery, he said, was already worth £1,500 a year '& in much better condition for the future advantage of itt, then it hath bin in mine or my Ancestors tymes, by reason of the New Collyery, and wagon way now layd'.[36] The Ravensworth way, with a final length of four miles, never left the Liddell family.

The next line in the area was the Dunston waggonway. The Hutton and Northbanks collieries, in the Gibside estate, were leased in 1692 by Sir Francis Blakiston to the Hon. Charles Montague, with a wayleave to take the coals to the Tyne. Though Montague worked the pits along with his Wortley relatives, it was not until 1699, when William Blakiston Bowes (still an infant) inherited Gibside from Blakiston and renewed the lease for twenty-one years, that they built the waggonway on a meandering route across Whickham Fell and down to Dunston.[37] It did not, however, last long. From 1723 the Hutton and Northbanks coals were carried down a new route to Derwenthaugh (the Western Way). In the same year the top

* A ten at this date was a measure of 10 chaldrons of coal. The history of this unit is explained on pp. 188-9.

half of the Dunston way was lifted and the track used to build a new line to Burdonmoor and Blackburn collieries, which were shared between the Montagues and Wortleys and the Liddells, and had previously sent out their coal down the Ravensworth way. This Blackburn line still fed into the lower half of the Dunston way, but it survived no more than a year or so in this form. After 1724 or 1725 only the top mile was retained, as a branch of the newly constructed route that was soon to become famous as the Tanfield waggonway. All the rest of the old Dunston way was lifted.[38]

Meanwhile, the first line down to Derwenthaugh had appeared. Sir John Clavering and Thomas Brumell wanted an exit for their coals from Byermoor, Lintz and Bucksnook collieries, but as they were not on good terms with the Bowes and Montagues they could not use the Dunston way. In 1710 they set about collecting wayleaves to build a line down the eastern side of the Derwent to its mouth, a very difficult route, but the best they could do.[39] Their right to some of the ground, however, was shaky in the extreme, and there was trouble even before the waggonway was opened with pomp and ceremony on 6 June 1712. The track on the most debatable stretch was torn up by a hostile crowd of Bowes partisans, led by Lady Bowes in person. One of her accomplices was William Cotesworth, Lord of the Manors of Gateshead and Whickham and the greatest wayleave tyrant in the area, whose settled policy it was to have a finger in as many waggonway pies as he could. In the course of the lawsuits which followed this riot, Lady Bowes managed gracefully to extricate herself, leaving Cotesworth to make reparation, a slight which it took him many years to forgive.[40] Sir John Clavering was an even more implacable enemy of his, and was therefore discomfited when Cotesworth bought a share in Bucksnook Colliery and began to use the waggonway himself. Sir John and, after his death, Lady Clavering also had their rights on it, so they shared it for a while in uneasy truce, along with George Pitt, who owned the Tanfield Moor collieries and used the way from 1714.[41]

When William Bowes came of age, he set about building a line to rival the Bucksnook waggonway. Usually known as the Western Way (the same title was occasionally used of the western lines in the Stella complex), it was opened on 17 April 1721 and ran parallel to the Bucksnook way, slightly nearer the Derwent. At the same time, the Wortleys' and Montagues' lease of his Gibside collieries having expired, he worked them himself and carried the coal down his new waggonway, with the result that the top of the Dunston way fell out of use.[42] Lady Clavering and Pitt immediately transferred their traffic to the Western Way to escape the contagion of Cotesworth on the Bucksnook, and the line was soon extended southwards to Burnopfield and Dipton collieries to accommodate Richard Ridley and

John Simpson. William Bowes died in October 1721 at the early age of twenty-four, and was succeeded by his brother George.[43]

Cotesworth was still in theory at loggerheads with Bowes over the disputed Bucksnook wayleave, inherited from Brumell and Lady Clavering, until in April 1726 the two men signed an agreement to drop all suits against each other and to work together in future to their mutual advantage.[44] At the same time, George Liddell and the Wortleys were also hand in glove: they were partners in Stella Grand Lease and in Blackburn, and in 1724 or 1725 together completed a brand-new waggonway to take their Blackburn coals to the staiths at Redheugh and Dunston, which finished off the old Dunston way for good. In July 1725 this new line was extended to Tanfield and Causey, two recently acquired collieries of the Wortleys, by way of the famous Causey Arch.[45] These two partnerships united in June 1726 to form the celebrated Grand Alliance: Sidney and Edward Wortley, with their agent Thomas Ord; George Liddell and Sir Henry Liddell the fourth baronet (still a minor); George Bowes; and William Cotesworth as Secretary. They agreed not to raise any lawsuits involving each other; to pool their wayleave rights and existing waggonways for their common use; to work their coal in quantities to be agreed; and generally to act co-operatively and have done with the destructive rivalries that had dogged them before. Cotesworth was to be paid 5*s*. a ten for all coals led from the partnership collieries, whether over his lands or not, as his price for not granting mining rights or wayleaves to other parties. All the major collieries belonging to the partnership were to be covered by the agreement, except Bowes' and Liddell's family collieries at Gibside and Ravensworth respectively, and the Liddell-Wortley share in Stella Grand Lease.[46] Thus constituted, the partnership completed the Tanfield way which Liddell and the Wortleys had begun, and extended it to Beamish, Shield Row and South Moor in stages. By 1738 it was over eight miles long, for a time the longest waggonway on Tyneside.[47]

Even before the Grand Alliance was formed, rumours spread that they were going to close the Western Way in order to concentrate its traffic on to Cotesworth's Bucksnook way, which was easier to work. Sir Francis Clavering, however, though he had a part share in the Bucksnook, was enjoying a fat income from wayleave rents paid for the Western Way, and to avoid losing them he retaliated by closing the Bucksnook instead.[48] This, as a partner, he had power to do. He had not safeguarded the Western Way, however, for the Allies persuaded Pitt to build a cross-line from Burnopfield to join the Tanfield way at Marley Hill, as an alternative outlet. Several times thereafter the Western Way was threatened with closure, but Bowes' own Gibside Colliery was much more conveniently served by Derwenthaugh than by Dunston, and so the Western Way was reprieved for a while, used by Ridley and Simpson with Burnopfield and Dipton coals, by Lady Clavering with Byermoor coals, by all of them with Bucksnook coals (for Cotesworth had died within a month of the Grand Alliance agreement), and by Bowes in his private capacity. The future had seemed so uncertain that Lady Clavering, Ridley and Simpson began planning a private, alternative, route in 1727, but as the danger of closure appeared to recede, the scheme was shelved.[49]

In the event, however, the day of reckoning was not long postponed. Bowes' pits at Northbanks and Gibside were worked out by 1738, and with them the Grand Allies' last reason for keeping open the Western Way was gone. Ridley, Simpson, Lady Clavering and a newcomer, Lord Windsor, who was working Pontop Colliery, were prevented from using it, and Sir James Clavering, speaking for his kinswoman to a committee of the House of Commons in 1739, complained that no wayleaves were possible except by courtesy of the Grand Allies; he might have to fall back on wains and carts. The solution was found in the same year, when a new partnership of Ridley, Simpson and Lord Windsor, retaining the top part of the Western Way, which was not in Bowes' control, resurrected their plan for a new line down the west bank of the Derwent, thus by-passing Bowes' domain.[50] At about the same time they built southward extensions to their new Pontop and Bushblades collieries. Some time later, another branch from Dipton southwards to Pont Head increased the maximum length of the line to a good ten miles.[51] The partnership ended after fifty years in 1789, when the whole Western Way was sold to the Tanfield Moor and Pontop Pike collieries. It survived complete for a short time, but Dunston was finally found to be a more convenient staith than Derwenthaugh, so, somewhere between 1801 and 1812, the central section was lifted, leaving the collieries with Pitt's cross-line of 1726 to the Tanfield waggonway as their sole exit. A branch, however, had already been laid from the Thornley and Garesfield collieries lower down on the west bank of the Derwent, so the bottom two miles of the Western Way remained in use solely for these mines. It was in this state that the system entered the era of the iron rail; most of the main routes were converted to iron, the rest sooner or later died the death.[52]

Proceeding eastwards again along the Tyne, very soon after the Redheugh staiths of the Tanfield waggonway, we come to the lower end of the Parkmoor, Sherriff Hill and Gateshead Fell waggonway, which belonged to Blackett, Bell and Wilkinson, and was first built, it seems, in 1694. Connected with it was a line serving Sir Ralph Carr's Bensham Colliery, dating from 1699.[53] Beyond Newcastle and Gateshead proper was a short line from Fawd Colliery to the river at Saltmeadows, where a waggonway may have been laid as early as 1656 (see p. 95), and another from Cotesworth's Gateshead Park mine. Then a second leg of Blackett's Gateshead Fell waggonway came down, after uniting with the Brandlings' Felling colliery waggonway, which was in existence by 1694.[54] At Pelaw was the Tyne staith of Washington and Usworth collieries, which for long had relied on a short connection to the River Wear, but in 1782 or 1783 were given this second exit, four miles long. A waggonway to Hebburn staiths had appeared by 1794.[55] Lastly, on the south bank we come to Jarrow staith, where Cotesworth and Dean Montague leased exclusive wayleave rights, together with leave to build waggonways from adjoining parishes, in 1705 – Cotesworth's first venture into the wayleave business. Whether any waggonways were laid to Jarrow at the time is not clear; it may well be that the first did not come until soon after 1800, when Simon Temple was opening up Jarrow Colliery.[56]

The waggonways of the Wear, which were the only begetter of Sunderland's commercial importance, were both fewer and more concentrated than those on the Tyne. In 1787, fifteen lines debouched on to both banks within a stretch of little more than two miles. Their history was just as complicated as that of the Tyne waggonways, but extraordinarily little information about them has survived. Of those on the north bank, tradition says that the earliest was that of Thomas Allan from his Flatts Colliery, just north-west of Chester-le-Street, to the Wear at Fatfield. The date is said to have been 1693. By 1746 it had been extended to just west of Pelton Fell, and by 1787 had reached Pelton Moor Colliery, six and a half miles from the Wear and no great distance from the southern end of the Tanfield waggonway, though there was never any physical connection. In the 1770s it was in the hands of the Shafto family, and in 1789 was inherited by General Lambton. At various dates it sprouted several branches.[57] Its next-door neighbour at Fatfield was built by Dean Hedworth before 1710. It ran westwards, parallel to Allan's, to pits in the Pelton area, and by 1776 had a branch to Ouston. Some time during the next ten years it was elongated westwards to Beamish South Moor, a mine close to Shield Row on the Tanfield, but again unconnected with it, so that the Pelton and Beamish lines ran more or less parallel for their whole length. By 1776 the waggonway was the property of Sir John Eden.[58] In 1779 William Joliffe opened a new colliery at Deanery Moor or Waldridge Fell, south-west of Chester-le-Street, and laid a third and heavily engineered line to Fatfield which shared Hedworth's old way for part of its route. A fourth waggonway came down from South Birtley, a mine of William Peareth's, using a small section of Allan's line.[59] At Washington to the east, although waggonways are mentioned in 1704, the two-mile line to Cox Green did not appear until after 1737.[60]

On the south of the river, three waggonways had been built by the Lambton family before 1737, from collieries at Lambton Main, South Biddick and in the Bourn Moor area. The latter line was augmented, it seems, in 1770 by a parallel route surveyed by William Brown, and was lengthened about 1780 to four miles by linking it to Rainton Ducks mine to the south.[61] Along very much the same route, from Newbottle to the Wear, a wayleave was granted in 1697 to Mrs. Jane Wharton, whose name long survived in Wharton Main Colliery. This waggonway was probably built immediately. Certainly by 1734 Messrs. Hilton and Nesham's colliery at Newbottle had a rail connection to Penshaw staith, where there debouched two more short waggonways belonging to John Tempest and one Smith.[62]

*

Although the coalfields of the seventeenth and eighteenth centuries were to a certain extent self-sufficient, it was inevitable that the larger ones should come to exert some influence on the practices of the smaller. Once the Tyneside waggonway had become firmly established, its rumour was spread about, and coal-owners elsewhere began to copy it (fig. 32). They might invite a Newcastle viewer or wright to build them one, or solicit details by post, or send a minion to inspect these marvels at first hand, or even go to see for themselves. In many cases, as on Tyneside, these waggonways were built to harbours or to navigable rivers. Once the canal system got under way, especially in the Midlands, it likewise attracted the railway to areas where it had been pointless before through lack of long-distance transport routes. Occasionally a waggonway which fed an ironworks or supplied a large town might have no water communication, but purely land-locked lines were exceedingly rare.

The first certain export of the waggonway from Tyneside was to Sheffield. It is said that a line was built here in 1722-3, but no details are known until 1729, when the surviving series of accounts of the Duke of Norfolk's colliery begins. On 13 December, Robert Littlewood was paid 'his Expences in going to Newcastle about the Wagen Way', and at the same time details of the waggonway expenditure begin to appear, and continue for ten years or more. Though they become fewer thereafter, the line survived into the 1760s.[63] It ran to a yard in the town, for supplying local industrial and domestic needs.

The second recorded case of Tyneside influence need not detain us long. When Ralph Allen, the Post Office reformer, built a railway from his stone quarries above Bath to the River Avon, it was described as

> such a Road . . . as the Gentlemen in the North of *England* had made between their Collieries and the River *Tyne*, that heavy Carriages

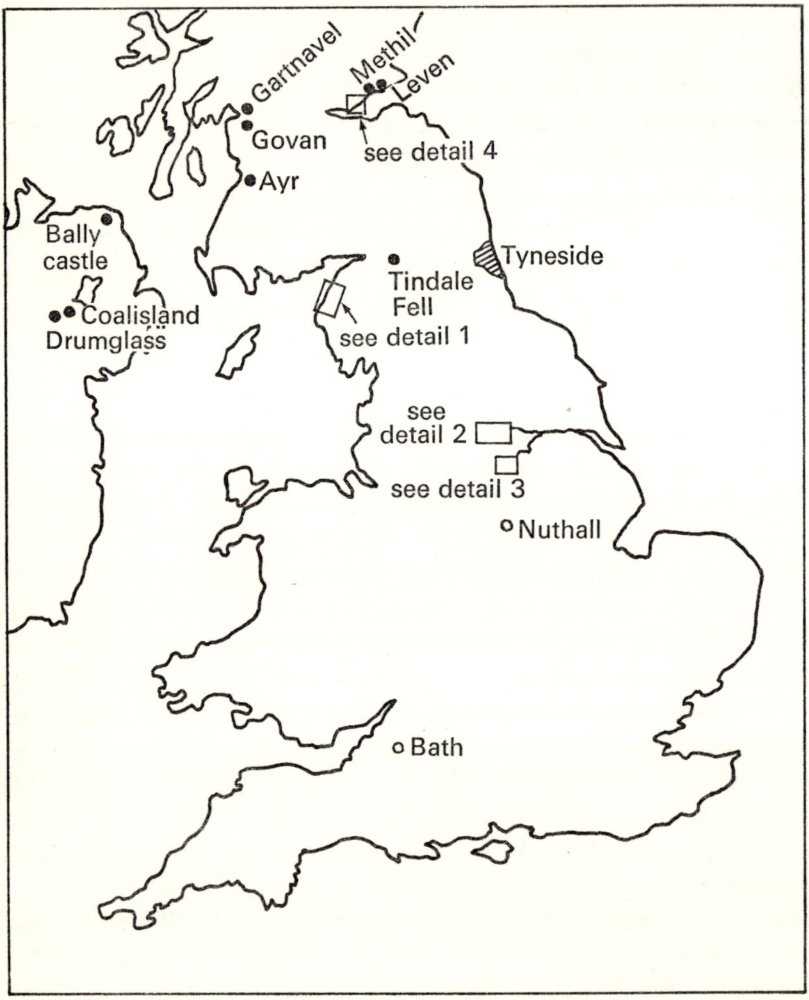

FIG 32 Distribution of Tyneside-type waggonways. Details on fig. 33.

> might be drawn along it with such little Strength, as would reduce the transportation of the Stone to the Water Side, to half the Price of carrying it down in common Waggons. Of the roads and Carriages in the North the late Mr. *Hedworth* sent proper Models to Bath.

Allen may have first got the idea when in the north on postal business.

But there was also Shropshire influence on the waggonway, unacknowledged, but so obvious in the design of the waggons, which can owe little to Dean Hedworth's models, that the details of this railway will be discussed in chs. 13 and 14, among other Shropshire railways.[64]

Tyneside's nearest neighbour, the Cumberland Coalfield, boasted a number of waggonways of which those leading from the Lowther family's pits to Whitehaven Harbour were the most important, though not the most extensive. Coal was carried down by packhorse and pannier until 1682, when an improvement in the transport became obviously necessary. One suggestion was for

> Waggons in imitation of those at New Castle, but our Case being different from theirs, where their Waggons are only Used to a Staith, 'twas objected that nothing would do here but what must go to the Ship Side, which a Waggon was most incapable off, as being only fit for a Staith.

John Gale, Sir John Lowther's agent, proposed that carts be used – a novelty in the area – and next year he built a 'coalway' from Woodagreen pit to the harbour, a 'Causey' or causeway bounded on each side by wooden baulks on which the cart wheels could run.[65] This type of construction, halfway towards a railway with its prepared track but no means of guidance, is rarely met with; but its success defeated Lowther's rivals and gave him command of the Whitehaven coal trade. By 1735, however, the problem of loading sea-going ships by waggon had been solved by the building of unusually tall hurries or staiths beside the harbour, and in that year Carlisle Spedding, Sir James Lowther's energetic coal agent, laid the first waggonway proper, from Saltom pit in Howgill – the southern part of the coalfield. It was apparently first used on 15 November, when John Spedding, Carlisle's brother and Lowther's general agent, told his master that 'we led Coals & carry'd the wagn^{s} round the Stage in the Steath & the Empty Waggons the Backway'. The nearby line from Parker pit to the Harbour, usually considered the first at Whitehaven, followed three years later, being opened on 4 August 1738.[66] Whingill, the northern sector of the coalfield, was only developed later. A waggonway to serve it was under consideration in 1738, but it was not built until 1754. Another of Carlisle Spedding's works, it was necessarily steep, and it ended a short distance from the quays, so that the coal had to be carted the last few yards. The building of a bridge to carry the line to the Harbour edge, halted by Spedding's death in an underground explosion in 1755, was not completed until 1803.[67]

Henry Curwen laid a waggonway from his pits to Harrington Harbour

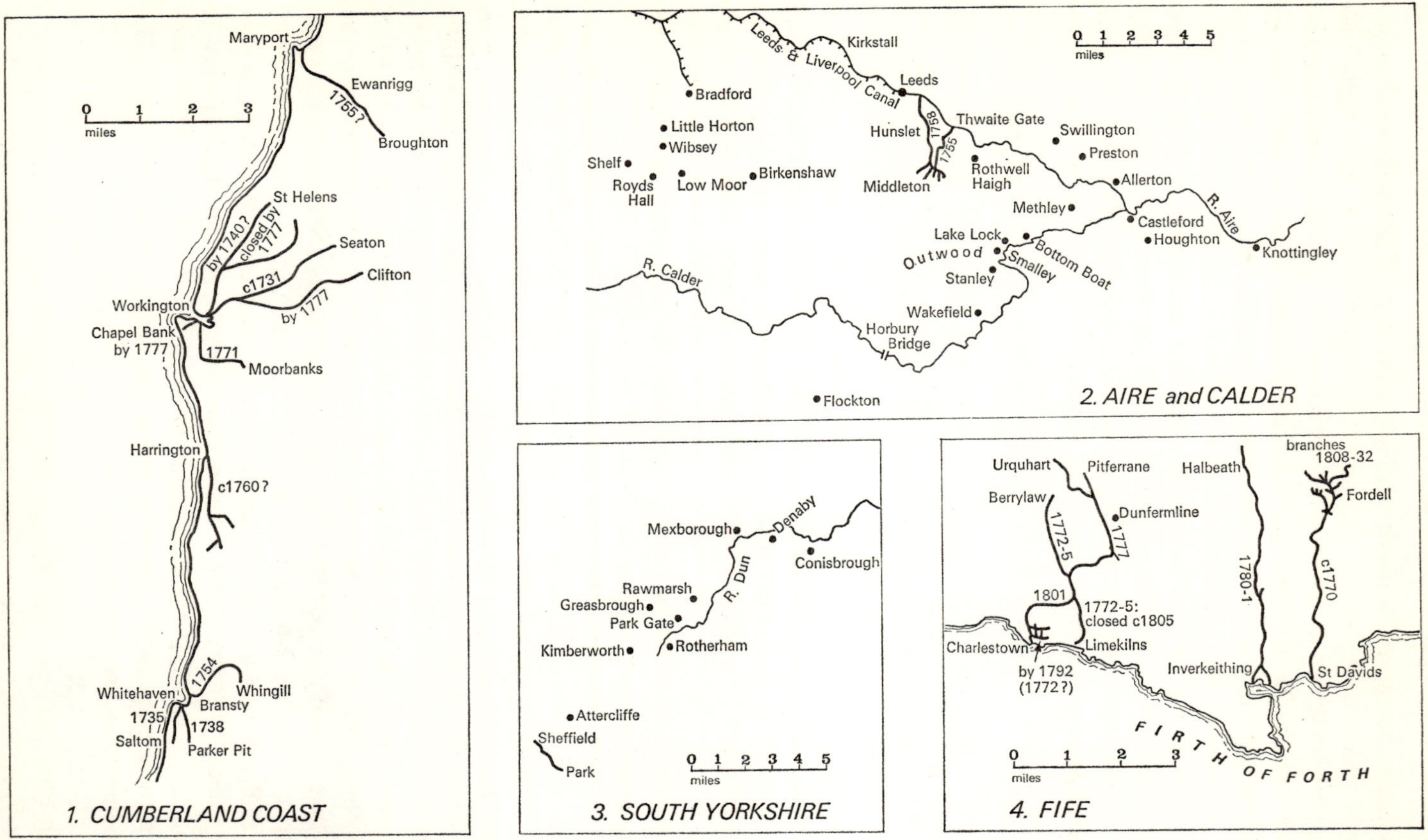

Fig 33 Detail maps of Tyneside-type waggonways.

soon after it was built in 1760, and the south side of Workington Harbour, three miles to the north, was likewise served by two short lines from Curwen collieries at Moorbanks (1771) and Chapel Bank (between 1771 and 1777).[68] On the north of the harbour, however, there were longer and earlier lines built by Sir James Lowther. The three-mile Seaton waggonway preceded even the Saltom line at Whitehaven. In 1751 William Brown told Carlisle Spedding that he would lend him an experienced waggonway wright, who had practised at Heaton and Long Benton: 'The man's Name is Ral. Douglas he tells me he was in your part about 20 years Since and laid a waggon way from Seaton to Workington.' This line probably dates to 1732, when Lowther took over the lease of Seaton pits; it was certainly at work by 1734. Two other Lowther lines, from the St. Helens Colliery to the north and from Clifton to the east, appeared a little later; but the first was closed before 1777, the second in 1781.[69]

To the north again, a two-and-a-half mile waggonway was built in 1755 by the local coal-lord, John Christian, from Broughton pits to Maryport Harbour, although in 1786 his wayleave for the lower portion expired and he was compelled to transport the coal for the last three-quarters of a mile in carts.[70] In the small and isolated coalfield south-east of Brampton, the Tindale Fell waggonway, which was to become famous for its early use of the wrought-iron rail, is said to have been built in 1775, but contemporary evidence is lacking. It seems to have been in existence by 1797, and probably began life with wooden rails.[71]

The coal exported from Cumberland went mainly to Irish markets, so it is not surprising that the earliest waggonways in Ireland imitated those of Whitehaven. Coal was mined on a small scale at Ballycastle on the Antrim coast, and in 1737 Hugh Boyd, the lessee of the mine, was given a grant by the Irish Government to build a harbour which would foster the Irish coal trade. Boyd's son and two others spent the winter of 1739 visiting parts of England and Scotland for ideas (Whitehaven would certainly be one of their ports of call), and in 1740 Boyd laid the first stretch of a short 'timber road' or waggonway for carrying the building stone from quarry to harbour works. Its longest extent was 486 yards.[72]

When a proposal was mooted in 1752 to develop the Drumglass colliery in County Tyrone, it was boasted that, in spite of difficult communications, it could supply coal as good and as cheap as that from Whitehaven. A critic of the scheme, however, argued that it was not nearly as good, and its cheapness was doubtful:

> In the first place, Sir, I do not find that the Proprietors have any proper Descent for a Waggon way, as they have on the other side of the Water, where, with so many Tun in a single Waggon, steer'd

> only by a single Man, from the Pit to the Ship, at the most trifling Expence. But, if there be such a Descent, I am confident, it must be form'd by an infinite Expence of Banking, and raising and levelling of Grounds, and tho' we should suppose the Proprietors able or willing to bear such an Expence,

how much will it raise the price of coals? As it happened, a short wooden line was built in 1754, but four years later it was reported to be in bad repair and was soon closed.[73] Then in 1757 Davies Ducart began building an ill-fated canal to the pit from the Tyrone Navigation at Coalisland, consisting of a series of level pounds connected by inclines which the boats were to be let down on rollers. The fall of the inclines varied from 55 to 70 ft. When the canal was finished, but not the inclines, John Smeaton was called in to report, and recommended 'a rail road of wood, in the manner of those at Newcastle, and Whitehaven' in preference to a canal of any sort. But it was too late to change, and Ducart simply laid rails on the inclines, with wheeled cradles for carrying the boats, which measured 10 ft. by 4 ft. 6 in. by 2 ft. 6 in. The system was completed in 1777. The whole concept was misguided, and it was out of use by 1787.[74] In spite of all that has been said to the contrary, Smeaton's suggestion was now followed up. In 1789 a Tyneside waggonway wright, Barnabas Brough, was called in to make an estimate, and the line was soon built. In 1793 the colliery was offered for sale, along with 'a wagon road which leads to navigation' (the Tyrone, not Ducart's canal), and the waggonway was still apparently going strong in 1830.[75]

The next main sphere of Tyneside influence was in West Yorkshire. Waggonways are found here from 1745 onwards in the Wakefield Outwood area, followed in turn by Middleton near Leeds, the Aire collieries below Leeds, the pits south of the Calder, and lastly those south of Bradford. Treating these lines in a geographical rather than a chronological order, we come first to the Bradford area, where a waggonway was carrying coal in waggons of a Tyneside type from Little Horton, Wibsey and Royds Hall down to the town by 1783.[76] By 1785 the Low Moor area was linked to Bradford by waggonway, perhaps by a branch of the same line. It was later connected to the Low Moor Ironworks, and carried coal both for local Bradford use and for transfer to the Bradford Canal. In the same area, the building of 'Newcastle Waggon Ways' by John Elwell and John Crawshaw of Shelf ironworks was permitted by an agreement of 1793.[77]

Just south of Leeds lay Middleton Colliery, the property of the Brandling family, who owned Felling and other pits on Tyneside. It was in 1755 that

Charles Brandling and his agent, Richard Humble (one of a Tyneside family of viewers), laid the first part of what was to become the famous Middleton Railway, a short wooden waggonway to carry coal north and east from the pits to staiths at Thwaite Gate on the navigable Aire, for transport to the agricultural east.[78] In 1758 they acquired wayleaves for a second route, due north to the centre of Leeds, to supply the town; and to safeguard these wayleaves against future controversy Brandling obtained a private Act of Parliament

> for Establishing Agreements made between *Charles Brandling* Esquire, and other Persons, Proprietors of Lands, for laying down a Waggon-Way, in order for the better supplying the Town and Neighbourhood of *Leeds*, in the County of *York*, with Coals.

The Act later described the waggonway 'such as is used for and about the Coal Works and Coal Mines in the Counties of Durham and Northumberland'.[79] It was duly opened on 20 September 1758,[80] and, with various branches, it continued until in 1807 and 1808 the wooden rails began to be replaced by iron,[81] and in 1812 the first commercially viable locomotive in the world, designed by Matthew Murray and John Blenkinsop, began to run on them. The Middleton was a waggonway of great importance, exerting considerable influence on the development of railways in Yorkshire, and, in the nineteenth century at least, on railways in other parts of the country.[82]

Moving down the Aire from Leeds, we come to a batch of waggonways serving collieries that were all worked by members of the Fenton family, either by themselves or in partnership with Timothy Smith. On the south bank, a line from the Glasshouse on Rothwell Haigh was there by 1774, with a longer one from the higher parts of the Haigh dating perhaps from 1783.[83] Three waggonways on the left bank, from the Swillington, Preston and Allerton Bywater areas, are first mentioned in 1773 and 1774, though the latter at least may date back to the 1760s.[84] By 1796 there was a line from Houghton to Castleford, and by 1778 one from a pit at Knottingley, several miles below.[85] Returning up the Calder from Castleford, we find a waggonway at Methley by 1796.[86] Next comes a series of lines descending to the river in Stanley from the Wakefield Outwood. The first recorded in West Yorkshire was in operation by 1745, from pits on the Outwood to the river at Bottom Boat.[87] It may be to this colliery, or to another in the adjoining parish of Rothwell, that a single pay-sheet refers, recording the output of 192 waggons of coal in a week in 1748.[88] The Fentons had a waggonway down to Lake Lock by 1774, which continued on the same or a similar route until the 1790s.[89] One other Fenton line ended at

Smalley above Lake Lock, built under a wayleave of 1783 as a 'new Castle Waggon way or road'.[90] Above Wakefield, there were only two known wooden lines, from Timothy Smith's and Richard Milnes' collieries at Flockton to the Calder at Horbury Bridge. Both were two miles long or so and belong to the early 1770s.[91]

The Tyneside influence of Brandling's waggonway is indisputable; Fenton's Smalley line, like another Fenton waggonway on the Don (see below), is called a Newcastle way; Swillington belonged to the Rev. Sir William Lowther, a relative of the Whitehaven family (see p. 127); Wibsey used waggons of the Tyneside type, and so did the Allerton line (pl. 52). In fine, there can be no doubt that all Yorkshire waggonways belong to the Tyneside family. There is never a hint of Shropshire railway influence in Yorkshire in the wooden rail era.

In South Yorkshire we have already looked at the line of 1722 at Sheffield Park Colliery, Tyneside's earliest known export. It was out of use before the 1770s, when John Curr became the colliery viewer. He was a native of Tyneside, born at Kyo: either that near Bushblades pit on a branch of the Western Way, or the Kyo near Greenside on the Stella Grand Lease line.[92] The Duke of Norfolk offered Messrs. Townsend and Furniss a lease of the Park Colliery on condition that the existing coal transport – carts – was improved, and Curr reported in 1773: 'The site of the Colliery in Question naturally points out a Waggon Way to be laid to a Coal Stage near the Town, there to deliver the Coals for the Regular supply of a great part thereof.' The work, described as 'the Newcastle method', was finished in November 1774 at a cost of £3,280, and sales of coal at the pits were forbidden. The waggonway reduced the cost of carriage to the town from 2*s*. 8*d*. to 1*s*. 2*d*. per ton, but the poorer customers, who had bought their coal at the pits and carried it home themselves, were hard done by. They rioted and burnt some of the line.[93] The story that Curr rebuilt the track with less combustible materials – iron rails and iron (later stone) sleepers – is quite apocryphal, for the colliery accounts show that the waggonway was still using wooden rails in 1787, and probably later.[94] It was closed before the end of the century.

The opening of the River Dun Navigation up to Rotherham in 1740 seems to have prompted the building of several waggonways in that area. The highest up the valley was that from Kimberworth Colliery, opened between 1742 and 1745. The lessee of the pits, John Bowden, was also working coal at Park Gate, near Greasbrough, just to the north, whence another line was built to the river in 1747. A second Greasbrough waggonway (a 'New Castle Road') was laid by the Fentons before 1763.[95] A little further downstream, the Rawmarsh Inclosure Act of 1774 stated that the lessees of Rawmarsh Colliery had staked out the ground over the common for 'making a Cut or *Newcastle* Road for conveying of Coals and Cinders ... to the River *Dun*'.[96] Much lower down, between Mexborough and Conisbrough, a small waggonway from Denaby Colliery is mentioned in a sale advertisement of 1769.[97] Far away to the south, William Brown of Throckley surveyed a four-mile route for a waggonway from Nuthall Colliery to Nottingham in 1764, a scheme which came to nothing.[98]

In Scotland there had existed since 1722 a waggonway from Tranent to Cockenzie Harbour on the south of the Forth. As we shall see, it was utterly un-Tyneside-like in character, and was probably based on Shropshire practice. In turn, it seems to have influenced two other Scottish lines, at Carron and Alloa. Once the Tyneside waggonway came to Scotland, it was inevitable that the two breeds should mix, and after about 1760 probably every Scottish railway was to some degree a mongrel, although a predominant strain, whether of Salopian or Northumbrian, can usually be traced.

Although there were a number of early technical contacts between the Tyneside and the Scottish coalfields, the Tyneside waggonway, it seems, was not introduced to Scotland until John Dixon arrived in the Glasgow area from Sunderland. Soon after 1750 he laid a wooden line from his Gartnavel and Knightswood pits, north-west of Glasgow, to Yoker on the north bank of the Clyde, whence the coals were shipped to his Dumbarton glassworks. This waggonway survived into the 1790s. Between 1775 and 1778 John Dixon's son William built a second line, from Govan pits to Springfield on the south bank.[99] Further down the west coast, two short coal railways had been laid to the Harbour of Ayr by 1775; from their large waggon capacity, they would seem to belong to the Tyneside rather than to the smaller Shropshire breed.[100]

On the south of the Firth of Forth, William Brown of Throckley was consultant to the Bo'ness Colliery, whom he honoured with his presence in 1754, building waggonways underground. The next year, John Burrel wrote to him from Bo'ness:

> As to the Waggon Ways above ground being now Determined to Raise all our valueable Coales by Pitts at the Sea Side we shall have no need for Waggon Ways but from these pitts to the Different Salterns and Harbour which is all on aplace. the only obstruction that lyes in our way from doing these instantly is some houses standing in the way which must be purchased so soon as this happens – I shall be likewise obligd to you for your Man Skilled in these Matters.[101]

An early instance of a process not unfamiliar today.

A series of waggonways ran down to harbours on the north shore of the Forth. Moving from east to west, we first come to Leven and Methil. Richard Pococke, Bishop of Ossory, called in 1760 at Leven, 'where there are salt works and a harbour for ships. . . . We passed some great Coal pitts and the waggon roads from them to the sea.' It was also reported in 1795 that 'A waggon-way of 2 miles from the pits [at Kirkland] to the harbour of Methil is now completed and everything promises an extensive trade'. Just after the turn of the century, if not before, these pits were employing Newcastle engineers.[102]

A little south of Cowdenbeath and some four miles from the estuary were the Fordell collieries of the Henderson family. Their produce was carried by carts along a 'coal road' to Inverkeithing until, in 1752, Sir Robert Henderson built a harbour at St. David's, fed by a branch from this road. About 1770 it was linked to the pits by a waggonway, which was destined to remain in use for some 175 years. Although nothing is known of its building, the large size of the waggons and the construction of the track reveal its Tyneside ancestry.[103] A very short distance to the west, on the east side of Dunfermline, lay Sampson Lloyd's Halbeath Colliery, whose waggonway to Inverkeithing was projected in 1779 and opened in 1780 or 1781. Both its engineer, Alexander Thompson, and its consultant, William Brown of Throckley, had rejected a suggestion that Halbeath coal should be carried down the Fordell line.[104]

The last waggonway to be noticed here had a most confusing history. The Earl of Elgin owned collieries to the north and north-west of Dunfermline, and two harbours on the Forth, Charlestown and Limekilns, both of which had limestone quarries and impressive arrays of lime-kilns. To carry the stone from the quarries and the coal from the pits to the kilns, horse and cart had long been employed. But in 1765 George Chalmers of Edinburgh began to pester the Earl, trying to persuade him to grant a wayleave for a waggonway to replace them. He had been to great trouble to discover the best Newcastle method, and wrote: 'I suppose we can have people from Boness or Carron who can make the Waggons of a proper size and probably one which will hold 2½ tons of coals may be sufficient for about 3 Tons of Limestone'. He would consult the ubiquitous William Brown of Throckley. Brown did not come in person – at least not yet, for in 1769 Chalmers wrote to Ralph Carr of Newcastle saying that a five-mile waggonway over rough terrain had been surveyed, but that he wanted Mr. Brown's advice. Could he come? Chalmers' ambitions were partially realized when a short line was built at Limekilns to serve the kilns themselves. Pennant in 1772, calling in during one of his tours, noticed that 'the limestone . . . is conveyed into them by a variety of rail roads'. Perhaps by

1777, and certainly by 1792, Charlestown kilns had waggonways too. The waggonway from Berrylaw coal pits to Limekilns, a distance of three miles, was laid between 1772 and 1775. In 1777 the Pitferrane pits, and later those at Balmule and Urquhart to the north-east of Berrylaw, were linked to Limekilns by a lengthy branch built by a Newcastle engineer, George Johnson. Another branch was put in to Charlestown in 1801. Though the original Berrylaw-Limekilns waggonway survived for a time, most of it was finally removed, thus converting the two branches into the main line.[105] With Lord Elgin's, the tale of the Tyneside waggonway empire is complete.

A word about terminology. To the Tynesider, until a very late date, a railway was always a *waggonway*: it was so in Huntingdon Beaumont's day, and the word is still by no means dead. The clear-cut distinction between railway waggon and road wain has already been observed – only once is a waggonway called a wainway;[106]* never is a coal wain called a waggon, or a road a waggonway. In the Tyneside colonies, the vocabulary was not so simple. *Waggon road* was popular in Scotland, Cumberland and Yorkshire, and visitors sometimes applied it to Tyneside too. The explicit *Newcastle Road*, *Newcastle Waggon Road*, *Newcastle Rail Road* occur in the north Midlands. A peripatetic Scot in Northumberland and Whitehaven called the waggonways *casseyways* – causeways.[107] *Railway* and *railroad* are Shropshire terms – they will call for closer consideration later on – and were used of Tyneside and of most Tyneside colonies only by writers brought up inside the Shropshire sphere of influence. Thomas Pennant, for example, the Flintshire traveller, restricted himself to *railroad* whatever part of the country he was in. The one exception is Yorkshire, which used *railroad* from at least 1774,[108] and John Smeaton, a Yorkshireman by birth, nearly always talked of *railroads* and *railways*. It was only the spread of the canals and their feeder lines, whose engineers knew not the Tyne, that brought the words *railway* and, to a lesser extent, *railroad* into general use. William Chapman, apparently, was the first Tyneside native to talk of *railroads*, in 1797,[109] and thereafter the usage spread. A law report on Wallsend Colliery, for example, spoke of 'Rail-ways or waggon-ways' in 1798.[110] With the locomotive, *railway* pushed into first place, with *railroad* at first a short head behind, and *waggonway* becoming more and more an also ran.

There remain two words which were sometimes used to describe the track. The first, *frame*, was reasonable enough, for that was what the rails and sleepers amounted to. It appears first at Broseley in 1606,

* In a Newbottle lease of 1697, permitting 'Railes, Sleepers . . . for the making . . . sufficient and convenient Wainway or Wainways'.

'Engin, fframe or Rayles', and occasionally recurs in the eighteenth century, used by such non-technical writers as William Stukeley and Bishop Pococke in the form 'frames of timber'.[111] Three of the descriptions of Allen's waggonway at Bath include the word,[112] which prompts the suspicion that there was some publicity hand-out which inspired these writers. The second term, *groove*, is a misunderstanding of the principle of the flanged wheel. Tennyson's line in *Locksley Hall*, 'Let the great world spin for ever down the ringing grooves of change', is often quoted in this respect. Tennyson later explained that it was dark when he first travelled by railway, and that he assumed the wheels ran in grooves. Even the *Westminster Review* could speak of the 'grooves of a railroad' in 1835.[113] In the earlier days, the same misconception arose. The hypothetical hand-out at Bath could well be the common source of the four authors who wrote that Allen's waggons ran in grooves, though contemporary pictures clearly show that they had flanged wheels.[114] John Loveday, a tourist, said in 1732 that 'two ranges of wooden grooves are fix'd in ye ways' at Tanfield; underground trucks in the Ecton copper-mines in Staffordshire were said to run in grooves in 1769; and an American visitor to Heaton on the Tyne in 1795 likewise spoke of 'wooden groves'.[115] Tennyson was far from being alone.

9

The Newcastle waggonway: wayleaves and engineering

Lo! here a lifeless waggon at thy will,
True to its ambit circles round a hill.
Down the descent, so vast the power of art,
The waggonway retains the flying cart.
So far from being cumbered with its freight,
Like virtue – see! it grows beneath a weight.
Down planes inclined the self moved engines fly
To lad the ships which near the Hurries lie.
Through wooden spouts descends the sooty ores,
An export grateful to Hibernia's shores.

'Poetical Prospect of Whitehaven', quoted in
The Cumberland Pacquet, 10 August 1802.

The right to grant wayleaves belonged basically to freeholders, but very occasionally to copyholders. In Durham much of the coal-bearing land south of the Tyne belonged to three powerful owners: the Bishop, the Lord of the Manors of Whickham and Gateshead, and the several rectors of parishes. In 1723 a Bill was petitioned in Parliament to confirm the Bishop's rights to mine in all the leaseholds and copyholds in his see. But behind its modest clauses, said its opponents, there lurked a claim to powers which the Bishop had never enjoyed before, including the granting of wayleaves and the laying of waggonways across such lands without paying any compensation for damage done:

> The claim of mines under inclosed copyholds and leaseholds, is modern and set up within the memory of man; and as to any claim of right for laying waggon-ways, fixing frames of timber, laying stones or rubbish upon the soil, cutting the soil, making trenches or levels for waggon-ways, erecting buildings, or hovels for workmen upon inclosed copyholds, was never pretended to; and indeed waggon-ways are of a modern invention.[1]

Although this Bill was rejected, the Bishop's attempt to extend his powers may well have been prompted by the fortunes made by William Ramsay, who leased the Manors of Whickham and Gateshead from him, and particularly by Ramsay's successor in 1716, William Cotesworth, who 'played the tyrant over the coal owners'. He was said, for example, to have received £3,000 a year for some land on Whickham Moor whose annual value was about 2*s*. an acre; and 'not so much as a whole acre was damaged by the way'.[2] To own wayleave rights was indeed a painless way of raking in money, and it is not surprising that men quarrelled over them. Again in 1723, a dispute arose over the Lordship of the Manor of Winlaton. Joseph Banks wrote to Roger Anderson complaining that Anderson, who had an eighth share in the Manor, had granted a wayleave for the Western waggonway and built new staiths at Derwenthaugh. Why hadn't he told Banks, who owned three-eighths of the Manor? Anderson replied that all Winlaton Manor was his, and 'that those lands through which the Waggon Way is laid and upon which the new Staiths in Darwenthaugh are built have belonged to me & my Ancestors'. Banks retorted that he had bought his three-eighths in 1719 from the Commissioners of Forfeitures after the 1715 rebellion, and that this entitled him to pull up any waggonways over roads, wastes and bridges for all that Anderson could do about it.[3]

Fortunately, it did not come to this: tearing up waggonways was generally resorted to, illegally, only by rival coal-owners. One exception, however, occurred in 1789, when the landowners lifted half a mile of the Crawcrook Stella way because the lessees' rent was overdue; and at the beginning of the century the Gateshead parish vestry was not averse, if the coal-owners were dilatory in paying wayleave rents, to pulling up their waggonways wholesale.[4] The tendency of rectors to be as arbitrary as their prelates was revealed once again in 1723, when it was proposed to replace the old Dunston waggonway, which passed through the Whickham glebe lands, by the new Tanfield line. One big advantage of the new route was its straighter and gentler course, but an even stronger argument for the change was that 'By this new way you would miss the Rector, and so not lye at the mercy of every New one'.[5]

It was wayleave questions of this sort in particular, and the cut-throat competition of the coal trade in general, that led to the forming of combinations of coal-owners. An important one appeared in 1709, with the intention not so much of easing the wayleave problem as of regulating the coal vend by laying down quotas for the various collieries. But by the 1720s things were rapidly getting worse, and the Grand Alliance of 1726 was formed largely to share existing waggonways and to grant new wayleaves for the mutual benefit of the members. They also indulged in negative wayleaves, the practice of paying a landowner rent on condition that he did not give a wayleave to a rival in the business.

Another question which sometimes arose was whether wayleaves originally granted for carts and wains, before waggonways became regular practice, could be deemed to include waggonways. At Maryport it was decided that they could;[6] but in a complicated series of actions between

George Pitt and Lady Clavering over wayleaves on Tanfield Moor, it was decided first one way and then the other. The Court of Chancery said in 1721, in Lady Clavering's favour, that waggons on one waggonway inflicted much less damage on the soil than carts and wains, with their ruts and frequent changes of route. In a return match in 1729 the Exchequer Chamber decided against Lady Clavering, although the Lord Chancellor

> thought it was far from a difficult question, whether it was less prejudice to the soil to make a way in this manner [a waggonway], than that such heavy carriages [wains] should be dragged all over the moor, sometimes in one part of it, sometimes in another.

But the finding was reversed on appeal to the Lords.[7]

If a waggonway was laid without a wayleave, the culprit was liable for trouble. When one of the Ridleys, however, built a branch to his Plessey waggonway in 1776, Sir John Delaval, the aggrieved landowner, merely wrote

> remonstrating against having my fences broken down and a waggon way being made through my Estate without my permission. You observe that I know by the present Way leave lease you have the same right to use the Waggon Way through Horton for Cramlington and Bassington as Plessey and Hartford, and therefore hope I will wave all further objection that the branch may be perfected. . . . I will give my consent for the new branch to be made across my field as it was staked out when I saw it provided you . . . do not claim it as a matter of right.[8]

The high cost of obtaining wayleaves was a continual and bitter complaint of the coal-owners. Lord Guilford, when on the Northern Circuit in 1676, was amazed by what he heard:

> Another thing that is remarkable, is their Wayleaves; for when Men have pieces of Ground between the Colliery and the River, they sell Leave to lead Coals over their Ground; and so dear that the owner of a Rood of Ground will expect 20 l. *per Annum* for this Leave.[9]

Wayleave rents had become so generally excessive by 1696 that the Company of Hostmen of Newcastle sent a petition to Parliament, saying

> that of late years the Colyeries adiacent to the river of Tine are soe much wrought out that the Coalowners have been forced to advance their workes severall miles further from the river of Tine into the

> Country, to digg for fresh supplys . . . by means whereof they are under a necessity of carrying their coales through other mens grounds . . . who for their damages occasioned by granting such liberty exact such extravagant rates from your peticoners for wayleaves and staith-roomes as is noe small preiudice and discouragement to the trade and will tend to the enhancing of the rates and price of sea coles in the Kingdom.[10]

Time and again Parliament received petitions of this sort, or set up committees to discover the reason for the high price of coal,[11] but nothing was ever done, although later in the eighteenth century, when the competition settled down, things grew somewhat easier. But the wayleave and the waggonway were so inextricably linked together that, to a non-technical mind, the word 'wayleave' seemed to be a synonym for 'waggonway'. Thus Joseph Banks, M.P., told the Commissioners of Forfeited Estates in 1719:

> Wayleaves are wayes from such mines to the said staithes (or wharves) laid with good Frames of Timber to fitt the waggon wheels, which being laid levell and descending, one horse will draw 3 or 4 [London] chaldron of coal several miles from the mines to the staithes. These wayes or wayleaves are so valuable that they are often farmed on the River Tyne at ye Rents of £100, £500 or £1,000 pr annum, and without such wayleaves, staithes are of no manner of use, nor will yield any profit.[12]

The same theme recurs in the diary of Lord Harley, who was travelling northwards from Chester-le-Street in 1725:

> We pass over two wayleaves which cross the great road. These way leaves are an artificial road made for the conveyance of coal from the Pit to the Steaths on the riverside; whereby one horse shall carry a greater burden than a whole team on a common way, and as they generally pass through the grounds of several proprietors, are very expensive to the coal owners, who pay very high prices for their trespass on that occasion.[13]

The main provisions of a typical wayleave, granted in 1692 to the Hon. Charles Montague, comprised leave

> to make lay a Waggon Way . . . (the same to Containe sixteen Yard in Breadth) from ye Pitt or Pits [of Gibside] . . . along . . . any of the lands . . . of . . . S^{r} ffrancis Blakiston . . . to ye said Rivers of Darwent

> or Tyne; and to fix, lay & Place in and along ye said lands & Grounds . . . Railes, Sleepers or other wood Iron or Timber, for ye making ye said Waggon Way . . . and to lead & Carry all other Materials & to Imploy Workmen & Workhorses in & about ye Amending and Repairing of ye same, as Occasion shall be, or Require, and to Cutt & Make Trench or Trenches, for ye Carrying away ye Water from ye said Waggon Way or Waggon Ways, and to make Bridge or Bridges, with Timber Earth or Stone, where ye said Waggon Way is Uneven by too great Descents, or Ascents, upon ye said Waggon Way, Grounds, as there shall be Occasion.

At first the coal was to be carried by wain, at a rent of £490 for 700 tens per annum plus 14*s*. per ten, with another 14*s*. per ten above 700 (counting 7½ bolls to the wain and 42 fothers or wainloads to the ten, which weighed about 36 tons. When the waggonway was built, the rent would be £630 for 900 tens per annum plus 14*s*. per ten, with another 14*s*. per ten above 900 (counting 15 bolls to the waggon and 22 waggons to the ten). If Montague gave up the lease once the waggonway was built, he was to assign Blakiston all the other wayleaves he had obtained for that waggonway.[14] Different wayleaves had slightly differing conditions – sometimes the lessee undertook to keep the gates, or was given permission to lay sidings and so forth – and the terms naturally varied too, according to the distance involved and the rapacity of the landowner. Newcastle Corporation was generally moderate in the rents it demanded; naturally enough, for it had to make a show of encouraging the expansion of the coal trade. The rent charged for the Willington waggonway in 1775 was a mere £100 per annum plus 2*s*. 6*d*. per ten.[15]

It will be quite obvious by now that all, or practically all, of the waggonways under discussion were purely private, built by the coal-owner or lessee at his own expense for his own purposes – they were, in effect, part and parcel of a colliery's equipment. This does not mean that Parliament was by any means unacquainted with the railway in the early eighteenth century and even the later seventeenth. The proprietors of the Stour Navigation had been given statutory powers to build railways as early as 1662 (see p. 244); the Committee on Privileges investigated Sir Humphry Mackworth's case which involved waggonways in 1705 (p. 250), and thereafter, as has been mentioned, Tyneside waggonways cropped up in several Bills and committee proceedings connected with the coal trade. Then a number of Acts on other subjects incidentally protected existing waggonways. The Whitehaven Harbour Act of 1739 laid down that the surveyor could dig gravel provided that he did not damage any colliery or 'the Waggon-ways or Causeways leading to and from the same'.[16] Several Acts turnpiking roads on Tyneside forbade the trustees

> to pull up, alter, damage, obstruct or prejudice any Waggon-way or Ways, laid or to be laid along or cross any Part of the said Road, or to stop or obstruct any Horses going or returning with Waggons laden with Coals, or empty, passing along the same,

and they authorize coal-owners to lay 'a Waggon-way or Ways along or cross any part of the said Road'.[17]

Brandling's Middleton Railway was the first to be specifically covered by an Act. It did not, however, empower Brandling to build his line; it merely ensured that the wayleaves he obtained would be honoured.[18] Likewise many canal Acts, from the 1760s onwards, laid down that coal-masters within a certain distance of the canal might build railways to it, provided the landowners involved were suitably recompensed. All these were private lines. The idea of the public railway, available to anybody with freight to carry upon payment of a toll, came in first as an extension of canal theory and practice. The earliest public railways were built by the canal companies in places where a canal would have been cut had the terrain, or finances, allowed; but the public railway that was completely independent of a canal company did not appear until the 1790s.

Once the precious (and probably pricey) wayleaves had been obtained, the way was clear to build the line. There were two main methods by which the coal-master could get this done. If his viewer (colliery engineer) was sufficiently experienced, he could put him in charge of the work, and either supply labourers from his own employees or hire them for the purpose. In this case much depended on the viewer. Some, like Amos Barnes and William Dent, specialized mainly in pit work, but would lend a hand on waggonways if required. But the large-scale coal-owners tended to employ different men for each purpose: viewers, who spent all their time in colliery management, and master waggonway wrights. In the latter class there fell William Cotesworth's Francis Baker, the Wortleys' Ralph Fetherston, and the Liddells' Albany Baker (later the Grand Allies'). They would simply build a waggonway at their master's orders at the lowest possible cost consistent with ease of working. On the other hand, if a colliery was just opening up, or if it was away from Tyneside in an area where the necessary know-how did not exist, the coalowner would call in outside help. He might be lucky enough to find a man like William Brown, who, though a lessee of mines himself, specialized in advising others about their waggonway problems; and if Brown could not help directly, he might be prevailed upon to send one of his experienced staff.

Otherwise, the coal-owners would invite tenders – or proposals, as they were known – from waggonway wrights, which would quote either a price of so much per yard run, plus extras for engineering works, or a single figure for the whole work. The coal-owner, normally, supplied the material himself. A typical proposal, for building a waggonway to replace the ill-fated Coalisland Canal in Northern Ireland, is worth quoting in full. It was tendered by Barnabas Brough, of Tubman Brough and Partners, a Newcastle firm of jobbing builders, waggonway wrights, and general contractors.[19]

Drumglass Colliery Septr 1789

> I propose to lay the Main & Bye Ways, Branches Checks Switches False Rails, Turn Rails, Shutts or Spouts making & fixing all Geers for the use of s^{d} W. Way & Branches, to make & fix all the Stand rails over Bridges, Staith &c that may be wanted for s^{d} way: to saw the Timber for the above purposes, beat up & make firm all the rails & Sleepers, and to stone and Ballast the same so as to be fully prepared to the Waggons to lead the coals down the same as soon as the nature of the business will permit f^{m} the Staples Pit at Drumglass Colliery to the Staith at Coal Island. Messrs Staples & Caulfield finding all Timber for Rails &c Iron for every purpose that may be wanted for s^{d} way & sleepers to level & form the Road or Way so as to be ready for proceeding with the above business of laying the sleepers & Rails. To find Horse or Horses for leading the Ballast, but not to find the Man to fill or drive, & to lead & lay down all the Sleepers rails &c as near or in the most convenient place for the W Way as may be. And I do futher propose provided that the s^{d} Messrs Staples & Caulfield perform the above & pay me or my Order the sum of 470£ of lawfull money of Great Britain, to lay the Sleepers on s^{d} W. Way so as not to exceed the distance of 18 In^{s} from each other middle to middle, and to do and perform every matter & thing touching or concerning the said business of laying said way to the Staples Pit at Drumglass cooliery from the Coal Island in a workmanlike manner.
>
> Barn^{s} Brough.

The coalowner would, of course, already have worked out the approximate route, both to obtain wayleaves and to provide details for the wrights. How the general line was decided is shown in a letter from George Liddell to William Cotesworth in 1718. They were thinking of leasing Newbottle Colliery, near the Wear, and sent William Dent and Albany Baker to view the land between it and the staith. It would be a tolerable way, he said, and not above three and a half miles long, if Mr. Bates' lease could be set aside: the cost of leading would be 1*s*. a waggon. If they could not pass through Bates' land,

> then we must Climb the hill more, so that the way will be not only ¼ of a Mile longer, but a much stiffer way, & cannot be mastered without a prodigious Cutt, or else Marrowing [using two horses] near half a mile: after having made a Considerable battery [embankment] over a Swelly. This is before you come to Panshea [Penshaw] hill. When you Come there; go wch of the Wayes we will, we shall have a Considerable pull for 300 yds, or a great Cutt, of near 600 yds.[20]

This particular idea was dropped because the engineering would be too heavy. Normally the coalowner would point out the general route, and the waggonway wright would survey it in more detail, and stake out the line. Wayleaves generally granted a maximum width of from 10 to 16 yards, much more than even double track would require on the level, but necessary to accommodate cuttings and embankments.

The ideal waggonway consisted of double track, the main way for loaded waggons, the bye way for returning empties (occasionally called the waggonway and side way respectively); and it was on a regular falling gradient from colliery to staith so that the waggons could run down by gravity, but not so steep that it was dangerous and difficult to pull back the empties (pl. 40). Naturally enough, not all waggonways fulfilled these conditions. Early lines were single-track – there is no clear evidence for bye ways until 1712[21] – with passing loops ('sidings' or 'bye-stands') at intervals.[22] A considerable number of waggonways remained single, and later ones were built so if their traffic was not heavy. Thus we find that Allan's waggonway to the Wear was still single in 1725, although its near neighbour, Hedworth's, had been doubled.[23] Away from Tyneside, where the coal trade was not so highly developed, single-track was perhaps the norm. This is true of almost all the Yorkshire waggonways, though the Middleton was a notable exception. On the Sheffield waggonway of 1774, a short length was double-track, but the rest was single, with sidings every 200 or 300 yards. In Scotland the Halbeath railway of 1781 had 'bye-stands' at intervals of 550 yards.[24]

On double-track waggonways, the bye way might wander some distance from the main way, especially if it was a later addition to an earlier single way. The most direct alignment and the most regular gradient were given to the main way, so that the bye way sometimes had to meander deviously about, not infrequently crossing over the main way to find a practicable route and an easier gradient. On the later line of the Western

Way, where it descended a very steep valley from Bryans Leap to the Derwent, there was just no room for double track, and main and bye were perforce united for a stretch.[25] Only once is triple track met with. The Grand Allies' Tanfield way, only a year after it was opened, was found to be carrying so heavy a traffic, especially in its lower reaches, that it was ordered 'the Owners to lay a new Main Way from ye furthest Branch to Burdon Moor Pits down to ye Partings of ye Ways at Lobley Hill', where the Dunston and Redheugh staith lines diverged. Even with this relief line, the older main way was worn out in four years.[26]

Although a continuous fall, sufficient for the waggon to run by gravity all the way from pit to staith, came to be the object of the later waggonways, the early ones were much more careless, with level stretches or even slightly uphill grades alternating with steep descents or 'runs'. This arrangement entailed hitching on and unhitching the horse at frequent intervals, a waste both of energy and of time.[27] As the years went by, an uphill grade against the loaded waggon was regarded as something to be avoided at all costs, and very steep gradients with the load were often eased. When the Tanfield waggonway was proposed in 1723 to replace the old Dunston way, its promoters said: 'We expect from the straightness of the way & Easy Runns, that the keeping the Way & Waggons will be cheaper, & the leading also cheaper then the present by 6 or £800 perhaps £1000 p. ann.'[28] The Tanfield was indeed a better way, but it included Albany's Run (later renamed Baker's Bank, both after its engineer) at Fugar Bar, with a quarter of a mile at 1 in 12. It was a fact that on steep runs the waggons played havoc with the track. Busty Bank, the great run on the Western Way, where the line dropped 500 ft. in a mile from Bryans Leap, was notorious. The waggonway keeper, submitting his accounts in 1789, explained, 'The Bryans Leap Run, being very steep, required extra attendance and Materials!!'[29] Henry Liddell told Cotesworth in 1710: 'As to our Team Waggon Way you know we have improv'd itt much w^th^in these 3 years last past, by raising where necessary & by easing of ye runs as much as ye situation of ye ground w^d^ admit off.'[30]

By 1750 a gradient of 1 in 10 down was considered the absolute maximum for the main way, and 1 in 30 up for the bye way.[31] Carlisle Spedding was worried about the gradients on the Whingill waggonway he was building at Whitehaven, and asked William Brown's advice:

Whitehaven June 16 1754

Dear Sir

As you have lately laid a new waggon way & must be well Skill'd in these metters, I should be oblig^d^ to you for Informing me what is

> the greatest Decent you Generally give in your Runs or what is the Greatest Decent your Convoys [brakes] will Comand them in Comon Weather. To be sure in frost Raggs or in Sleet Rain &c you use Some means to Ruffen the Railes with Ashes Small coales and Suchlike. My enquirey is for Streight way for our new waggon way from whin Gill is so circumstanced that we Canot lessen the Decent by windgs (which both lessen the Decent & create more friction by Rubbing against Check Rails) & Having a hill on one Side and a Gill or Dean on the other for 4 or 5 Hund^d Yards in length I desire youll be pretty particular about your Several ways of Manageing the Runs and waggons in them our way falls about 2½ or near 3 foot in Ten Yards in some places.

Brown replied:

Throckley 30 of July 1754

Dear Sir

> Your kind favour of the 16th Ins^t I rec^d for which am much obligd. As to the Runns in our Waggonways I know of none that exceeds 3 Inches in a yard and very few that is so much for we find that our Convoys will not hold a Large Waggon when the Runns is so much Notwithstanding we lay Ashes and sometimes Cynders to Ruffen the Railes. Therefore in the Collys. where Sharp Run happens are forced to go with lesser waggons than Comon tho knows of none so little as yours which as I Remember Contains only 18 Bools . . . I think there is no Deficulty at all in getting your wg. Down Tho your way be Streight as your greatest Descint is no more than 3 foot in Ten yards.

He went on to describe an improved type of brake which, together with ashing the runs, solved Spedding's problem.[32]

Steep gradients occurred elsewhere in the Tyneside empire – 1 in 13, for example, on the Fordell railway. George Chalmers, however, passing on information about Tyneside waggonways in 1765 to Lord Elgin, was more cautious.

> The outmost Descent that a Waggon may goe with safety on a Straight line is 2 inches in a Yard [1 in 18] but if the Way Winds round a Hill 3 Inches fall [1 in 12] may be dispensed with – Thirteen Inches ascent in 20 yards [1 in 55] with a loaded Waggon is enough for a single horse and they doe not use more about Newcastle.[33]

The practice of letting a waggon down a steep gradient on a rope was used to some extent in the Shropshire coalfield, but was not known on Tyneside and its colonies. The self-acting incline, whereby the down-going load pulled up the empty by means of a rope passed round a drum at the top, was used on the Continent at an early date (see pp. 72-3). In Britain it was patented by Michael Meinzies, a Scottish advocate, in 1750 as

> a Machine for the carrying Coals from the Coal Walls where they are dug to the Bottom of the Pit or Shaft and from the Mouth of the Pitt or Shaft to the Heaps at some Distance therefrom and in some Cases of carrying them to the Staith or Place where they are Put on Board Keels or Ships, without the Assistance either of Horses in drawing Empty or Loaded Waggons.

He suggested a double waggonway on an incline, with an endless rope passing round pulleys at each end. His patent expired only in 1778.[34] Although the Shropshire empire used the balanced incline early on, it was not apparently applied in Tyneside colonies before Ducart's ill-fated railway inclines on the Coalisland Canal in 1777. There may have been another early example on the Thwaite Gate line of the Middleton waggonway. In March 1781 the entry appears in the accounts 'Paid J. Broadbent for Makeing & Keeping in Repair the offside Waggon Wheels in Woodhouse hill Lane for Two Years, £16.16.0.' Similar entries recur annually – one instance is 'Robt Nunns the remainder of his acc^t^ for upholding & keeping in Repair the offside wheels sills &c in Woodhouse Hill Lane £18.18.0' – and cease in the 1790s.[35] Since it seems unlikely, to say the least, that a man was stationed on the hill to repair the waggons as they passed, and only the wheels on one side at that, it is possible that the wheels in question were side-rollers to guide the rope on a curved incline, where they would be on the inside of the bend, the offside or right-hand to a descending waggon.

The balanced incline first appears on Tyneside in 1784, when a French traveller, Faujas de Saint-Fond, noted that

> where the local circumstances have permitted, the weight of the load, and the accelerated movement have been combined in such a manner that files of loaded waggons run down the inclined plane and at the same time cause the empty waggons to reascend without the assistance of horses, along another road parallel to the first.[36]

This antedates by some years Thomas Barnes's self-balanced incline with iron rails at Benwell, built in 1798, which is usually said to have been the first in the North.[37] After 1800 self-acting inclines became common, but until then the gravity-worked main way and horse-worked bye way remained the norm.

The waggonway wright and his labour force would prepare the ground by levelling a strip perhaps 20 ft. wide for a double-track way, at a cost which varied from 2*d.* to 1*s.* per yard. To maintain a reasonably steady gradient, it was necessary to excavate cuttings and raise embankments. Every waggonway, except in the very smoothest of terrains, would have small cuts and batteries, as they were called (embankments were sometimes known as 'mounts' or 'bulwarks' as well). Often they would be no more than 1 yard or so high or deep, but entailed extra work and were charged for at a special rate, 3*d.* per cubic yard for cuts, 3*d.* to 6*d.* for batteries. An easy line of this sort was envisaged by John Christian when he was opening up Broughton Colliery above Maryport in 1759, 'as wo'd any waggon way from ye remotest part of my Common scarce require a battery of two yards high to come into the main road and in 9 parts out of 10 of ye length no battery at all'.[38] Some major waggonways, however, followed particularly difficult routes, and extremely heavy and expensive engineering work was the result. Sanderson said in 1767:

> Some of these Ways are from one to eight Miles in length; and great Hills have been cut away, Vallies filled up, and high and distant Mountains joined together by Arches, at an immense Expence, to procure an easy Descent from the remotest Inland Collieries.[39]

However deep some of the cuts, tunnels were perhaps unknown on wooden waggonways. A 100-yard tunnel on the Flockton Railway in Yorkshire has been ascribed to the opening years of the line,[40] but it is in fact on a subsequent branch of uncertain date, maybe of the early 1800s, when tunnels were beginning to appear on iron railways elsewhere.

George Liddell's account of his suggested line for Newbottle has already been quoted (p. 144), with the choice between two unwelcome alternatives – either 'a prodigious Cutt, or else Marrowing near half a mile: after having made a Considerable battery over a Swelly'. When the Western way was completed in 1721, William Bowes was inordinately proud of it:

> Last Munday we begun to lead down ye new Waggonway, which is ye beginning of my profitt: it is a work of such great importance, & crosses so many Mountains & Vales, which are all levelled that I can compare it to nothing more properly than to ye Via Appia.

None the less it was not free from dangerous runs, and in wet weather the more prudent waggon-men deserted it in favour of the rival but gentler Bucksnook way.[41] Even when Bowes' line was replaced by the new one on the other bank of the Derwent, the earthworks were still heavy, especially on Busty Bank. Hutchinson wrote in 1794:

> Immense works are frequently made of masonry or soil across valleys

> and rivulets . . . not far from Chester[-le-Street] are others executed by Mr Jolliff at an immense expence, and several more are to be seen in this county: Darwenthaugh waggon-way, extending from Burnopfield and Brian's Leap to the river Derwent, was the most difficult and expensive work of the kind hitherto executed in the coal trade.[42]

Sizeable earthworks are still to be seen along the routes of the Bucksnook and Western ways, and some of the Stella lines (pls. 41-2).

The Tanfield waggonway of 1725 was another of the wonders of Durham. Its main way could not be purely gravity-worked, as it had to surmount the watershed between the Beamish Burn and the Tyne, but its designers did their best to keep the uphill haul to a minimum by means of massive cuts and batteries. To some these works were a blot on the landscape and a threat to the established *status quo* of the coal trade. James Clavering, protecting the interests of Sir Francis Clavering, wrote to his lawyers a week before the Grand Alliance was formed in 1726:

> What's suspected is a project to stop all the western collierys, Beckley is one of them, if this succeeds Mr Wortley & Coll. Liddell is then to carry Mr Bowes to Tyne down their new way, & that I truely hope & believe may also be in S^r Fr: power to prevent, otherwise will be an imense loss to him in many articles. This new way is laid adjoyning to S^r Fr: hedges, where the herbage of the Comon belongs to his Ten^ts in right of their Farms, & the vast cutts and damages done by laying the way is a very great detriment and loss, this you must duely weigh.[43]

The massive works of the Tanfield way were also quoted by those who sought to show that the Grand Allies had raised the cost of coal. Thomas Stokoe, under examination by a Parliamentary Committee in 1739, said that Dawson's and Davison's mines at Tanfield could not have been worked twenty years ago,

> Because the Expence was so great, they being brought down by a Waggon Way and there being two great Bridges built at a vast Expence.
> *Q.* How large are these Bridges?
> *A.* Tanfield Bridge Gentlemen come to see, on purpose, who say it is the largest they ever saw and must be a very great Expence to the owners.
> *Q.* Is not the Price of the Carriage of Coals much increased?
> *A.* Yes, – there are two Bridges built which cost £7000 besides Waggon Ways, Staiths, Cutts and Batteries.
> *Q.* What might they cost?
> *A.* I believe it could not be done for less than 50000£.[44]

To the tourist, however, these works were something to marvel at. William Stukeley passed this way in 1725:

> We saw Colonel Lyddels coal-works at Tanfield, where he carries the road over valleys filled up with earth, 100 foot high, 300 foot broad at bottom: other valleys as large have a stone bridge laid across: in other places hills are cut through for half a mile together; and in this manner a road is made, and frames of timber laid, for five miles, to the river side.

John Loveday visited the Tanfield way in 1732: 'The Proprietors have been at great charge in levelling hills, raising valleys, &c. Nigh y^{s} bridge [Causey Arch] is also another, a long Arch, und^{r} a vast Hill of Earth.'[45] All these sources mention the Causey Arch, of which more in a moment, and all of them regard as equally vast and wondrous the other 'bridge' or 'hill of earth' – the embankment over the Causey Burn, which passes through it in a culvert, 350 yards below Causey Arch. It still exists, and carried trains until 1962. We are so accustomed nowadays to the achievements of engineering that it takes some imagination to understand the contemporary astonishment at these works, which would be a mere flea-bite to a nineteenth-century railway engineer. But in the 1720s nothing had been built on this scale since Roman times, as Bowes realized when he compared his waggonway with the greatest Roman road of Italy. Small wonder, then, that people came from far and wide to gaze open-mouthed at man's mastery over Nature in the cause of trade and industry. Though they did not know it, they stood on the threshold of the Industrial Revolution. Romans apart, there had been few precedents for a work of this magnitude: the French canals of the seventeenth century, a number of docks and harbours, but little else. It might be said that it was on the Tanfield waggonway that British civil engineering on a grand scale first saw the light of day; while at the same time mechanical power, in the shape of Newcomen's steam engine, was wheezing into life. A century later the marriage of the two was consummated and the modern railway was conceived. It is not recorded who designed the Tanfield line, but all the honour, it seems, is due to the waggonway wrights Albany Baker and Ralph Fetherston, under the aegis of Wortley and Liddell.

Like cuts and batteries, bridges were also inevitable. Timber was the normal material. For crossing the steep and deep valleys with which Northumberland and Durham abound, a design was evolved which seems, to eyes accustomed to modern railway bridges, exceedingly slender and fragile. But both the speed and the weight of the waggons were low, and

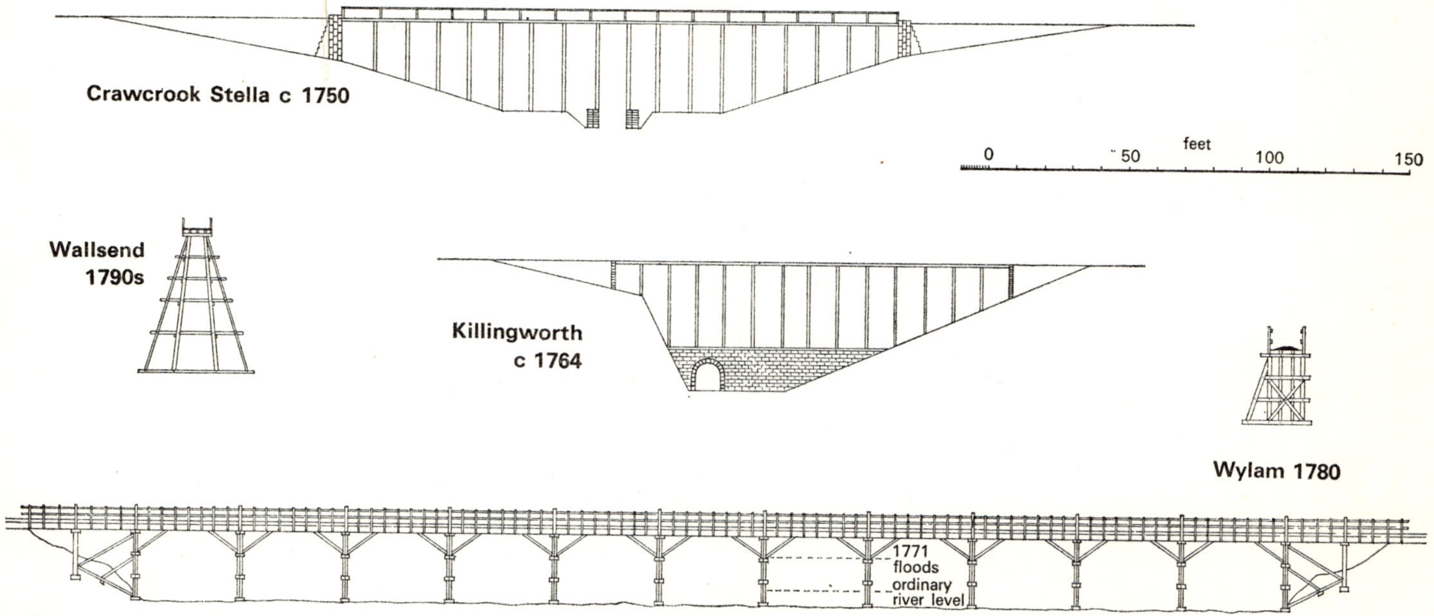

FIG 34 Tyneside timber bridges.

the bridges fulfilled their purpose well. The simplest form is illustrated by the plan for a proposed bridge over Red Burn in Guards Wood, to carry an extension of the Crawcrook Stella waggonway to Mickley pits, dating from somewhere in the middle of the eighteenth century (fig. 34). A timber platform carrying a single track is supported by seventeen pairs of uprights or 'geers' set 10 ft. apart, and each 1 ft. square. The length of the bridge itself is 196 ft., the height above the stream 36 ft. The approach batteries, which add another 164 ft. to the length, end in retaining walls or 'breasts'. The estimate was this:

To 8 pair of Geers 33 Feet long each	£89	4	0
To 9 Pr Do 24 do	83	14	0
Two Pillars 6 feet high each 4 feet thick & 10 yd long	6	0	0
Two Breast Walls 5 yds high each and 5 yds wide, 7 Feet at Bottom and 3 ffeet at Top at 4s p yd Heywork	10	0	0
40 yards of Flank at 2s 6d p	5	0	0
950 yds of mettle [earth or rubble] at 3½d p	13	17	1
Wood	10	0	0
20 Feet of Covering at the ends Barrowy Deals &c	2	0	0
	£219	15	1

The cost of a battery of the same length (360 ft.) and height, and 76 ft. wide at the base, with an arched culvert for the stream, was calculated as £224 10*s*. 6*d*.[46] It was probably the bridge that was built, although the maintenance costs of a battery, one imagines, would have been much less. A similar but larger bridge was built by John Buddle in the 1790s at Wallsend Colliery over Wallsend Burn (fig. 34). It was 494 ft. long and 50 ft. high in the middle. Each of the forty-four sets of geers consisted of two legs and two side braces, resting on a transverse sole and supporting a transverse crown tree. The crown trees carried the longitudinal sills, to which were fastened the floor, rails and hand-rails. The cost was kept extremely low (£178 3*s*. 4*d*.) by using cheap and bad timber, with the result that the bridge had to be strengthened within a few years by the addition of transverse binding balks – four to each set of geers – and of longitudinal stretchers.[47]

A development of this arrangement was to build a low battery or even a stone foundation across the bottom of the valley, and set the timber work on top, thus saving on the most expensive item, the long and massive geers. Such a bridge was designed for the Grand Allies' Killingworth (Long Benton) waggonway in about 1764, with stone walling 15 ft. high and 30 ft. geers above. The length, including the approach batteries, was 208 ft.

To 8 pair of gears at £17. 10s p pair Riga Timber	£140	0	0
To 5 pair of D° at £10 p D°	50	0	0
To Winning & Leading Stones &c 600 y^ds at 9^d p	22	10	0
To Working & Walling the Arch [32 feet long]	18	9	0
To Walling the Flank Walls 354 y^ds at 18^d p	26	11	0
To 2 Breast Walls Each £5	10	0	0
To Mettle work 900 yards @ 4^d p	15	0	0
	£282	10	0

The cost of building a battery instead was estimated at so much less, £194 3*s*. 9*d*., that one wonders why the bridge was considered at all. We do not know which was decided upon.[48]

A similar plan had been proposed in 1738, when a branch of the Heaton waggonway was about to be laid. The estimates include two bridges, a timber one founded on an earth battery over a gill (the 'Dean'), and a plain wooden one over Benton Lane. The first two entries refer to both bridges; the last entry to Benton Lane bridge; all the rest to that over the Dean.[49]

To Casting the foundation for ye 4 Land Breast wall	0	10	0
Winning stones leading & walling the 4 Land Breasts Walls – 2 at Benton Lane & 2 at the Dean containing about 10 Yards each & 1 yard thick one part with the other at 4s p	8	0	0
To Casting a foundation for the Arch or Cunduit 31 y^ds at 12d p	1	10	0
To Winning stones leading walling & turning an Arch over the Cunduit a Yard wide and a Yard high & 31 y^ds long at 12s p	18	12	0
To filling up the Dean with earth to ye height for setting the timber on 5670 Square Y^ds of earth att 3^d p	70	17	6
To 600 feet of firr wood for 11 pair of Geers 16 feet in height att 10^s p foot	25	0	0
To 160 feet of firrwood Sills at 10^d p foot	6	13	4
To 2½ Double Redd wood Deals att £10 p	25	0	0
To Carpenters work for D°	12	0	0
To Nails & Spikes for D°	2	10	0
To Timber Deals & workmanship for crossing Benton Lane	23	14	5
	£194	7	3

These bridges were evidently no rarity, for the French traveller Gabriel Jars noticed them in 1765, and a Swede, Bengt Qvist Andersson, mentions in 1776 the strong and tall wooden bridges over deep valleys.[50]

When a waggonway engineer was faced with a river of some size in a wide, shallow valley, rather than a steep and narrow one, his bridge was designed on a different basis. He would make the spans as wide as possible to reduce the risk of damage by flood-water and flotsam, and the resulting struts which carried the lengthened spans were somewhat reminiscent of Brunel's timber viaducts. Such a bridge was envisaged for the Wylam waggonway in 1780 (fig. 34) in circumstances which are not recorded.[51] The river it was intended to cross, 407 ft. wide, can only be the Tyne. An extension from Wylam to the West Wylam Colliery on the south bank may have been proposed; but it was certainly never built. The bridge which carried Bowes' Western Way across the Derwent at Swalwell doubtless belonged to the same family. Barely a year after it was built, its approach battery was swept away by floods, and William Cotesworth, now the sole user of the rival Bucksnook way, was gleefully told about the ensuing events in a series of letters from his agent, Thomas Sisson.[52]

> (10 July 1722) The great raine . . . has done very great damage by the flood that came down Derwent and Tyne. . . . At the New bridge . . . the water has drove down the Bulwork at the west end and carryed away at least 30 yards of ground into the field where the end of the bridge was fixed so that it will be full as much to add to the bridge as the bridge contained in length before. . . . The water was so high that . . . it tore up part of the waggon way in the field where the west end of the bridge is fixed – drove away several waggons that were standing by the new waggonway side in the Haigh.

The trouble was caused, Sisson thought, by the bridge itself – its timbers held back the sludge washed down from the mill-dams at Crowley's iron-works, which reduced the channel; the river then built up until it burst its banks. Bowes promptly set a gang of men to repair it, with much timber for 'the Gears of the Bridge'. They removed a quantity of gravel from the river-bed to repair the approach batteries before Sisson, on Cotesworth's authority as Lord of the Manor, forbade them; and they built a coffer dam behind which they remade the stone abutment:

> (29 July 1722) The New Bridge is so repaired that the waggons go now, they began last Fryday tho there is still a great deal to do at before it will be finished, but as the waggons can go, that answers their End.
> (3 August 1722) The Newbridge is now very near finished, and they

> lead currantly now they have made a sack and seame way [embanked horse track] thro' my lady & Mr. Anderson's grounds.

Stone bridges were few. The first, as far as we know, and certainly the greatest was Causey Arch, sometimes called Tanfield Bridge or Dawson's Bridge (pl. 43). It was built by Wortley and Liddell to carry the Tanfield waggonway from Thomas Dawson's Causey and Tanfield collieries, which they were leasing, over Causey Burn. The line came into operation in July 1725, and it is said that the first bridge here collapsed almost as soon as it was finished. Its successor, the present structure, was begun in August 1725, and as it took well over a year to build, the stone bridge that Stukeley saw in the summer of 1725 (p. 150) must have been the original one. Loveday viewed the Causey Arch in 1732:

> Another day We roade . . . to ye Arch belonging to Blackburn Colliery [a misapprehension], 7 miles from Newcastle; it is ye highest Arch I ever saw, perhaps has not it's Equal; it has not been built many years, 'tis a little Stream y^{t} flows under It; ye occasion of making it so high was, y^{t} ye Coal-Carts might go on plain ground.[53]

Hutchinson gives the first detailed description in 1778:

> Near Tanfield is a great work called Cawsy-Bridge, carried over a deep gill or valley, to obtain a level for the coal-waggons to pass, constructed at the expense of . . . the Grand Allies: The span of the arch is one hundred and two feet, which rises in abutments or land piers about nine feet in height, and being semicircular, makes the elevation about sixty feet: the level is from thence kept up by forced bankings of earth, in some parts forty feet in height, and one hundred and forty paces in length; much more expensive than mason-work. . . . The valley is wooded, wild, and romantic.[54]

By 1789, when John Brand wrote, myths were beginning to form:

> There still remains a remarkably large arch standing over Causey-Burn . . . which appears to have been finished, after having once fallen in for want of weight A.D. 1727. It was built by a country mason for a waggon-way to a colliery of the late Lord Ravensworth [Sir Henry Liddell], which was set on fire, and has been long unwrought. It is said to have cost 1,200 l. The architect involved himself in many difficulties concerning it, and is reported to have destroyed himself at last from the dread of its falling a second time. . . . I copied the subsequent inscription from a sun-dial on this stupendous arch:
>
> 'Ra. Wood, mason, 1727.'[55]

This is clearly one of the 'sacrifice stories' proper to ancient bridges.

though here the architect himself was said to be the victim. The tale of his suicide is probably apocryphal, for Wood survived at least four months after the bridge was finished – long enough to receive a gratuity from the Grand Allies.

There are various versions of the cost. Andersson in 1776 said £10,000;[56] Brand £1,200; and most sources from then on, probably misreading Brand, give £12,000. All are wrong. A set of summary accounts among Bowes' papers shows that the total cost was £2,252 16*s*. 1¾*d*., including gratuities. A clause in the Grand Allies' agreement, signed ten months after work had begun, ordained that Bowes should contribute 'his proportionable part of the Charge already Expended or to be Expended in makeing & Erecting ye Bridge called Dawson's Bridge', and accordingly an account was submitted to him when the bridge was finished. Bowes, Liddell and the Wortleys each paid a third of the cost.[57]

For ⅓ of the charge of building a bridge between Mr Clavering and Mr Dawson^s ground (vizt)			
For ⅓ of £1.1.– given to Ra: Wood who undertook to build that bridge in earnest	0	7	0
For ⅓ of £907 p^d Mr Wood on acc^t of building that bridge from 26 August 1725 to the 11th of June 1726 both days included	302	6	8
For ⅓ of £747.18.3½ p^d Hugh Boagg on acc^t of building that bridge from the 9 July 1726 to 5 Novr 1726 both dayes included	249	6	1¼
For ⅓ of £197.12.10 for timber sent from Team to that bridge before 1st Novr 1726	65	17	7½
For ⅓ of £121.8.– for 303½ fothers of lime and leading at 8^s p fother sent from Team to that bridge before ye 1 Novr 1726	40	9	4
For ⅓ of £207.5.0 for Rag had of Mr Cha Atkinson to that bridge as p. 2 notes	69	1	8
For ⅓ of £12.15.–¼ p^d Tho Surtees the 2^d of August 1726 for Oak timber had to that bridge	4	5	0
[later] To Hugh Boag £21 To Geo Airey £10 and Mr Davison £10 for their care and inspection at the bridge to be paid in ⅓^ds	[13	13	4]
[later] By ⅓ of a guinea given John Bourn a Mason the 21 October 1727 for directing &c in building a bridge Between Mr Claverings and Mr Dawsons grounds	0	7	0

In addition, the Grand Allies minuted on 18 March 1726/7: 'Agreed. To Give Ra: Wood 15 Guineas.'[58]

Nothing is known of Ralph Wood. The payments to him are so small – the £907 is clearly for workmen's wages – that Brand may well be right in describing him as a country mason, self-taught perhaps, like William Edwards, who built the famous Pontypridd thirty years later. Brand says that Wood was responsible for the failure of the first bridge, but that, like Edwards, he persevered until he succeeded; but the building account's phrase 'Ra: Wood who undertook to build that bridge' rather suggests otherwise. Hugh Boag, who took over the construction in 1726 when Wood for some reason had given up, had started his career – fascinating touch – as Colonel George Liddell's gardener.[59] He later went on to become a respectable colliery viewer. John Bourn, the mason who received a small and late gratuity, was apparently employed in finishing off the work. Of the other two men who received fees for supervision, George Airey was probably a relative of the Liddells' agent, Thomas Airey; and Mr. Davison was none other than William Davison, Lord of the Manor of Beamish, in which the Arch stood, who acted as treasurer for the building of the waggonway. So it seems that purely local talent, some of it from a surprising source, was responsible for this magnificent structure.

The magnificence of Causey Arch is emphasized by its span. For thirty years, as far as can be discovered, it boasted the longest single span of any bridge in Britain. There were several greater bridges in France and Italy, but in this country its nearest rival, it seems, was the medieval Twizell Bridge in Northumberland, with a span of 90 ft.; and it was only surpassed in 1755 by William Edwards's 140-ft. masterpiece over the Taff at Pontypridd.

The dimensions quoted by early writers are slightly wrong. In fact, the span is 105 ft. The arch is semi-elliptical, not semicircular, and rises about 35 ft. from springing to crown. The deck, 10 ft. above the crown, stands rather over 80 ft. above the stream, and is 22 ft. 7½ in. wide. It is drained by hollowed stone spouts (pl. 44). The approach embankments were retained by massive walls – that on the north-west included a retaining arch – and the arch is flanked by buttresses which rise to parapet level. The sundial adorned the top of the south-western buttress before it was removed to Urpeth Hall for safe keeping: it has now been lost.[60] Though much of the retaining walls of the embankment has collapsed, the bridge still stands, scheduled as an ancient monument, but in urgent need of consolidation (pl. 45).

It is always said that the waggonway over the arch ran only to the Causey pits, a few yards from the bridge end. In this case, why was so elaborate and expensive a bridge, with accommodation for double track, built for a mere branch? In 1725 the only two collieries worked in this area were Causey, immediately west of the bridge, and Tanfield, also on the west bank, but nearly a mile upstream (fig. 31). It has always been assumed that the route of the Tanfield waggonway, as it was up to its closure in 1962, was the original one, staying on the east bank until just short of Tanfield Colliery, where it swung over the burn by a very easy crossing. But a substantial and long-disused waggonway is traceable from the west end of Causey Arch along the *west* bank to Tanfield pits, which is quite possibly the original Tanfield line of 1725. If this is so, Causey Arch was built to carry a main line, not a branch; it was more difficult to bridge the burn here than further south, but in compensation the route on the west bank demanded much smaller earthworks than that on the east. The eastern route must date in part to about 1730, when the extension to Beamish was laid, and the rest to some later date, perhaps when the subsidence of the approach embankments to Causey Arch had reached a dangerous level. Causey pits were closed between 1733 and 1738, but an engraving of the 1770s shows the arch still carrying traffic.

After the Causey Arch, the thin tale of other stone bridges must come as an anticlimax. The cost of building 'A bridge over Horton Burn of Stone & Mettle [earth]' on a proposed new alignment of the Plessey waggonway in 1787 was estimated at £400.[61] At Whitehaven, Whingill coal was carried by waggonway to Bransty, whence it was carted the short distance to the harbour. Carlisle Spedding had planned and begun a bridge over the turnpike road to continue the line to the water's edge, but its progress was halted by his death. It consisted of a solid masonry arch 27 ft. high and 20 ft. in span, flanked by pedestrian ways; the waggonway on each side was supported, not by embankments, but on wooden geers in the ordinary manner at staiths (fig. 35). The arch long remained unfinished and unused. John Smeaton recommended its completion in 1775, but it was not opened for use until 1803.[62] The reverse occurred at Maryport, where John Christian and the Broughton Coal Co. built a waggonway in 1755 to the Harbour, across the 'arches' at Ewanrigg – a viaduct of brick arches 18 ft. high and 24 ft. wide. In 1786, however, the wayleave for the line below them lapsed, and henceforth the coal was carted from here to the Harbour.[63] Lastly, Ducart's inclined railways on the Coalisland Canal crossed several bridges: semi-elliptical arches with fine voussoirs, more than worthy of the regrettable line they carried.[64]

All these bridges were to carry a waggonway across an obstacle, and only one example of the opposite has come to light, in the Ravensworth accounts of 1727: 'Repaireing a bridge Crossing y^{e} Wagg^{n} Way 2/8.'[65] Roads

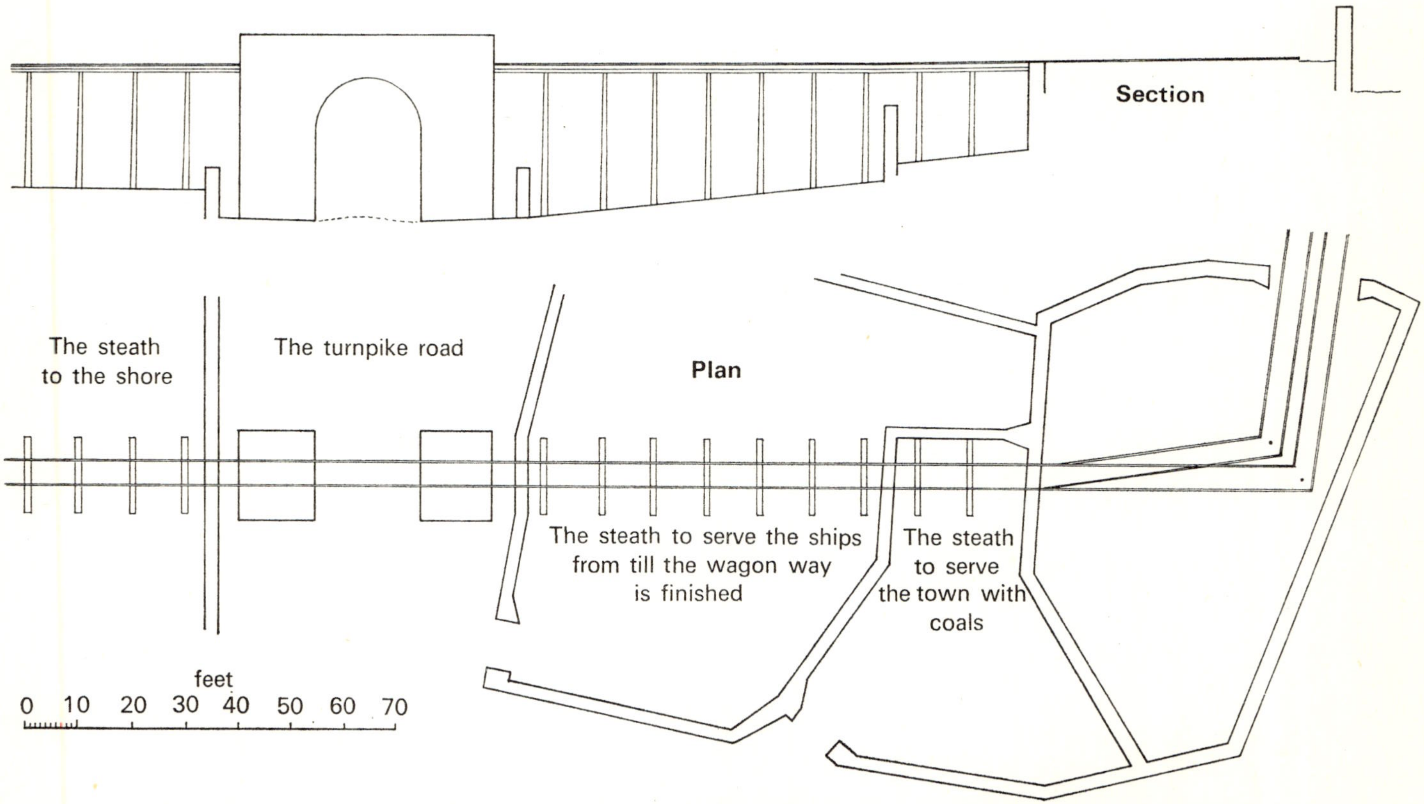

FIG 35 Plan and section of Bransty Arch and approach to staith on Whingill waggonway, Whitehaven, 1755?

and paths were usually crossed on the level. The legal position of level crossings was challenged in 1714 by some influential but unknown opponents, who included an alderman (Richard Ridley?). Trouble had been brewing for some time before they brought a test case in March of that year, on the grounds that waggonways were a common nuisance in public highways. It was heard at York, and

> ye success of ye day is entirely on ye aldermans side . . . whether or not they have ye courage to execute their threat & pull up all ye other waggon ways in ye country y^t^ lye or cross ye common high ways.

The campaign was carried on to Parliament, where a petition by keelmen to throw open all wayleaves was being sponsored by a Newcastle Member. Sir Henry Liddell was busy lobbying against it:

> My son Harry's indisposition deprived us of his assistance this day agst the petition for waggon-ways from Newcastle. . . . Mr Headworth wishes it [the Bill] cou'd be so model'd as to secure such waggonways as cross highways from being pull'd up as he apprehends his way may be the earliest of any. . . . J. H[edworth] and L[ambton] seems apprehensive that his way will be pulled up and severall others, but I propose that those wch are not, should contribute to ye charge equally, being from a publick advantage. If such an offer were made to him [Hedworth] it wou'd induce him to remain firm in opposition.

The petition was lost, and no actual removal of level crossings seems to have taken place.[66] Indeed, when the earliest turnpike Acts for County Durham were passed in the 1740s, the free use of the roads by waggonways was permitted.

The track bed of a waggonway was hedged or fenced on both sides, an almost invariable rule from the very earliest days. The object was both to mark out the sideways limit of the wayleave rights, and especially to keep men and animals off the track for their own safety and to prevent delays to traffic. On the very opening day of the Bucksnook waggonway in 1712 there was confusion because 'all y^e^ beast in y^e^ pasture had unluckily got on y^e^ way'.[67] Sometimes a plain wooden fence served – at Ravensworth in 1729 they used superannuated rails and sleepers as posts – and sometimes regular hedges were planted. Ralph Fetherston sent in a bill to George Bowes: 'To $\frac{1}{3}$ of y^e^ Charge of planting Quick & repairing Hedges of y^e^ Waggon Way in Mr Bowes Ground in y^e^ following years [1731-7] . . . £46 4*s*. 5$\frac{3}{4}$*d*'.[68] The price of this protection was usually small. On a realignment of Allan's way to the Wear in 1769, hedges cost only 2*d*. a yard for both sides of the track.[69]

At its top end, a waggonway would often divide into several branches to serve the various shafts of a colliery. Tanfield Moor, which was riddled with pits, had a bewildering fan of branches which were constantly changing as the years went by. In 1772, for example, the branches totalled nearly a mile in length, against seven-and-a-quarter miles for the waggonway to Redheugh.[70] *Branch* was the ordinary word for such a subsidiary line; *siding* was reserved practically always for passing loops on single track. At the bottom end, some means were necessary for transferring the coal from waggon to boat or ship. Sea-going ships could not sail above Newcastle Bridge, or any distance up the Wear, so craft called 'keels' were used as lighters to ferry the coal from the staiths or wharves higher up-river down to Shields or Sunderland. They were broad-beamed and propelled by oar, with a capacity of some eight waggon-loads, or 21 tons. In the earlier days of waggonways the keels apparently tied up at special quays – in 1666 one of the Stella waggonways ran down to a 'Stone stayth containing fower keelroomes' – and the coal was transferred from the waggons by wheelbarrow.[71]

Before long, however, a more efficient device was invented. This was still called the staith, or sometimes the 'trunk', and consisted of a platform, with or without a roof, projecting out into the river and standing on wooden legs or geers (pl. 40). The waggons reached it by means of wooden bridges from the bank, usually two in number, the ongate and the offgate, whose names explain the one-way system of traffic. At intervals between the rails on the platform were square openings, down which the coals were emptied through the bottom of the waggons. Some led to diagonal spouts which ended 5 or 6 ft. above high-water level, so that the coals could be shot directly into keels stationed underneath them. Other chutes, leading down from the platform to the store-house below, were used if there were no keels waiting to be loaded when the waggons arrived. When the time came, these coals were barrowed from the piles in the store-house and shot down smaller spouts to the keels.[72] These roofed wooden staiths with spouts had appeared by 1700, and soon became general on the Tyne and Wear. They sometimes achieved a very great size. The Western Way in 1721 had two adjoining staiths, one on the Tyne and one on the Derwent, which were respectively 390 and 360 yards long, with berths for twenty-six and twenty-four keels.[73] They appear on a map of Derwenthaugh made in January 1723/4 (pl. 46) – one of the earliest known maps to mark waggonways in any detail – which also shows the Bucksnook staith on the other side of the Derwent.[74] A staith storehouse might hold 30,000 chaldrons, or 79,500 tons, of coal.[75]

The Cumberland coal harbours used similar devices, though they were usually higher than on Tyneside, since they had to load large sea-going ships. The Howgill staith or 'steer' had five spouts (here called 'hurries'), and its platform stood 37 ft. above high-water mark, with the store-house floor 25 ft. below.[76] Thomas Pennant described it in 1772:

> Close to the shore, on the south side, is another [quay], covered with what is called here a *steer*, having in the lower part a range of smith's shops, and above an extensive floor capable of containing six thousand waggon loads of coal, of 4,200 lbs. each. But this is only used as a sort of magazine: for above this are covered galleries with rail roads, terminating in large flues, or hurries, placed sloping over the quay, and thro' these the coal is discharged out of the waggons into the holds of the ships, rolling down with a noise like thunder.[77]

The next stage in the development of the staith, however, whereby the waggon was lowered bodily to the ship and emptied there, did not appear until after the wooden waggonway era. Although it was invented by William Chapman in 1800, it was first applied only in 1812 by Benjamin Thompson at Wallsend.[78]

The Newcastle waggonway: track

But should the Frenchmen shew their face
Upon our waggon ways, man,
Then there upon the road, you know,
We'd make them end their days, man.

'The Pitman's Revenge against Buonaparte', quoted by John Bell, Jun., *Rhymes of the Northern Bards*, Newcastle (1812), 37.

With the track-bed cut and levelled, the track itself could be laid. Let Gabriel Jars describe the general arrangement as he saw it in 1765:

> When the way has been laid out, with a width of six feet, and the gradients have been fixed, an excavation is made of the same width as the said way, deeper or shallower as the levels and the solidity of the ground require. The whole length of this excavation is then laid with pieces of oak, four, five, six or eight inches square; they are placed crossways, at intervals of two to three feet apart. These beams need to be squared only at their extremities, on which other well-squared and sawn pieces of wood, some six or seven inches wide by four or five thick, are fixed with wooden pins. These pieces are placed on both sides of the way for its whole length; they are generally laid four feet apart, which forms the inside gauge of the way. It will be seen that these new ways are nothing but a wooden framework. All the space between the pieces of wood is filled with stones, compressed as much as possible to strengthen the way. The whole is covered with sand and gravel, which is placed between the longitudinal bars and only about two inches deep above their base. The buried wood is thus preserved, and the way is made very solid.[1]

It is in fact to be doubted whether the track was ever actually sunk into the ground except in cuttings. The Tyneside engineers were fully aware that the solidity of the way depended on adequate drainage, and they therefore tended to raise the track above ground-level. The ballast played an extremely important part, for it had to carry the weight of the traffic, drain water away from the wood, bind the whole framework together, and protect the sleepers from the elements and, if possible, from the hooves of the horses as they walked along between the rails. In a modern railway, the weight of a train is transmitted through the rails to the sleepers, which in turn rest on the ballast. On wooden waggonways, however, some of the weight of traffic was transferred to the ballast directly by the rails. The sleepers rested on a thin layer of ballast, which was primarily to assist the drainage, and they served as much to keep the rails to the right gauge as to transmit the load. So, once the sleepers were laid on their bed of ballast and the rails pinned on to them, 'the ashes or material . . . were then beat firmly against the under surface of the rail, for the purpose of strengthening it, and making it more rigid'.[2]

The rest of the ballast was then filled in between the rails, up to the level of the sleeper-tops or slightly higher. Outside the rails, it was raised nearly as high as the rail-tops, and was extended at this level for some 20 in. on either side to form the paths or footgangs for the waggon-men to walk along. Outside these again, the ballast dropped away to allow the rain-water that had fallen on to it, and percolated through it, to drain off into the two gutters or ditches.[3] The total width, outside the gutters and inside the fences, might be some 14 ft. for a single-track way, or 22 ft. for a double-track one. Waggonway ballast was no doubt originally reused ships' ballast, with which Tyneside (and the bed of the Tyne) was liberally provided by the colliers. At Whitley in 1704, when the word first occurs in a railway context, clinker from the salt-pans was used 'for the ballast of the waggonway',[4] a material much used thereafter, along with coke, cinders from the colliery pumping engines, gravel and even small coal. Clinker and cinders were perhaps not the ideal for drainage, but as they were often available close at hand the expense was saved of transporting better material from a distance. Broken stone, such as is general nowadays, was little used in the days of wooden rail, although the Clavering Stella waggonway accounts of 1711 contain entries for 'breaking ballast', and 'stoning' appears at Sheffield in 1780.[5] No special surface was normally provided for the horses' hooves to grip on other than the packed ballast. The route of the Western Way, however, is still strewn with large, rounded stones, which suggests that sometimes the horse-track was cobbled. Ballasting a yard of single track varied in cost. At Throckley in 1754 it was $1\frac{1}{2}d.$, on Allan's waggonway in 1769 it was *5d.*[6] During the Napoleonic Wars it rose to $6\frac{1}{2}d.$, $8\frac{1}{4}d.$, and even *10d.*,[7] while away from Tyneside the cost was often a great deal more.

The sleepers, as Jars remarked, could range in size from 4 to 8 in. square: 6 in. was probably about average, and they tended, if anything, to be wider than they were deep. Often they were squared only at the ends, leaving the centres untrimmed. Their length depended to some extent, naturally, upon the gauge of the rails, but with Tyneside gauges mostly falling into the area of 4 ft. to 4 ft. 6 in., 6 ft. was a rough average. None the less, the length could vary considerably even on one waggonway. William Brown, ordering sleepers for Throckley in 1765, wrote: 'I will take one Thous^{d} Sleepers deliv^{d} at Shields providing they are one foot in measure one with another', and the 4-ft.-gauge Elgin waggonway had 5-ft., 6-ft. and 7-ft. sleepers in stock in 1796. At Ballycastle, where the gauge is not known, they were 5 ft. long.[8] Modern practice is to make sleepers of a length equal to nearly twice the gauge (a present-day wooden sleeper is usually 10 in. by 5 in. by 8 ft. 6 in.) in order to spread the weight of trains over the ballast outside the rails as well as inside. In the eighteenth century this principle had not yet appeared: sleeper lengths were rarely more than one and a half times the gauge. There was also a considerable difference in the intervals at which they were laid. Main ways, which were almost always more solidly built than bye ways, generally had their sleepers set at 1-ft. 6-in. centres, sometimes at 2-ft., and very rarely indeed at 1-ft., 2-ft. 6-in. or 3-ft. On the bye ways and small branches, 3-ft. bearings were not uncommon, though 2 ft. was perhaps the most usual distance. None more than 3 ft. is known.

On to the sleepers were pinned the rails. Their dimensions were remarkably constant. No details of seventeenth-century ones survive, but in the eighteenth both height and width of rails in Tyneside and its colonies generally fell between 4 and 5 in. If the rail was not square in section, it was always higher than it was wide so long as the old single rail survived. Perhaps the most usual dimensions were 5 in. by 4½ in. No recorded example agrees with Jars' statement that a rail's width exceeded its depth, nor that its larger dimension was as much as 6 or 7 in. Only, it seems, in Cumberland were lighter rails favoured, though Whitehaven began with 5-in. square ones, and they never went below 4 in. by 3 in.[9] Very little information survives about rail lengths: 6 ft., the figure given by Nicholas Wood, was probably the norm, though at Heaton in 1795 the rails varied from 5 to 8 ft.[10] The Seaton waggonway was supplied with uncommonly long ones – 12 to 15 ft. – in 1749. By 1776, however, a Scottish supplier had to tell a Cumberland agent that he had a quantity of round timber suitable for rails

> which would square 4 inches. . . . You mentioned in your last that you could have none but what is 7 foot long or upwards. Now I cannot agree to that for it is a grett lenth; I think 6 foot is a good lenth and perhaps some chance ones may be less, but as to lenths, I believe it will please you.[11]

From the 1770s the upper surface of the rails was sometimes planed to a slightly convex shape, apparently in an attempt to reduce the friction of the wheels.[12]

The pins, or trenails, which pegged the rails down were usually of young oak and 6 or 7 in. long. Holes were bored through the rail and half through the sleeper to receive them, one per sleeper per rail. Rail joints were made on sleepers, rather than between them, the two ends simply butting up against one another and each held down by a pin. A 6-ft. length of track, with sleepers at 1-ft. 6-in. intervals, would thus need ten pins. A typical price was 15*s*. per thousand in 1794.[13]

Thus constituted, a waggonway was ready for traffic. But, with traffic or without, it would not last for long. The effects of rot in the buried sleepers, alternate sun and rain and wind on the exposed rails, and especially the wear and tear caused by waggon wheels, all combined to shorten its life. Under only moderate traffic, a rail might last for three years before it was worn dangerously thin; under very heavy traffic it might succumb in a year. As the sleepers generally outlived the rails, this meant replacing the old rails with new, and selling them off as firewood or the like. The sleepers also suffered:

> To secure the free passage of the flanch of the wheels, along the sides of the rails, little or no road material could be laid above the level of the top of the sleepers; the action of the horses' feet soon, therefore, wore the sleepers through: added to this, the frequent perforation and insertion of the pins, through the same part of the sleepers, rendered them unserviceable. These defects led to the adoption of what is termed the 'double way'.[14]

The double way (fig. 36) involved pinning a second layer of rails on top of the first. Whereas in the single way the rail joints necessarily occurred on sleepers, in the double way the joints of the upper rail could be anywhere and the lower rail, with a much-extended lease of life, offered a continuous surface for pinning down replacement top rails. The upper rails, in their turn, lasted longer, since they could be worn almost completely away before the track lost its strength; and the sleepers benefited because they could now be covered with ballast to a depth of 5 or 6 in. Whilst still leaving plenty of room for the wheel-flanges, this ballast protected the sleepers from the horses' hooves.[15] The top rails on a heavily used line might still last only a year, but it would be at least four years before the sleepers and the bottom rail need be disturbed. John Buddle wrote of Washington in 1789: 'Tyne Way having now been in use 6 or 7 years, the Bottom Rails (of Firr) are failing, and will require some additional labour.'[16] It was not unknown for them to have a life of ten years.

The construction was identical to that of the single way, with the addition of the extra rail on top, except that the rails tended to be slightly wider than they were high, for stability: a typical double way would have bottom rails 4½ in. high by 5 in. wide, top rails 4 in. high by 5 in. wide. The idea probably arose in the 1750s, for the first known example of its use was in 1760 on the Shiremoor waggonway.[17] It spread fast, and by the 1770s the double way was the norm. But only on main ways, where the

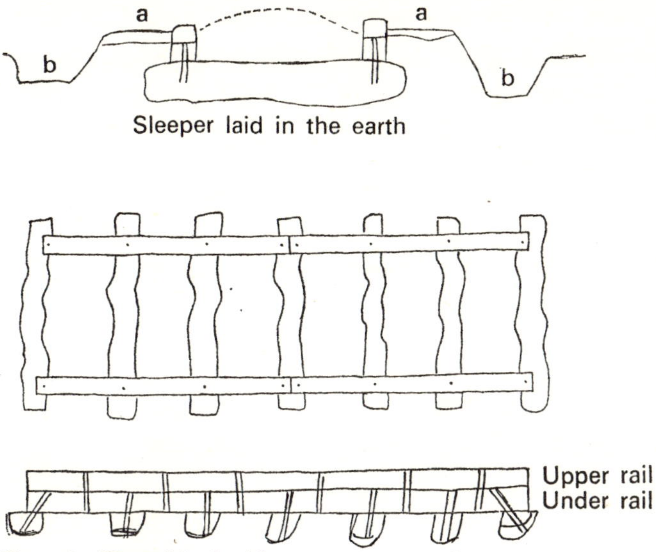

FIG 36 Tyneside double way, 1807. *aa*, foot gangs 20 inches wide; *bb*, gutters to drain the way.

loaded waggons were liable to break the rails and wear them out quickly; there seems to be no case on record of a double bye way.

It was early recognized that wood was not the ideal material for rails. It wore badly, and created a great deal of friction:

> The yielding nature of the material, especially when saturated with wet, would create very considerable resistance to the wheels, which, by sinking into and compressing the rails, would always form a rising surface, and thus impede the progressive motion of the carriages. . . . In some parts of the road, where occasional acclivities occurred, which could not be levelled, or where sudden windings of the road were obliged to be made, thin plates of wrought iron were laid upon the surface of the rails, and fastened down with common nails, to diminish the resistance opposed to the wheels, and equalize the draught of the horse. This, no doubt, would be found a great improvement, not only in diminishing the friction, but also in preventing the Rails from wearing. Yet we do not find the use of them much

> extended, beyond the above-named instances; probably, from the difficulty of keeping the plates fast upon the Rails, as the nails, by the elasticity of the wood, would be constantly working loose, and occasioning a continual expence in keeping them right. Upon the whole, however, the use of such plates would, in many cases, be attended with considerable benefit. . . . This plate was generally about two inches broad, and half an inch in thickness; and was fastened to the wooden rail, by nails counter-sunk into the iron bar.[18]

When and where this improvement was introduced is not at all clear. The *Penny Cyclopaedia* claimed in 1841, without quoting any sources, that 'it became usual at an early period, at least as early as 1716, to nail thin plates of malleable iron upon the surface of the wooden rails, wherever a steep ascent or a sharp curve rendered the draught harder than usual'.[19] It may be so, for Tyneside wayleaves usually included a reference to iron, which seems to have had no part in track-work except as plates on top of the rails. Even so early a lease as the Gibside one of 1692 gives leave 'to fix, lay & place . . . Railes, Sleepers or other wood Iron or Timber, for ye making ye said Waggon Way'.[20] A number of subsequent leases repeat these words, with variations, but it is not until Brandling's Middleton Act of 1758 that we first find specific mention of iron rails. It lays down that

> it shall and may be lawful to and for the said *Charles Brandling* . . . to carry, convey, fix, lay, and place, Wood, Timber, Bricks, Stone, Earth, Gravels, Iron Rails, Sleepers, and other Materials, unto, in, and upon, the said Fields, Lands, Wastes, and other Grounds.

It may be that a comma has been accidentally omitted between Iron and Rails, which would alter the meaning, but there are a few hints that the Middleton used thin iron plates at an early date. The accounts record the purchase of large quantities of iron bars from 1762, specified as hoop iron in 1767 and as 'Waggon Plates' in 1789.[21] In the same part of the world, the Greasbrough waggonway was plated with iron in 1766, when a colliery employee was indicted for stealing it. A witness said that he saw George Meller, the defendant, 'rip and break the said piece of plated Iron from the said Waggon Road and found there were ripped up and carried away above thirty foot of Plated Iron from and off the said Waggon Road'. Another witness deposed that 'Meller seeing this Informant going up the aforesaid Road threw down the aforesaid piece of plated Iron and run away across a Close adjoining the aforesaid road and secreted himself under a Hedge'. As the total weight of stolen iron was 60 lb., the plates must have weighed about 6 lb. per yard.[22]

On Tyneside, Jars remarks in 1765 that 'all rails must be carefully aligned at their ends; strips of iron are sometimes placed at this point' – presumably to reduce the wear at the rail-joint.[23] We find iron plates at Throckley in 1771,[24] but it was not until Cort's invention of the puddling process in 1784 that wrought-iron plates really came into favour, and even then they were used only on curves and gradients, it seems, never on the whole of a waggonway. A valuation of the Western Way in 1789, for example, divides the whole line up into 220-yard stretches for costing. Many of these stretches had no iron on them at all, some had short lengths of 7 or 9 yards of iron plate, and none had more than 29 yards. But where iron was laid, it was always on sharp curves or steep runs, generally on the main way, but sometimes on the bye.[25] Hartley waggonway, a short but steep line, had no less than £59-worth of iron plates on its wooden rails in 1808, a fair length with the general price of plates standing at 1*s*. a yard. Killingworth Colliery invested in some Swedish wrought-iron plates in about 1806, which gave great satisfaction; Jarrow owned 284 yards of 'waggonway plates' in 1809, which again weighed 6 lb. per yard; and the rails on Kenton staith were covered in 1812 with iron plates at 8 lb. per yard.[26] Specimens of this type of track survived in use well into the 1840s, and one isolated example, in a timber yard at Sunderland, lasted until 1965. Its iron plates, 2 in. by ½ in., lay on massive oak rails 8 in. high by 5½ to 7 in. wide. Recent excavations on the route of one of the waggonways to Ayr Harbour have also revealed wooden rails bearing rust-marks for iron plates 1½ in. wide.[27]

The Cumberland Coalfield also adopted the idea. Some of the Whitehaven waggonways were laid with iron strips measuring 2½ in. wide by ½ in. deep by some 10 ft. long, and at Maryport in 1837, on a short wooden waggonway much dilapidated and shaken by wear, 'some of the wooden rails are covered with thin strips of iron. . . . Only one waggon is brought down at a time which runs the greatest part of the road itself, and is drawn up the incline by a horse.'[28] Cumberland has also been credited with the introduction of cast iron for the same purpose, as an armour-plating for wooden rails. Longstaffe's history of Darlington, followed by a number of later writers, claimed in 1854 that 'Rails wholly of iron are said to have been unsuccessfully tried at Whitehaven in 1738'. But if the ancestry of this remark is traced back, it will be found to rest on several misunderstandings. One source was probably T. G. Cumming, who wrote in 1824 that wooden railways were adopted at Whitehaven in 1738, an observation immediately followed by an account of the substitution of iron for wood.[29] Cumming did not mean that iron appeared as early as 1738 at Whitehaven, but a quick reading of his words could well give that impression. Secondly, Alexander Scott said, again in 1824:

> With regard to the introduction of cast-iron railways, a late anonymous author (1821) says, without advancing his authority, 'that, in 1738, cast-iron rails were first substituted for wooden ones: but, owing to the old waggons continuing to be employed, which were of too much weight for the cast-iron, they did not completely succeed in the first attempt'.[30]

This late anonymous author, whoever he was, seems to have been reading, or misreading, John Farey's account of railways in Rees' *Cyclopaedia* (1819), which, like Cumming's, says that Whitehaven waggonways dated back to 1738, and continues with the story about early cast-iron rails being too weak.[31] In fact, there is no evidence of cast-iron rails in Cumberland until the turn of the century. Nor is there any support for the oft-quoted story that, when John Curr's 1774 waggonway at Sheffield was burned by the angry populace, he replaced the wooden rails with his own cast-iron angled variety.

Indeed, it seems clear that the invention of the cast-iron bar to lay on top of wooden rails took place at Coalbrookdale in 1767, and that its use spread only to Shropshire colonies. No example has come to light on Tyneside or in its empire.

The type of wood used for rails depended on supply and on the sort of line it was wanted for. In the days of single ways, oak was naturally the most favoured wood for rails, although ash, occasionally birch, and later fir, beech, alder and maple were also used. It often happened that the main way was made of oak and the bye of fir or ash. With the advent of the double way, the bottom rail was usually of fir, though oak was by no means rare for the purpose; the top rail was almost always of beech, a wood capable of acquiring a smooth gloss ideal for reducing friction. Oak, fir and sycamore are very occasionally found as top rails. The sleepers, both for single and double ways, were nearly always of oak or ash.

In the early days of waggonways the timber for rails and sleepers and waggon wheels came largely from local sources. The Sheffield colliery accounts of 1732–3, for example, include a number of such entries as:

paid Mr Philip Smilter for Rail &^{cc} out of the Little Wimse	30.16.4
Robt Union for hewing & Sawing Rail in the Roe	2.13.9$\frac{3}{4}$
Paid ffrancis Glossop for sawing 200 sleepers at 16^{s} p hund	1.12.0
p^{d} Mr Smilter for 7 Tun of Realle 2^{l}. 5^{s} p Tun	15.15.0
p^{d} D^{o} for sawing 12 Tun of Rail at 7/6 p Tun	4.10.0
pd for felling wood for sleepers in Tom Cross wood near Crookmore	13.0[32]

Tyneside coal-owners likewise relied on woods near their own collieries. Lionel Vane, for example, one of the chief partners in the Stella Grand Lease collieries, owned enough woods in the late seventeenth century to supply his waggonway's needs as well as producing plenty of rails, sleepers and planks for sale at a very handsome profit. Some of the accounts of his agent, William Iley, have survived to show how trade of this sort was managed:

An accompt what I have Disbursed for Cutting sawing Hueing and leading in of the wood I sould for my mast Lionell Vane Esqr use.

	ll s d
Imp. pd the wrights John Wright and nic Andrew for workeing & sawing 1200 Yards of rales at 7^d p [10 yd]	03:10:0
pd for Cutting & Huing 2 for sleepers	00:02:0
pd for Cutting i fother of barrowwood	00:01:0
pd Anth Stepson for leading in the Rales was 20 foder at 2^s 6^d p fod	02:10:0
pd him for leading 3 foother of Sleepers & wood at 2^s 6^d p foder	00:07:6
In all	06:10:6

An accompt what wood I have sould for the use of my mast Lionell Vane Esqr was cut in Chopwell wood the 4th July 1684

	ll s d
Received from Anthony Stepson in all of waggon Rales the number of 1200 yards was delivered to Anthony Armstrong for Mr Nath: Midletons use as 3^d p yard is	15:00:0
D^r to him two foother of sleepers is	00:13:4
D^r to Ralph Chambers 1 foder of barrow wood is	00:08:6
In all	16:01:10
pd as p. other side in all	06:10:6
p. Will Iley Rests [profit]	09:11:4

Mr. Iley thus contrived a profit of nearly 150 per cent. on his immediate outlay.[33]

As the coal industry, and especially the waggonways, grew, local timber became scarcer and scarcer. Some oak and beech forests were specially planted in Northumberland to supply the demand for waggonway materials, the benefit of which was reaped for a time;[34] but the bulk of the forests planted from the 1740s can hardly have matured before the wooden rail was ousted by the iron. The fitters, or commercial agents of the coal-owners, bought this timber from local merchants. Some of the correspondence survives of William Dobson, Sir John Delaval's fitter at Hartley. Here is a selection:

Mr Wm Dobson S^r.

This to aquent you I Have 2000 yards of oak Rales at Alemouth [Alnmouth] If you Have a Call I wood be glad to Deale with you I allsa Have oake Slepers & props Side Sheth Convows & Brest Wood [waggon parts: see p. 191] If you want any thing In that way I wood be glad to sarve you. . . .

I Reemane your vere Hble Sarven^t
Jno Prissick

Newcastle 21th March 1773
NB prises $7\frac{1}{2}$^d pr yard for Rale Slepers 9^d Sheth 6^s pr Doz.

*

Alemouth 23th June 1773

Mr Wm Dobson S^r

I have loaden Mr Gubb with oak Slepers Rals & props as follows

To 630 Oak Slepers at 9^d
To 90 ash Slepers at 6
To 5 foots props & sum 6 foot at $2\frac{1}{2}$
To 4 foot do at 2^d
To Oak Rals 110 yards at /$7\frac{1}{2}$
50 Ends

	£	s
You will be so good as Deelever Him & ples to pay Him His frite	5	0

Jno Prissick.

*

Yarm, 2^d Octr 1769

To Wm Dobson, Seaton Sluice

S^r In the 22 of Septer last I Shipt for you one Board of the Rose Captn Richd Parkinson master the perticklers as below and I think the Best Sleepers I ever See but thay will come at Sum think higher price the Mr Manchester Spock of when he was at yarm but when I went to see them I thout them worth the monney I have charged.

thare is severl of them that has two foot of Timber In them a good many of them will make you Timbers for your keels. From your Hum Ser John Merryweather.

	£	s	d
To 329 Sleepers at 1^s p Sleeper	16 –	9 –	00
To 274 Yards of Rails at 11^d p yard	12 –	11 –	00
	£29 –	00 –	00

PS I was from home when the Rails and Sleepers was Shipt but me Brothers mesered them of which understand those things very well. [Dobson's hand] The Sleepers to be charged 9^d p s. The Rails at 8^d p yard.

By 1807 Dobson's successor was having to try still further afield. He was told then by a Southampton merchant: 'In consequence of there being scarce any Beech Timber cut in this Neighbourhood this Season I have it not in my power to furnish you with any Beech Rails. . . . When I have Rails shall be happy to serve you.'[35]

By the 1720s large quantities of rails and some waggon wheels were supplied to Tyneside from the woods around Sheffield and down the Don Valley, being carried by river to the Humber and thence by sea. This trade continued until at least 1750. Meanwhile, still more distant sources had begun to be tapped. William Scott, father of the first Earl of Eldon, was Bowes' fitter at Newcastle and a dealer in waggon wheels and rails in his own right. His letters give an interesting glimpse into the fluctuations of the market. Although he bought most of his stocks from Sussex and the New Forest, with some from Northamptonshire, there was still some local timber available.[36]

18 February 1745/6. Wheels are at present a great drug from so many y^t came last year. Rails will be wanted, but the people pays so badly for them that wo^d weary anybody to serve them, a years credit is nothing with them.

28 February 1745/6. You say you have bo^t a parcel of rails you formerly wrote ab^t, if you mean ash rails theyl come too late, for the gentleman that wanted them is already servd. Your timber is not yet arr^d. I find the best oake rails will scarcely give 6d. p yd this year, as there will be a great many cut in this countrey and led to the wag^n ways at 6d p yard. This parcell of rails proved very indifferent and were badly squared.

24 July 1746. To Mr Richd Chitty, Singleton, Medhurst, Sussex. 'The Ark of Scarbro': Mr John Masterman arriv'd here last Sunday and bro^{t} me (without any advice from you) about 20 load of rails and 185 waggon wheels, wch rails and wheels are both greatly decayed. I wont say but the rails may some time pay the fr^{t} but the wheels will contribute very little towards it, for I fear they will never sell at any rate but for firewood, considering the great numbers of wheels that are here unsold and many of them fresh. I am perswaded it wont be worth any bodys while to send any here for some years.

26 May 1747. To Daniel West, Slyndon, Arundel, Sussex. Being now resolved to receive no more goods such as w^{n} wheels, rails, and such like from anybody, finding I no sooner receive them, than soon after, the money is expected, tho I can make it appear I often don't receive the money for them in less than sometimes 12, 18, nay for some 2 years and upwards, so I am now determin'd to employ myself in making up accounts with such people as I have dealt with in that way, and amongst the rest yours wch I propose to send in July; in the mean shall not pay any more bills, but shall be ready to pay you whatever may appear due to you, not doubting if on a fair account you may be found in my debt, you'l do the same by me.

20 October 1747. To Mr West. It may not be amiss to inform you that some people begin to want wag^{n} wheels, tho some others have too many, and here is 500 or 600 yet unsold, tho I've none to sell, and my friends Mr Shafto and Mrs Bell begin to own they want, and there will be beach rails wanted.

8 January 1747/8. To Richard Chitty. Beach rails will not be wanted as formerly. I mean not so many, the long wag^{n} ways being on the decrease.

Not long after this, Baltic fir began to make its appearance on Tyneside, with the extra heavy expense of shipment. On top of this was the local freight, for transport in keels from Shields or Newcastle to the colliery staiths. Such items crop up in many accounts:

Clavering Stella 1712: Keel-Hire of 203 Wheeles 10/-.

Plessey 1715: 3 Tun of Rail wood at 50^{s}: 7^{l}:10^{s} & for 102 wheels p. recept 32^{l}:10^{s} Wherry Fright 1^{l}:19^{s}:4^{d}.

Bigges Main 1785: Sundrys for Keel dues of Rails & Sleepers with Expencies £1.19.6.[37]

Cumberland, cut off from Tyneside's timber sources, had to find its own. A trade with Loch Moidart in Scotland was in full swing in the 1770s, and by the 1790s Curwen at Workington was being supplied with timber for rails from the Mawddach Estuary behind Barmouth, but the forests here were soon depleted.[38] Charles Brandling supplied his Middleton colliery by purchasing woods near Leeds, but was very soon forced to buy further afield – Snaith, Worksop, Gainsborough, Easingwold, or even the Baltic – with corresponding increases in price.[39] The following figures, culled from thirty-one different sources on Tyneside prices, five on Cumberland and two on Yorkshire, show the general trend of rising costs:

	1600-1700	1700-25	1725-50	1750-75	1775-1800	1800-15
oak rails p yd	3*d.*	4½-6*d.*	3¾-7*d.*	7-8¼*d.*	6-8*d.*	10-14*d.*
beech rails p yd				4¾-6*d.*	5-7*d.*	7-10*d.*
fir rails p yd				5*d.*	6*d.*	7-13*d.*
ash rails p yd				4-5*d.*	4-6*d.*	
sleepers each (oak or ash)		5-8*d.*	6-10*d.*	6-10*d.*	8-12*d.*	12-18*d.*

In out-of-the-way places, prices were very much greater. At Ballycastle in 1740, for instance, oak and fir rails cost 13½*d.* per yard and sleepers 22½*d.*[40] Rails were normally bought by the yard, but Yorkshire waggonways and a few others often reckoned by the ton, which made about 120 yards of rail.

The total cost of material and workmanship inevitably depended upon the scale of the engineering works. In the early days of waggonways this was not excessive, although James Clavering, who was thinking of building a line from Urpeth Colliery to the Tyne, wrote complainingly to Henry Liddell in 1712:

> We think laying ye way computed at 6000 y^{ds} to be highly rated at £1000, for where there are few or no difficulties 2^{s} p y^{d} is ye usual rate, In ye most substantial ways 2^{s} 8^{d}, & I have been informed Mr Headworths way . . . wch is laid as strong as possible, & had many difficulties to encounter did not cost him 3^{s} . . . whereas here as P computes ye demand amounts to 3^{s} 4^{d}.[41]

Prices rose hereafter. William Brown wrote in 1754 to Carlisle Spedding telling him the usual Tyneside dimensions and prices, which came to about 2*s.* 10*d.* per yard without levelling:

> We Generally Lay our way of about 4½ Inch by 5 [rails] at about 4 foot or 4 foot 2 inches Inside and Inside [gauge] and Sleepers at about 15 or 16 Ins Distance but this Breadth is for a Waggon of 24 Bools these Rails cost laid down at the way Comonly about 7d p. yard and

sleepers 7^{d} or 8^{d} pp. piece. The Carpenters has for a new way of laying and filling Sleeper Height $3\frac{1}{2}^{d}$ p. yard End ways and the filling with Stones &c Comonly Cost us about $1\frac{1}{2}^{d}$ p yard as to leveling the Ground That varies as the ways is to be laid is regular.[42]

Table 3

Cost per yard of building new wooden waggonways

Date	Place	Rails	Sleepers	Pins	Ballast	Laying	Total	Type
1712	Tyneside	—	—	—	—	—	2/0-3/4†	single main
1717	Bucksnook	—	—	—	—	—	2/7	single bye
c1726	Tyneside	1/0	1/4	—	—	2½*d.*	(2/6½)	single main
1740	Ballycastle	2/3	1/10½	—	—	1/0	(5/1½)	single main
1745	Tyneside	1/2	2/6*	1*d.*	2*d.*	3*d.*	4/2	single main
1754	Throckley	1/2	1/3	—	1½*d.*	3½*d.*	2/10	single main
1762	Maryport	—	—	—	—	—	3/11	single main
1769	Allan's	3/0	2/7½	2½*d.*	5*d.*	9*d.*	7/0	double main & single bye
1769	Throckley	1/8	1/8	2*d.*	3*d.*	6*d.*	4/3	double main
1773	Sheffield	2/3	1/8	2*d.*	1/0	8*d.*	5/9	double main
1780s	Wallsend	3/9		—	—	4½*d.*	(4/1½)	double main
1780s	Wallsend	2/8		—	—	3*d.*	(2/11)	single bye
1787	Plessey	—	—	—	—	—	3/6	single main
1792	Twizell	1/10	11¼*d.*	2*d.*	8¼*d.*	4½*d.*	4/0	double main
1792	Twizell	1/0	9*d.*	1*d.*	6½*d.*	3½*d.*	2/8	single bye
1803	Willington	2/4	2/6	1*d.*	2*d.*	6*d.*	5/7	double main
1803	Willington	1/2	1/10½	1*d.*	2*d.*	3*d.*	3/6½	single bye
1805	Tyneside	3/3	3/0	3*d.*	10*d.*	8*d.*	8/0	double main
1805	Tyneside	1/8	2/3	1½*d.*	6*d.*		4/6½	single bye
1806	Tyneside	3/4	2/8	2*d.*	—	9*d.*	(6/11)	double main
1806	Tyneside	1/8	2/0	1*d.*	—	6*d.*	(4/3)	single bye
1806	Sherriff Hill	2/9	2/6	2*d.*	—	7*d.*	(6/0)	double main
1806	Coxlodge	—	—	—	—	—	8/0	single branch
1808	Middleton	—	—	—	—	—	7/6	single bye

* Sleepers at 1-ft. centres.
† Including cuts and batteries.
Bracketed totals are incomplete.

Table 3 has been compiled from sources such as this, omitting the highly variable cost of levelling. Costs away from Tyneside, it will be noticed, tended to be high. When all the factors are taken into account – cuts, batteries, bridges, levelling and fencing as well as the materials and workmanship – a list of individual costs per mile will be seen to fluctuate somewhat, but also to show a general rise as time went on:

		Cost per mile
Wollaton	1604	£ 82
Broseley	1605	133
normal Tyneside	1712	176
Hedworth's	1712	264 maximum
Urpeth charge	1712	293
Tanfield	1725	c535 without Causey Arch
Sheffield	1773	880
Halbeath	1780	c750 including waggons
Heaton	1786	780 including staiths
Plessey	1787	c410 very flat ground
Twizell	1792	1129 very rough ground
Willington	1803	876

One of the main results of the increase in the price of timber during the Napoleonic Wars was the spread of the all-iron rail and the stone sleeper-block. This development will be discussed in ch. 15.

When a railway line curves sharply, the flanges of the outer wheels tend to climb up on to the rail and derail the vehicle. To counteract this, an extra rail, the check rail, is laid nowadays just inside the inner rail on the curve, so that the back of the flange of the inside wheel bears against it, and prevents the outer wheel from climbing the outer rail. Check rails, or false rails, are surprisingly old, being first heard of in 1729 at Ravensworth.[43] The evidence for their form in the days of wooden rail, however, is slight and contradictory. Two maps, of the Throckley and Tanfield Moor waggonways, drawn by William Brown in 1771 and 1772, mark false rails by dotted lines, always on the inside of curves, where one would expect to find them if they were the same as modern ones.[44] On the other hand, a narrow flange-way between running rail and check rail would be a perpetual and dangerous trap for dirt and stones kicked up by the horses, and our only description, written by the Swede Bengt Qvist Andersson in 1776, says that they were applied to the outside of the outer rail:

> Where the way bends, the rail is provided with what is called a false rail or jack rail [probably a corruption of 'check rail'], which serves to guide the waggon on curves. This false rail is nothing but a baulk fixed tightly to the outer edge of the rail, nailed to the cross sleepers and raised some four to five inches above the rail, with which it forms an angle. It rarely has to be more than four or six feet long, depending on what the bends in the way demand.[45]

It is clear enough that this type of check rail was a raised ledge which restrained the outer wheel on the curve from climbing the rail and going off at a tangent.

Whatever the normal type of false rail, Carlisle Spedding of Whitehaven knew all about them in 1754, when he said that the steep gradient of his waggonway could not be decreased by bends in the route 'which both lessen the Decent & create more friction by Rubbing against check Rails'.[46] After 1770 false and check rails find frequent mention on Tyneside in estimates and valuations of waggonways. The Clavering Stella line, ancient, tortuous and steep, was the subject of a report in about 1800:

> This Waggon Way has many Runs, lies badly for Ballast, only small Coals and Pan Rubbish, many Check Rails, Crossed in many Places by the Lead Way [the Greenside-Chopwell road], crouded on the Sides [i.e. narrow cuttings]; but much opened the last 3 years. Sandy Ground destructive to the Sleepers.[47]

The Western Way, another serpentine and precipitous route, was divided up for valuation purposes in 1789 into 220-yard stretches. Just as the more curved of these stretches were provided with wrought-iron plates to reduce wear on the rails, so they were given false rails for safety, sometimes as much as 131 yards of false rails on a 220-yard stretch of track. Nor were the false rails confined to the main way: here, as elsewhere, the bye way also had them in profusion. Over the whole line, with 11,071 yards of main way and 10,740 yards of bye, there were 2,569½ yards of false rail.[48] The same sort of proportion is found on the Elgin waggonway in 1796, which had 8,230 yards of main way, 2,800 yards of bye, and 1,370 yards of check rails.[49]

Not infrequently 'double false rails' are mentioned, sometimes fitted to double main ways and therefore ostensibly made double the usual height in order to be effective with two-tier running rails. But as double false rails were also attached to single bye ways, the term must mean a false rail fitted to each running rail. Single false rails cost 8*d.* per yard in 1729 plus 1¼*d.* for laying; in the 1780s single false rails cost 1*s.* 1*d.* per yard (laying 2*d.*), double ones 1*s.* 7½*d.* per yard (laying 3*d.*).[50]

The surviving evidence about points is meagre in the extreme. This is the greatest of pities, for their construction can be a useful guide to technical ingenuity. One thing does seem certain: that they were not like modern points. The two tapered blades, anchored firmly at one end and capable of being moved at the other end under the elasticity of the iron, appeared first in 1839. Their immediate predecessors, two short blades pivoted at one end, go back probably to the beginnings of all-iron rails, and were used in the dying days of wooden rails as well. The only illustration of point-work in timber shows just such an arrangement (pl. 71) in use at Königsgrube in Silesia in 1819.

In the heyday of the wooden waggonway, however, the evidence which survives shows that there was but one 'switch rail' to a set of points, not two, as the ordinary arrangement demands. Apart from two early references to 'ye parting of ye ways', which tells us nothing, nearly all the mentions of switches occur in valuations of existing waggonways. Most of them talk of a bolt, presumably the pivot for the switch rail, and some talk of iron plates and check rails as well. Thus at Walbottle in 1771 we find 'Iron plates upon a Switch Rail $3\frac{1}{2}$ yds=4/-'. The valuation of the Western Way in 1789 contains many references, such as '20 yards siding: Check & Switch: 5/- and 1/4'; '9 yards of Points @ 5^d=3/9' (the only early use of the word 'points' that we have); 'Switch and plate'; 'Switch & Bolt' at various prices from 10*d.* to 1*s.* 6*d.* – the latter at one of the most important switches on the whole line, the junction of the main and bye ways for the precipitous Leap Run.[51] Similar items are found for other waggonways. The Montague Main way in about 1810 included '5 Switch Bolts 10 lb. @ $2\frac{1}{2}$ [*d*]: 2*s.* 1*d*'. Bigges Main had 'Switch and Bolt 1. 11d' in 1787, and in 1809–10

950 y^{ds} & 2 feet single Bye Way	1/6	£71. 5. 6
19 yds double checks on do	9^d	14. 3
2 Switches with Bolts on do	2/-	4. 0
6 yds of Iron plates on do		6. 0

This apparently meant one switch at each end of the bye way.[52]

Because the switches always seem to come singly, it is likely that most of these early points were on the single-shifting-rail principle, as used for guide-wheel trucks on the Continent and for the crude flanged-wheel vehicles of Transylvania (fig. 27 and pl. 37). The design is indeed primitive, turnouts are necessarily sharp, and the angles would no doubt give the waggon a rough ride – the switch rails seem to have been only 3 or $3\frac{1}{2}$ yards long. But they had the merit of simplicity, and the overwhelming advantage that the horse-path between the rails was never obstructed by the switch rail. The only known drawing which even hints at the design of English point-work, a large-scale plan of Eighton Colliery at the head of the Ravensworth Way in 1783, does suggest that this was the pattern used. Examples of this type, made of iron, were still to be found underground on Tyneside in 1867.[53]

All these early points were no doubt set by hand, presumably by the waggon-man. No surviving accounts record wages being paid to special pointsmen, although in some places there would be a real need: on the Tanfield waggonway at Lobley Hill, for example, where half of the very heavy traffic branched off to Redheugh and half to Dunston staiths. There were places, too, on single lines where a self-acting point would have been desirable, such as the junction of ongate and offgate at the entrance to a staith, and at passing loops. In later days, in cases like this, the switch rail was attached either to a spring or to a counterweight, so that it always remained set correctly for a waggon approaching from the facing direction; waggons trailing over it would open the switch automatically against the spring with their wheel-flanges.

This self-acting point is always supposed to have been first applied to the Sherriff Hill waggonway in 1826, and the Stockton and Darlington of 1825 is generally credited with the first use both of catch points, to carry runaway waggons off the main line and out of harm's way, and of rodded points worked by a lever from a distance. In fact, the Elgin waggonway had sprung points worked by remote control in the early 1820s, with iron rails.[54] More interesting still, Carlisle Spedding intended to use all three devices on his Whingill waggonway at Whitehaven in 1755, although history does not relate whether he actually did so: probably not, for he was killed by a fire-damp explosion less than two months later. Fearful for the safety of the waggons on a gradient of 1 in 11, he wrote:

> To prevent great Mischiefs by the Running of Waggons I intend at the Bottom of the Run to have a Branch laid and a switch Rail to be Shutt by an Old Man to sit in a Cabbin and with an Iron Rod thro' a Conduit fixt to a small Leaver (when he Sees a Waggon got loose) he may Shut the Switch Rail and turn the Waggon into the Branch which Riseing till it Runs against a Battery of Earth and Turf will take no harm and will fal back again gently and I'll have another Switch Rail which shall Shutt of itself and turn the Waggon (in its coming back) into another Branch where it will rest till brot into the Main Way below the first mentioned Branch by which means if another Waggon or more shou^{d} follow they will Escape breaking one another.[55]

On bends, the track of a timber waggonway would normally be laid in a series of short straight sections, which, with 6-ft. rails, produced a reasonable result. Slightly curved, or in severe cases specially short, rails might be used. When a bend was too sharp for a waggon to get round at all, a turntable had to be provided. There were several of them on the bye way of the Howgill line at Whitehaven, which also had some serving the function of points at junctions.[56] On Tyneside they were much rarer: Jars remarks, understandably, that they were avoided on main lines if humanly possible. But their use on staiths was a different matter, where the waggons had to be turned through a right angle in a very confined space. We first hear of them in 1714 as 'staith-diskes' at Stella,[57] and thereafter they become a regular part of a staith's equipment, under the name of 'turnrail'.

Jars' drawing (pl. 47) shows a turnrail of nearly 10 ft. in diameter; at Whitehaven they were much smaller, as a description of 1801 shows:

> A turn-rail is a frame of oak wood, about 5½ feet long, and of the same breadth as the waggon way. It has a large oak block in the centre, about a foot deep, with a hole through the middle, of 3 inches diameter, cased with steel, to receive an iron bolt nearly of the same diameter, which is fixed to a large block of wood, below the turn-rail. The bolt has a collar, or shoulder, of iron upon it, for the case of the turn-rail block to rest upon, and is always kept well-oiled. By placing the waggon upon this moveable frame, one man can, with great ease, turn the rail and waggon round; or as far as appears necessary.[58]

A frequent entry in waggonway accounts is 'To Greace for the turnraile 7^d' (Ravensworth, 1726), or 'Oyle for the Turn Rails 0.6' (Tanfield, 1767). Jars says that the waggon-man operated the turnrail himself, but the Ravensworth waggonway's Team staiths employed a special man for the purpose: the accounts regularly record entries like 'To Turning ye Raile 43 Gates at 1^d p: 3/7'. A century later, in 1831, a turnrail-man was employed on the Derwenthaugh staiths of the still-wooden Garesfield waggonway at 12*s*. a week.[59] As to their cost, Bucksnook waggonway bought 'Two Tunn of Oak Timber for Turn railes, 06:00:00' about 1715; the Ballycastle line in 1740 needed 'Nine turning Railes at the several Angles of the Timber Roads . . . at 1l. 10s. a-Piece'; and they were valued at £1 each at Shiremoor staith in 1760, and at £1 10*s*. each at the Kenton staith in 1810.[60] Until the coming of the locomotive, if not later, waggons had a downhill and an uphill end, and were never turned round. Turnrails were intended only for overcoming impossibly sharp corners.

Another subject where the contemporary engineers, clerks and tourists were regrettably reticent is that of gauge. There are not many surviving references, and most of them are late; and they show that on Tyneside and in its colonies the gauges fell between the limits of 3 ft. 10 in. and 5 ft. The Bath line, according to Labelye's excellent drawings, had a gauge of 3 ft. 9 in., but this comparatively small figure may well be due, like other features at Bath, to its Shropshire sources rather than to its Tyneside ones. 4 ft. was perhaps the most common, although paradoxically we can pin down few individual waggonways to this gauge. Gabriel Jars, in 1765, remarked that 'the rails are usually laid four feet apart, which forms the inside width of the way', and the *Encyclopaedia Britannica*'s 1778 edition likewise stated: '*rails* . . . are laid . . . four feet distant from each other'.[61] William Brown of Throckley, writing to Carlisle Spedding in 1754 about Tyneside practice, said, 'We Generally Lay our way . . . at about 4 foot or 4 foot 2 Inches Inside and Inside'. Since the Lambton waggonway, built by Brown in 1770, was 4 ft. 2 in. gauge, it is likely that his Throckley way was the same.[62] It is often said, on the basis of Jars' account, that the great Tanfield line had a gauge of 4 ft., but while it is true that the Tanfield is the only waggonway he mentions specifically, his reference to gauge is quite clearly general and not particular. The Tanfield's predecessor at Dunston staiths, the 1699 line from Northbanks, was 2 in. smaller. A note of 1721 records, 'The Breath Dunston Waggon way Inside & Inside – 3 foot 10 Inch', together with the Ravensworth way: 'Colo Liddell Waggon . . . the Waggon way 3 foot 10 inches Broad'. Since the Blackburn branch of the Tanfield way took over all the Dunston way waggons, the Tanfield was no doubt of the same gauge.[63] The same is probably true of the Western Way, which replaced the original Dunston line. Alexander Scott said in 1824 that the 'breadth of track' of Lord Elgin's waggonway was 4 ft. 3 in., but he probably meant the gauge over the iron rails, since original drawings of 1821–3 show that the distance between the rails was only 4 ft.[64]

Two independent foreign observers in 1825-7 both give the gauge of the Middleton waggonway near Leeds as 4 ft., although later authorities prefer 4 ft. 1 in.[65] The drawings of the first all-iron rails used on Tyneside, dated 1797, show that the Lawson Main gauge, and consequently the Byker and the later Heaton gauge, was 4 ft. 3 in.[66] Hedworth's waggonway to the Wear and its later extension to Beamish had a 4-ft. 4-in. gauge in 1832; and the Fordell in Scotland was the same. Sir John Clerk described the Plessey waggonway in 1724 as 'a cassey way of about 4 feet $\frac{1}{2}$ broad', meaning about 4 ft. 6 in.[67]

The group of lines which debouched at Wallsend and Willington shared, or nearly shared, a common gauge. The Killingworth way was 4 ft. 8 in. when George Stephenson first worked there, and presumably was so when first built in 1763–4.[68] The Willington way of 1776, in physical contact with the Killingworth, must have been either 4 ft. 8 in. or 4 ft. $7\frac{1}{2}$ in., for when the Kenton and Coxlodge railway was laid in 1808, incorporating a mile of the Willington track between Bigges Main and Wallsend staith, it adopted a 4-ft. $7\frac{1}{2}$-in. gauge.[69] The $\frac{1}{2}$-in. difference, however, whether the break occurred at Bigges Main or at Willington, was not enough to prevent through running. Again, the independent Walker waggonway, immediately upstream of Wallsend, had 'Breadth of Waggon Way within Rails 4 f 8 is'.[70]

The first Whitehaven line from Saltom – and doubtless its successors – had its rails '4 foot 10 Inches Assunder'.[71] The last line to mention, and perhaps the widest, was the Wylam waggonway, which is usually said to have been 5-ft. gauge, although the Trevithick locomotive destined for it in 1805 was built, according to the drawings, to a gauge of 4 ft. 10 in.[72]

The number of different gauges need cause no surprise. At a time when most waggonways were completely self-contained and isolated there was no reason to have a standard, although a connected group as at Willington would naturally tend to use the same gauge. Why the Dunston way, to take an example, should have adopted 3 ft. 10 in. instead of 3 ft. 9 in. or 3 ft. 11 in. or any other figure is a question that nobody now can answer; even to this day, industrial railways are notorious for the variety and inconsequence of their gauges. It is often tempting to suppose that rails were laid to fit a waggon built at random, rather than that waggons were built to fit a deliberately chosen gauge. There is no real mystery, however, about the basic reason why the Tyneside waggonway gauges fall into the range of from 4 to 5 ft. Charles E. Lee has pointed out that a figure of this sort is the optimum axle-length for a one-horse vehicle; if the axle is much shorter, the vehicle is small and the horse does not pull an economic load; if it is much longer, the cart becomes too heavy for one horse.[73] Times change, but the horse does not. This seems to be the simple reason why Greek and Roman and even pre-classical vehicles, as evidenced by the ruts they left behind, were of much the same gauge as the Tyneside waggons of the seventeenth and eighteenth centuries and British Railways of today.

For this reason there is a mite of truth in the oft-repeated explanation of our standard gauge of 4 ft. 8½ in., that George Stephenson copied it from Roman chariot ruts found on Hadrian's Wall: both were following the same natural law. But to assert that there is any direct connection between the two is nonsense. Stephenson did not build Killingworth waggonway; indeed, when he first went there at the age of twenty-three the 4-ft. 8-in. gauge was already at least forty years old. When he built his first locomotives for Killingworth, they were of necessity of 4-ft. 8-in. gauge, and when, with reputation established, he went on to build new railways elsewhere, he not unnaturally gave them the gauge he was accustomed to. The Hetton and Springwell colliery railways of 1822 and 1826, the Stockton and Darlington of 1825 and the Liverpool and Manchester of 1830, all were built to 4-ft. 8-in. gauge. The extra ½-in. which makes up our present standard gauge seems to have crept in during the 1830s, through the desire for greater side-play for the flanges. If Stephenson had found his mission in life at Heaton, our standard gauge would probably be 4 ft. 3 in.; if at Wylam, about 5 ft. It was largely a question of chance.

The Newcastle waggonway: waggons

My name is William Robinson,
  That same I do advance.
The first little galloway I ever drove,
  They called him Little Chance.
Chancey had two greasy legs,
  Likewise a kettley back,
And going down the waggon way,
  He made the waggons knack.

'Little Chance', a Tow Law song published by A. L. Lloyd,
Come All ye Bold Miners, London (1952), 21.

If the complexity of Continental measures is enough to make the historian weep, the system of weights used on Tyneside is guaranteed to make him tear his hair. The ton and the hundredweight were virtually unknown there, except to writers who were trying to explain the local system to outsiders. For the transport and sale of coal in bulk, four main measures were in use: the *boll*, the *waggon*, the *chaldron* and the *ten*. The confusion arises largely from the fact that there was little standardization. The chaldron was fairly early established as a weight, while the rest remained basically dry measures. Their content varied widely, and so did their relationship one with another. All four tended to grow in size as time went on, but not at a regular rate over the whole area. Thus in 1713 the Stella Grand Lease collieries counted 17 bolls to the waggon and 30 waggons to the ten; on the Bucksnook waggonway nearby, in the same year, there were 19 bolls to the waggon, but only 21 waggons to the ten, which was some 22% smaller than Stella's, always assuming their bolls were identical.[1] The subject is so complex that it is not intended to discuss it in detail. What concerns us primarily is the capacity of the waggon.

The *boll*, or *bowl*, was originally the tub or barrow used for loading coal into keels. It was legally standardized in 1678 at 22 galls. and 1 pottle, or 11¼ pecks, but a number of eighteenth-century colliery leases defined it as 35½ or 36 galls.[2] Such evidence as there is, however, is quite clear and consistent in showing that in practice the boll was always smaller than in theory – about 16 galls. or 8 pecks. A gallon of coal as a measure was evidently deemed to weigh up to 15 lb. on Tyneside, although it was lighter in other areas. As a general rule, based on the fact that a 19-boll waggon held 42 cwt., and a 24-boll waggon 53 cwt., it is safe to assume that a boll of coal weighed about 2·2 cwt. during our period. As has been seen, the wain-load, or fother, usually contained 7 or 8 bolls. There is no indication of the size of Beaumont's waggons, but after the Restoration the standard seems to have been set at 15 bolls, or 33 cwt. From about 1700 the 19- and 20-boll waggon (42- and 44-cwt.) gained ground, and remained the norm, until the 24-boll (53-cwt.) waggon, which first appears in the 1750s, became established as the standard late in the century.

This is, however, only the broad outline of the picture. Waggons holding 18 and 17 bolls were favoured at various times and on various waggonways, and even 10-boll waggons were envisaged, though not apparently used (see Table 4). To a certain extent the waggon size depended on the nature of the waggonway: if it included steep gradients, then the capacity would be kept low. But the more coal that could be carried in one trip, the cheaper; and the upward trend in capacity reflects the improvement in engineering of the lines and perhaps in construction of waggons. An interesting case of stepping up the capacity occurs on the Dunston way from Northbanks and Gibside. The line was opened in 1699 with 15-boll waggons; the next year they were enlarged to hold 18 bolls, and in 1702 to hold 19 bolls, where they remained until the way was closed in 1725.[3] The usual mode of increasing the capacity was to add extra boards or ledges to the top. When the Grand Allies leased the Stella Grand Lease waggonway in 1728, they minuted: 'The Waggons to be Enlarged, to the Size allowed by the Lessors'. Nothing was done immediately, for in February 1729/30 they

> Order'd that Ledges be imediately put upon the Waggons at Stella to make 'em carry 20 boules. That the Steathmen do every one build a 19 boule Measure at some convenient part of the Steath & Every Waggon not Carrying the quantity, no payment to be made for that Gate [trip].[4]

Again, the Walker Colliery wayleave of 1751 specified 20-boll waggons, but two years later the lessees petitioned the landowners to be allowed to use 24-boll ones.[5] Waggons did not always, however, grow larger. The Tanfield ones were measured at 20 bolls in 1735, but in 1737 Bowes was pressing the Allies to agree 'That the Partnership waggons be measur'd and that they carry full 19 Bolls and no more'.[6] Glaring inconsistencies appear elsewhere, too. When wayleaves were obtained for the Willington waggonway in 1773, it was agreed with one landowner to pay rent on the basis of a 20-boll waggon; with another, on the basis of a 19-boll waggon, even though waggons with a capacity of 21 to 26 bolls were permitted by the same document.[7]

The precise capacity of a waggon was a frequent source of dispute between the coal-master and his customers. Complaints had been made in 1722 that John Wilkinson used a very odd waggon-load on his Gateshead Fell way, and though he had difficulty in selling his coals as a result, he refused to change it. There was still confusion in 1734 when four viewers were sent to measure the loads:

> The waggons not being on this day wee went to ffellon [Felling] Staiths & the Empty waggons being standing there wee got a waggon Driver wch belong'd to Nicho Gally Staithman at ffellon and desired him to fill an empty Waggon with coal sclates (there being no Coales at the Staiths) in such manner as he & the other waggon men usually did at the Pitts, wch he did, and wch we afterwards measured and it Contain'd 17 Bowles.

A few months later they went again, waited till two waggons were filled at the pit, 'and after they were come from the s^{d} pitt upon the Waggon way at some distance, we mett them & told them our Errand'. One, they found, contained 17 bolls even though 'upheaped', the other 18 bolls. When the dispute finally came to court in January 1736/7, William Row, an aged waggon wright, deposed that the waggons he built to specified dimensions were

> said to carry 15 bowls and made for that purpose & allowed & admitted both by Landlord and Ten^{t} to be a fair & just measure & were never objected to till by Mr Cotesworth: has orders from Owners to keep waggons to the same Gage ever since 1694; loose ledges of no Service to the loading of the Waggons at the pitts being only put up after the Waggon is loaded to save the Coles on runs: Mr Wilkinsons constant order not to fill with the ledges up & when done otherwise the Overman pull'd them down.

In 1722 the waggons had been unfairly measured for the Dean and Chapter, the landlords, by treading the coal down; a second measuring in 1725 showed that only one waggon held 15 bolls. Everything depended on how they were filled, and 'when wagons grow old they will hold more'. A driver of long standing said that he had carried nine or even ten 13-peck corves of coal in a waggon, but never ten 15-peck ones. Various measurements in 1735-6 produced various answers, ranging from $16\frac{3}{4}$ to 19 bolls, but averaging out at $17\frac{1}{2}$ bolls.[8]

As a result of squabbles of this sort, an Act of Parliament was passed in 1738. A previous Act (1695) had laid down that keels, wains and carts be measured and marked, but since then 'divers Persons concerned in the Coal-Trade have used and employed Waggons, Barrows and other Carriages, not mentioned, specified or comprized with the Directions of the said Act,

Table 4a

Capacities of Waggons and of Tens, Tyneside

Date	Bolls to waggon	Equivalent in cwt.	Waggons to ten	Bolls to ten	Waggonway
1663	15	33	20	300	Crawcrook Stella
1676-83	—	—	20	—	Stella Grand Lease
1692	15	33	22	330	Dunston proposal
	or 10	22	33	330	" "
1694	15	33	—	—	Felling
1698	15	33	21	315	Elswick
1699	15	33	—	—	Dunston
1700	18	40	—	—	"
1702-23	19	42	22	418	"
1703	15	33	25	375	Winlaton?
c1710	15	33	20	300	Winlaton
c1710	20	44	20	400	"
1713?	19	42	21	399	Bucksnook
1713	17	37½	30	510	Stella Grand Lease
1713	17	37½	30	510	Bensham
1714-16	—	—	22	—	Bucksnook
1717	—	—	21	—	Kenton
1723	19	42	22	418	Benwell
1724	20	44	22	440	Heaton
1726-9	—	—	22	—	Ravensworth
1728	—	—	22	—	Western Way
1730	20	44	—	—	Stella Grand Lease
1734-7	16¾-19	37-42	—	—	Gateshead
1735	20	44	22	440	Tanfield
1735-6	19-20	42-44	22	418-40	Long Benton
1737	19	42	24	456	Crawcrook Stella
1737	19	42	—	—	Tanfield
1743	19	42	22	418	Byker
1751-3	20-4	44-53	22	440-528	Walker
1754	19-24	42-53	—	—	(Tyneside)
1755	20	44	22	440	Brunton
1773	19-20	42-44	22	418-40	Willington
1789	24	53	20	480	Washington
1804	24	53	—	—	Wylam
1809	20	44	22	440	Kenton

Table 4b

Capacities of Waggons, Yorkshire

Date	Capacity in cwt.	Waggonway
1748	38	Wakefield Outwood?
1773	44	Sheffield*
1774	46	Middleton
1774	38	Swillington
1774	37	Glasshouse
1796	48	Milnes Flockton
1796	44	Stanley Main
1796	44	Swillington
1796	40	Methley
1796	40	Allerton
1808	45	Middleton

* 20 bolls to waggon, 22 waggons to ten.

whereby divers Frauds, Deceits, and Abuses have arisen'. After 1 June 1738, it was ordained, all waggons and barrows were to be measured, weighed and marked with their capacity.[9] As a result, entries appear in waggonway accounts, such as these from the Tanfield way in 1767: 'Expence in Measuring the Waggons, 3/4', and 'Expence in Bolling the Waggons, 4/-'.[10]

The *chaldron* was basically a measure used in the shipping of coal, and appeared rarely in connection with waggonways. The Newcastle chaldron, not to be confused with the London and other chaldrons, which were considerably smaller, gradually increased in size, and for a very long time was larger than the largest waggon. In 1616 it weighed about 43 cwt.; in 1678 it was fixed by statute at $52\frac{1}{2}$ cwt., and in 1695 at 53 cwt., where it remained.[11] It was thus only in the second half of the eighteenth century that the waggon caught it up, which no doubt eased the fitters' and merchants' arithmetical problems. It was then that the expression 'chaldron waggon' first appeared, later abbreviated so that 'chaldron' became a synonym for 'waggon'. The keel, the river lighter, usually held 8 chaldrons.

The unit used to measure the output of a colliery, or the quantity of coals delivered by waggonway to the staith, was the *ten*. This began life as 10 chaldrons – it was still so at Ravensworth when the waggonway was built in 1669[12] – but soon became considerably bigger. It could contain from 20 to 30 waggons, but since the waggon varied so much, the ten varied even more (see Table 4), from 300 to 528 bolls. However, the ten most often consisted of 20 or 22 waggons, and so grew from about 300 bolls (33 tons) with 15-boll waggons to 480 bolls (53 tons) with 24-boll waggons. The ten was just settling down at this figure in the nineteenth century when it was superseded by the ton. The divergence in the contents of the ten from colliery to colliery must have caused intense difficulty in the trade; and even more so since at least one colliery used different tens simultaneously. At Winlaton in about 1710 a ten was counted at 20 waggons, whether the waggons held 15 or 20 bolls, and so contained either 300 or 400 bolls.[13]

Away from Tyneside, weights and measures were often different again. Yorkshire used the waggon or the chaldron, which were mercifully the same, ranging from 37 to 48 cwt. For smaller units, it either adopted the Tyneside boll, or else reckoned in *corves* (about 2 cwt.) or in *bushels* (from $\frac{1}{2}$ to $\frac{3}{4}$ cwt.).[14]

Whitehaven waggons were similar, adhering to Newcastle measures: they increased from 18 bolls, or 40 cwt., in 1754 to 44 cwt. in 1794 and 45 cwt. in 1803.[15] The waggonways immediately north of Whitehaven, however, clung to the obscure *Cumberland ton*. It was defined about 1700 as $15\frac{1}{4}$ ordinary cwt., consisting of 192 galls., each of which weighed 8·9 lb., much less than on Tyneside. Thus the regular waggon load at Workington in the 1730s and 1740s was three Cumberland tons, or $45\frac{3}{4}$ ordinary cwt., slightly more than at Whitehaven, but roughly equal to contemporary Tyneside loads. It may have grown over the years, since Hutchinson in 1794 describes it as 3 English tons, but he may have specified the English ton, wrongly equating it with the Cumberland one, to avoid confusion with the rather larger Dublin ton, which also appears in Cumberland. Here too there were difficulties about the method of loading. In 1738 John Spedding realized that heaping the waggons increased the capacity by one-eighth, which meant that one waggon load in every nine was being given away free; so he gave orders 'to alter the Wagns, so as to hold 3 Ton Strik'd [level] measure, there having been odd practices Carried on by the uncertain way of Heaping them'.[16] At Maryport the waggons were originally much smaller – only 2 Cumberland tons, or $30\frac{1}{2}$ cwt. – the ton being defined again (in the colliery lease of 1755) as 192 galls., i.e. 12 corves of 16 galls. each. But by 1773 the corf had grown, it was discovered to the coalowners' horror and amazement, to 25·8 galls., increasing the ton to nearly 25 cwt. Although the true Cumberland ton was restored, the waggons had evidently grown with the corves, and hereafter (it is recorded in 1778 and 1792) were built to the same capacity as those at Workington – 3 Cumberland tons, or $45\frac{3}{4}$ cwt.[17]

Scottish waggons were often larger still than those on Tyneside. At Halbeath, the capacity was from 41 to 42 cwt. in 1795, but on Lord Elgin's line it grew from 50 cwt. in 1784 to 60 in 1796, and likewise at Fordell from

48 cwt. in 1795 to 60 cwt. early in the nineteenth century. So, too, the waggons at Ayr held 60 cwt.[18] Robert Stevenson remarked that 3 tons was too much; it shook the track and necessitated heavy rails. These tons and hundredweight were apparently of normal size.

Tyneside waggons were generally very similar, but there were two

FIG 37 Waggon at Plessey, 1724 (Clerk).

varieties of design. In one, probably the rarer, the body was hopper-shaped, with both ends sloping upwards and outwards symmetrically. In the other, the rear or uphill end was vertical, and only the front end sloped outwards. The earliest representation, a quick ink sketch of the symmetrical type, occurs in the journal of Sir John Clerk in 1724 (fig. 37). A delightful map of the Long Benton pits and the extension of the Heaton waggonway, drawn in 1749, includes two non-symmetrical waggons, both being pulled

FIG 38 Waggons at Long Benton, 1749.

by horses.[19] One is going down full and the other is returning empty (fig. 38). The first detailed drawing and description were published by the *London Magazine* in 1764, under the *nom de plume* of T. S. Polyhistor of Chester-le-Street. Neither illustration nor account is very satisfactory. But later the same year the *General Magazine* printed a much superior description and drawing (pl. 48) from John Buddle, Sen., the father of the great coal-viewer.[20]

Description. A represents the body of the waggon, loaded with coals,

B, drawn by a single horse C by means of the traces *a*, *a*, and conducted or drove by a single man, called the *Waggon-man*, whose most common action on the road is, enticing his horse forward with a bit of hay in his hand, which he supplies from under his arm *b*. The quantity of hay sufficient for a day being kept in the Hay-poke D. The body of the waggon is in form of an inverted prismoid, composed of pretty strong fir deals, having strong pieces of oak or ash wood at the bottom called Soals, at the corners and sides called Sheths, and at the top, called Overings, which are occasionally raised with Ledges to make the waggon hold any additional quantity.

The top of the fore part of the waggon projects farther out than the bottom; therefore the greatest part of the loading is supported on the fore-wheels, and consequently the waggon, when in motion, is drawn with less force than if the fore part did not project at all: The draught is also lessened by the fore wheels being considerably larger than the hind. . . . E is the convoy, being a strong curving piece of Alder-wood, whose end (*c*) is kept indifferently tight in the Ring-iron (*d*) by little wedges called Scotches, deposited in a box (*e*) called the Scotch-box, and kept from touching the wheel by hanging its end (*f*) in a loop of leather (*g*), fastened to the hind Corner-Sheth of the waggon; Its use is to regulate the motion of the waggon down the sides of the hills (called by the Waggon-men, Runs) making it uniform, which, by the laws of motion, would be accelerated was no such regulation to take place; the Waggon-man, taking the end (*f*) out of the loop, lets it down upon the wheel, and, placing himself astride upon the end (*f*), with one foot on the waggon-soal, he presses more, or less, according to the nature of the declivity of the Run; the Convoy acting at that time as a Leaver of the 2d kind, whose fulcrum is in the Ring-iron (*d*), and the friction of the Breasts (*h*, *h*), (either of which may be used at pleasure) regulates the motion; the large iron (*k*) is called the Hind-iron, and hinders the convoy from slipping off the side of the wheel. . . . When the waggon arrives at the staiths, he lets down his Bottom-board, which is the bottom of the waggon, having Hinges on one side, and a Hasp on the other; and the coals run down an opening in the waggon-way under the waggon. . . . As the waggon is going to the pit, the convoy is prevented from jumping out by a pin put through it, between the Ring-iron, and its end, called the Loiter-pin.

A number of drawings of waggons, of varying quality, appeared during the next twenty years or so. That of Jars was copied from the *London*

Magazine, and is much inferior to the *General Magazine*'s version. Three engravings depict waggons on gravity runs (pls. 40, 49 and 50). That by William Beilby for Morand's *L'Art d'exploiter les mines* (pl. 49) is astray in that it omits the wheel flanges, although Morand later acknowledged the error.[21] Several descriptions were also published, none of which adds much to the admirable account quoted from the *General Magazine*.

Practically all waggons on Tyneside and in its colonies were designed, of course, for carrying coal. In the coalfields, the only exceptions were the ballast waggons, of which most waggonways had a small number, used in connection with track maintenance. They were evidently similar to coal waggons: in 1729 the Ravensworth wright was 'altering 3 Ballass Waggons into Coale Waggons at 4/- p.'[22] There were besides a few lines for the transport of stone in one form or another: at Limekilns on the Forth for broken stone, and at Ballycastle for building stone. What their waggons looked like we do not know. The Clavering Stella lead traffic was probably accommodated in ordinary coal-waggons (p. 119).

The earliest waggon wheels on Tyneside were of solid wood. Occasionally a wheel might be made from only two pieces of timber, one disc for the main part of the wheel and another, of larger diameter but narrower, nailed on to form the flange. The normal practice, however, was to build wheels up from several small pieces. Andersson wrote in 1776:

> The wheels consist of thick beech planks, 6 inches wide on the rim, larger in diameter on the inside than on the outside, built up of two, three or four close-fitting pieces, joined with pegs and S-shaped clamps of iron driven in across the joints.[23]

Such an arrangement can be seen in pl. 48. Elm, ash, oak, birch and fir were also used. The supply of ready-made wheels was very similar to that of rails and sleepers (see pp. 170-5) – either from local carpenters or from more distant sources – and the cost varied very considerably with size and with the fluctuations of the market. On average, it increased from 5*s*. a wheel in 1679 to as much as 14*s*. in 1765,[24] although lower prices are very occasionally found for one reason or another – 3*s*. 6*d*. at Sheffield in 1731, 4*s*. 4*d*. at Ravensworth in 1729, 4*s*. 8*d*. at Leeds in 1762.[25]

The first categorical evidence for flanged wheels is the drawings of the Bath waggon of 1734 (pl. 62), but there is no doubt whatever that the flange dates back to the Restoration, if not to the very beginnings of the English waggonway. Lord Guilford said in 1676:

> The Manner of the Carriage is by laying Rails of Timber, from the Colliery, down to the River [Tyne], exactly straight and parallel; and bulky Carts are made with four Rowlets fitting these Rails; whereby

> the Carriage is so easy that one Horse will draw down four or five Chaldrons of Coals, and is an immense benefit to the Coal Merchants [an exaggeration, even if he meant London chaldrons].[26]

These rowlets can hardly mean anything but flanged wheels. Lists of waggon parts bought and in stock, which survive back to about the same date, show no material difference over a period of 150 years. If guide-wheels, for example, and not flanged wheels, had been in use in the 1670s, one would have expected the stock lists to have distinguished between them and the running wheels. The flange was often called the 'creese' or 'crease', a word perhaps derived from 'crevice'. 'Creesers', the workmen employed in clearing out dirt from the flangeways inside the rails, are found regularly from 1715;[27] crease plates or nook plates were iron strips nailed to the vertical part of the flange to reduce wear. Several entries in the Ravensworth waggonway accounts of 1726–9 refer to these flange-plates: 'Thom. Burrow for Crease Plates for ye Ballass Wag^n 2^d', and 'Thom Burrows mending ye Nook plates on Ballass Waggons 2^d'.[28] It has been inferred from a few descriptions of wheels as 'hollow', 'grooved' or 'pulley' that double flanges were sometimes used.[29] In fact, all the authors in question were non-technical ones who perhaps did not fully understand the principle; and there is no good evidence for double-flanged wheels until iron rails were well established.

The flange was usually $1\frac{1}{2}$ or 2 in. deep, and the overall width of the wheel was between 6 and $8\frac{1}{4}$ in., the tread having about the same width as the rail. In diameter, wooden wheels varied from 27 to 46 in., with an average of roughly 3 ft. It seems that they were hand-cut, not turned on a lathe. Some arrived at the Bucksnook waggonway in 1722 and necessitated payment of £1 'to rounding 120 wheels as above @ 2d p', and Nicholas Wood remarks:

> From the very probable inaccuracy of the workmanship, it is not likely the periphery of the wheels would be perfectly circular, which would cause considerable jolting, or an undulatory motion, to the load, and thus increase the draught.[30]

Ravensworth was supplied in 1727 with a number of wheels that were too small, for a wright spent eight days 'parteing Waggon Wheels' – presumably separating the built-up pieces – and then comes the entry, 'to Jos Greenwell & Pn^{rs} for Enlargeing 54 Wheels from 27 & 28 Inches to 38 & 40 Inches at 2^s p. Wheel'.[31] Throughout our period, the front (downhill) wheels of a Tyneside waggon were invariably of greater diameter than the rear ones, the difference ranging from 1 to 6 in. The Ravensworth wheels just mentioned are a case in point, while at Hartley in 1758 we find diameters of 38 and 32 in., and at Sheffield in 1773 of 44 and 40 in.[32] The reasons given, both of them unsatisfactory, were that unequal wheels kept the top of a waggon level, and so prevented spillage, when it was free-wheeling (and no doubt lurching) down a steep gradient; and that larger front wheels reduced the effort needed to keep a waggon in motion. The wooden wheels at Leeds in 1762 weighed nearly 1½ cwt. each.[33]

If wooden wheels were allowed to become too dry, they were liable to crack and warp. No doubt they were usually doused with buckets of water to keep them damp, but some large waggonways had special ponds in which waggons were parked overnight, half-submerged, in the summer. A tradition that there was such a pool at Lintz on the upper reaches of the Western Way is confirmed by the mention in a valuation of 1789 of '78 yards of siding to Pond'. The 'Wash-holes' at Dunston on the Tanfield line and at Throckley likewise had special sidings running through them.[34]

The wear on wooden wheels, both from the rails and from the pressure of the convoy, was considerable. From the few surviving indications, it seems that a year was a long life for a wheel. It was inevitable that means should be sought for reducing this expense, and the solution was identical to that found for the wooden rail. First the tread of the wheel was covered with thin wrought-iron plate, and then the whole wheel was made of cast iron. On 20 May 1731 a patent was granted to Elias Thornhill of Sunderland, who

> having observed the great distruction of the best timber trees in severall parts of our dominions, and finding the same mostly to proceed from the cutting and vsing the same in large quantities for coal waggon wheels, which are made out of solid timber trees, and are about a yard and ten inches diameter and seven or eight inches thick, which waggon wheels seldom last above a year or two . . . hath . . . invented . . . 'A New Way or Method of Making the Rim or Edge of the said Wheels with Iron or Steel, and with Iron Ribbs or Tabbs, and Iron Bolts, Rivets and Screws for the Fastening the same, which . . . will preserve the Wood of the said Wheels and the Rails which they go upon from Wearing, and will last many Years longer than those now used'.[35]

This clearly means iron plates on the wheel tread, held on, perhaps, by nails or screws through projecting lugs at the side. John Loveday in 1732 found an application of Thornhill's patent on Allan's or Hedworth's waggonway: 'We saw some Coal-Waggons to day, whose Wheels were straked with Iron, a new Invention, & will save much Wood.'[36] It was not, however, so new a device as Thornhill and Loveday imply, for just the same thing was fitted to Ravensworth waggons in 1726 and 1727, under the name of Trod plates, the counterpart of crease plates:

> Thos Borrow for 2 Trodplates & Nailes 6d.
> Tho Borrow for a Trod plate & Nailes for a ballas waggon as p Note.
> Thos Borrah for 2 New Hoops wt 2 pound at 3d p 6d.

These plates remained long in use. Even the waggons on the Halbeath waggonway in Scotland, it was proposed in 1780, should have 'Iron shod fir Wheels'.[37] Matthew Dunn records an alternative method of armouring the wheels:

> In order to prevent the wear of the wheels (which were extremely expensive to maintain) they were studded thick with nails driven up to the heads; but the wear was proportionately great upon the breasts of the convoys, which was a source of great labour and expence.[38]

The cast-iron wheel first appears in this country in Shropshire and underground, but it was quickly adopted for waggons in the colonies of Tyneside, though not for a while on Tyneside itself. Ralph Allen's waggons at Bath, built in 1731 under Shropshire influence, boasted cast-iron wheels. The very next year the Sheffield colliery accounts record, 'By Charges going about Sough Metall [sow metal=cast iron] Wheels, £1', and in 1733 'Pd Jonathan Ragg for fetching the Saugh Metall Wheels, 5/-'. These entries probably refer to waggon wheels, since they occur alongside others relating to wooden ones, but they may have been only a brief experiment.[39]

Whitehaven started off in 1735 with cast-iron wheels (fig. 39) of a surprisingly modern appearance when contrasted with the clumsy examples at Bath (pl. 62). Their diameter was 22 in., their weight about 1 cwt. 2 qr. 25 lb., their treads were slightly coned, and all four wheels on a waggon were the same size. In time the weight increased a little – by the 1770s the average was 1 cwt. 3 qr. 21 lb. apiece, at 12*s*. a cwt. Sir John Clerk wrote in 1739 that 'these waggons run upon 4 little cast iron wheals', and all subsequent evidence confirms that Whitehaven wheels were always of iron. The early use of this material may be due to influence from Shropshire, which had been using iron wheels since 1729. There were strong links between the two areas in connection with steam engines. In the end, a foundry was built at Saltom about 1786 to supply waggon wheels for the Cumberland coalfield. At Seaton Colliery near Workington, however, the later Tyneside practice of using one pair of wood and one of iron wheels was already to be found in 1741.[40] The Middleton waggonway at Leeds began to convert its wheels to cast iron in 1767.[41]

Around Newcastle the wooden wheel lingered on for many years:

Bailey's engraving of the Parkmoor waggonway in 1783 shows a waggon with all four wheels still made of wood (pl. 50). Waggon wrights were at first reluctant to adopt iron wheels, partly because they were brittle and partly because they tended to cut the rails. Tyneside, moreover, was poorly

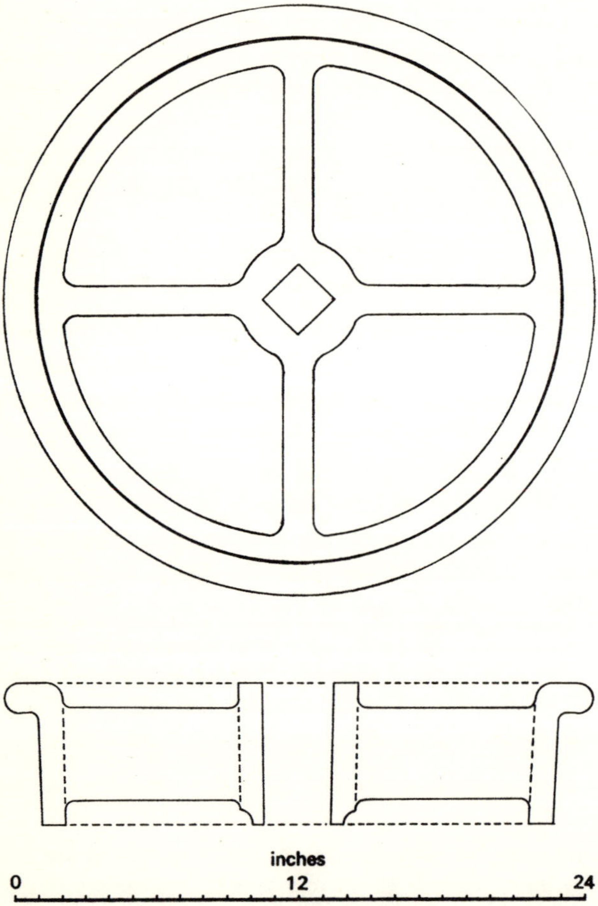

FIG 39 Cast-iron wheel, Saltom waggonway, Whitehaven, 1735.

off for foundries. From the 1750s, however, it gradually became normal practice to fit iron front wheels and wooden rear ones, as it was considered that the convoy could more easily stop a wooden wheel. William Brown wrote to Spedding at Whitehaven in 1754:

> The Hard thing upon you will be your mettle wheels some of which indead is in our Country but these is used as fore wheels for we

> allways have wood for the Hind wheels which with a Convoy well Breasted will totally Stop or Trail the Hind Wheels.[42]

This arrangement remained the norm for thirty years, until in the 1780s the wooden wheel was gradually ousted for good. In the drawings of waggons at Long Benton in 1749 (fig. 38), all the wheels appear to have four spokes, which suggests iron; but they are too small and sketchy to place much reliance on. The first time we can be certain that all four wheels were of iron is in 1772, when the Bedlington Foundry began to supply Hartley Colliery with two sizes, 38 and 42 in.[43] Front and rear wheels were still made unequal, and generally remained so until the locomotive replaced the horse (pl. 51). The size of iron wheels varied between limits similar to those of wooden ones, and their weight and price naturally varied too. At Hartley in 1772 a 42-in. wheel weighed 3 cwt. and cost £2 2*s*. 0*d*. (14*s*. a cwt.); at Wallsend in 1794 a 32-in. wheel weighed 1 cwt. 3 qr. 9 lb. and cost £1 9*s*. 4*d*. Thereafter, in contrast to the ever-rising cost of wood, cast iron stayed at this level of 16*s*. a cwt., or even less, throughout the Napoleonic Wars, a fact of prime importance in the rise of the iron rail. Wheels were generally cast locally, even on Tyneside. Whitehaven had its own foundry from 1786; the Middleton accounts show their wheels were supplied by a Hunslet foundry; and the inclusion of 'Wheels for Newcastle waggons' in a price-list dated 1782 of the Birkenshaw Foundry near Bradford suggests that this firm provided for the needs of other West Yorkshire waggonways.[44]

A typical Tyneside cast-iron wheel – one is illustrated by Jars (pl. 47) – generally had eight spokes (sometimes ten) and was narrower on the tread than a wooden one. Towards the end of the wooden rail era, John Buddle recorded a few dimensions:[45]

	Percy M[ain], in.	Willing^{n}, in.	Walls-end, in.
Breadth of Trod . .	$3\frac{7}{8}$	$3\frac{1}{4}$	$4\frac{1}{8}$
Ditto – Margin [flange] .	$\frac{3}{4}$	$1\frac{1}{4}$	$\frac{7}{8}$
Height of Ditto . .	$\frac{7}{8}$	$\frac{3}{4}$	$\frac{3}{4}$
Diameter . . .		to 34 inches	

In the course of time, it seems, the flange had become shallower. The Whitehaven drawing of 1735 and Jars in 1765 give the depth as $1\frac{1}{2}$ in. The play allowed between wheel-gauge and track-gauge was only minimal.

Axles were of wrought iron and always revolved with the wheels. They were usually square in section throughout, except at the bearings – the Sheffield accounts record in 1787 'to 32 Axeltres turning £1/12'. A square hole cut or cast in the wheel centre fitted over the axle end, where it was held by wooden keys or wedges. At Bucksnook in 1723, 3*s*. $4\frac{3}{4}$*d*. was paid for 'a Wedgeing Maull for the Waggons'.[46] When Hartley waggonway first adopted iron wheels, Sir John Delaval received this advice from the manager of the foundry:

> Bedlington Furnaces 3 Oct 1772
>
> Sir
>
> Inclosed you have an Acco^{t} of six Waggon Wheels Cast from the New Pattern made to your Direction. I hope they will be found agreeable to your Intention . . . I cannot pretend to give any directions for the Bushing of these Wheels only would beg leave to recommend moderation in driving your Wood especially if its very dry and of all things to avoid striking the Metal in any Part when tight drove, but if any of these should [break through] fair Play I shall without hesitation replace them. We hope in less than a Fortnight time to sending you some more and shall I doubt not be able to supply you with all you may want
>
> I am
> S^{r} Y^{r} most hble serv
> Jno Charlton.[47]

Linchpins were sometimes also inserted in the axle ends, so that if the keys worked loose the wheel would not fall off.

Axles were bought in 1723 for the new Burdonmoor way at 16*s*. 6*d*. apiece. Those on the Middleton waggonway in 1762, made at the Kirkstall Forge at 23*s*. 4*d*. a cwt., weighed 134 lb. and cost 28*s*. each; three years later on the Killingworth (Long Benton) way, where Lord Ravensworth's smith made them at 32*s*. a cwt., their weight was 88 lb. and their price 25*s*. 2*d*.[48] The axle bearings were simple castings attached to the main frame. In the Sheffield accounts for 1733 we find 'P^{d} for Soughmetall for the Axletrees to run in 4/-', and in 1729 'P^{d} for grease for the Waggins 1/4' (though Whitehaven in 1775 ran to best butter for lubrication at 4*d*. a pound).[49] The axles were held in the bearings by hinged straps, bolted down at the free end, so that when repairs were needed the bolts could be undone and the body lifted off the axles.[50] Brass bearings did not come in until after 1800.

The construction of the body, or coop, was simple, a mere development of cart-building, from which most of the technical terms came too. The description in the *General Magazine*, already quoted, gives the main outline, which was identical from 1679, the date of the earliest surviving stock-list of waggon parts (at Stella) down to and beyond the end of our period. The main frame consisted of two soles, hefty beams running fore and aft, about 4 in. wide and 8 ft. long, though in later days they extended further out at each end to form dumb buffers. They cost 1*s.* 4*d.* each early in the eighteenth century, and 3*s.* 6*d.* at the end of it. They were connected by the bottom sheths or sheaths, two smaller transverse beams (6*d.* to 2*s.* each). Into this frame were mortised the ribs or side sheths, one at each corner and two on each side and each end (2*s.* 6*d.* to 7*s.* per dozen), which were held together at the top by the two overings on the sides (8*d.* to 1*s.* 4*d.* each) and the howtrees at the ends (4*d.* to 10*d.*). At Whitehaven overings and howtrees were called 'overends' and 'ends'. Inside these timbers were nailed the deals, or cleadings, the horizontal planks which actually contained the coal. Extra planks or ledges, a few inches high, might be added to the top of the front and sides to increase the capacity. The base of the waggon was largely, and often entirely, occupied by the bottom board, attached on one side by two hinges and on the other by a hasp, which was released at the staith to empty the coal. The axles were set under, or nearly under, the corner sheths, to allow the bottom board to be as large as possible. Even so, the wheelbase was rarely more than 5 ft., and was often less than the track-gauge. In the earliest Tyneside waggons, it is said, there was no bottom board, and the coal had to be shovelled out from the top, a method preserved in later ballast waggons.[51] As far as is known, most Tyneside colonies followed their parent's lead, although the Fordell waggons, in later days at least, had top-hung doors at one end and were emptied by tipping forwards. This device may, however, have well been copied from Shropshire practice, which had some influence in Scotland.

The operation of the ordinary brake is obvious from the various drawings and from the *General Magazine* description. The convoy (sometimes called the 'tiller' or 'rudder' by non-technical writers) was a bar of wood, usually curved and between 5 ft. 6 in. and 6 ft. 6 in. long. It was wedged at the bottom end into a pivoting ring, the ring iron, which was attached to the sole, spare wedges or scotches being kept in a small box on the centre panel of the waggon-side. The convoy was pierced at its very lowest point by a loiter-pin, which prevented it from shaking free of the scotches and rising up and out of the ring iron. The lever was held off by a leather or rope loop, or an iron hook, supporting its upper end. When this was taken off, the convoy was free to move in the vertical plane only, for it was restricted on one side by the corner sheth and on the other by a bar, the hind iron or long iron, attached to the waggon-side. Nailed to the underside of the convoy at each end was a breast, or brake-block, usually made of beech; when one was worn out the convoy was turned round and the other breast applied. Breastwood was bought by the yard at about $4\frac{1}{2}$*d.*, trimmed to shape by the waggon wrights, and fitted to the convoys at a rate of 2*s.* 8*d.* per dozen. The convoys themselves cost a mere 1*s.* a dozen in 1679,[52] but during the first quarter of the next century had risen to 6*s.* to 9*s.* a dozen. To judge by the numbers purchased, their life was very short. Most waggonways were content with this simple, single, brake operating on one wheel, and applied by the waggon-man sitting on the lever on long gradients.

When heavy waggons ran down steep slopes, however, the simple convoy was not enough. One solution, first mentioned by Jars in 1765, was to duplicate the convoy on the other side, and link the two upper ends

FIG 40 Whitehaven (Howgill) waggon, late eighteenth century.

by a horizontal bar.[53] But all four wheels could be braked by a more elaborate method. William Brown told Spedding in 1754:

> We have lately contrived what we call a long Convoy which goes to the fore wheels & by the Help of a little Contrivance at the Hind End of the Wagn. the Same man Manages both so That now can trail all the 4 wheels.[54]

Here the two levers were pivoted right at the front of the waggon, with breasts bearing on both wheels on each side, and linked by a bar across the back. An extra advantage was derived from the greater leverage made possible by the longer convoys. Several variations on this theme were developed, notably by Carlisle Spedding, who was bedevilled by breakneck gradients at Whitehaven. He wrote to Brown in 1754:

> By our long Convoys find we can Govern the waggon pretty well in Dry weather. I hope may do it when the Railes are wett after the Way is worn and rough with ashes &c. I fear them most in Raggs or frost. I have fix^d a chain to lock the Hind Wheels as they do the Spake Wheel Carriage Wag^y [?] but fear the Carelessness of the waggoners may occasion many accidents by draw^g the waggons Too near the Runn head before they lock them.[55]

This pair of long convoys bearing only on the front wheels remained in use at Whitehaven until the early 1800s (fig. 40). Another variation is described by John Farey, writing in 1805:

> Some waggons which we have seen, had the convoy fixed to a moveable joint at the front of the waggon, and had a large block of wood thereon, which, when the convoy was let down, wedged in between the fore and hind wheels, and acted most strenuously as a brake.[56]

Another very widespread method was apparently invented at Shieldrow on the Tanfield waggonway in 1790. It consisted basically of a simple

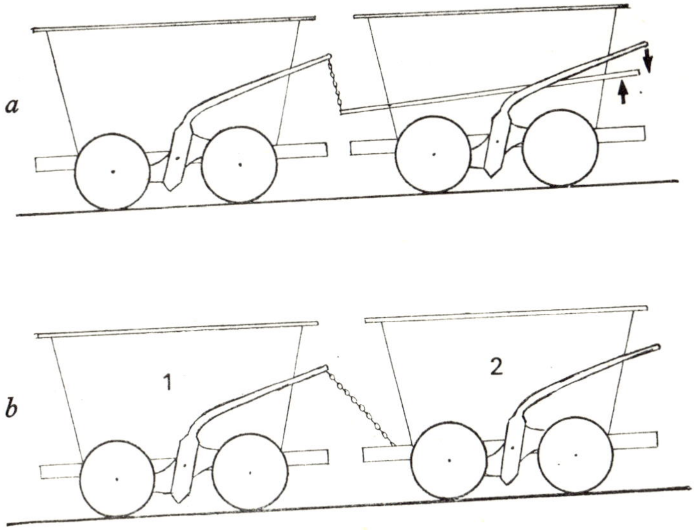

FIG 41 Long convoys or continuous brakes on Tyneside waggons.

convoy of the old type, but extended down beyond the fulcrum to form a compound lever, bearing on the front wheels with a breast below the fulcrum (fig. 41). Then, in 1793 on the Western waggonway, the ancient principle of one horse, one waggon, one man was broken. Two waggons were sometimes coupled together on runs. The brake on the first one was put on, and held down by hanging a hay-poke to its end, while the waggonman controlled the rear waggon in the usual way. Many accidents, however, resulted until in 1795 one Hall of Pontop colliery invented the long brake (fig. 41*a*). The second waggon was equipped with a long bar pivoted at the front, whose forward end was attached by a chain to the upper end of the ordinary convoy on the first waggon. The waggon-man looked after the second waggon's brake and also pulled the back of the long bar upwards to apply the brake on the front waggon. Even longer trains could be controlled and in a simpler way, by chaining the brake handles down to the frame of the next waggon back (fig. 41*b*). If waggon No. 1 tried to run away from waggon No. 2, its brake would automatically go on. The waggon-man would of course always operate the brake on the last waggon.[57]

Two other means of braking are possible: the sprag, which is often bad for the wheels, and the drag, which can harm the track. Scotland shows signs, inherited perhaps from Shropshire, of having used both, as well as ordinary brakes. Fordell waggons were spragged and braked in 1798, and Halbeath waggons had beams of birch to pull behind as drags on gradients.[58]

The wrought ironwork in and on a waggon amounted to a considerable weight. As many as 378 nails were needed to hold it together, and sometimes screws were used, though they were expensive. The Ravensworth accounts of 1726–7 include such items as

> Thos Borrow 55 pound of nailes @ 3¾d
> D° for 8 Longe wormes for a ballas wag^{n} 2/-
> D° for 2 long wormes and 4 short 10^{d}

Table 5

Internal Dimensions of Waggons

Place	Date	Length Top ft.	in.	Length Bottom ft.	in.	Width Top ft.	in.	Width Bottom ft.	in.	Height ft.	in.	Capacity Bolls	Capacity Cwt.
Felling	1694	6	11	4	3½	4	5	2	2½	3	6	15	33
Gateshead	1725	6	11½	4	4½	4	3	2	2½	3	6½	15	33
Whitley	1721	7	0	—		4	8	—		3	9	—	—
Blackburn*	1721	7	0	—		4	6	2	3	3	10	—	—
Dunston	1721	6	11	4	0	4	5	2	2	3	8½	—	—
Ravensworth	1721	6	11	4	2	4	6	2	4	4	1	—	—
Tanfield	1730	7	0	4	11	4	1	2	4	3	10	—	—
Byker	1749	7	6	5	11	4	1	2	8	—		—	—
Tyneside	1778	7	9	5	0	5	0	2	6	4	3	—	50
,,	1797	7	0	5	6	6	0	3	0	3	9	—	—
,,	1797	7	6	—		6	0	—		4	0	24	53
,,	—	7	1	4	4	4	9	2	6	3	6	19	42
,,	—	7	5	4	5	5	6	2	8	3	9	24	53
Willington	*c.* 1815	7	5¾	4	4¾	5	5½	2	10	3	10	24	53

* Waggons going out via the Ravensworth way, before Blackburn was linked to the Dunston or the Tanfield waggonways.

Table 6

Cost of Waggons

		£	s.	d.	
1679	Stella Grand Lease .	4	5	0	wooden wheels
1716	Plessey . . .	6	10	0	,, ,,
1723	Burdonmoor . .	4	5	0	,, ,, second hand
1724	Tyneside . . .	7	1	2	,, ,,
1758	Hartley . . .	7	15	0	,, ,,
1765	Killingworth (Long Benton) . . .	9	14	2½	,, ,,
1765	Tyneside . . .	10	0	0	,, ,,
1776	,, . . .	10	0	0	,, ,,
1778	Moorbanks (Workington)	8	0	0	iron wheels? valuation
1778	Chapel Bank (Workington) . . .	6	0	0	,, ,, ,,
1778	Harrington . . .	7	0	0	,, ,, ,,
1780	Halbeath . . .	20	0	0	wooden wheels
1784	Elgin	13	0	0	2 iron, 2 wooden wheels
1793	Middleton . . .	6	4	3	iron wheels, second hand
1796	Wallsend . . .	13	9	0	,, ,,
1799	Middleton . . .	10	0	0	,, ,, second hand
1802	Middleton . . .	18	19	4	,, ,,
1805	Tyneside . . .	23	11	0	,, ,,
1806	,, . . .	22	5	2	,, ,,

At Killingworth in 1765 'Robt Yelloley Smith his Bill for Nails & Worms' was 5*s.* 3*d.* per waggon.[59] Then there were the two axles, all the iron connected with the convoy, and the two hooks at each end for attaching the horse's traces to. The total weight of iron on a waggon, excluding the wheels, ranged from 2 to 4 cwt. A complete waggon in 1807 weighed 17 cwt. when empty and 3½ tons when full.[60]

The dimensions of a waggon body varied over the years, between one colliery and another and even between waggons of the same colliery, and they were naturally related to the desired capacity in bolls. Table 5 lists some surviving measurements.

The cost of a complete waggon at various dates is set out in Table 6. The entry in the Stella accounts for 1679, which simply reads, 'I built two new waggons this yeare 005:00:0', probably represents merely the workmanship and ironwork involved; the wooden components were worth about £1 15*s.* 0*d.* per waggon.[61] At a later date, the advent of the iron wheel was responsible for raising the overall cost (although the subsequent repair bills were much reduced), and later still the sharp increase in timber prices and labour charges in the Napoleonic Wars raised the cost still further. The Wallsend figure of £13 9*s*. 0*d*. in 1796 consisted of: timber, £2 8*s*. 8*d*.; workmanship, 12*s*. 0*d*.; wheels, £6 8*s*. 4*d*.; other ironwork and smith's labour, £4 0*s*. 0*d*.[62] Second-hand prices naturally depended on condition. The whole stock of 109 waggons on the Dunston way was bought on its closure in 1723 by the new Burdonmoor waggonway at about two-thirds of their value when new. Middleton Colliery at Leeds bought three second-hand waggons from Thomas Barnes at Walker in 1793 and another from Brandling's Felling Colliery in 1790 at about half their first cost. The Walker waggons, if not the Felling, would have needed re-gauging to fit the Middleton track. The Cumberland figures for 1778 are from the valuation of Curwen's effects after his death, and allow for depreciation.[63] The cost of ballast waggons was only half that of coal waggons. The waggons for stone at Ballycastle in 1740, in sharp contrast to the expensive track, were cheap at £3 2*s*. 7½*d*.[64]

The number of waggons on any waggonway depended entirely on the output of the colliery and its distance from the staith. Brockwell (Winlaton), a small concern near the Tyne, had 12 waggons in use and 9 laid up in 1734; Hartley 18 in 1758; Stella Grand Lease 23 in 1680; Felling 26 coal and 2 ballast waggons in 1734; Benwell 28 in 1711; Plessey 50 in 1717; the Dunston way 85 in 1711 and 109 in 1723; and the Tanfield waggonway with its branches, by far the busiest line, had no fewer than 490 waggons in 1727, of which Bowes, the Wortleys and Liddell owned 120 each, George Pitt 70, and Sir Francis Clavering 60.[65] Away from Tyneside, Brandling's Middleton waggonway was using 93 coal and 5 ballast waggons in 1808; Fordell about 30 at the end of the eighteenth century; Halbeath a mere 14 in 1795.[66] Although the illustrations very rarely show it (see pl. 51), most waggons were probably marked with their number for identification. Those at Hartley certainly were.[67]

12

The Newcastle waggonway: operating and maintenance

Saw ye owt o' ma lad
Gannin' doon the waggon-way,
Wi' his pocket full o' money,
And his poke full o' hay?
  Aye, but he's a bonny laddie
    As ever ye did see,
  Tho' he's sair pock-brocken,
    And he's blind of an e'e.

There's ne'er a lad like ma lad
Drives to a staith on Tyne;
Though he's coal black on work-days,
On holidays he's fine.
  Ma lad's a bonny lad,
    The bonniest I see;
  Wiv his fine posey waistcoat
    And buckles at his knee.

'The Waggoner', anonymous song first published by John Bell, Jun., *Rhymes of the Northern Bards*, Newcastle (1812), 300.

The traffic on waggonways could rarely be called heavy. It will be seen from Table 7 that the number of waggons run per day was often less than 100. On an average day at Long Benton, for example, during daylight hours, waggons would pass at about three-and-a-half-minute intervals. The Tanfield waggonway, however, would have had a waggon every three-quarters of a minute, by far the heaviest concentration of traffic on any line – 'the way is almost always covered with waggons', wrote Jars in 1765. For several years after the formation of the Grand Allies in 1726, the coal carried down the Tanfield accounted for almost half the total vend from the Tyne, and nearly a third of the whole product of the Great Northern Coalfield.[1] But only occasionally does the evidence allow us to see how many gates or journeys any one waggon made in the course of a day. The number was obviously greater on a short line than a long one: the Benwell waggonway, for example (see Table 7), was very short, while the Plessey was comparatively long.

It was an almost universal rule that one waggon should be pulled by one horse, with one man in attendance, the waggon running by gravity wherever possible. Oxen very occasionally, it is said, replaced horses in the eighteenth century,[2] and on the staiths the waggon-men unhitched the horses and manhandled their own waggons – horses could all too easily

Table 7

Traffic on Waggonways

		Waggons per annum	Average waggons per day[a]	No. of waggons	Average daily gates per waggon
Stella Grand Lease	1676	12,674	58	—	—
” ” ”	1679	15,080	68	—	—
” ” ”	1680	13,873	63	23	2·8
Stella Clavering	1711	14,434	66	—	—
Ravensworth	1710	*c.* 26,400	*c.* 120	—	—
Dunston	1711	—	85	85	1[b]
Benwell	1711	—	70	28	2·5[b]
Plessey	1715[c]	6,080	56	44	1·3
Bucksnook	1716	7,565	34	—	—
”	1717	8,165	37	—	—
”	1718	12,296	56	—	—
”	1719	13,674	62	—	—
”	1720	13,380	61	—	—
”	1721	7,332[d]	33	—	—
”	1722	13,520	61	—	—
”	1723	10,290	47	—	—
Western Way	1721-2[e]	57,230	260	—	—
Dunston & Burdonmoor[f]	1723	49,426	225	109	2
Benwell	1723	12,372	56	—	—
Tanfield	1727	204,600	930	490	1·9
Seaton (Workington)	1734	—	21	—	—
” ”	1741-7	ave. 11,103	37	—	—
Whitehaven[g]	1740	103,610	345	—	—
Long Benton	1751	48,181	219	—	—
Middleton	1762-4	ave. 6,811	38	—	—
”	1801-8	ave. 35,000	194	93	2
Maryport	1780-1	ave. 16,401	53	—	—
”	1786	21,202	71	—	—
Harrington	1785-6	ave. 13,627	45	—	—
Sheffield	1781-8	ave. 12,305	41	—	—
Workington	1794	—	*c.* 150[h]	—	—
Wylam	1804	—	—	—	3[i]

These figures are taken from specific reports on the number of waggon-loads carried down the various ways. Reasonably accurate figures can also be obtained, for any colliery which had no appreciable landsale, by translating the annual output, whether in tens or chaldrons, into waggon-loads.

Notes to Table 7

a. Tyneside: based on 220 working days per year (mid-February to late October, excluding Sundays).
Middleton: based on an arbitrary 180 working days per year, mainly in spring and autumn.
Cumberland and Sheffield: based on 300 working days per year. See p. 210.

b. Restricted vend.

c. July-December only.

d. Traffic reduced by opening of Western Way.

e. May 1721 to April 1722.

f. Waggons and some of track transferred from Dunston to new Burdonmoor way in course of year.

g. Nearly all by waggonway from Howgill; a very small proportion came by cart from Whingill.

h. 100 waggons a day on Curwen's two lines, 50 on Lowther's Seaton way.

i. In trains of two waggons.

fall down the spouts. Very rarely, too, when an adverse gradient against the load was unavoidable, Tyneside coal-owners had to resort to 'marrowing' with two horses to one waggon.[3] For bringing back the empties, the Halbeath and Fordell waggonways sometimes used two horses per waggon.[4] Needs must, perhaps, but they were not steep lines. It may be significant that Shropshire influence was to be seen in Scotland, for Shropshire often used more than one horse to a waggon. The concept of the train – two or more waggons coupled together – was practically unknown, except for occasional instances where the waggonway was level. On the Wylam line, running for five miles along the Tyne, one horse regularly pulled two waggons[5]; and the Garesfield waggonway in 1831, when it was still laid with wooden rails, was divided into two independent sections. On the upper two and a half miles from the colliery to Winlaton Mill, where the gradient was steep, there was one waggon to a horse as usual; but on the two-mile stretch from Winlaton Mill to Derwenthaugh, the last surviving and almost level remnant of the Western Way, one horse coped with four waggons.[6] Tyneside only began to adopt the train widely at a late date, for economizing on manpower as much as on horse-power. From the 1790s two horses might pull two waggons together, with one man in charge (see p. 201), but it was the iron rail, which engendered so much less friction than wood, and the locomotive of several horse-power, which finally killed the single waggon.

Tyneside waggon-men or drivers, just like the miners, banksmen and all other colliery workers, were bound servants of the owner. In other words, they contracted to serve the same master for a year, and could not leave his employ till the period was over. Waggon-men's bonds lay down the terms of service and the wages to be paid, as well as penalties for improper conduct. On the level Wylam way in 1804, for example, where there was no gravity working, the fine was 1*s.* for riding on a full waggon, and 2*s.* 6*d.* for leading an empty waggon up the main way instead of the bye.[7] Waggon-men came under the authority of the staith-man, who was responsible for seeing that the staiths were sufficiently supplied from the colliery. The horses might belong to any of four different owners. Very often they were provided by the coal-owner's tenants, under a clause of their lease – a profitable business, but in these cases the rent was pegged at an unusually high level.[8] Often, too, the horses belonged to the waggon-men themselves. As the drivers' 'occupation is not much more profitable than that of a common pit-man, it may be supposed they are not *excellent* cattle, nor in high order'. But subsidies were not unknown. In 1715 the Wortleys gave their agent on the Dunston way £60 'to lend the waggon-men to enable them to buy horses which he is to repay'.[9] Thirdly, the horses were occasionally supplied by the coal-owners themselves, especially when for some reason the usual sources were insufficient – George Bowes in 1723 'sent his fine horses to the waggons'.[10] Finally, somewhere around 1800, the outside contractor appeared on the scene, who agreed to lead the whole of the traffic for a fixed sum. In Yorkshire, however, small-scale contractors had led the waggons since the 1760s – there were seven such firms on the Middleton in 1804 – and in addition the collieries undertook a small proportion of the leading themselves.[11]

Waggon-men and horse-owners were paid for their services by the gate or waggon-journey. The sum naturally varied according to the distance of the pit from the river, but for most of the eighteenth century in both Tyneside and Cumberland it was based on 3*d.* to 4*d.* a mile for the downward trip, with nothing for the return journey. In about 1716, for example, the 'leading charge' on the Bucksnook waggonway, five and three-quarter miles long, was 1*s.* 8*d.* Of this sum, by an arrangement apparently universal on Tyneside, three-fifths went to the owner of the horse and two-fifths to the waggon-man.[12] From the slight surviving evidence, it seems that there were about the same number of waggon-men as of waggons. In 1727 on the Tanfield, for example, out of a total of 490 waggons, the Liddells were responsible for leading and repairing 170 (120 on their own account, and 50 of Sir Francis Clavering's). To work them, the Liddells employed in 1728 (we do not know about 1727) 160 waggon-men.[13] As the waggons averaged nearly two gates a day during the working season, and the average leading charge on the Tanfield was 1*s.* 7*d.*, the waggon-man who did not own his horse would earn about 1*s.* 3*d.* a day, a sum very similar to the wage of an ordinary miner. The traffic was, of course, almost entirely one-way, though occasional return loads of timber, bricks or hay for the colliery were carried, known as 'bye-work' and paid for at half the rate for coal. A horse, it was said, could travel up to forty miles a day; but it seems that it was rarely called upon to do so.[14]

If the waggon-man did own his horse, then he kept all the leading charge – an average of about 3*s.* 2*d.* per day on the Tanfield – but out of it he had to supply the harness, and feed, shoe and eventually replace his horse. He was at the mercy of rising prices, especially during the Napoleonic Wars, when the cost of horses and fodder soared, although some coal-owners were generous in times of crisis. In 1800 John Buddle at Lambton 'Fixed to give the Waggon Carriagemen 10 Shillings in the Pound earned as a Consideration for the Dearness of Corn and Hay; with which they were satisfied'. On the Garesfield in 1831, worked by a contractor, the agreement provided for a rise in the leading charge should the cost of fodder rise above a certain level.[15]

The average price of three- to six-year-old horses, mares and geldings bought for the Middleton waggonway rose from £14 in 1771-89, through £21 in 1790-9, to £26 in the first five years of the nineteenth century. Old horses were sold by Sheffield Colliery to the knacker for a pound or two in the 1780s. Horse harness cost £1 5*s.* 6*d.* a set in 1799.[16] Detailed costs for keeping horses, however, do not appear until later still. The Heaton owners, when they were considering investing in a locomotive in 1821, worked out the annual expense of a waggon horse which they kept themselves. It was this[17]:

Cost of 1 Horse – say £40 – to work 5 years	£8.	0.	0
Hay 4 Tons @ 70/- p	14.	0.	0
Oats & Beans – allowing 1¼ Boll of Oats and ¼ Boll of Beans for each horse – say Oats 6/- p Boll & Beans 10/- p boll will be 10/- p week for 52 weeks	26.	0.	0
Shoeing	2.	0.	0
Wear and Tear in Trapping	1.	10.	0
Horse Tax	1.	1.	0
Horse-keeping	3.	0.	0
	£55.	11.	0

Since the waggon-man of a century earlier would not be saddled with the horse tax, and would look after his own horse, the Heaton total can be reduced to £51 10*s.* 0*d.* By comparison, three-fifths of the average daily earnings on the Tanfield in 1727, over 220 working days, works out at about £20 a year, a sum which included a measure of profit for the horse owner, be he waggon-man or not. In a century, therefore, the cost of keeping a horse had been multiplied by at least two and a half.

Coal-owners sometimes ran small farms attached to the collieries, both to provide fodder for and to breed their own underground ponies. The coal-owner rather than the waggon-man was entitled to collect 'the manure bred on the waggonway'.[18]

When waggon-men were difficult to get and to keep, they were sometimes given retaining fees at the beginning of a year, to ensure their return for the next. George Liddell wrote in 1729, 'I am sorry to hear the Western Waggon-way have begun so indiscreet a president as giving Retainers to new carriage[-men] at their coming on and especially when corn is cheap'.[19]

The coal trade on Tyneside was to a large extent seasonal, since work in the mines and on the waggonways was dependent on the movements of the collier fleet in the coastal trade. Other areas differed. In Cumberland the shipping, though reduced by about a third in the winter, still kept going, and the waggonways, like those at Sheffield, which catered for a purely local landsale, remained largely unaffected. At Yorkshire collieries on the Aire and Calder the largest vend came in the autumn, between the summer droughts when the rivers were too shallow to navigate, and the freezing of the cuts in winter. During the summer, miners and waggon-men would return to agricultural pursuits until trade picked up again. But on Tyneside the collier fleet was generally laid up from November to February to escape the winter gales. The waggon-men would have to stock up the staiths in readiness for the spring arrival of the fleet, but, as George Liddell said in 1729, they worked only eight and a half months in the year, and the constant threat hung over their heads that the start of the shipping season would be postponed until March, leaving them only seven months of work.[20] This was not, however, a hard-and-fast rule. On the Stella Grand Lease way in 1679, 1,918 waggons were led between 8 November and 26 December, compared with 15,080 for all twelve months, which represents an almost regular flow of traffic over the whole year.[21] But in general it seems that the winter was a lean time for waggon-men, with no pay coming in; and even in the spring and summer they were paid only at long and erratic intervals.

For these reasons, along with the keel-men, who suffered the same disadvantages, but were more vocal about it, the waggon-men occasionally came out on strike. The spark that set off a dispute was sometimes economic, as in 1710, 1740 and 1789, when shortage of corn and high prices led to strikes and riots by pit-men, waggon-men and keel-men. On the first of these occasions some of the Clavering Stella waggon-men went so far as to run their waggons off the staiths; but on the second, while most of the coal-workers of the area were busy sacking Newcastle Guildhall, the Stella waggon-men continued placidly at work, a tribute to the enlightened outlook of Sir James Clavering. Sometimes, too, strike action was prompted by industrial disputes. A localized example happened at the beginning of the 1713 season, when the Clavering Stella pits were going through a difficult phase and output was restricted. The waggon-men apparently objected to the lack of trade, and refused to carry what little coal there was. James Clavering wrote to his agent, 'If they will not fetch their waggons today, yo will do well to go among them tomorrow, & pray Endeavour to discover ye Ring leaders'. The matter was somehow settled, but three weeks later 'ye waggon men were again in rebellion; now we have no coles upon ye Steath'. This time they stayed out for ten days. A larger strike, this time involving the whole of Tyneside, took place in 1765, lasted over a month, and resulted in much violence. The cause was an attempt by the owners to enforce labour beyond the limits of the miners' bonds; the result was a decided, though not complete, victory to the miners.[22]

The waggon-men were usually drawn from the local farms. Sometimes they even leased their own smallholdings and tried to indulge in both waggoning and agriculture, an unhappy combination in which the agriculture suffered. A report on the Claverings' farms in the area of their Stella waggonway in 1748 emphasizes the point. One tenant 'relys too much upon the Coal Trade and by that Neglects the main thing [his husbandry]'; another 'is very Careful but none of the best Husbandmen; he follows the Waggons himself & has nothing but Children to manage his farm'. At Bucksnook, leased by Anthony Buck,

> There has been formerly a vast of damage in it by Pitts sinking and Waggon Ways but now none save only Silvertops Way through part of it. During the whole time of his [Buck's] Farming that Ground I believe he has not so much filled up one Pitthole nor improved one Inch of the Waggon Ways that has been laid off for many Years. He's what we tearm in this Country a downright Sloven. . . . Buck is so much taken with the Coal Trade and Silvertops fine promises that he's Wholly neglected his Husbandry and Spoild his Farm.

Sometimes, though, there was a real economic need for a smallholder to work on the waggons. Of Follyhill near Greenside the report says:

> This Farm bounders on each side the Lead Way and Waggon ways. It is not above 35 acres, the Ground hill & Dry Sandy . . . & cannot be improved, nor do I believe would let for the Rent it now gives [£13 1*s.* 0*d.*] if it was not for the Coal Trade wch those that farm it must raise the Rent by; & no other Way.[23]

As the cost of living rose towards the end of the eighteenth century, so too did the waggon-men's wages, though more slowly than the miners'. At Wallsend about 1794, $3\frac{1}{2}d.$ was paid for a 1,040-yard gate, and $4\frac{1}{2}d.$ for one of 1 mile 330 yards. In 1804 on the Wylam way, the waggon-men (who did not own their horses) were paid *6d.* per waggon, which may seem a small reward for a five-mile journey. But with two waggons to a train and an ordained three gates a day, it meant a daily wage of *3s.*, and from November to February the rate was raised to *7d.* a waggon as slight compensation for the lack of trade. The Kenton and Coxlodge contractor in 1809 was paid $6\frac{1}{2}d.$ per waggon per mile on wooden rails, or *4d.* on iron.[24]

In the early days, the coal was raised from the pits in corves or baskets by horse-gins, and simply emptied on a heap by the pithead. It had then to be shovelled by hand into the waggons, sometimes by a special gang of men, but usually, it seems, by the waggon-men themselves as part of their duties, without any extra pay. From the 1760s onwards, however, most pits were equipped with screens or sloping grids for sieving the coal. The pithead gear was now elevated some distance above the ground, the corves were emptied on to the screen, and the coal fell through into waggons underneath. The waggon-man attached his horse's traces to the hooks on the end of a loaded waggon, and set off for the staith. In Jars' words,

> When the way is almost level, the horse pulls the waggon. . . . When it reaches a point where the slope is more pronounced, where the waggon would go too fast, and where the horse becomes useless, the driver unhitches it and puts it behind. At the same time he climbs up behind his waggon, releases the end of the lever or the two levers, and by pressing it on one or both of the rear wheels he stops them . . . and thus reduces the waggon's speed. He varies the pressure according to the gradient and the speed at which he wishes to travel. . . . When he is past the point where the waggon will move by gravity alone, he attaches the horse again, and continues in this way until he reaches the staith.[25]

On gravity runs, the waggon-man usually sat on the convoy (pls. 49 and 50), but a few illustrations show him actually standing on it.[26] At Whitehaven a traveller commented on the driver, who

> fearlessly rattles down the steep, and regulates his speed at pleasure. . . . [One sees] the cool and careless faces of the helmsmen, some smoking their pipes, and others accompanying the rough music of the wheels with a song.[27]

The greatest danger came in rain or frost, when the braked wheels slipped on the wet rails and the waggon continued to accelerate. In such weather,

William Brown said, 'we lay Ashes and sometimes Cynders to Ruffen the Railes' – ashing was the duty of the waggonway wright – and Spedding at Whitehaven went even further:

> When the Railes are wett we lay Small freestones about the bigness of a hen Eggs which Crushed to Sand upon the Railes and makes the Convoy take more hold of the Mettle Wheels. So that we can do pritty well in wett weather. Frost Raggs will be the worst for if we could hold all the Wheels the Waggon would [go] faster than when the fore wheels Run and Crushes what we lay on the Railes.

By this he probably means that in frost the waggon would move faster with the brake on than off.[28]

In really bad weather, steep waggonways were compelled to suspend traffic altogether, such was the danger of runaways. Sir Henry Liddell said of his Ravensworth line in 1710, 'We . . . shall not lead but in Cases of necessity w^{n} ye weather is indifferent', and Ralph Sowerby of Tanfield Moor, when asked by a Parliamentary committee, 'Do you never leave off?', replied: 'Never but when it is convenient for our Colliery, when the Weather is wet.'[29] In 1721 Bowes, Pitt and Lady Clavering took their traffic off the Bucksnook waggonway and sent it down the new but much steeper Western Way. In rainy weather, however, they continued to use the Bucksnook with its easier runs. They made no payment for the liberty, much to the fury of William Cotesworth, who wrote to Thomas Sisson, his agent, in 1722:

> Pray let Mr Bell & Mr Baker lay their heads together to get us a plump affidavit about Mr Pitts Lady Clav & the rest if they do it, that they come down the old way in the wett weather . . . set forth, as truth is, that they have during the late wett weather forsaken the new way thro' Mr Bowes & Hardings grounds & led down the intire old way till they tooke off to go over Derwent . . . & let Mr Baker set forth that notwithstanding he was appointed the general agent for keeping the way, & that they damnify it much by useing it in wet weather, they will not pay him for repairing it.

Within a fortnight Sisson reported back:

> Yesterday was a most severe day of Thunder and Rain in the afternoon among wch were Hailstones, and Mr Dent saies that when the Raine came on the Waggons that were in the way going to Steath were turned all off the New way by Ra ffenwick who is an agent for the Undertakers of the New way who stood at the partings of the New and Old way in Byermoor who shut up the gate that goes down

> the New way and ordered all to go down the Old way, Mr Bakers man Mr Bindless was present and see this and can make affad^{t} of it if you think it of service.[30]

The spectacular results of brake failures on steep runs are mentioned by many contemporary writers. Nicholas Wood said this:

> In wet or damp weather, the wheels, by *licking* up the dirt and mud from the rail, became so slippery, that the action of the brake was almost destroyed; and the attendant having thus no power over it, it frequently got away, destroyed everything in its course, killed the horses that were upon the declivity, and finally was dashed to pieces. . . . To obviate this inconvenience, in wet weather, boys and men were employed, strewing ashes upon the rails down the steep declivities . . . to cause the brake to take effect . . . . Frequently, on these very steep descents, for many days the work was laid off, on account of the weather; a sudden shower of rain occurring, when any of the waggons were on the declivity, set the whole away; and men were stationed to draw ropes, as booms, across the line of road, to stop their progress. If the ropes could be drawn across before the momentum became very great, the damage was less; but if they broke the ropes, then the most disastrous effects followed.[31]

The *General Magazine* adds:

> Waggon-men, in going down very steep Runs, commonly take their horses from before, and fasten them behind their Waggons; as they would inevitably be killed was the convoy to break (which frequently happens), or any other accident occasion their waggons to run *amain*; nor is this fatal consequence only attendant on the horses, but the drivers often receive broken bones, bruises and frequently the most excruciating deaths: – Indeed, in some places, a most humane custom is established, which is, when any Waggon-man loses his horse, the other Waggon-men go a Gait for the poor sufferer, which is little out of their profits, and purchase him another horse.

It was not unknown for from ten to fourteen horses to be killed in one accident, because the waggon-men had no time to free them before abandoning ship. Pressing too hard on a convoy carried another risk: fire. This sometimes destroyed the whole waggon, sometimes only the convoy, which left the waggon-man just as helpless. As Jars remarked, when a waggon went out of control the least that could happen was derailment. The *London Magazine* adds:

> When a waggon happens to be off the waggon-way, if laden, it will take

> two or three horses to way the waggon again; sometimes the waggon axle breaks, then, if laden, it causes a great stop to the other waggons before they get the coals and waggon clear from stopping the way, called by the waggon-men a caud [cold] pye.

In a case like this, if the driver 'can but get a *shovel-full* of Coals carried in his waggon to the Staiths, he is paid for his Gait, otherwise not'.[32]

Tyneside burial registers sometimes mention causes of death, such as 'killed with a waggon' or 'killed by upsetting a waggon',[33] and children were often the victims. We have seen two such accidents at Whickham in 1650 (p. 95). Another example is contained in a note (referring to the Bucksnook way in 1714) which asks for counsel's opinion on a nice legal point[34]:

> A man driving a waggon full of Coals through Swallwell suggested to be in the Highway but in the mannor of Whickham leaves his horse & waggon going on the way wch is a way laid for the use of Coalwaggons only. The Horse goes on & draws his Waggon over a childes head by wch it was Instantly slaine – a coroner pretending to have a deputation from the proper Coroner appointed by Comission from the Bishop sitts an enquiry & finds the wheele that Run over the Childes head a deodand. The inquisition being held by a deputy is void. May the Lord [of the Manor] recover by action a deodand?*

At Sheffield in 1733, however, there was no inquest, no deodand – just a paltry item of expenditure in the accounts: 'P^d for the Woman that was killed with the Waggon 0.10.6'.[35]

The Bucksnook waggonway was a line with an extraordinary capacity for attracting misfortunes; James Clavering once said of it: 'This famous waggonway occasions such variety of proceedings & adventures that really Could we have its history writ by a good pen, w^ld outdo Don Quixot.' Nobody obliged, and we have now to be content with occasional glimpses, mostly from the pen of James Clavering himself. One such incident – it can hardly be called an accident – cannot be passed by. He wrote in 1712:

> Shall next give you a pleasing narrative. It was just hinted last post to ye Governour [Henry Liddell] how civilly my cheife [Sir John Clavering] entertained ye judges, who were by justice Wrights cunning managem^t brought to Swalwell under ye pretence of seeing S^r Am. Crawleys works . . . They came by water. At their landing ye Bar^tt had his chariott wth a pair of horses that had never drawn at ye wheels before & ye postilion who had never driven before for coachman; that

* *Deodand*: an object that has caused accidental death, forfeited to the crown for pious uses.

> they might have an exact view of y^{e} [waggon]way, ye vehicle [drove] all upon it, till at last in ye midst of Swalwell towne they came upon a kind of bridge wch had on each side a descent of about 3 yds at least, this proveing too narrow over went one of ye horses & hung in ye harness & if ye great help had not been at hand ye vehicle in wch was ye worshipful ye mayor & ye two judges had followed, this was very unlucky for both ye judges declare if their opinion should be asked whether waggon ways be common nusances in publick high ways, they must give it in ye affirmative; ye country say ye Bar^{tt} did this purposely, for their judges usage of him at ye Durham assizes, when ye Bar^{tt} had not success in one point.[36]

To return to our waggon-man. Assuming he made the journey to the staith without succumbing to any of these hazards, he stopped at the entrance to the ongate, unhitched his horse, pushed his waggon by hand over the bridge on to the turnrail, revolved it until his waggon was pointing along the staith, and reported to the staith-man. This official marked one gate against our waggon-man's name on his list for the pay clerks (in Cumberland he gave the waggon-man a token, redeemable for cash at the pay office[37]), and told him which spout to use. The driver pushed his waggon to the requisite opening, released the bottom-board catch and emptied the coal down the chute. This done, he closed the board, and took his waggon via another turnrail to the end of the offgate, where he attached his horse and drove as fast as possible up the bye way, luring the horse on with handfuls of hay.

Track and waggons were maintained, in the early days especially, by the coal-owners themselves, or rather by their agents. Thus in the 1670s we find William Iley accounting to his masters, Sir George Vane and Lionel Vane, who owned Stella Grand Lease.[38] But soon the system grew up of entrusting the 'keeping' to contractors, and by the end of the eighteenth century it was almost universal on Tyneside. The keepers or undertakers were usually gentlemen of some standing and a commercial bent, who employed master wrights as their agents on the spot. Sometimes the master wrights kept ways and waggons on their own account. A typical deed of contract is that whereby Joseph Weatherburn, gentleman, convenanted with Charles Atkinson on behalf of the lessees and proprietors of Gibside, Marley Hill and Northbanks collieries, that for nine years from 25 December 1712 he would uphold and maintain the Dunston waggonway and its branches from the several pits with sound oak rails and sleepers, and repair all bridges, banks, risings and mounts. He would make sufficient waggons with good and sufficient wheels and iron axle trees, lead all the coals from pit to staith, and lay down branches from the main waggonway to every heapstead as necessary. He would maintain at Atkinson's cost the bye way between pits and staiths in good repair, and keep the main way and bye way well filled with good hard ballast or pan rubbish, and clean or dress the same as often as needful to preserve the same dry hard and clean, and more convenient for the horses to be employed thereon. Atkinson was to pay Weatherburn 13*s.* per ten of coal led.[39]

This tentale – so much per ten – was the usual method of payment, the figure itself varying both with the length of the waggonway and with the difficulty of the route. Steep runs wore out waggons and track far more quickly than a gentle gradient did. When Henry Liddell was trying to find a keeper for Ravensworth in 1710, he asked Cotesworth to

> talk this matter wth Cozen Baker, & see if y^{o} can article wth him. I think this proposall a reasonable one to keep ye Way & waggons &c in Repair att 10^{s} a Tenn in ye low ground, that I call fr ye Waggon Bridge as farr as ye ditch wch is above ye Engines, & all beyond that att 15^{s}. I think wth ffranks [Baker's] good menagem^{t} he may save handsomely by ye year.[40]

In the sixteen months after the Tanfield and Burdonmoor (Blackburn) waggonways were opened, their keepers made no profit at all, so in November 1726 the Grand Allies increased the tentale from 14*s.* and 12*s.* to 15*s.* and 13*s.* respectively.[41] Recorded way- and waggon-keeping tentales range from 6*s.* per ten at Lands colliery in 1754 to 16*s.* at Collierly in 1727, or roughly from 3$\frac{1}{4}$*d.* to 8$\frac{3}{4}$*d.* per waggon run. The one was a short distance, the other a very long one. In Cumberland, a quite exceptionally high rate was paid to John Westray in 1741: 1*s.* 1*d.* per waggon on the three-mile Seaton line at Workington.[42] Occasionally ways and waggons were kept on a different basis: £2 per week, for example, on the short Wallsend line in 1795, the owners providing and leading the materials; or 1$\frac{1}{2}$*d.* per ton per mile on the Halbeath in 1780.[43] Sometimes, too, way- and waggon-keeping were kept separate. At Maryport about 1769 the waggon-keeping was let at 6*d.* per waggon run to one Jackson, who made a profit of £1,500 in nine years, while the way-keeping was given to another man at $\frac{1}{4}$*d.* to $\frac{1}{2}$*d.* per waggon, depending on the distance of the pits.[44]

For waggon-keeping, the keeper employed specialist waggon wrights and smiths, usually paying them by the ten carried (1*s.* 4*d.* to 2*s.* 6*d.*), plus all materials. In the accounts of Henry Ellison, who kept the Ravensworth way and waggons from 1726 to 1731, there are regular entries, such as, 'Repaireing ye Waggons to ye Smiths & Wrights for 397 Wag^{ns} led at 1^{d} p D^{o} £1 13s 1d' – that is, 1*s.* 10*d.* per ten. On the Bucksnook in 1721, however, the wrights were paid by the day (1s. 6d.), the smiths by the ten (1s. 3d.).[45] When waggons were kept by the owner, the cost appears greater because materials as well as workmanship are included in it. Thus at Stella in 1679–80, waggon repairs cost 2s. or even 3s. per ten; in 1680 £67 17s. 7d. was spent on keeping 23 waggons, or nearly £3 per waggon over the year.[46]

The work of way-keeping was done by expert waggonway wrights, unskilled labourers and creasers. They were normally employed by the keeper (either individually or, paid by piecework, in companies), but sometimes directly by the coal-owner. Ravensworth had 15 to 18 wrights in 1726–9, paid between 8d. and 1s. 4d. a day, plus a fluctuating number of labourers at 8d. Washington in 1789 employed six wrights at 9s. a week and five labourers at 6s.[47] By the turn of the century a good wright was commanding 2s. 6d. a day. Brand says of the creasers:

> Many of the superannuated workmen earn a little livelihood by what they call 'creesing'; that is, crevicing, or cleaning out the crevices of the rails in these waggon-ways, upon which in steep places they are employed in laying cinders, to prevent them becoming too slippery, and the waggon from running amain.[48]

A further duty is mentioned at Wallsend in 1795: 'The men to Shovel Snow'. In 1792, the collieries which supplied the Wear vend employed 40 waggon smiths, 100 waggon and waggonway wrights and 60 creasers for the ways, compared with 500 waggon-men, and a total of 15,000 in the whole coal trade.[49]

If the owners employed their own staff for track maintenance, there were several ways of reckoning the cost: at a flat rate for the whole line for a year, as at Heaton in 1820 (£300 for three miles) and Fordell in 1798 (£560 for four and a quarter miles); or, in a general Tyneside example of 1797, at $1\frac{1}{2}$d. per waggon per mile, which equals £93 15s. 0d. per mile per annum on a total traffic of 15,000 tens; or by the ten over the whole line as in waggon-keeping – at Sheffield in 1773 it was 3s. 6d. per ten.[50] There is a very nice instance of this, where two men were tendering their proposal directly to the coal-owners:

> Poaposell for keeping Kenton waggonon way in repare viz. from the Ingen Pit to the woding Bridg By the ten of 440 Boulls to the ten the Ownears finding all materll Exsept asshen the runn Reparing all the Raylling By the waggn way siead at 2s 0d pur ten and the ouners to lead Balles and all materills the men inploied on the Waggon way to Be found in Hous and fiearing.
>
> Agus 15th 1809 — John Hogget
>
> Rowland Crozer[51]

This is not an isolated case of houses being provided for the track workers. A wayleave for Plessey waggonway in 1709 includes

> liberty to erect and build upon the said lands and grounds near the said Waggon ways Houses Lodges or Shedds for the workmen to live in for the better repaireing and mending of the said Waggon ways so that the Houses lodges or Shedds do not exceed ffive in number.[52]

The actual work done by the waggonway wrights and labourers included sawing new rails and sleepers, leading them to site, replacing worn-out track with new, clearing ditches and repairing fences. Within limits laid down by the contract, they also built branches to new pits. Perhaps most of their time was spent in replacing the ballast and packing it ('backing and filling') – it was constantly compressed by traffic, and not infrequently washed away by rain and floods. The best way of describing their multifarious tasks is to quote from various waggonway accounts. Those of Sheffield colliery do not give much detail:

6 Nov. 1731	Paid for repairing ye Waggon Way	11s
29 Apl. 1732	paid for laing ye Railes & mending the Waggons	11/-
Jan-July 1732	By the house to lay Rail &c Carpenter work about ye pitts	27. 9. 3¾
9 Sept. 1732	pd John Kent for mending the Way & Getting up Rails	3/6
10 Feb. 1732/3	pd Mr Smilter for Sawing 12 Tun of Rail at 7/6 p Tun	4. 10. 0[53]

When a short branch of the Clavering Stella way was built in 1710, 'An Count of ye Workmens Days att ye moore gaite' shows that some dozen men were employed for a month at 10*d.* a day. A few entries are more particular: 'Wm. Brown for leading Rails & Sleepr 9 days at 6d', and 'Jno Gray 15 day att ye bulwork 12/6'. The total cost was

Railes DD [delivered]: 2331 yards [at] 05¼ [d]	50 : 19 : 9¾
Sleapers Do Do 1058	26 : 09 : 0
Lab.	18 : 18 : 11
	96 : 07 : 8¾

At the same time an even shorter branch was laid – 'faws Brench – This

Brench was old Wood' – so only the labour was charged for, at £1 18*s.* 0*d.* In 1711–12 on the same waggonway, we find items like

Wrights. Wm Anderson 2 days breaking ballast at 10^{d}	1/8
Jos. Pallister 92 Waggons of Ballast bo^{t} of him at 4^{d} p Wagg	1 : 10 : 08
$Widd^{w}$ Parkinson 86 ditto	1 : 08 : 08
Carrying up 80 Waggon Wheeles	6/4[54]

The most detailed accounts are those for the Ravensworth way in 1726-31, kept by Albany Baker, the agent on the spot of Henry Ellison, the keeper. Apart from the entries for purchase of materials and payment of wages, there are many such entries as these:

Leading:	
Dec. 1726	Robt Dickinson Cart 2 days leading sinders.
Mar. 1726/7	$Anth^{o}$ Stephinsons Waggons leading 30 Gates of Railes 10^{d} p gate; Do $10\frac{1}{2}$ Days leading 43 Gates of Ballas @ 30^{d}.
Laying:	
Dec. 1726	Mat Moffitt & Pn^{s} for pouleing up leavelling & laying backeing & ffilling 100 yards of maine way at $4\frac{1}{2}^{d}$ p yard
Nov. 1726	Casting a Gutter at y^{e} bogg [3 men at 5 days each]
Jan. 1726/7	Jno Barnes 11 Days Ballasting 828 Yards of way at 30^{d} p Day.
Mar. 1728/9	$Matt^{w}$ Moffitt & Pn^{rs} for 46 Yards of ffalse Railes laid $1\frac{1}{4}^{d}$ 4/$9\frac{1}{2}$
	Ditto for makeing a Sideing in D^{o} way 7/6
Woodwork:	
Nov. 1726	Jos Greenwell & Pn^{s} for Bresting 45 Dosen of Convoys at 2^{s} 8^{d} p.
Mar. 1726/7	Greenwell 1 Day Repaireing ye Ballas Waggon
Aug. 1727	Thos Wheatley & P^{nrs} for Sawen 10 Deales for Cleathing y^{e} Waggons @ $2\frac{1}{2}^{d}$ p Cutt 2/1
Ironwork:	
Dec. 1726	Thos Borrow 55 pound of nailes for Convoys @ $3\frac{3}{4}^{d}$
Sep. 1727	Thos Borrah for mending a Axeltree fr a Ballas Waggon 8^{d}
Oct. 1729	Thom Burrow for 8 pound of nailes for y^{e} $Wagg^{n}$ ledges.

It will be seen that much of this work was carried out during the winter, when the traffic was light or non-existent.

Way- and waggon-keeping was not necessarily profitable. Henry Ellison made a loss of £300 in 1727 and 1728; a profit of £278 in 1729; a loss of £266 in 1730 and of £17 in 1731. His income from the tentale paid by George Liddell, the owner, was £1,300 in 1728, plus a little from George Bowes, who carried some Blackburn coals down the Ravensworth way. But he paid out £732 in wages and £672 for materials. The detailed breakdown of these purchases in the following year was:

10,180 Yards of oake raile at 6½ & 2,414 at 5ᵈ	326. –. –
149 Yards of False raile at 8ᵈ	4. 19. 4
532 Sleepers at 9ᵈ & 1078 at 6ᵈ	46. 18. –
257 Waggon Wheels at 7/6, 115 at 9ˢ & 2 at 4/4	148. 11. 2
54 Soales at 1/4 & 29 at 8ᵈ	4. 11. 4
108 Overings at 8ᵈ & 36 Howtrees at 4ᵈ	4. 4. –
14 dozen of Side Sheths at 4/4, 34 10/12 doz. at 4ˢ & 4½ Dozen at 2/6	10. 11. 3
20 Bottom Sheths at 16	1. 6. 8
1050 Yards of Breast wood for Convoys at 4½	19. 13. 9
49 dozen of Convoys at 9ˢ	22. 1. –
20 deales at 15ᵈ, 20 at 13 & 140 at 4½ᵈ	4. 19. 2
380 deals at 4½	7. 2. 6
	600. 18. 2[55]

Not all of them would be used in the same year that they were bought, for keepers always held a stock of materials in their yard or store which were listed in the annual accounts. An early and simple example from the Stella Grand Lease waggonway is dated 1679[56]:

I have of wag wheels in new Storehouse 80 at 5s. p wheales is	020 00 0
In Store House are hundredd Dailes	007 00 0
Of waggon Soales sheths & offerings	009 00 0
Of Convoys 15 Dusen at 12ᵈ is	000 15 0

Inevitably, way- and waggon-keepers did not always give satisfaction. In 1729, for example, the Grand Allies complained 'that the Partnership Waggon Ways are in very bad repair Viz one of the Main Ways not useable, the other the Rails thin & Wants Ballast & the By Way very Bad particularly Albany's Run'.[57] Again,

> Memorand^{m} That this 10th of ffeb^{y} 1725/6 Stephen Chambers & George Chambers Waggon-Wrights has carefully view'd that part of Winlaton or Brockwell waggonway which Mr Albert Silvertop kept for Sr Edwd Blacket Mr Ridley and Mr Simpson . . . the same being soe much out of Repaire that Scarce one Sleeper in the whole length of Silvertops way is whole and Nothing to hold the Rails together but as they pin them to the Ground or put in halfe Sleepers And that by a Moderate Computation that part of the waggon way which Silvertop keep^{t} . . . will Take Seven hund and ffifty Yeard^{s} of Rails and Above ffour hundred Sleepers . . . before the same can be made a Currant Goeing Waggon way.

The branch to Winlaton Colliery was so out of repair that 'no Waggon can lead Coals thereon'.[58]

William Cotesworth kept the Clavering Stella way. In December 1712 James Clavering wrote: 'Could I once get his accts, I think in ye main we should have no reason to complain, am sure ye waggon way, & waggons in a much better state than ever I see them.' Alas for first impressions; little more than a year later he felt otherwise. He had at last received Cotesworth's accounts, charging more than £1 per ten, a monstrous figure indeed, excluding staith repairs. He wrote to Henry Liddell:

> His fine tongue pleads many Excuses, as ye ill repair ye way was in w^{n} he enter'd Aprill 1710, to w^{t} it's now, ye severall long branches laid, & ye small quantitys we have yearly led; were this really so, much might be said, but between you and me . . . ye way is not in a better condition. . . . As for ye branches, some of them were laid wth old railes pulled up fr branches that have been laid again, & this for want of Stock, nay at present I believe Scarce either a Sleeper or raile, that ye 2 new branches that are now laying to Whitefield & moorgate, are forsed to be laid in ye same manner.

Cotesworth's reply was that, though he had been appointed keeper or undertaker, his proposal or contract had never been accepted and signed. He was thus free to charge what he thought fit. It was the many new branches which put up the cost:

> none that ever undertooke to keep a Waggon-Way in ye River in my memory tooke it without limitation of length of Branches, and I thinke you can satisfy yoselfe There never was any Colliery in the River that had either so many Branches or of so great a length occasioned by yor being obliged to seeke out the best coal in the different partes of your Grounds to help to support a sinking Trade:

> so that yo[r] Pitts lay at Great distance from each other viz[t] at Moor Gate from Leadgate, Buck's Nooke from Colebournes and even ye Colebourne Pits at a Great Distance from ye Main Way. You may see by ye vast Number of Sleepers I have put in ye Way since I entered, what a Condition it was then in – not one Waggon save 6 in ye way when I enterd but were stoppd with Turfe, Strawe or Ling – Pray see if there be now one Waggon that will loose a Coal as big as a Nutmeg. . . . I have put the Staith Waggons & Way into good Order wch were run to a ragg before I enterd.

Finally, he said, the stock of materials his predecessor had left behind was

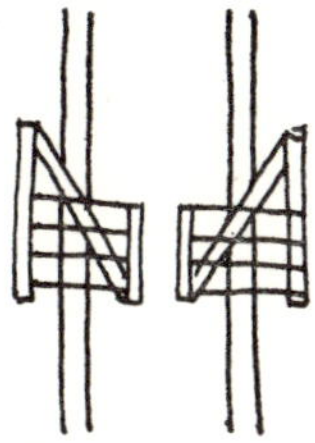

FIG 42 Tinkler Gate on Tanfield Moor waggonway, 1772.

worthless. Clavering was forced to pay, but was still rumbling about it in 1717, when he told Henry Liddell:

> I also remark'd y[e] present condition of y[e] Wagg: way [under a new keeper, compared] to what it had been for some years past, notw[th]-standing y[e] great sume Mr Cotesworth got by it, I had like to have said either cheated y[e] family or a friend of y[rs] of.[59]

At level crossings, traffic on a public road was usually protected by gates, with a gatekeeper living in a cottage alongside. Gates were occasionally erected where there was no level crossing, in cases where for some reason the waggonway was not fenced in and livestock had to be prevented from straying from one field to another. Pitt's Tanfield Moor waggonway in 1772 had three gates: Crookgate, where the main way crossed the bye way, and some sort of traffic control may have been practised; Tinkler Gate in the middle of nowhere, apparently on a field boundary (fig. 42); and Maugham Gate, immediately short of the junction with the Grand Allies' Tanfield way, where there must have been some organization to prevent collisions at what was the second busiest junction of Tyneside. The real Clapham Junction of the day also had its gate – Lobley Hill Gate on the Tanfield itself, where the Dunston and Redheugh staith lines diverged.[60] This gate was indeed on a road – in fact, on the exact site of the 1645 Waggon Gate of the Whickham waggonway where it crossed the wainway from Ravensworth (fig. 30), but one suspects that its main function was akin to that of the modern railway signal.

Gates were occasionally built and maintained by the coal-owner: the Heaton wayleave of 1724 includes the clause, 'Gates to be hung and kept by the Lessees and Damages for Trespasses of Cattle by neglect of keeping the Gates to be made good by Lessees'. Sometimes, too, the coal-owners built the gates and houses, but went halves with the way-keepers on the cost of repair and on gate-keepers' wages. Two gatehouses built on the Tanfield way in 1738 under this arrangement cost £27 17*s.* 3*d.*[61] Most often, however, the way-keeper alone was responsible for the gates. Albany Baker's Ravensworth accounts mention in 1727, 'Jam Wallas for makeing Haspes & Steuples &c for y^{e} Gates as p. Note $12/1\frac{1}{2}$'.[62] Gate-keeping was a light job, well suited to the elderly. We have met 'wedow Howborne at waggon yate' at Whickham in 1645; another appears at Stella in 1712 – 'To Cash pd Widdow Richardson for keeping Stella Freehold Gate 7/-'.[63] Payment was sometimes by the day – the cost on the Washington way in 1789 was a regular £1 2*s.* 6*d.* a week – but it generally varies so much that it was probably based on, and fluctuated with, the number of waggons which passed. Thus at Ravensworth, there was one bulk payment in 1729, 'For 3 Years Gatekeeping near ye bowbridge at Team £4 10. 0', but for the rest the weekly sums paid to four or five different men and women range from 6*d.* to 3*s.* Likewise at Plessey in 1715 the weekly cost varied from 5*s.* to £1 6*s.* 0*d.*; a typical entry reads: 'Gates keeping in the Way, p^{r} Thos Ellot & Sand. Knight £1.3.6.'[64]

It may be of interest to relate the transport charges to the cost of mining and raising the coal and to the selling price. A typical set of figures (though the profit was perhaps on the high side) refers to George Pitt's Tanfield Moor Colliery in 1712, estimating the costs if it were connected to the Bucksnook waggonway:

21 Waggons to a Ten: Working	£ 1. 12. –
Leading at 15^{d} p Waggons	1. 6. 3
Repaires of Wayes & Waggons	0. 14. –
Wayleave & Steathrooms & all other Charges at Staith	6. –
	3. 18. 3

Since the coal would sell at £9 4*s.* 0*d.* per ten, the profit would be £5 5*s.* 9*d.*

The document also compares the cost of waggon and wain transport. By road, Pitt could only move 350 tens a year, at a total profit of £656; by waggonway, he could move 1,200 tens a year, with a total profit of £6,345. Whereas by waggonway the cost of transport was nearly one and a half times the cost of the actual mining, by wain it was three and a quarter times the mining cost.[65]

It remains only to describe some of the highlights of a waggonway's life. In some cases, it was opened with the minimum of fuss. There is no record, for example, of festivities when the first waggons went down William Bowes' great new Western Way on 17 April 1721, and the completion of the Moorgate line of the Clavering Stella way in 1711 is commemorated by a laconic entry in the accounts, 'Jno Bainbridge for opening ye Bars when ye Waggons first went on' – in the sense of removing some barriers which had prevented premature traffic, rather than providing alcoholic refreshments.[66] But some openings were celebrated with no half-measures. Brand records in 1789:

> There is a tradition among the people belonging to the coal-works, that the first waggon that was used for this purpose in the vicinity of Newcastle, was lined with tin, and filled with the liquor called punch. It is easy to conjecture that the unlading such a waggon would prove a very grateful task to the thirsty workmen.[67]

This is almost literally true. We do not know about the first waggon on Tyneside, but the punch-waggon seems to have become the traditional way of celebrating an opening in the coal trade. The Willington waggonway came into use on 6 March 1776, but some months before, on 2 November 1775, the sinking of the first Willington pit was completed. The owners contributed a roasted ox, a sea of ale and a waggon-load of punch to mark the occasion. Similar festivities took place at the opening of William Joliffe's Waldridge Fell waggonway to Fatfield on the Wear in 1779, when

> some thousands of people attended, preceded by a band of music, colours flying, &c. In the afternoon they returned to the houses near the colliery, where an excellent cold dinner was provided, consisting of a sheep roasted whole, six sheep in quarters, and half an ox, which was washed down with 18 barrels of good ale. The bells at Chester were rung at intervals during the day.

But when the Middleton waggonway at Leeds was opened in 1758, there is no record of a beanfeast; merely that 'the Bells were set a ringing, the *Cannons* of our Fort fired, and a general joy appear'd in every face'.[68]

It is the opening of the Bucksnook waggonway, however, on 6 June 1712 which gives us the best foretaste of the brass band, cold collations, and seemly junketings which were *de rigueur* at any sod-cutting or railway-opening ceremony in Victorian Britain. But before describing the opening, a word must be said about the background, for nothing connected with the Bucksnook was ever clear-cut or simple. Sir John Clavering and Thomas Brumell, with Gilbert Spearman and William Wright, J.P. (a 'great waggonway merchant'), had obtained what they thought were sufficient wayleaves for building a waggonway from their Bucksnook and Byermoor pits to Derwenthaugh; but the land-owners on the route thought otherwise. Thomas Shafto was one; another, even more militant, was Lady Elizabeth Bowes, châtelaine of Gibside, whose sons William and George were not yet of age. She took personal and violent action against Clavering and Brumell. As early as March 1712 her brother-in-law, a Prebend of Durham who might have known better, advised her 'best to bring as many persons as you can into ye Riot . . . I see no reason for your Ladyship paying them any compliment in forbearing to beat up theire quarters'.[69] Under strong pressure from Lady Bowes, the 'Five' also came out against the waggonway. They were a cartel of five of the most powerful coal-owners on Tyneside – Henry Liddell, Alderman White, John Ord (the Wortleys' agent), John Wilkinson and James Clavering, a kinsman of Sir John's (the family was divided against itself), with William Cotesworth as secretary. On 29 April they agreed to incite the copyholders along the route to destroy the new Bucksnook way, 'and if they are opposed that then a joint force be raised to pull it up by force'. This is exactly what happened. Sir John caught Lady Bowes, staunchly supported by Cotesworth, red-handed in Fawdon Field, and a bloody brawl ensued. Sir Henry Liddell wrote in her support: 'The cause I hear well of; these wish it success and that you may ride triumphant with your Female Heroine who fac't the Baronet with courage beyond her sex and humanity becoming it.'[70] In the subsequent legal action, Lady Bowes adroitly contrived to wriggle out of the case – a mere injunction was issued to prevent her from interfering again – and Cotesworth was left, most unwillingly, holding the baby. We later find him sending a rueful note to his agent:

> Mr Bell I do heereby Desire you to pay Mr Fra Baker the Comon Agent for keeping Bucksnook Waggon Way in Repair Seventy One pounds five Shillings being two thirds of the money laid out by him for repairing the Waggon Way in Fawdon feild wch was destroy'd & demolished by Mrs Bowes, Deducting out of the said Sume seven pounds four shillings for Sleepers furnished by me towards repairing the said Way.[71]

Sir John and Brumell would not be put off, though the opposition was still on the warpath. James Clavering wrote to Henry Liddell on 3 June:

> Ye new waggonway will do harm. Lady B has got ye injunction I am told dissolved & T.S[hafto] obtained an injunction for keeping him in ye peaceable possession yet at ye same time ye other side talk as big as ever & have appointed Whittson Monday [June 9] for ye procession of ye punch waggon, therefor time must try if force can overcome right . . . a waggon way wthout a colliery is a pretty thing. . . . (later) I mett my cheife [Sir John Clavering] their who told me had intercepted severall of Lady B notes of Summons to our tenants and Servts to attend as yesterday in order to pull up ye waggon way; that he had sommoned ye posse com[itatus], & if any did attempt to pull up, thay wld be taught their duty; so I expected att my return to have heard of a battle in Fawdons feild.

His next letter to Liddell, on 8 June, describes the final outcome:

> Ye waggon way successfully on, last Fryday ye punch waggon down in triumph, ye whole country almost was invited. Ye procession too large for a letter, therefore yo must be content wth only a few hints. Parson Walker led us, followed by ye cheife justice [William Wright] on one hand of sr John, Jack Spearman on ye other, ye fidlers & pipers intermixt wth ye whole company. Ye first waggon was filled wth Byermoor coles, ye 2d punch, & ye 3d &cra Bromells. Wn they came to Tho Shaftoes ground he stoped them & behaved himself like a hero, ye cheife justice commanded ye bayliffs to take him into custody, wch they not readily obeying, 4 of Sr John servants was ordered to do it, & did carry him of a prisoner. My Lady was their I must not omit telling yo, & that yr humble servt was not invited, because ye justices precept to ye Sheriff for raising ye posse com was grounded upon an information agt him, who is very ignorant of ye matter, but shall resent it in due time. . . . I had like to have forgot mentioning after ye punch &c was demolished att Steath, great many of ye company went up wth Sr John to White House. In ye midst of their jollity & cups, they were allarm'd wth tydings that Lady Bowes was wth 200 men in Fawdons Feild pulling up ye way. Ye whole company to horse immediately, & messingers dispatch'd to all ye neighbouring villages for aid, their was such ryding & tumbling as ye like has not been known. Wn they arrived att Fawdens feild all ye beast in ye pasture had unluckily gott on ye way, wch at a distance appeared to confirm ye report, a halt was made & a councel of war called how to

> attack these horn beast; I w^{ld} gladly have a ballad made of all this; put it into some good hand, print a few & send us them down.

Celebrations over, and Lady Bowes having shot her bolt, it was time to get down to business. James Clavering wrote on 13 June: 'Y^{e} new waggon way continues to make a great noise & will be perfected in a few days now, I am told ye waggons were to be on yesterday.'[72]

Tynesiders were not yet much enamoured of law and order. In 1713 various stewards were instructed by the Gateshead Vestry

> forthwith to use the most proper methods and likelyest means to oblige the Bensham Owners to pay what money is due and in arrears for wayleave . . . and that the Grassmen be secured from all inconveniency that may attend them in Pulling the wayes up or using any other means that may be thought necessary by Councel.[73]

The victim, whose name does not appear, complained to Lady Carr:

> I was no sooner begun to worke & lead coals to the River wch I laid on a staith belonging to Redhugh Estate & close adjoining to staiths farmed by Sir Ra Carr but I was immediately made sensible that I was too near y^{r} gentleman, for his tenants & agents encouraged by I know not whom gave me great disturbance in leading my Coals through Redhugh Grounds, they stopd my Carriage, turned me out of the way that I had agreed with Mr Errington for, locked up my Gates in the Milbanks Lands . . . indicted me for asserting my Right of makeing a way through the said Milbanks grounds. Sir Ra Carrs staithman an intimate of Mr Erringtons came upon me with above 20 Rude fellows & they got me down in the feild Bet & bruised me sore, spilt my blood & left 5 marks of their wounds on my body wch I bear this day.[74]

Another method of stopping a waggonway, apart from pulling up the track and assaulting the owner, was attempted at Plessey in 1720. Between 4 and 5 a.m. three waggon-men discovered two labourers of Robert Wright and John Spearman, the owners of West Hartford colliery,

> trespassing or digging on or about the south side of the Waggon Way in West Hartford Moor belonging and appertaining to Plessa Colliery in order to put down (as they said) a Post or Rail there and another opposite as these Deponents believe in order to put up a Barr to stop the coal waggons which pass that way from Plessey Colliery to Blyth.

After being evicted, the intruders returned and began to sink a pit in the middle of the bye way. On being questioned they said that they

> were ordered to throw all the Rubbish and Earth that came out of this intended Pitt upon this s^{d} Waggon Way . . . in order to stop the leading of Coales. . . . They digged up some earth and threw the same with a shovell into the s^{d} Waggon Way betwixt the Rails with a design to hinder the Waggons from going along the same.[75]

Disturbances of a different sort arose in Sheffield in 1774, on the building of the new waggonway. A contemporary letter describes the event.

> A few days since [in late November] we had a very riotous meeting of the populace of this town, occasioned by an imposition which some persons belonging to the Duke of Norfolk had the cruelty to put in practice, with respect to the price of coals got in the Duke's estate, adjacent to the town. These merciless wretches formed a scheme the beginning of last summer to bring all the coals got in the Duke's land about the town into a yard at the outside of the town, by means of large carriages on low wheels, which run on a road made of timber, in imitation of one at Newcastle. This road, together with their carriages, being ready for use, they have begun to bring coals to the intended place, where they exact from the buyers almost double the price they were sold for before at the pits; and in order to compel the people to comply with their exorbitant demands, they have totally put a stop to the usual delivery of coals at the pits, and also refuse to sell less quantities than a horse load, which is a great hardship to the poorer sort of people, unable perhaps to spare more than a penny at a time to buy coals with. These were alterations so disagreeable to the populace, that they assembled in a prodigious number, and destroyed several of their carriages, totally pulled down a watch-house and compting-house in their new coal-yard, and set fire to all the timber machinery erected for discharging their loading, brought one carriage in triumph through the town, afterwards kindled a fire in, and sent it flaming into the river.[76]

John Curr, the engineer, is said to have been 'reduced to the necessity of concealment in a wood, for three days and nights, to escape the fury of the populace'.[77]

Waggonways were also at the mercy of striking miners. In February 1733/4 the pitmen at Newbottle struck over an alleged increase in the size of corves, and threatened to pull up the waggonway, destroy the fire engine, and tear one of the partners, John Nesham, limb from limb. To defend the engine-house against siege, a supply of men and muskets was smuggled in, secreted in a ballast waggon to avoid being spotted by the pitmen. In the resulting affray a miner was mortally wounded.[78]

A gang of twenty-two roughs – possibly disgruntled waggon-men – was charged in 1766 with

> a Riot and hindring and obstructing Seven Coal Waggons [on the Tanfield way] . . . from passing along the Waggon Way . . . and for throwing Wasting Scattering and destroying two waggon load of Coals . . . and for riotously assembling . . . armed with offensive weapons for the space of half an hour to the Great disturbance and terror of the liege subjects of our Lord the King.

Each was fined 1*s.* 8*d.* and jailed for three months. A similar hold-up occurred on the Western Way in 1775, again for no obvious reason. One Robert Beverley was charged with unlawfully stopping a number of waggons and preventing the waggon-men 'from loosing their Horses from the said Waggons for the space of Eight Hours and upwards and for breaking and destroying such waggons and consuming the coales therein'. He was sent into the Navy.[79]

Waggonway materials were a standing temptation to thieves. The Clavering Stella accounts in 1712 record: 'To Cash pd for a Warr^t^ & prosecuting 3 people for Stealing Wood off ye Way 4/-.'[80] We have already seen a colliery worker at Greasbrough in Yorkshire accused (but acquitted) in 1766 of removing wrought-iron plates from the rails.

When the end came for a wooden waggonway, death might be due to one of several causes. Most often, the colliery it served was exhausted or flooded or burnt out, and the way had no further object in life. The track would be lifted – it still had a value of between 7½*d.* and 1*s.* a yard – and the rails and sleepers sold as firewood or fencing posts. If the earthworks were slight, as they usually were, the land would revert to agriculture, and today only the larger embankments can still be traced. Sometimes a waggonway would close because a better one had been built, as in the case of the Dunston Way and the old Western Way. Only once, it seems, was a line closed for 'political' reasons – the unfortunate Bucksnook again, which was put out of its misery on 28 June 1726 to spite the Grand Allies, the very day after the Alliance was formed. The actual closing was done by James Clavering, who described his action to his lawyer:

> I went on Tuesday and stopped Bucks nook way, the waggons were going wch obliged me to pull up in two places, to stopp both the full and empty waggons & whats most remarkable the two foremost of each belonged to Cotesworth. The full waggons in the evening

> standing all in a row made a fine figure and barricade; wch I have released this morning on a petition, suggesting the danger the Coles standing there were in of being stole; I considered the admitting them to pass could be no prejudice, & a refusall might look like malice prepense; if Sr Fr[ancis Clavering] has entered into no agreements, we're safe. Then dined with Bowes at Gibside – gave consternation to his new allies.[81]

Most frequently, however, wooden waggonways survived long enough to be converted into iron railways. This process had begun on Tyneside in 1797 at Lawson Main Colliery on the Heaton way and at Benwell, but it was a gradual one. Wooden waggonways were still at work in their original state in the 1840s. The story of how the iron rail drove out the wooden will be pursued in ch. 15.

The Shropshire railway: I

. . . Hence is seen
The New Made Road, and wonderful *Machine*,
Self-moving downwards from the Mountain's height,
A *rock* its Burden of a Mountain's weight.

Mary Chandler, *A Description of Bath*, 3rd ed., London (1736), 16.

Wooden railways in Lancashire, North and South Wales, and at a few places in Scotland, owed little allegiance to Tyneside practice, but were apparently inspired by the example of the west Midlands, and especially of the coal-mines and iron-works around Broseley and Coalbrookdale in Shropshire. Sometimes the link between colony and motherland is recorded in documents, but more often it has to be inferred from the distinctive features of the railway. Although our knowledge of this group is patchy and unsatisfactory, a dim picture does emerge of a line of evolution springing perhaps from the same roots as the Newcastle waggonway, and occasionally influenced by it, but none the less remaining independent for a long while. The differences are most marked in the terminology (Shropshire invented and preferred the words *railway* and *railroad*), in the smaller gauges, and often in the smaller waggons. One reason for these differences was the different situation of the coal in the Severn field. Here, unlike most of Tyneside, it lay close to the surface, often outcropping in the steep banks of the river, and was won as much by level drifts or adits as by vertical shafts. Railways laid underground often ran straight out of the adits and down to the river without interruption; and gauge and waggons therefore had to be small for underground work. We have seen that this was true as early as 1608 (p. 99), and just such an arrangement was noticed a century later at Madeley by Francis Brokesby, where, he wrote in 1711,

> are considerable Cole Mines into which they descend not, as in other Places, into Pitts; but go in at the Side of a Hill, into which are long Passages, both strait forward, and from thence on each side; from whence they have dug the Coles: which, by small Carriages, with four

39. *Above* Flanged wooden wheel from Broseley.

40. *Below* Loaded and empty waggons on the main and bye ways. On right, staith. Tyneside, 1787 (Gibson).

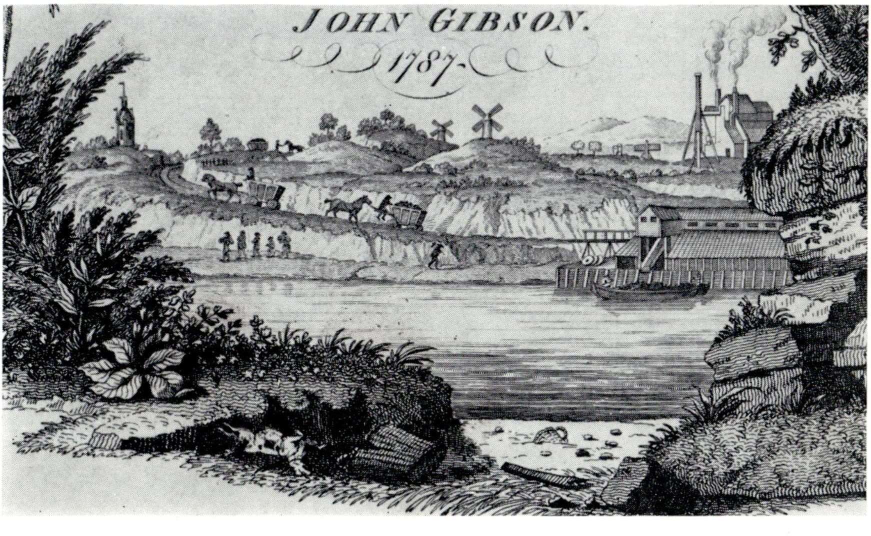

41. *Above* Cut on the Bucksnook waggonway of 1712.

42. *Below* Battery on the Bucksnook waggonway of 1712.

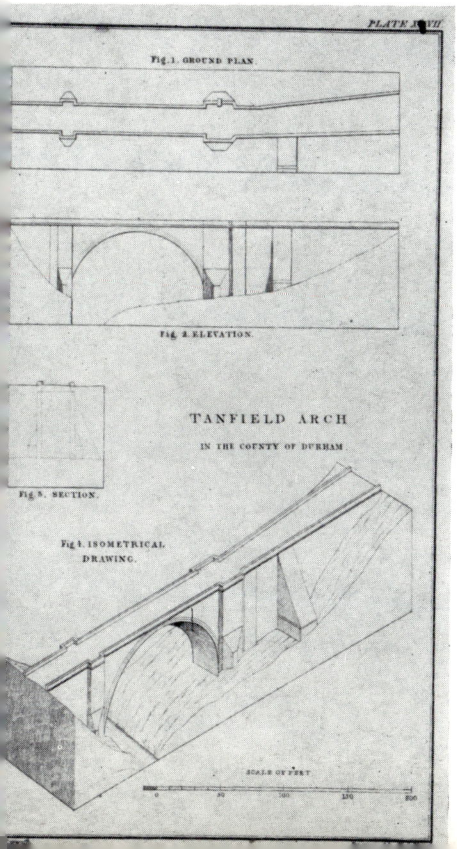

43. *Above* Causey Arch in 1804.

44. *Left* Causey Arch in 1834.

45. *Left* Causey Arch in 1967.

46. *Below* Plan of Derwenthaugh staiths, 1724. North is at the bottom; River Derwent (stippled) down centre, River Tyne across bottom. Western waggonway (n) to staiths Z and &, Bucksnook waggonway (4) to staiths 5 and 6.

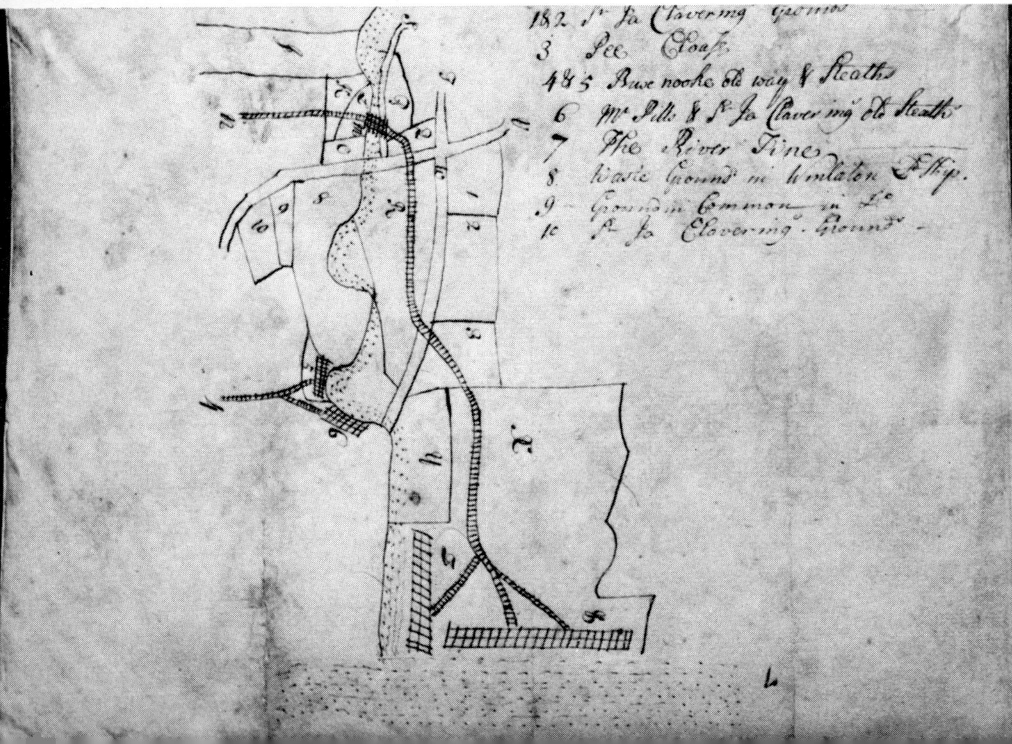

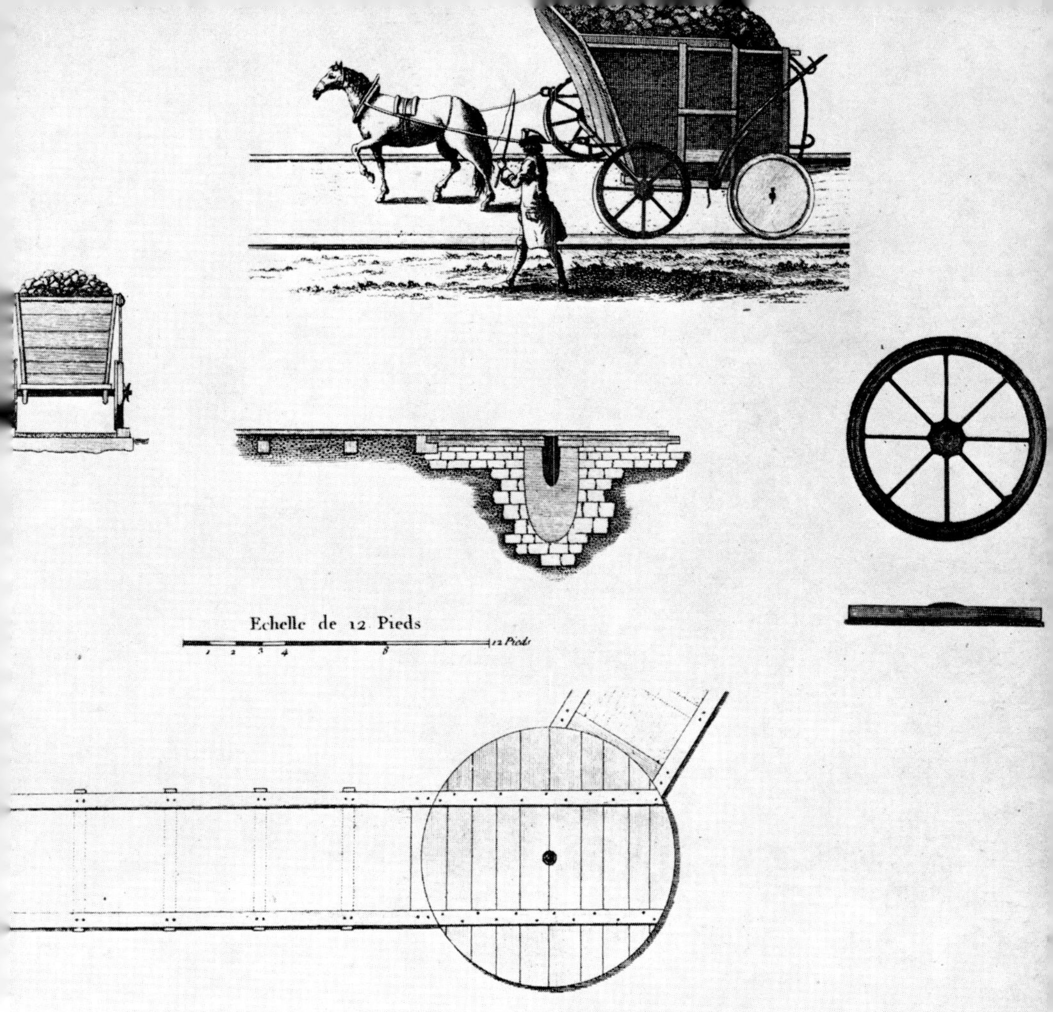

47. *Above* Waggon, cast-iron wheel and turnrail, Tyneside, 1765 (Jars).

48. *Below* Tyneside coal waggon, 1764 (*General Magazine*)

49. *Above* Tyneside waggon on gravity run, 1773. Staith in background (Morand).

50. *Below* A run on the Parkmoor waggonway, with Newcastle in the background, 1783 (Brand).

51. *Above* Coal waggon, probably on the Willington waggonway, *c.* 1815.

52. *Below* Horse gin, Newcomen engine, keel and waggonway, Allerton, Yorkshire, 1786.

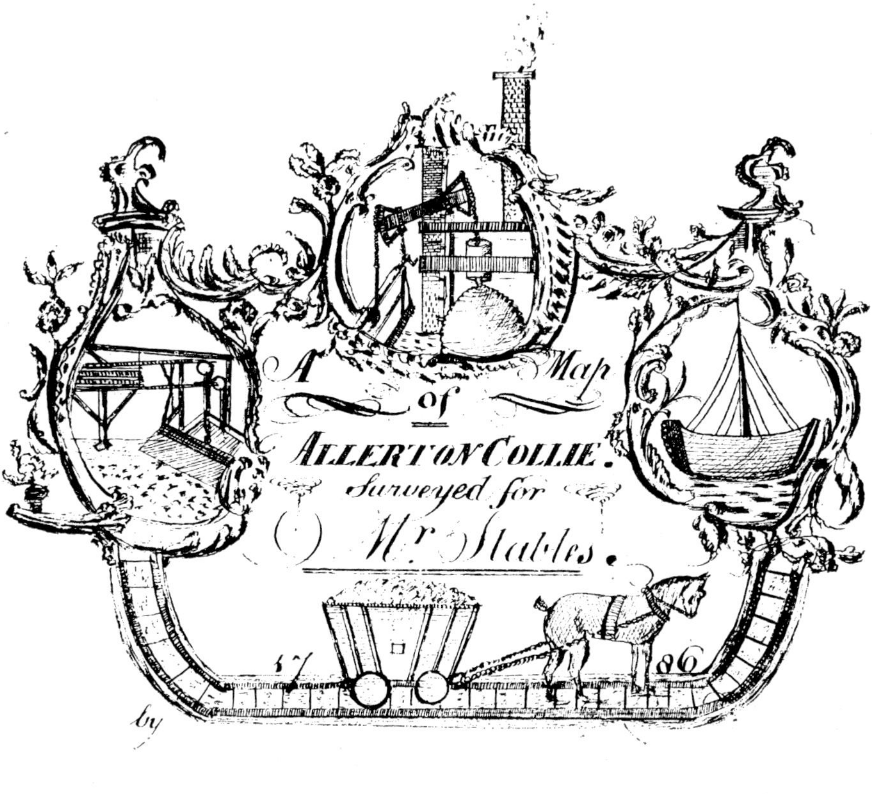

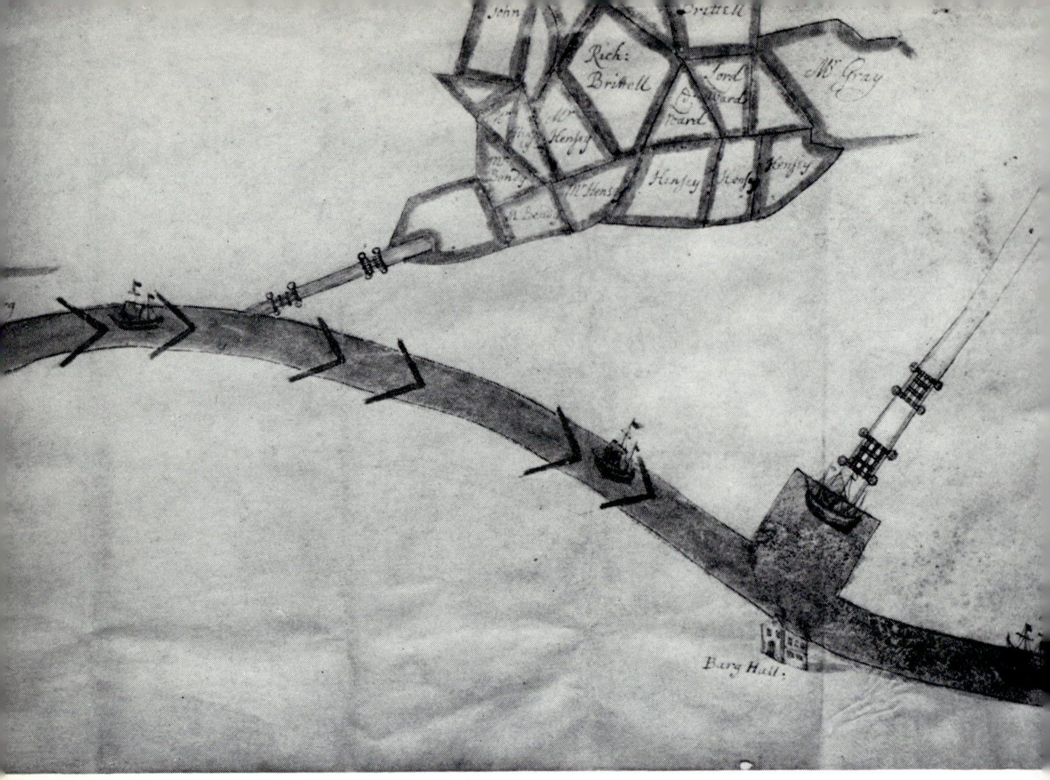

53. *Above* Map of upper end of proposed Stour navigation, *c.* 1662, showing two footrails each with two waggons (Yarranton).

54. *Below* Coal-pit and train of waggons, Neath Abbey, 1805.

55. *Above* Coal staith at Landore, with ox and waggons, 1796 (Ibbetson).

56. *Below* Ralph Allen's waggonway and Prior Park, Bath, 1750.

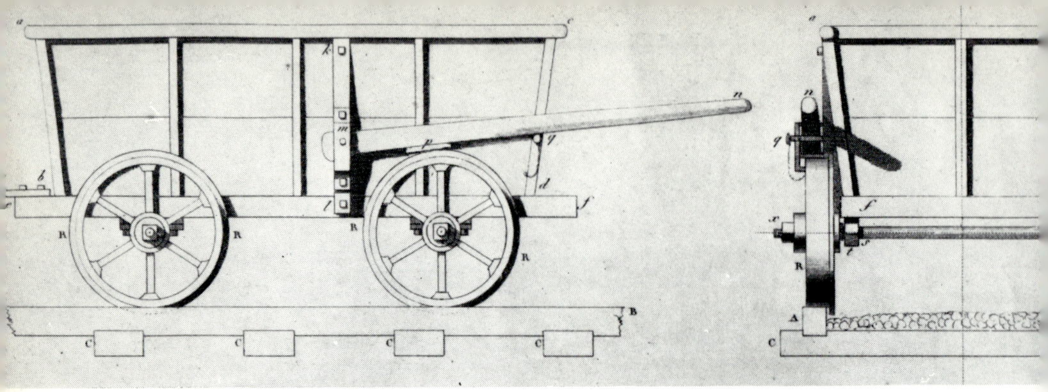

57. *Above* Small Shropshire waggon on wooden rails (Baader).

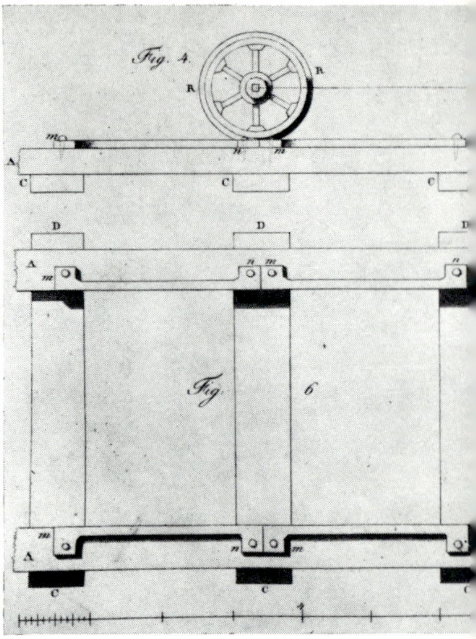

58. *Right* Cast-iron rails on wood (Baader).

59. *Below* Mouth of a coal-pit near Broseley, 1788.

60. *Above* Large waggon and team on Jigger Bank, Coalbrookdale, 1777 (Williams).

61. *Below* Small waggon and team, Coalbrookdale, 1801 (Loutherbourg).

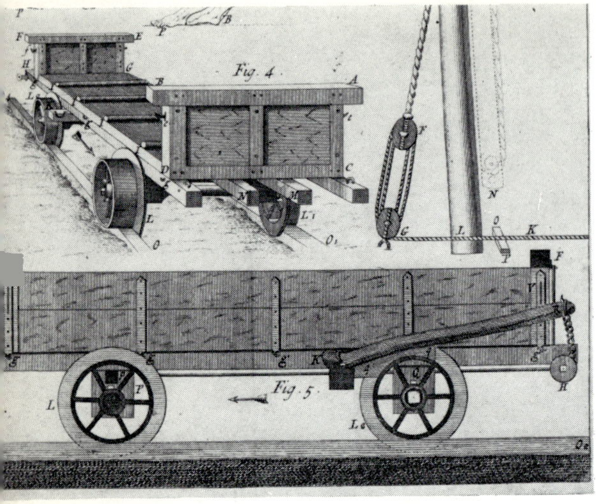

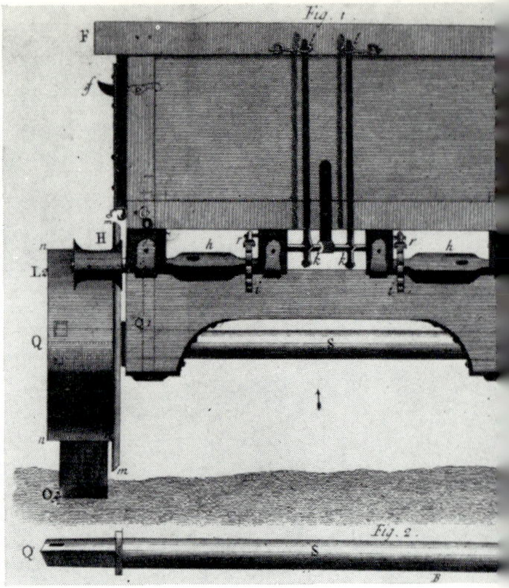

62/63. *Above* Bath waggon, 1734 (Labelye).

64. *Below* Shropshire Canal incline head, 1797.

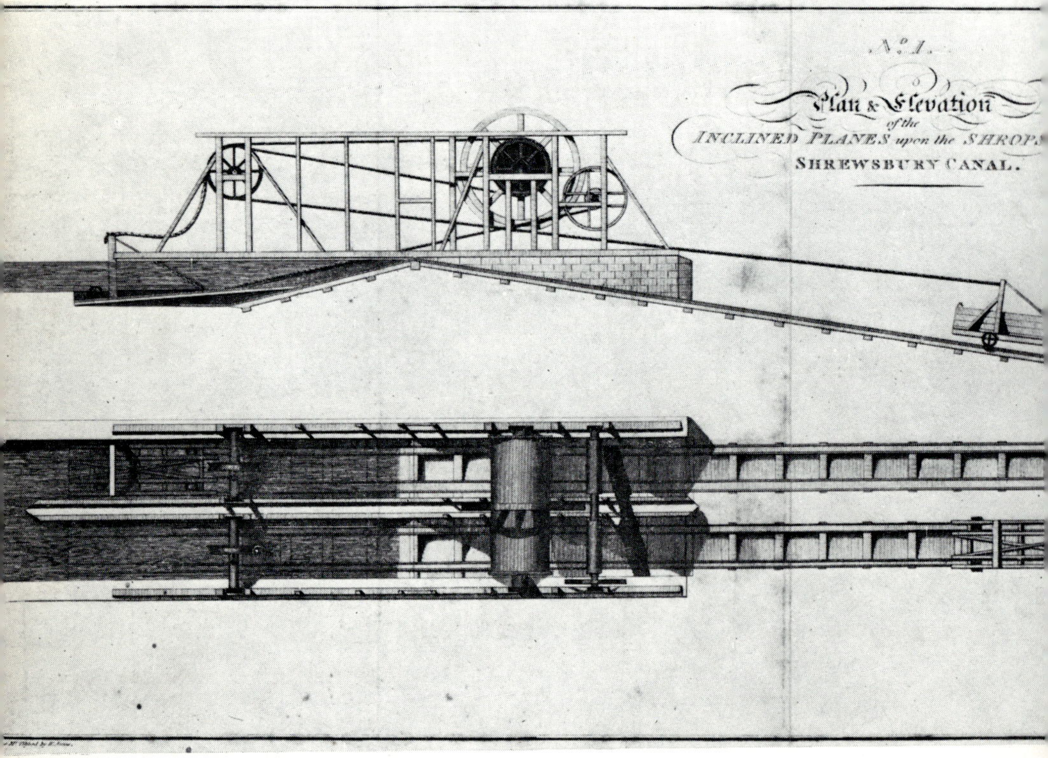

65. Wooden edge rail and sleepers from Groverake mine, Blanchland (York Railway Museum).

66. Remains of wooden rails, Deeside railway, Glyndyfrdwy, 1960.

67. Contractor's wooden railway in Midlands, 1848 (Tait).

68. Kibble and English waggon from the Harz in Zellerfeld Heimatmuseum.

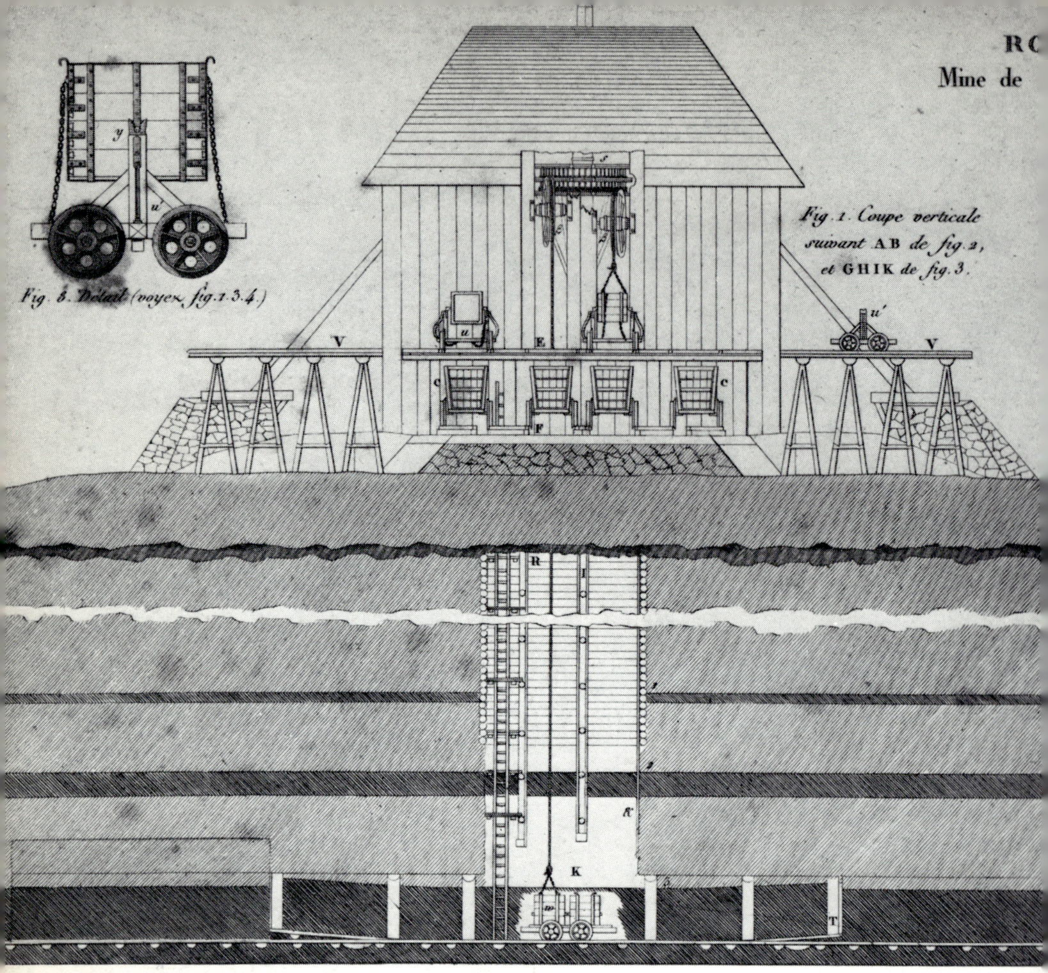

69. *Above* Königsgrube gallery, shaft-head and (top left) container on truck, 1819.

70. *Below* Underground English waggon with four axles, Königsgrube, 1819.

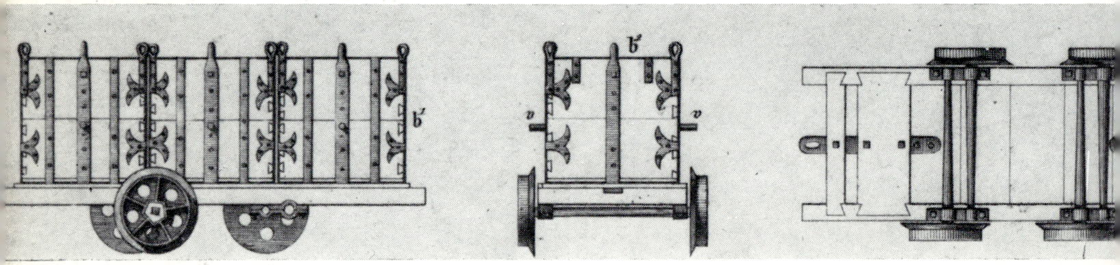

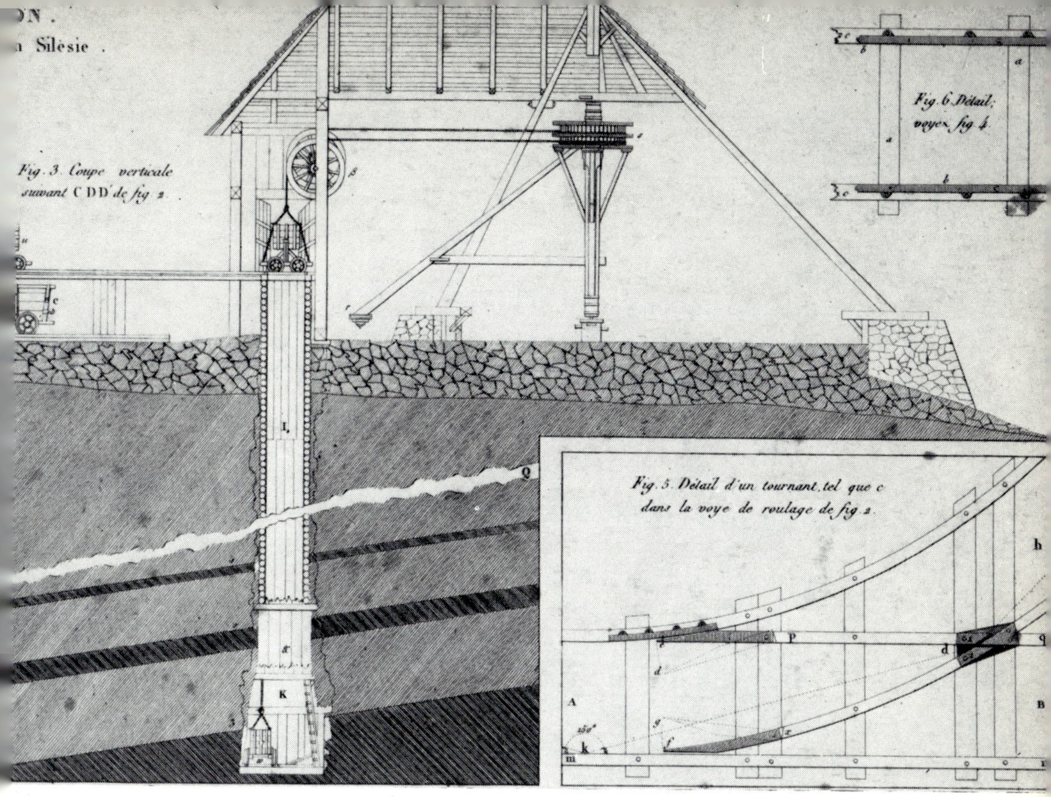

71. *Above* Underground points, shaft-head and surface cast-iron rails, Königsgrube, 1819.

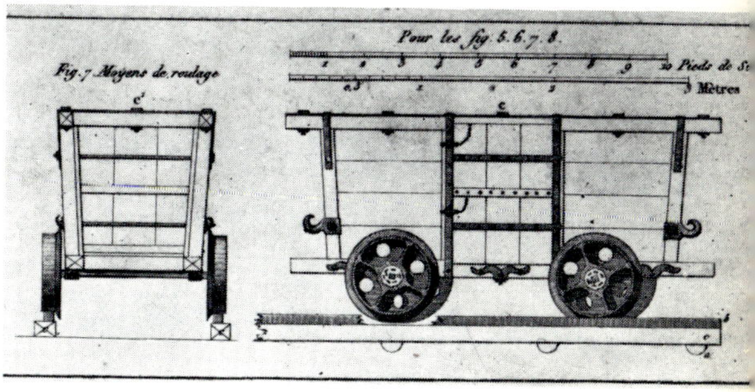

72. *Right* Surface waggon, Königs-grube, 1819.

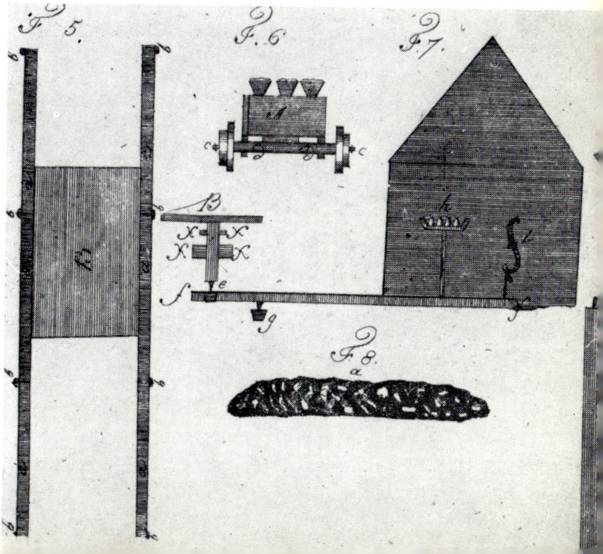

73. *Right Lower* English waggonway, waggon and weighbridge, Le Creusot, 1789.

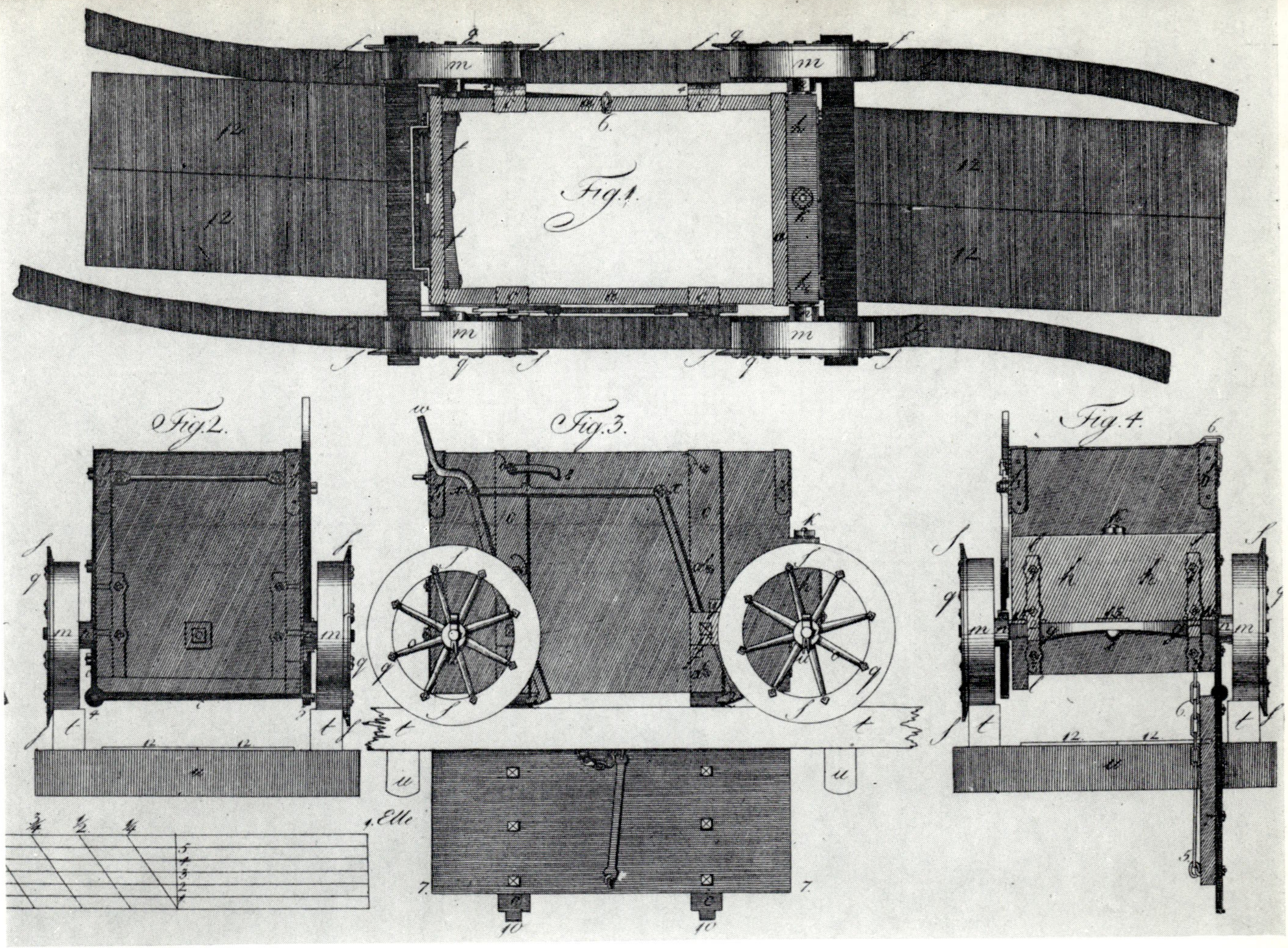

71 Outside flanged waggon, Schneeberg, 1794

> Wheels of above a foot Diameter, thrust by Men, they convey not only out of the long underground Passages, but even to the Boats which lye in the Severn ready to receive them; a sight with which some years ago I was not a little pleas'd.[1]

The word originally used to describe these drift-mines was *inset*, which began to be replaced in the 1660s by *footrid*, or its variants *footrail*, *footroad* or *footridge*. This is derived from *rid* or *ridding*, the overlying soil removed in opencast work. The Wednesbury Parish Register for 1577 records that 'Christopher Daly was buried: he was killed in the ryddinge, in the cole-pit';[2] and *footrid* originally meant an opencast working. Dud Dudley described it in 1665:

> Where the Coles is deep and but little Earth upon the measures of Coles, there the Colliers rid off the Earth, and dig the Coles under their feet; these Works are called Foot-rids.[3]

By extension, the word was also applied to horizontal drifts. Robert Plot wrote of Staffordshire mining in 1686 that sometimes

> the Coaliers can work a *cuniculus* [adit] into the side of the *Hill*, according to the course of the *row* or *streak* of the *coal*, for some hundreds of yards in a direct line upon a level, through which they bring forth the coal as easily altogether as in a *flat open work*; as they doe at the *Footerill* at *Apedale* [near Newcastle-under-Lyme] . . . [which] is driven in them upon a level about 200 yards.[4]

Footrail is still used in this sense in Staffordshire. By a further extension, *footrid*, like *inset*, came to mean a railway laid in an adit, and even a railway unconnected with an adit. A wayleave agreement of 1670 gives liberty to

> carry and convey to the River of Seaverne all the Coles that can shall or may be gott in the land of the sayd William Crompton and William Jerman situated in Broseley aforesaid through the lands of him the said ffrancis Huxley Situate in Broseley aforesaid by way of Insett or footerid.[5]

The railway had first taken root in Shropshire on the south bank of the Severn around Broseley, Willey and Benthall (fig. 43); we have already seen how a fair number of lines were built or authorized between 1605 and the Restoration. During the later seventeenth century, and all the eighteenth, the process continued. The evidence is fragmentary and mostly in the form of legal documents, and much research has yet to be done before a connected history of the Broseley and Madeley coal-mines and their railways can be compiled. Moreover, the nature of the material is less tractable than on Tyneside, where the mines tended to be large and

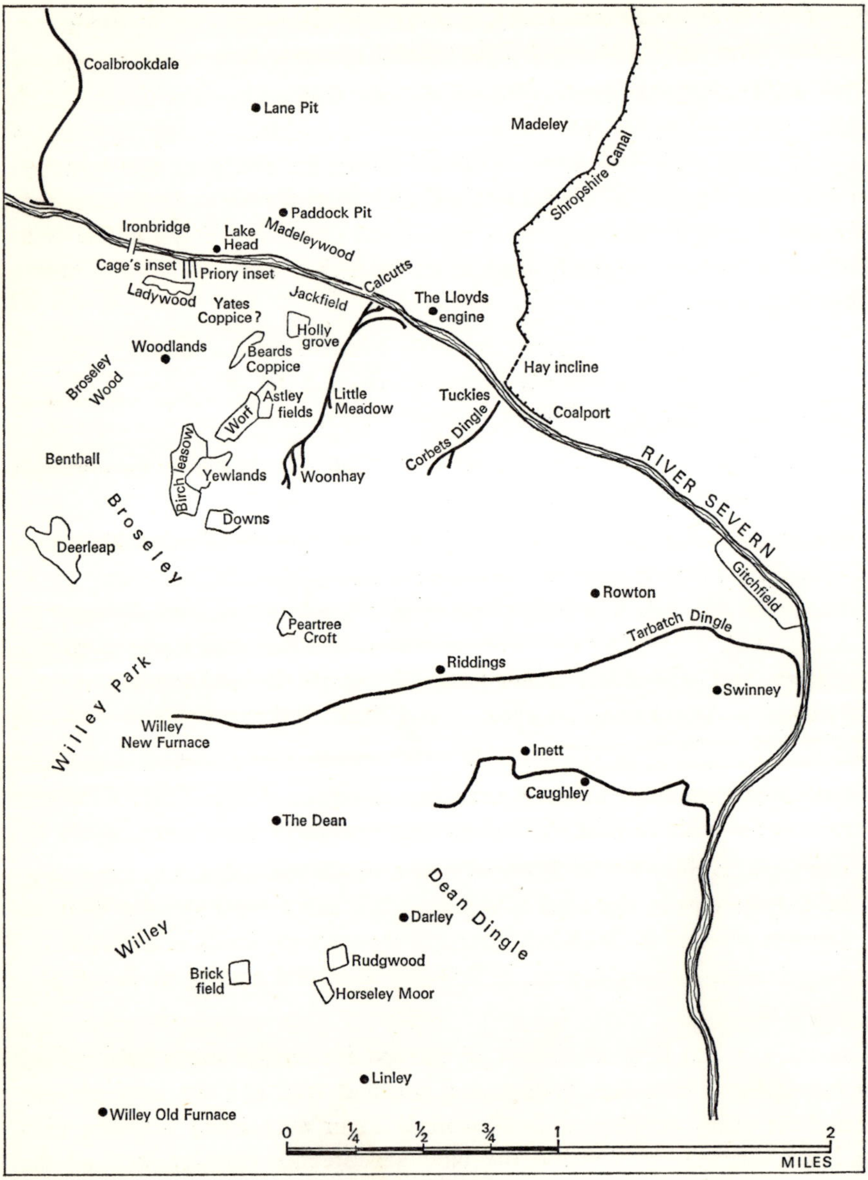

FIG 43 East Shropshire coalfield, southern half.

their waggonways of a certain length. Here there was much mining on a very small scale in a restricted area. The documents often give no precise indication of where a railway was, and, unlike Tyneside, there is an almost complete dearth of early maps that mark railways. But as far as can be disentangled at the present, this was the development of railways in the district after the Restoration.

At the west end of Broseley there were still a number of insets in the Ladywood area. In 1686 John Huxley leased his mines in Yates Coppice and Ladywood Meadow to Sampson Crompton, who 'Within and from either of the mouths of the said Insetts shall have liberty to fix and sett one Rayled Waggon Way thourowe the said parcell of land to the River of Seaverne'. Crompton might fell three oaks which obstructed the making of the waggonway, and at the end of his twenty-one-year term might 'take up and carry away All such Timber Rayles and other materialls'.[6] The colliery called 'Foot Road near Crumptons Boat in Broseley' in 1761 may have been a survival of this mine;[7] but it does not seem to have lasted much longer.

Whether the lines of 1605 and 1607 down to Calcutts enjoyed a lengthy life we do not know. Certainly a lease of 1673 by Francis Huxley gives leave to sink pits in Calcutts and Peartreecroft, which had been demised to him by Richard Young, 'and to lay Rayles and stacke Coles More Barges and leade the sayd Coles soe gotten'. Later in the same year Samuel Langley, still hammering his claim to the land, made a complaint that the lessees had

> sunke pitts made or driven ffooteridds Insetts or Additts waggon wayes gate wayes [underground levels] or other passages for ye getting of Coales in ye sd peece or parcell or land or ground called . . . Calcotts, Astly Feilds, Perry furlong [Peartreecroft], Yew Lands and Burch Leasowe and Little Meadowe,

and had likewise 'leavled sevall waggon wayes gate wayes Comegate wayes or windways . . . in lands . . . called . . . Downes and Worfe'. This list of mines around and above Calcutts suggests that a main line existed on the same, or very nearly the same, route as that of 1605, with a complicated system of branches. The very next year Huxley authorized another such branch to Hollygrove footrid.[8] In 1697 the Calcutts railway achieved the distinction of being the first to appear on a printed (but exceedingly crude) map, its lower reaches being marked as 'The road from the Coal-pits to the *Severn*', passing and serving Martin Ele's pitch and tar works at Jackfield.[9] The Langleys' claims to Calcutts and much of the disputed Broseley Manor apparently triumphed, for the property was inherited by Edward Purcell through his wife, Herbert Langley's daughter. In 1726 Purcell is found leasing all his coal in Broseley to Michael Stephens, except those got through the New Footroad. From 1728, however, he began to sink into debt, and sold off parts of his estate. A view of his assets throws some light on the state of his mines:

> Purcell is also a partner in a Colework called the footrid, having a 3d share, his part of all tools, waggons and implements of the colliery, now worth £30. Also is a partner wth Wm Crompton and Wm Ashwood in several pits, his 3d share of Gins, Ginhorses, Ropes and Barrells, Waggns and implements of colliery there are worth at least £30. Also he hath a 4th share in a work called Woonhay Colework, his share of those implements of colliery are worth £30.[10]

An estate map of the same date, with additions up to the 1760s, shows the railway running down from six pits around Woonhay to Calcutts and Jackfield Wharf. The Calcutts line survived well into the nineteenth century.[11]

Half a mile downstream, the same map marks a short line down Corbetts Dingle to the Tuckies, serving some very ancient pits which had once belonged to James Clifford (see p. 108).

Further still downstream beyond Coalport was a wharf at Gitchfield where coal was shipped from a footrid immediately behind, and from Rowton, a mile inland up Tarbatch Dingle. Gitchfield and Rowton pits were originally worked by the Old family, with Michael Addenbrooke and Richard Edwards, who gave to Messrs. Manning and Taylor in 1702 a wayleave to the Severn from both footrid and pits, and leave

> to sink pits, drive levels, lay down way-boards, make railroads, and convey coal, ironstone and limestone, from Rowton, by their servants and workmen, to Gitchfield meadow, to trows and barges on the Severn, paying in addition to the Royalty to the owners, 3d for every usual ton or waggon of coal or ironstone; also to carry from the Severn, hay, corn, wood, &c, for the use of the works.

Before long Michael Stephens took over the lease. He sub-let Rowton pits to Thomas Sprott and Nicholas Harrison 'master collier', together with leave for winding water, repairing the rails for the conveyance of coals, and

> for getting coals in a certain place adjoining Rowton called Tar Batch Dingle, and the use of the rails from the said pits to the river Severn, as the same are now laid by Michael Stephens, paying him 2d per ton, out of their moiety of the proffitts.

In addition, Thomas Crompton was assessed in 1715 for a 'footridge £30, and for a footridge at Gitchfield £20'.[12]

This early Tarbatch Dingle railway was to become part of a more important line. In 1757 the Willey Company, which had built a new furnace some distance to the north of Weld's old furnace of 1618, laid a connecting line to the existing railway down Tarbatch Dingle. This was by virtue of a wayleave from George Forester (inheritor of the Weld estates) to John Wilkinson and others, with right to 'lay Rails and make a Waggon Way' over Forester's ground from the coal and ironstone mines to the furnace and from furnace to Severn. One track was evidently not enough, for in 1759 William Bromley granted the Company the right

> of their own costs and Charges to lay new Rails and make a double Railway adjoining the said Railways so that the whole double Railway does not exceed in breadth ten yards, with liberty of passage with Horses and all other Cattle and with Waggons and all other Carriages whatsoever.

The Company was to pay 'twelve pounds a year in satisfaction for the damage and trespass', and at the end of the term they were to remove all rails and sleepers on Riddings Farm and to level the land, but leaving the rails in Swinney Farm for Bromley's use.[13] This line survived, with modifications, for at least seventy years more.

Lastly on the south bank of the Severn, a railway was laid before 1780 from Caughley pits past Inett to the river half a mile below Gitchfield.[14]

On the north bank there was much mining in the mile-long stretch which embraces Madeleywood, the Lloyds and their hinterland. The earliest mention of railways here occurs in 1692, when Basil Fitzherbert, John Purcell and Basil Brooke leased out some iron- and coal-mines just west of Madeley. There was to be a flat payment for 'levelling Ground & laying Railes of saw'd Timber, for Waggons loaded with Coles to pass upon, from the Lane Works to ye River of Severn being in length by measure 1500 yards or thereabouts'. Liberty was granted 'to sink Pitts, make foot Railes & Heads to search for Coles . . . & Waggon Ways to Make & Railes to lay . . . for ye Carrage & Conveying of ye said Coles . . . with Carts Waggons Oxen & Horses & any other Carrage whatsoever'. Lane Pit is only half a mile from the Severn as the crow flies, due north of Madeleywood (fig. 43), but the line doubtless took a longer route to reduce the breakneck gradient. Another coal-mine, Holland's footroad, was also opened at an early date, nearer the Severn and possibly that which Brokesby described in 1711. In 1715 a dispute arose as to whether these works were in Madeley glebe land, as the Vicar, Jeremiah Taylor, claimed, or belonged

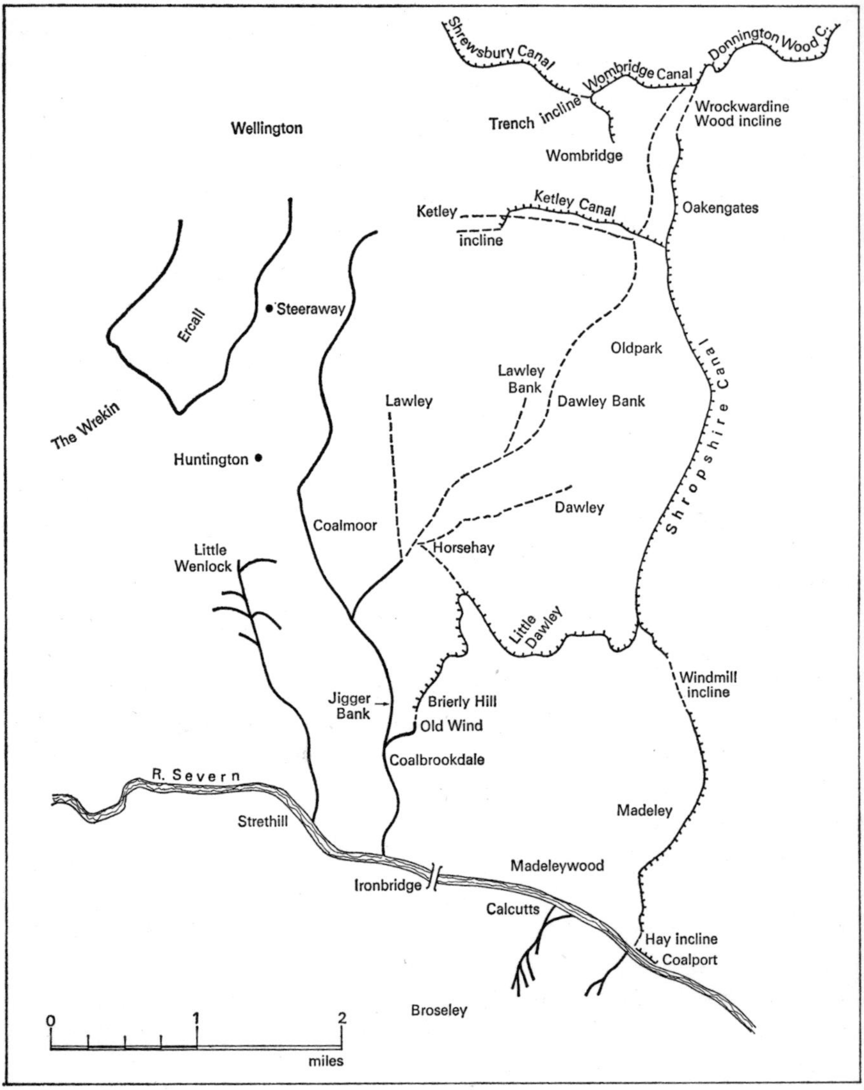

FIG 44 East Shropshire coalfield, northern half. Routes marked with broken lines are completely conjectural and are inserted only to give an idea of the extent of the system.

to the Manor. The lessees of Holland's footroad (including Nicholas Harrison, whom we have met at Broseley) pleaded that the coal they were working 'lyes a great way (Viz^{t}) six or seven hundred yards or thereabouts from the mouth or entrance of the foot Road', and was not under the glebe land of New Leasow; and that anyway their profit was negligible. They sold coal at 5*s*. 3*d*. per waggon, but the 'digging mining waggon wayes wood and other charges . . . where the Work is in good order costs these Defendants about four shillings three-pence p Waggon', in addition to the rent of 6*d*. per ton. But the case was decided in the Vicar's favour, and in 1722 we find him leasing Holland's footroad back to the original lessees. In the same way he leased in 1726 to George Weld and Basil Brooke a mine in the New Leasow glebe land, at Smiths Gin Pit in the Lloyds, 'where ye fire Engine now stands' – the earliest certain mention of a steam engine erected in Shropshire – with liberty to use an existing 'Waggon way or ways & Wind ways from ye said Smiths pitt'.

The Madeley Coalwork accounts for 1741 show that, out of a total of £895 spent on mining and transport, the waggoning cost £72 and repairing the rails £14. At the Lane and Paddock pits, both now in the glebe land, an inventory of the same year includes many waggons, wheels and axles, as well as 'horse gears and Rails from the pitts to Severn, £31.1.0'. Holland's footroad was still at work in 1747.[15]

Railways first appear in the accounts of the great Coalbrookdale iron company in 1748, not connected with the works themselves, but a mile away on the bank of the Severn, immediately below the site of the later Bedlam or Madeley Wood furnaces. Here the Lakehead footrid was being dug and provided with a railway, and a waggonway was being laid from the ironstone pits leased from George Weld. The expenditure appears in the accounts in such forms as 'Laying the Rails & making Waggons for the Lake Head footrid', 'levelling the Railway', 'driving Waggon Way', and 'laying Rails at Esqr Welds'.[16] These and similar items are entangled together, and it is not clear if they refer to two lines or to one. Weld's pits were at Amphlett's Leasow, whose site is unknown. But by 1757-8 three separate railways converged on the new Madeley Wood furnaces at Lakehead.

On land which indisputably belonged to Madeley Manor, J. Smitheman had evidently laid a railway by 1729, when he began to buy cast-iron wheels from Coalbrookdale.[17] In 1756 he and the Giffards, who shared the Manor with him, gave Thomas Brook and later Edmund Ford leave to lay rails and get coal and ironstone through footroads in Madeleywood, paying a royalty of 1*s*. 6*d*. per $\frac{1}{2}$-ton waggon and 3*s*. per whole-ton waggon, the weight of the ton 'to be the same as sold at the Old Jacobite Footroad in Madeley Wood.'[18] Also in 1756, the Giffards leased pits and wharfage at Lower Lloyds to William Hallen, to supply the Sambrook and Caynton forges near Newport. Hallen was to maintain and repair the 'said Pits Shafts Wagon Ways and requisites, Rails, Sleepers Wagons Gins Buildings and Railways already erected and laid or to be erected and laid.'[19] Lesage, a visiting Frenchman, found a footrid in the Madeleywood area in 1784, with an iron-plated railway in it.[20]

In Dawley, the parish north of Madeley, the only reference to railways before the Coalbrookdale system spread there seems to be a 1733 lease from Lord Craven to Thomas Barker (later sub-let) of coalworks in Little Dawley, with power to 'make drains ditches and ffootroads . . . and make and carry on all such Ways rails and passages . . . as shall be necessary'. He was given free access with 'Carts horses Waggons or other carriages'. In Wombridge, the next parish to the north again, the Charlton family built a railway from their coalpits in Horsepasture in 1747.[21]

Further west, leave was granted in 1727 'to lay railes' from pits around Little Wenlock to the Severn at Strethill Farm, with power to lay 'side railes' or branches up to 700 yards long 'from the main raile already made and laid'. The main line was about a mile and a half long, with a very steep descent to the river bank.[22] A little to the north, and not connected to the Severn, a railway ran in a great loop from Watling Street, over the col between the Wrekin and the Ercall, and back to Watling Street. It served a number of quarries and pits which fed the Steeraway lime-kilns, and references to rails and waggonways are frequent in the kiln accounts from 1737 to 1753.[23]

When in 1708 Abraham Darby took over the old furnace at Coalbrookdale, he laid the foundations of the great Quaker empire of iron-works which was perhaps to influence the course of the Industrial Revolution more profoundly than any other establishment. Here iron was first successfully smelted with coke instead of charcoal; here was built the first iron bridge; here cast-iron rails were first laid on a waggonway, and here the first railway locomotive ran. The Dale became the main supplier of cylinders for steam engines; and the Dale provided an outstanding example – not, alas, widely imitated – of social welfare and concern for the working conditions of its employees. Coalbrookdale and its satellite foundries were run by a succession of Darbys, with their relatives, Richard Ford, Richard and William Reynolds, and Richard Dearman, until 1851, when they left it as the largest foundry in the world.[24]

For forty years the furnaces were supplied with their raw materials – ironstone, coal and limestone – by pack-horses which carried about 3 cwt. each, and the advent of railways came surprisingly late. When they did come, they were evidently influenced to some extent by Tyneside, with which Coalbrookdale had business links in connection with steam engines. After the death of Abraham Darby II, who was responsible for the introduction of railways, his widow wrote: 'They used to carry their coal upon horse's backs, but he got roads made and laid with Sleepers and rails, as they have them in the North of England.'[25] Indeed, the Coalbrookdale system of the 1750s and the Willey furnace line of 1757 seem to have been the first railways of any considerable size in Shropshire, both in length of track and in size of waggons. One or two of the Madeley lines of the same date, though short, also used large waggons. Previously, Shropshire railways had been small in extent and in waggon capacity.

A series of wayleaves was obtained by Abraham Darby from Brooke Forester between 1749 and 1752. They empowered him, so that

> Coals may be conveyed in the easiest and best manner, to make a wagon way and lay rails on sleepers in such manner as is commonly used, and with coal wagons and horses and oxen to draw the same on or along the said railway to Coalbrookdale.[26]

The first line was laid southwards from Coalmoor to the Dale to carry coal and ironstone, and was opened in about 1750. It was extended north almost to Wellington some time later in the century. The railway down from the foundry to the wharf, for bringing up the limestone from the other bank of the river and for carrying finished iron for transport down the Severn, appeared only in 1756, an event noticed in Abiah Darby's diary: '1756, 1mo 31st. First Waggon of Pigs came down the Railway.'[27]

The early history of the railways to the Coalbrookdale outposts is hazy in the extreme. Their dates cannot be guaranteed, their lengths are uncertain, and their routes can be little more than guessed, so much has the area been churned up in later years. The following account is therefore not to be relied on in detail. The Horsehay furnaces were begun in 1753 and blown in two years later. To supply them, a railway was laid in 1754 with two branches to Lawley Bank and Dawley Bank, followed by others to Lawley in 1761-2 and to Dawley a little later. A proposed branch is mentioned in a letter to Forester by William Ferriday, his agent and spy on Coalbrookdale affairs, early in 1755:

> This morning I had a letter from Mr Darby, he desires the Esqr will be pleased to give him leave to lay a branch of railway from the Horsehays to the lymeworks. . . . I suppose the above is of the mushroom breed, only the produce of one night, the place where he proposes to lay the rails is Huntingdon, and neither he nor we are sure of good lymestone there, it will cost 130 l. full to lay the said

> branch, and their is nobody besides himself thinks the Horsehay furnace will come to anything of profit.

This branch may have been connected to the Coalmoor line, but probably came to nothing. To carry out Horsehay's produce, a link was built in 1756 to the original Coalmoor-Coalbrookdale line.

Horsehay, in spite of Ferriday's sneers, did extremely well, and it was followed by Ketley furnace, blown in in 1756 and supplied from Donnington Wood by the Bank railway via Ketley Bank, which was also joined on to the Dawley Bank branch from Horsehay under a wayleave of the same year.[28] The line from the Coalmoor route to Horsehay and perhaps as far as Dawley Bank was known as the New Railway, in contradistinction to the earlier line from Coalmoor to the Dale. The very route of another major branch, the Park Railway, is in doubt. It was laid in 1769, perhaps from Horsehay towards Old Park. With minor branches, the total length of railways belonging to Coalbrookdale was now approaching twenty miles, and in 1784 it was said to be thirty miles, most of them laid with the cast-iron plates first introduced in 1767.[29]

Moving away from Shropshire, the nearest railways, and the earliest after the Restoration, were close to Stourbridge in Staffordshire. They are remarkable for several reasons. A fair amount of documentary evidence for them – though not as much as we would like – has survived; they were the first railways authorized by Act of Parliament; and they remained unique for a century, in that although they carried coal they were built and worked by the proprietors of a river navigation, not by the proprietors of a colliery. This situation did not recur until the Canal Age was in full swing. The moving force was Thomas Lord Windsor, together with George Earl of Bristol and Thomas Smith as largely sleeping partners. The first step was taken on 13 April 1661, when John Addenbrooke of Bewdley and Richard Smart of Stourbridge agreed to deliver to Lord Windsor and his partners 15,000 tons of coal a year at a wharf near Wollaston Mills, Oldswinford (Stourbridge). Windsor was to obtain

> sufficient leave and freedom for the said John and Richard to lay and sett one ffootrayle as by the said John and Richard shall be judged the most convenient way for that purpose from the Coleworkes in Pensnet now in the occupation of William Winehurst and Thomas Smart to the said key or wharfe . . . for the better and more easie conveyance and carriage of y^{e} said Cole.[30]

The Addenbrooke family had coal interests near Broseley (p. 236). Pensnett is some two and a half miles north-east of Stourbridge.

The main part of Windsor's scheme, however, was to make the River

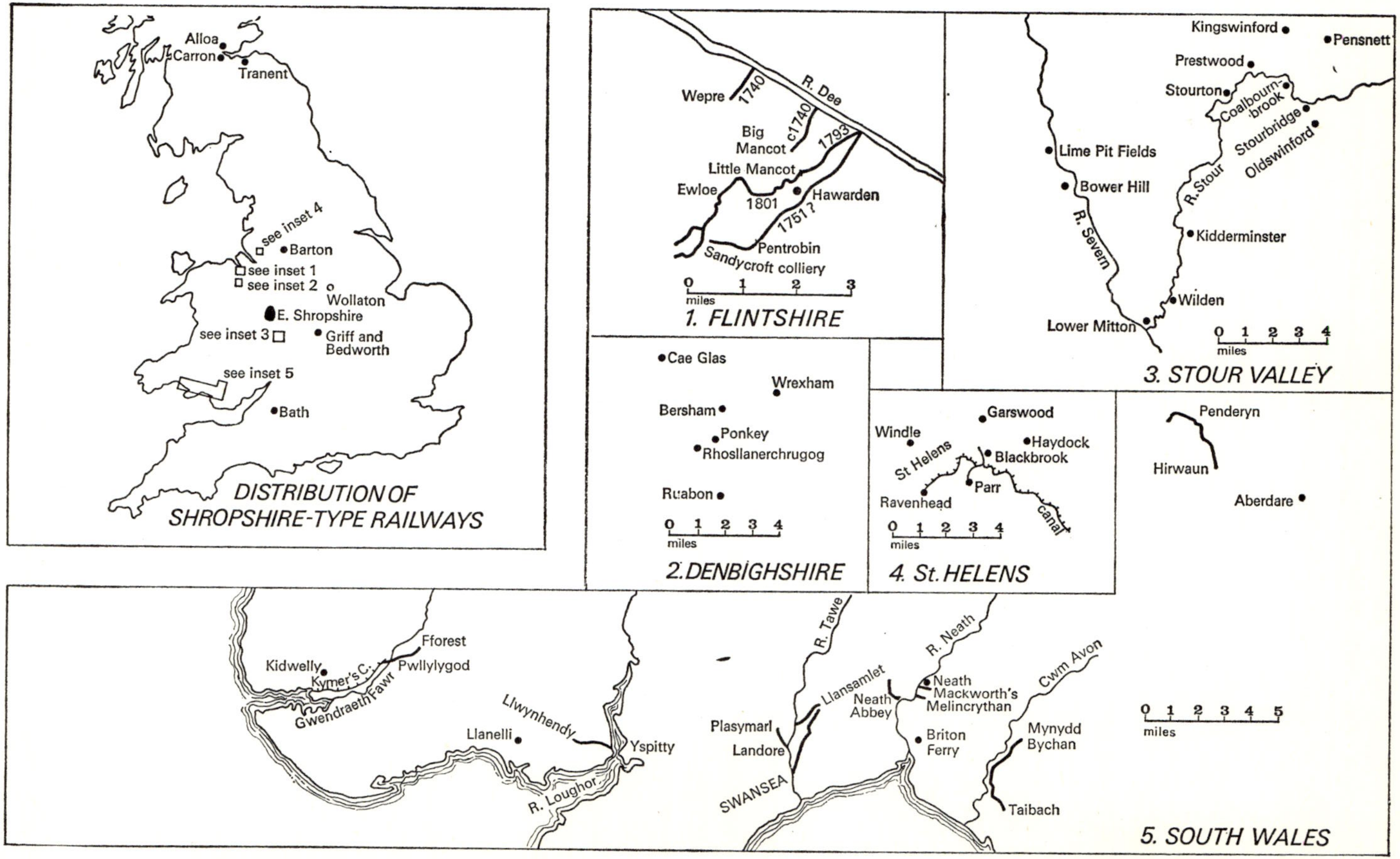

FIG 45 Distribution of Shropshire-type railways.

Stour navigable, for the better supply of the ironworks along its banks and to give the coalfield of south-west Staffordshire, hitherto landlocked, an outlet to the Severn and the markets downstream – Worcester, Gloucester and Bristol. A private Bill to this effect was submitted to the House of Lords on 11 May 1661. The scheme was in direct competition with the Shropshire mines, which had a virtual monopoly of the Severn coal trade, sending down 100,000 tons a year; and the coal-owners of Broseley, Benthall and Madeley, along with Shrewsbury Corporation, raised an immediate outcry.[31] The Bill, however, passed smoothly through Lords and Commons (though interrupted by a long adjournment), and received Royal Assent on 19 May 1662. Entitled 'An Act for the making Navigable of the Rivers of *Stower* and *Salwerpe*, and the Rivulets and Brookes running into the same, in the Counties of *Worcester* and *Stafford*', it empowered Bristol, Windsor and Smith to build locks and weirs, 'And to make any Wayes, passages, Foot-rayls or other conveniencies for the carrying and conveying of any Coals or other Carriages to the said Rivers', and to maintain them. The proprietors were to have sole use of barges on the rivers and of 'Carts, Waynes, Waggons and other Carriage upon the said Wayes, Passages or Foot-rayles . . . from the Coal-pits'.[32]

The engineer they employed for the work was Andrew Yarranton, a Worcestershire man, formerly a Captain in the Parliamentary forces, agriculturist, ironmaster, and industrial and commercial reformer. His first love was inland navigation. He was concerned, sooner or later, in many river projects, and had already worked with Windsor in trying to improve the Salwarpe, which flows through Droitwich, for the salt trade. His plan for the Stour survives (part reproduced in pl. 53), showing eleven locks down the river from Stourbridge, past Kidderminster, to Lower Mitton (now Stourport), and two footrails ending on the north bank between Stourbridge and Stourton. One, which runs off the map, is evidently that from Pensnett, ending at a wharf opposite Barge Hall near Coalbournbrook; the other is shorter, apparently disappearing underground in the Prestwood area. On the footrails Yarranton drew little pictures of waggons – the earliest we have in Britain – each with a rectangular framework and four wheels splayed out at the sides in defiance of perspective. Other drawings of wheeled vehicles by Yarranton in his *England's Improvement* have the same peculiarity.

Before long Bristol and Smith were replaced as proprietors by Samuel Baldwin and his brother Timothy (owners of a local ironworks at Wilden on the Stour, who had interests at Broseley as well), for a short time by Yarranton himself, and by Samuel Whyte, who acted as the Director of Works. Operations began late in 1662, but proceeded slowly. Over the next two years only £130 was made available to Addenbrooke to buy timber for the footrail. However, in 1665 the navigation was complete as far as Kidderminster and the footrail was finished to the Stour from Pensnett. Yarranton had estimated that 1 ton of coal would cost 2*s.* at the pit, plus 6*d.* for transport down the footrail and 3*d.* for its maintenance. But in practice the cost turned out to be 2*s.* 5*d.* at the pit and 1*s.* 2*d.* transport and maintenance, with an unfortunate effect on the profits expected. It was not until August 1666 that there were any profits to distribute, and things were going badly with the footrail because of the clerk's inefficiency. Yarranton wrote: 'I will not give 2*d.* for Rivers and foote-Rayle as they now stand and are managed but I can put thinges in A better way in Afortnights tyme', and later: 'As to the erers of the ffootrayle it is in the bad managment and I have provided the Remedy and can put it into order when I see cause provided that there may be monies constantly to pay the Collyers.' The proprietors therefore appointed Yarranton 'to manage the foote Rale', and once again all seemed set fair.

Although traffic was flowing, the lower river was still only navigable in favourable conditions, and the locks were often damaged by floods. In 1670 Yarranton contracted to complete the navigation by 1672, but little was done for want of money. A memorandum of 1671 urges 'That care may be taken for ye preservation of ye timber and carriages belonginge to ye ffootraile & allsoe of ye barges'; in 1673 Yarranton could not find the money for a 'bargain of wood that Mr Addenbrooke bought for him of Bendy by & for ye footrail'; and on 12 March 1673/4 Yarranton lamented to Baldwin: 'The fluds and the Runing in of the Land upon the footrayle hat near brook us; my sonns hart was almost could and Indeed Mr Whyte was to blame to see us in such straits and did not pay his parte.'

As a result of these disasters, the 1670 contract, long out of date, was scrapped and the iron-masters signed another agreement with Yarranton that he should complete the navigation by 1676 and build a new underground footrail. A year later it was reported that Yarranton

> is preparing in setting out a little Foot Rayle under ground out of Mr Bendyes Copice to lay Dry a pcell of Mr Henseys Cole which hee hath newly Bought and it is Beleved in laying out 50^{l} he will git Cole to Suply the cuntry and navigation for three or fore yeares.

This was within a quarter-mile of Coalbournbrook. 'The Stack of Cole containing 2600 wayt will be Delivered downe the footraile in the high way at 22^{d} the Stack.' By 1676 the surface section, at least, of this footrail was ready, but as always shortage of money prevented much work on the navigation. Yarranton complained bitterly in his *England's Improvement*, published in 1677, that he 'fell on, and made it completely Navigable from Sturbridge to Kederminster; and carried down many hundred Tuns of Coales, . . . and there it was obstructed for want of money, which by Contract was to be paid'. He severed his connection with the scheme (he died in 1684), but his son Robert took over.

Another contract was signed in January 1677/8 between Lord Windsor, Samuel and Timothy Baldwin (both were now knights), three iron-masters, and William Farnolls and Robert Yarranton, whereby Farnolls and Yarranton were, for £1,758, to make the Stour navigable for 12-ton barges from Stourbridge to Farm Bed just above Lower Mitton,

> and shall and will make a waggon Rayle of one hundred and forty yards or thereabouts into the ffarme bedd towards Mitton Bridge for the Cariage of coales and other materialls And from the lower end of the Waggon or footrayle to make the said River Stower navigable

into the Severn for 20- to 30-ton barges. In other words, to avoid the expense of an entrance lock where Stour met Severn, the navigation was to be replaced by a short railway whose waggons served as intermediaries between the small Stour barges and the larger Severn ones. Farnolls and Yarranton were also to 'make a footrayle from Ananias Henseys Coleworke to Aldens Locke [at Stourbridge]. And shall make Eight Waggons for this footrayle.' All this to be done before 24 June 1679. But once again the plans were foiled. Three of the iron-masters got cold feet and withdrew their support. The navigation continued in a dilapidated state, and the two footrails – that from Pensnett and Hensey's underground one – were still in occasional use in 1680/1, when an anonymous report was compiled:

> Jenowary 4^{h}: 1680: I tacking A suruay of the foote raile, as I haue often times made many Journyes to take anaccownt of what raiels ware mising, I goeing to the house of Nickles hodgkis in the parish of Kingswinfoorde inkeeper, wheere I found 1 great sill as did beelong to the locks at bargehalle 24 foot long and 13 inches brod and 9 inches thick. goodman hodgkis I could not speeck with. his wife tould mee that they bought that peese of Mr Robert Yarinton and neuer any more but that 1 peas. I loocking A Bout the backside found 17 neaw Joysts in a barne; thos Joysts wear cut at Kidderminster and carried up in A barge to repare the foote raile before wee gin of carying coole, and they wear leaft at bargehall. thar is as many tacken a way besides those as did lay 300 yardes. William Skelding 1 of Mr Hensys Colliears Hath tacken the gretest part of them away and I asked Skelding how hee durst take Awaye my Lord Windesors raiels

> and hee tells me that Robert Yarinton ordered him to take them for monys that Robert oued him and I bid him show mee that order but hee saied hit was by word of mouth. 22 pound will not make good the raile again, beeside the raile that Hogkis hath. I tould Skelding that my Lord Windesor would bee very angry with him for distroying his raile way and he tould mee beefoore 1 of mi oun men that hee cared not for my Lord Windesors anger. I often mising the railes I and my man wached seueral nightes but neuer could catch any for after that wee wear come awaye they found theyer time to take what they see good. the wagins being left at barge hall thay have stole the brases out of the wagins and the bords allsoe. I thinke hit not conuenient to tacke thos persons before A staffordsheere Justis for I thinke there will bee littell dun till my Loord Windesor is in the country and then his Lordship to have the hearing of hit.

Thereafter everything lay fallow until the Staffordshire and Worcestershire Canal was opened in 1772.

East of Birmingham there were two wooden railways in the Coventry coalfield, where the coal-owners had been among the first to realize the potential of Newcomen's steam engines and had close links with Coalbrookdale, where so many early cylinders were cast. By 1703 the Newdigate family had built a system of small canals in and around their collieries at Arbury and Griff, which fed and were fed by a number of 'waggon roads'. These were probably cart-roads rather than waggonways in the railed sense. But by 1760 railways had certainly appeared here, and more were built as a result of the opening of the Coventry Canal nearby. Just to the south, Bedworth Colliery had a railway by 1767, and maybe some time before.[33] The owner was Richard Parrott, a connection of the Stonier or Stanier Parrott who had built one of the first Newcomen engines in the district, perhaps in 1716, and who is no doubt the same man as the Stoneyard Parrot who was working coal pits at Calcutts by Broseley in 1715 and 1721.[34]

In 1697 the Shropshire railway first came to South Wales.[35] Sir Humphry Mackworth of Betton Grange near Shrewsbury had married the daughter of Sir Herbert Evans of Gnoll, near Neath, where the collieries had been abandoned for thirty years through lack of ventilation and transport. Mackworth was given a lease in September 1697 by the burgesses of Neath,

> whereupon he travelled into other countries to find Skilfull Workmen or Miners to assist him therein, and after great Expences, and by carrying on a level and Wind-way, commonly called a Foot-rid or Waggon-way, after the manner used in *Shrop-shire* and *New-Castle*;

> he recovered the said Coal-works, and at great Expences continued the said Waggonway on Wooden Railes from the face of each Wall of Coal twelve hundred Yards under Ground quite down to the Water-side, about three-quarters of a Mile from the Mouth of the Coal-pit; the said Coal-works without the conveniency of this Foot-rid and Waggon-way, could not be carried on to any Profit; but by means thereof, great quantities are brought forth and sold to Sea.

The source of Mackworth's railway was acknowledged more precisely still:

> These Waggon-ways are very common, and frequently made use of about *New-Castle*, and also about *Broseley*, *Bentall*, and other places in *Shropshire*, and are so far from being Nusances, that they have ever been esteemed very useful to preserve the Roads, which would be otherwise made very bad and deep by the Carriage of Coal in Common Waggons and Carts.

And again,

> where the Coalworks were lost for want of Air . . . and could not be recovered for want of Artists in that Country; whereupon Sir *Humphry Mackworth* got some out of Shropshire to instruct others, and to carry on an Artificial Waggon and Windway, for draining the Water, Circulating the Air, and driving out the Coal, which was never seen in that Country before.

About the same time as he built his coal railway, Mackworth bought from Sir Carbery Price a number of lead-mines in mid-Wales, and transferred them to the new Company of Mine-Adventurers of which he was Deputy Governor. He also transferred his smelting works at Melincrythan, less than a mile south of his own colliery, and greatly improved them during 1699 by building a short canal, flood-gates, cranes and a short 'artificial Waggon-way for carrying ore and Stones to and from the River at Neath'.

Mackworth's coal railway, however, attracted the jealousy of his rivals in the trade, notably Sir Edward Mansell of Briton Ferry Colliery, who was a J.P. and did not scruple to misuse his powers (according to Mackworth, in whose favour almost all the evidence is biased). In 1705 Mansell tried to entice away Mackworth's workmen with bribes, and forcibly impressed them into the Army if they refused. He sent his minions to steal coal and to sabotage Mackworth's works, with a level crossing as his special target. 'The destroying this Waggon-way', Mackworth said, 'is in order to destroy the Profit of the Coal-works, and consequently to prejudice the Undertakings of the Mine-Adventurers in that County, which depend on the Plenty and Cheapness of Coal from these Adjacent Works.' On 13 August 1705, at the Grand Sessions,

> an Indictment was presented and found by the Grand Jury against Sir *Humphry Mackworth*, for erecting a *Nusance* cross the Highway leading from *Neath* to *Cardiffe*, which was set forth in the said Indictment to be *Ten Foot high, to the great annoyance of her Majesty's Subjects, passing that way* . . . the Place presented was the said *Waggon-way for carrying down the Coal to the Water-side*, and has been made by Sir *Humphry Mackworth Eight Years* before, and upwards, and quietly enjoyed by the general Approbation not only of the Overseers of the Highway, and the neighbouring Justices, but of all others travelling that Road, it being formerly a *deep Pool of Water and Dirt*, and thus amended by the said Sir *Humphry Mackworth*, and by the Slag of the Smelting-houses, made *the best Spot of Ground for heavy Carriage between* Neath *and* London, the highest part of it being three Foot two Inches on one side, and two Foot five Inches on the other, with a gradual Ascent of thirty-six Foot and upwards on each side.

In spite of the favourable evidence of Mackworth's witnesses and the approval of the Jury and the Overseer of the highway, the case was adjourned to the next Sessions. Mansell's agents

> supposed that they might then justly pull down the said *Waggon-way*, and hinder the carrying of Sir *Humphry Mackworth's* Coal to the Waterside; and according on the *Eighth Day of September last* [1705], *John Morgan and Philip Williams*, two notorious wicked Fellows, were employ'd to break down the said *Waggon-way*, and the . . . Agents to the said Justices . . . were then present, countenancing and directing the said Illegal Proceeding.

Mackworth tried to get a warrant against the culprits, but Mansell refused it, and even arrested Mackworth's men who tried to prevent the damage.

Six times during September the railway was torn up by Mansell, six times it was repaired by Mackworth. For example:

> On the 17th of September following the Way being Mended, the said *John Morgan* and *Morris Morgan*, with an Iron Bar, two Hatchets, two Pickaxes, two Shovels, broak up the said Way, and *cut the saw'd Rails of Sir* Humphry Mackworth *in two Places*, which were laid by him to mend the Way, and support his heavy Carriage cross the same. . . .

The next day,

> the said Rails or Waggon-way were broak and cut in Pieces not only cross the High-way, but also *within the lands of Sir* Humphry

> Mackworth, *in his own Possession on each side thereof*; and that his Waggons loaded with Coal, and coming down from the Pits towards the Waterside (where Vessels waited for their loading) were actually obstructed, not only by Cutting in Pieces the said Rails; but when they saw that by strength of Horses, and Assistance of Men, the Agents of Sir *Humphry Mackworth* were able to draw the Waggons cross the High-way, without the Benefite of the said Rails, *they dug deep holes in the Ground to let in the wheels of the Waggons*, and to make the Way impassible for such sort of Carriage.

John Morgan's reply to Mackworth's charges was

> That Sir Humphrey, some time ago, for the convenience of carrying his Coal, rais'd a great Bank cross the Highway, and fixed upon it two Rails, distant about two Foot from each other, and three Inches above the surface; last great Sessions a Bill of an Indictment was presented, and found by the Grand Jury for this Nusance, and some of the Persons now complain'd of endeavour'd to remove it, but were hinder'd by the workmen of Sir *Humphrey* who assembled for that purpose, in a riotous Manner; however Sir *Humphrey* soon afterwards order'd his own Workmen to remove part of this Nusance, tho' he now Complains of others, for only Endeavouring it. For this Sir *Humphrey* now Prosecutes in the *Queens-Bench* and the *Leet-Court*.

Mackworth eventually won. The Sessions Jury presented the culprits 'for feloniously stealing and taking away the sawed Timber and Waggon Wheels of the said Sir *Humphry Mackworth*, found in their Custodies the first Day of this Instant *October* 1705', and Mackworth also complained to the House of Commons (he was Member for Cardigan) of breach of privilege. The Committee found eight of the offenders guilty, but by this time they had fled back to Wales and were apparently never brought to book. Mackworth's day of reckoning came in 1709, when he was accused of embezzling from the Company of Mine-Adventurers, and was found guilty by the House of Commons. The railway, however, survived under his descendants. It was plated with iron before 1791,[36] and disappeared with the closure of the pits in 1809-10.

Just over the river from Mackworth's line was Neath Abbey Colliery, which in 1767 boasted a waggonway with oak rails 'pretty much of the same construction as the present', as a witness told a Parliamentary committee in 1810, when the line had still not been converted to iron.[37] Neath Abbey Iron-works, founded in 1792, apparently used railways from the start.[38] An inferior illustration of the coal-works in 1805 (pl. 54) shows a train of waggons on an impossibly serpentine track. A little further west again, on the east side of the Swansea Valley, Chauncy Townsend had built a waggonway which is first mentioned in a lease of February 1743/4, from Llansamlet Colliery to the Tawe. He was also a lessee of pits at Neath and was doubtless imitating Mackworth's railway. His son-in-law and successor, John Smith, laid a longer line from Llansamlet to wharves near the mouth of the Tawe, which was later replaced by a branch of the Swansea Canal.[39] On the opposite bank of the river, the pits at Landore were worked by Sir John Morris, whose commonplace book records: 'Charged to Landore Colliery Acct – 1774: New waggon way from Plas y Marl now us'd on Coalwharf etc £243 17s. 7d.' Sir John's father had come from Bishop's Castle in Shropshire to Landore Copper Works, and he himself had dealings with Coalbrookdale, as we shall see, when he laid his underground waggonways with cast-iron plates. Joshua Gilpin, an American visitor in 1796, noted that 'The Coal Pits abt. here are very numerous & the railroads running from each of them down either to the woods or River form a great Number of paths'.[40] J. C. Ibbetson painted the Landore staith in the same year (pl. 55).

Still further west, a short wooden railway is said to have been built in the 1750s from pits at Llwynhendy to the east of Llanelli, running down to the Loughor River at Yspitty. It is significant that Chauncy Townsend of Llansamlet Colliery had considerable coal interests in this area, and built a canal about 1770 along much the same route, perhaps replacing the railway.[41] The westernmost of early South Wales lines was laid from Thomas Kymer's anthracite mines at Fforest to Pwllylygod on the Gwendraeth Fawr, at the head of his canal from Kidwelly. The canal was opened by 1769, and the railway was probably of the same date.[42]

East of Neath, a wooden railway was laid about 1750 from Mynydd Bychan pits above Cwmavon to the Taibach Iron-works near Port Talbot.[43] Further east still, the network of railways which filled the valleys from Aberdare across to Blaenavon did not appear till the Canal Age, and the age of the iron rail, were in full swing. The only recorded wooden railway – though there may well have been others – linked the Hirwaun Works above Aberdare with the Penderyn limestone quarries in the hills above. The furnace was started in 1756, and the railway was laid, according to different authorities, either about 1763 or after 1786, when Samuel Glover of Birmingham leased the works. In 1794 the line was absorbed into the Aberdare Canal's new railway from Hirwaun to Aberdare and converted to iron rails, although it was ordered as late as 1799 that small connecting railroads from quarries should have wooden rails.[44]

The next railway to be recorded was both famous and unusual. It was famous because it was on the outskirts of Bath, and a sight which every curious visitor to the spa had to see; it was unusual because it carried not coal or ironstone, but building-stone, and because, in sophistication, its design was more elaborate than anything Shropshire or Tyneside had seen. Its builder was Ralph Allen, the Post Office reformer who was making a fortune by farming the cross-posts, and who now intended to add to it by quarrying and selling Bath stone. In his early works on Bathampton Down he seems to have used wooden railways, but his *magnum opus* was the line built in 1731 from Combe Down Quarries past his mansion at Prior Park to the Avon (pl. 56). The quarries were 500 ft. above the river, the roads, as so often, were in a hopeless condition, and the blocks of stone weighed up to 4 tons apiece. In the circumstances, a railway was the only possibility. As we have seen (p. 126), Allen had contacts in the Great Northern Coalfield, and ostensibly copied the waggonways there. Dean Hedworth even sent him models of waggons. But although no acknowledgement is made to Shropshire, it is quite obvious that Allen's waggons owed much more to Shropshire than to Tyneside. This can be seen in the comparatively small gauge, in the small and early iron wheels, in the long wheelbase, in the sprags, in the cast-iron axles and brass bearings, and in the terminology. The actual construction was undertaken by the engineer John Padmore of Bristol, a city which had much closer links with Shropshire, by way of the great highway of the Severn and its heavy coal trade, than with the Tyneside coalfield.[45]

Many contemporary letters and descriptions of Bath contain references to the railway. John Evelyn, for example, great-grandson of the diarist, wrote to his father in 1738 that the stone was

> conveyed in a very clever manner down to ye town upon carriages with low broad wheels, covered with iron, which run upon a wooden frame made ye length of ye hill, so that when ye machine is sett agoing it runs down ye hill without any help, only one man behind to steer it, & in this manner above three hundred Tunn of stones are carried down at one load.

Defoe described it in the same year, in similar and again slightly inaccurate terms.[46] On Allen's death in 1764 the waggonway was scrapped, to the disgust of travellers in the area. The Rev. S. Shaw wrote in 1788:

> The former mode of conveying the large blocks directly down the hill to Bath, by machines running in grooves or frames of wood, such as we see in the collieries about Newcastle, is now no more; they carry them in common waggons, to the detriment of the roads, and inconvenience of travellers.[47]

A *cri de cœur* that is not unknown today.

In North Wales, Richard Ford and Thomas Goldney, two partners in the Coalbrookdale Works, had leased a furnace at Bersham, near Wrexham, in 1731, and they were soon joined in the venture by Abraham Darby II. They may well have laid railways to supply coal, ironstone and lime to the furnace, although nothing is certainly known until the works had been sold to Isaac Wilkinson, who was later joined by his sons, John and William. In 1757 Isaac leased all the coal and ironstone under Cae Glas near Minera, two miles from Bersham, with liberty of 'laying rails or making a railroad to the pits from the main or great road', and another railroad from the Ponkey to a colliery at Rhosllanerchrugog to the south. These lines were probably short ones, running merely from the pits to the nearest road, but in 1758 the Wilkinsons obtained land in Wrexham itself for building a railroad to carry coal direct to Bersham furnaces, and a year later laid another line in Ruabon parish.[48]

Ten miles to the north, the Flintshire Coalfield around Hawarden received its first short railway in 1740, when a Chester company built a line from the Latchcraft pits at Wepre, between Shotton and Connah's Quay, to the River Dee. About the same time, a mile-long railway was laid from Big Mancot to the river, referred to in later leases as 'Mancott Old Rail Way'. In 1751 John Dutton and Walter Stubbs of Beckbury, near Broseley, leased the right 'to level the ground and lay rails' from pits in Ewloe and from Pentrobin and Sandycroft collieries. The latter line at least seems to have been built, running past Hawarden on the east for four miles to the Dee, for in 1756 Sandycroft Colliery paid Isaac Wilkinson £28 7*s.* 0*d.* for waggon wheels and £5 4*s.* 3*d.* for waggons. But it cannot have lasted long, since later in the century Sandycroft coal was carted to Mancot and taken down the Mancot railway. Then in 1793 Beriah Botfield of Old Park furnaces near Coalbrookdale, who had leased both Sandycroft and Mancot collieries, laid an iron railway (no doubt plated in the Coalbrookdale manner) from Little Mancot to the Dee, thus superseding Mancot Old Railway; and in 1801 the Rigbys of Hawarden, who took over from him, pushed the line another four miles southwards to the Sandycroft area.[49]

In Lancashire there were a number of early but elusive railways. By a draft agreement, undated but before 1745, with Scroop, First Duke of Bridgewater (father of Francis, the celebrated Third Duke), the Company of Proprietors of the Irwell Navigation were to make a side-cut to Worsley Brook near Barton Old Hall, with provision for making railroads to enable the Company to collect coal from any pit within two and a half miles. The agreement was never signed.[50]

The building of the Sankey Canal from the Mersey to St. Helens, opened in 1757, prompted a number of colliery owners in the area to lay railways to it for the easier transport of their coal. The inspiration of these lines – and indeed of all early lines in Lancashire – is open to doubt. One might suppose a Tyneside influence, though at several removes. Yorkshire, just over the Pennines, was a Tyneside colony in the matter of waggonways; and Henry Berry, the engineer of the canal, had been in Ireland with Thomas Steers when he was supervising the construction of Ballycastle Harbour with its waggonway.[51] On the other hand, Sarah Clayton, whose industrial interests were unusual for a woman at that date and who laid the first railway to the Sankey Canal, was well acquainted with Ralph Allen's waggonway at Bath;[52] the nearest coalfield to Lancashire is Flintshire, an undoubted Shropshire colony; Shropshire miners emigrated to Lancashire, though not in any number until the 1760s;[53] and, finally, the vocabulary points to Shropshire. The Tyneside word *waggonway* is never found in eighteenth-century Lancashire; the neutral *waggon-road* sometimes occurs (as it does in Yorkshire); but usually it is *railed road* or *railroad*, good Shropshire names which Tyneside and Yorkshire, until a later date, knew not. All this, however, is only indirect evidence. There is no categorical reason for supposing that Lancashire railways copied those of Shropshire.

Sarah Clayton announced on 16 December 1757 that her new mines in Parr were at work: 'The Waggon Road and other conveniencies are fixed in such a manner that Flatts [barges] may be loaden in a few hours.' Twenty-one years later her mine was offered for sale 'with all the horses, waggons and other carriages, railed roads, ways, windlasses'.[54] An agreement of 1766 referring to Windle Colliery mentions 'Windey Engine, the Rail Road, the Waggons and the Hovel', and the sale advertisement of this colliery three years later includes 'a parcel of timber, suitable for railed roads, with all the utensils belonging, waggons and waggon wheels'.[55] The colliery agent at Haydock wrote in 1768: 'Whilst the Navigat^{n} is in cleansing we have a fine opportunity of paving our rail'd road.'[56] The Warrington Company of copper-smelters obtained leave to lay down 'a planked and railed road' at Blackbrook in 1772; but when Thomas Williams bought the Company in 1785 he found it engaged in 'removing the Rail Road'. Garswood Colliery had a railroad by 1766, the Ravenhead Plate Glass Manufacturers by 1792. All these lines ended at the Sankey Canal.[57]

Moving on to Scotland, we come to still more debatable country. The appearance of the first railway here was an indirect result of the 1715 rebellion, for his part in which the Earl of Winton was attainted. His estate was bought in 1719 by the York Buildings Company, whose primary purpose had been to supply London with water. They worked the Earl's collieries around Tranent (nine miles east of Edinburgh), built one of the earliest steam engines in Scotland, enlarged Cockenzie, or Port Seaton, Harbour, and linked it to the pits by a two-mile waggonway, which also served Preston salt-pans.[58] The traditional date of 1722 is probably correct: it was certainly before 1736.[59] The pits were leased out to tenants, one of whom, John Cadell, eventually bought the company out in 1779.[60] The railway hit the headlines, as it were, in 1745, when the Battle of Prestonpans was fought astride its route. To meet the Pretender's forces advancing from the west, Sir John Cope, the Royal General, drew up his troops along the 'narrow cart road', as he called it at the Court of Enquiry, and likewise stationed his twelve pieces of artillery on its slight embankment.[61] It availed him little, for he was caught half-asleep by the Highlanders; but it explains the allusion in Adam Skirving's derisive song:

Hey, Johnnie Cope, are ye wauking yet?
Or are your drums a-beating yet?
If ye were wauking I wad wait
To gang to the coals i' the morning.[62]

The Tranent line used small waggons pulled in trains, and its gauge was also small – features which cannot be paralleled on Tyneside waggonways. So unless it was a descendant of the hypothetical railway introduced to Scotland by Thomas Tulloch in 1606, its ancestry was presumably Salopian, to which these features do belong. The same arguments apply to a lesser extent to two other Scottish lines, although as the eighteenth century progressed the breeds became more and more mixed.

The Carron Iron-works just north of Falkirk were founded in 1759 by Samuel Garbett and John Roebuck, two Birmingham industrial chemists, and William Cadell, whom they had met at their Prestonpans sulphuric acid works. Several experts were recruited from Coalbrookdale, and Cadell's son, William Jun., was made first Manager of the iron-works. In 1760 he built a waggonway to bring coal from Kinnaird Colliery, a mile and a half to the north. Nine years later there was a three-mile line from Quarrole, yet further north.[63] The inspiration no doubt came either from the Dale or from the Tranent way, for it was not until later that Carron began to take an interest in Newcastle methods.

John Earl of Mar, the owner of the Alloa Colliery on the north of the Forth, had sent his Manager to Newcastle to inspect machinery as early as 1709. The visit, however, produced no Newcastle waggonway, and when a railway was eventually built down to the Forth, either in 1766 or 1768, it was a close copy of the Tranent line in gauge, waggon size and operating practices. Its track, however, rang all the changes from the single wooden way through the double way to the cast-iron and wrought-iron armour-plating, until the all-cast-iron rail finally appeared.[64]

The question of terminology needs some final clarifying. We have already seen how *waggonway* was the universal Tyneside word until about 1800. Shropshire and its colonies also used the word – indeed, its first recorded appearance is at Broseley in two independent documents of 1631[65] – but not nearly so exclusively. More often, in the earliest days, Shropshire records refer simply to *rails*, or in some contexts to *inset* or *footrail* and its variants. *Waggon road* is not infrequently found in both Tyneside and Shropshire colonies – notably Lancashire and Yorkshire – but apparently not in either of the motherlands. *Railway* (or *railed way*) was evidently coined in the Shropshire empire, and remained its monopoly until the spread of the canals disseminated the word and finally established it as the orthodox British term. *Railway* is first met with on the Pensnett line in 1681,[66] closely followed by the compromise *railed waggonway* at Broseley in 1686.[67] *Railroad* (or *railed road*) had much the same history, the earliest example being at Rowton on the Tarbatch Dingle line in 1702.[68] For a long time it ran neck-and-neck with *railway*, and it was not until the middle of last century that *railway* drew significantly ahead. In America, of course, the same race had a different result, for there *railroad* won. *Tramway* and *tramroad* hardly concern us here. In ch. 17 we will be meeting underground trams and tramways, but as terms applied to surface vehicles and railways they appear only at a late date and, to begin with, in South Wales. The first recorded use of *tramroad*, for example, on the surface was only in 1798, when the railway concerned had iron angled rails.[69]

The Shropshire railway: II

Thy fam'd Inventions, Mackworth, most adorn
The Miner's Art, and make the best Return:
Thy speedy Sails, and useful Engines show,
A Genius richer than thy Mines below.
Thousands of Slaves unskill'd *Peru* Maintains,
The Hands that labour still exhaust the gains:
The Winds thy Slaves their useful Succour joyn,
Convey thy Oar, and labour at thy Mine.

. . .

This, Mackworth, fixes thy immortal Name,
The Muses Darling, and the boast of Fame:
No greater Vertues on record shall Stand,
Than thus with Arts to grace, with Wealth inrich the Land.

Thomas Yalden, *A Poem on the Mines late of Sir Carbery Price, dedicated to Sir Humphry Mackworth*, 2nd ed., London (1698).

The way is now clear to deal with specific aspects of the Shropshire railways and their offspring. The evidence being unhappily meagre and erratic, nothing like so comprehensive a picture can emerge as did for Tyneside. In many respects, no doubt, Shropshire and Tyneside were similar; in some, they differed markedly; and on a number of points we just do not know.

As wayleaves were often leased together with the right to mine coal, it is difficult to disentangle the two charges. At Yates Coppice in Broseley in 1632, William Pacie paid 10*s*. a year 'for ye trespasse of the Insett'. At the same mine in 1686, John Huxley charged Sampson Crompton a royalty of 1*s*. a ton for coal and 1*s*. 3*d*. for ironstone, with no separate charge for carrying it down the line to the Severn; but he demanded for all coal and ironstone got elsewhere 'and wch shall be conveyed . . . downe the said Waggon Wayes as aforesaid fourepence for evry Tun'.[1] At Rowton in 1702, in addition to the coal royalties, the lessees paid 3*d*. per waggon; in the lease for the same line to Willey in 1759, a flat rate was charged of 'twelve pounds a year in satisfaction for the damage and trespass', plus an unspecified sum per waggon. On the Little Wenlock way in 1729 (and on others subsequently) the lessee paid 'two shillings and sixpence for each hundred yards of land in length whereon the waggonways and the necessary or convenient ditches' were made.[2] Sir John Glynne charged a rent of only 5*s*. a year, plus the free supply of 25 tons of coal, near Hawarden in 1751, but Thomas Dundas, lessor of a Carron waggonway in 1769, was more extortionate: £200 a year, plus £10 an acre on a three-mile line. Even this was a fleabite compared with some Tyneside wayleave rents.[3] Another cheap wayleave was granted at Bath in 1731 by Philip Bennet to Ralph Allen: 5*s.* a year, with the option of buying the ground outright for £13 9*s.* 6*d.* It was, however, probably a short length of land, and there was a snag. Bennet reserved to himself and his family, friends and servants the right

> with His and their horses coaches chariots chairs chaises to ride pass and repass in upon and through the said Ralph Allen's new made waggon-way . . . provided the same be not made use of by . . . Philip Bennet . . . with any carts cars drays waggons or any other heavy carriages.

Which was very convenient for Bennet, so long as his chair did not collide with a four-ton block of stone running down by gravity.[4]

Incidental references to ordinary cuts and batteries appear in the Steeraway accounts, such as:

> 1737 Paid Thomas Jarviss for cleansing a fall of earth and gravell that fell into the Wagen Way 3/4
>
> 1749 Paid Robt Haseldine for Cleansing ye Rails to ye Earth Mount 3/4
>
> 1750 Paid for building a wall to keep up the Rails in Machems Ground.
>
> Paid Barnett for Walling & Soughing [draining] in the deep of the waggon way being 30 yards at 4^d a Yard.[5]

But there is no record of any great engineering works. A single-arched stone bridge of about 24-ft. span which carried Kymer's Pwllylygod railway over the Gwendraeth Fawr above Kidwelly still survives. It is said to belong to the original line of 1769, and it may well do so.[6] By a coincidence, the only other bridge we hear of was built in the same year. It was on the Park Railway near Horsehay, was mainly of wood and cost £100.[7]

Gradients were sometimes terrifying. The Bath railway, dropping at 1 in 10 for a mile, was worked by horse-power up and by gravity down, though the waggons were equipped with specially effective brakes. The Little Wenlock line climbed up from the Severn at 1 in 8·2 for the first 192 yards, and at 1 in 9 for the next 300 yards, apparently without any ropes to control the speed of the downcoming waggons.[8] But this was unusual. Ropes or chains were very often provided in such cases. As early as 1607, on Clifford's Broseley line, the rioters 'cut the Roope that coales were lett downe w^th^all to Seaverne side', and there must have been a similar arrangement on most of the railways which plunged down the sides of the Severn Gorge. On the incline on the Coalmoor-Coalbrookdale railway, it was said in 1751, 'The Chain for letting down ye Wagons 112 yds long is 18 Hund. weight' – perhaps worked by hand or by horses rather than self-acting.[9]

The balanced incline, whereby the loaded waggon going down pulls up the empties on a parallel track, was patented in 1750 by Michael Meinzies, as we have seen. The Coalmoor incline, if it was not originally balanced, was probably made so at some later date, perhaps without knowledge of Meinzies' patent. It is said that this incline was not controlled by a brake on the drum at the top, as was normal later, but by the brakes on the individual waggons. In Shropshire parlance, the brake-lever was called the 'jig' and the brakesman the 'jigger'; hence the incline's name of Jigger Bank.[10] The road which once followed the incline, but has since been realigned, is still known by the name. Others followed in the Dale area. Arthur Young, who visited here in 1776, wrote that the railways 'made for the limestone waggons on the steep hills are so contrived that the loaded waggon winds up the empty one on a different road'.[11] Further afield, Richard Jones, Ralph Allen's foreman at Bath, built a short balanced incline in 1755 to carry materials for the Palladian Bridge at Prior Park. Quite independent of Allen's main railway, it was, Jones said, 'an exceeding good contrivance, and was . . . by *my* plan'.[12]

The unloading arrangements at wharves are not at all clear. On one early occasion at least, direct unloading from waggon to river trow was not envisaged, when in 1686 a lessee in Broseley was empowered to

> errect or inclose one penn or fold att the foote of the said Waggon Wayes and nere unto the River of Seaverne for the better ordering and disposing All such Coales as shall or may be gotten or brought.[13]

Later Shropshire-inspired waggons sometimes, if not normally, had doors in one end, rather than in the floor as on Tyneside, and the staith at Landore (pl. 55), where the rear end of the waggon was hoisted up to empty the coal out of the front, may be typical of Shropshire practice. At Bath the massive blocks of stone were removed from the waggon by crane. Another unusual device was to be found here. When the need arose in 1739 to transport stone from the south bank of the Avon, where it was normally transhipped, to the north bank, two barges were built which could each carry two waggons across the river – the first instance of a train-ferry.[14]

Double track was something of a rarity. The Willey lease of 1759 permits a 'double Railway' with a maximum width of 10 yards. De Givry, a Frenchman who visited the Dale in 1784, recorded: 'We have found railways near Coalbrookdale where the way is double, for the convenience of waggons passing each other.'[15] The drawing of Neath Abbey in 1805 (pl. 54) appears to show double track, but the two lines are so intertwined that it is difficult to credit the artist. Generally, in fact, single track with 'pass-byes' was used.

Various materials served as ballast: furnace slag, obviously enough, at Coalbrookdale, and broken stone at quarries like Steeraway. The track at Alloa was packed with clay, which can hardly have assisted drainage. The proprietors of Haydock Colliery in Lancashire paved their railroad, presumably for a horse-track between the rails.[16] Sleepers were laid at 1-ft. 6-in. to 3-ft. centres, and were normally about the same size as Tyneside ones. At Coalbrookdale their cost was slightly less – from 5*d.* to 7*d.* each in the period 1755-70 – perhaps because the gauge was smaller and their length less, and because they were sawn by direct labour, not bought ready-made from middlemen. Sometimes, however, exceptionally wide and thin sleepers were used. The Horsehay accounts record in 1771: 'Dawley Railway 63 Broad Sleepers @ 10d'; drawings show dimensions of 4 in. by 8 in. by 5 ft., and 3 in. by 10 in. by 5 ft. 2 in. (pls. 57–8); and of the Alloa line it was said that 'the sleepers are very broad'.[17] Pins were made at 1½*d.* a score (Horsehay, 1769), but Mackworth's railway at Neath, most unusually, was buying 'waggonway nails' at 5*d.* per pound in 1744.[18]

The rails tended to be a trifle smaller than Tyneside ones: 4½ in. high by 3½ in. wide, 4 in. by 3 in. (both Coalbrookdale area in the 1760s), or 4 in. square (Mackworth's in 1705). Those at Bath, with unusually heavy weights to carry, measured 6 in. by 5 in.[19] The normal length was 6 ft., although 7 ft. was known, and 10 ft. is recorded at Madeley in 1741.[20] Oak and ash, from local woods, were the only types of timber used in Shropshire. The wood was usually bought by the ton of 120 yards, at £2 in 1767, and sawn at 2*s.* 4*d.* per 100 ft. The price per yard varied from 3½*d.* (ash) to 5*d.* (oak) in the period 1741 to 1770. The life of the rails at Horsehay seems to have been about six years, for the railway built in 1754 was completely re-railed in 1760-1 and the old rails charcoaled.[21] Alloa was unique among Shropshire railways, as far as we know, in using the double way with two tiers of rail – one strongly suspects Tyneside influence from other Scottish lines. The first form of track consisted of single rails of Scotch fir, 5 in. by 4 in., and before long a second layer of fir was fixed on top. This wore so badly that the upper fir was replaced by beech, which became smooth and glossy and unusually free of friction.[22]

In 1767 there took place at Coalbrookdale an event which was to have an immense effect on the history of railways: the casting of the first iron rails. Richard Reynolds, the Manager of the Dale works, was the man responsible. His granddaughter wrote in her memoir of his life:

> For the conveyance of coal and iron to different parts of the works, and to the river Severn, wooden rails had been in use, which from the great weights carried upon them, were not only soon worn out, but were liable to give way and break occasioning loss of time and interruption to business, and great expense in repairing them. It occurred to him that the inconveniencies would be obviated by the use of cast-iron. He tried it at first with great caution, but found it to answer so well, that very soon all their railways were made with iron. He did not attempt to secure by patent the advantage of this invention, and the use of cast-iron in the construction of railways was afterwards generally adopted.[23]

The Coalbrookdale books of the period are now lost, but the date of 1767 is provided by a letter of 1823, when they were still extant. Robert Stevenson, the great Scottish engineer, wrote to ask for details of the origin of cast-iron rails. Barnard Dickinson replied that he had

> had many dusty old Books examined & much enquiry made relative to the introduction of Cast Iron Rails, most of the Old Men here that know anything on the subject agree that it is upwards of 50 years since Iron was substituted for Wood rails, but in our Ledger we find but one *old* entry of Iron Rails which is on 13th of 11mo 1767 – 100 Iron rails 111c. 2q. 11 lb. @ 7/- but no other Ledger entry do we find respecting them until 1781 – Stephen Hughes an old workman who is very particular in his acct of past events says it is 56 years since he first worked for the Compy & that Iron rails were used about 2 years after, and we have found a Moulders acct. in which the first entry of Iron rails is in 8mo. 1768, made by what are called French Men, 232 rails at 1½ each & from that date there is a regular entry every Month, so there is no doubt but they were used here in 1768, & continued from that period, but we have nothing to prove their invention here, or that they were imported from any other place . . . In the Moulders Accts for about a year from 8 mo. 1768, we find near 3000 rails moulded at 1½ each.[24]

Stevenson concluded – and there is no evidence to the contrary – that the rails of 1767 were the first cast anywhere.

Dickinson was Manager of Ketley works, but his letter was written from Coalbrookdale, and his researches were apparently in the Coalbrookdale books. So, after the first experiments in 1767, rails were in full production at the Dale from August 1768. Horsehay began casting rails in October

1768 – the Horsehay Waste Book is the sole contemporary source that survives – and by the end of 1774 had made 16,067 rails, enough for nine miles of track. Here the standard rail was 6 ft. long (a few were 7 ft.), and weighed 124 lb. (average of thirty-six rails) compared with the 125-lb. average of the first 100 cast at Dale in 1767.[25] The actual cost of the iron was about 3*s*. 6*d*. per cwt., plus 1*d*. for moulding and casting, but, if the rails were sold, they were priced at around 8*s*. per cwt.[26] It is not certain how the rails were cast. The casting charge of 1*d*. per cwt. is very small – for axles it was 10$\frac{3}{4}$*d*. – which suggests that the moulding was simple, without a top cover. But this would leave the underside of the rail (the top side in casting) irregular and difficult to seat firmly on the wood.

As there is no evidence that the Coalbrookdale system ever used the double way, with two layers of wooden rail, and the single rails were distinctly on the small side, very little wear was necessary before the wood became dangerously weak. The iron rails did not, as is often said, replace the wooden rails, but were laid on top, to provide extra strength and especially to protect the wood from wear. This is clearly shown by the evidence given by Jabez Hornblower to a Parliamentary Committee on Turnpikes in 1809, although he gives a misleading reason for the introduction of iron:

> Railways have been in use in this Kingdom time out of mind, and were usually formed of scantlings of good sound oak, laid on sills or sleepers of the same timber, and pinned together with the same stuff. But it was not until the proprietors of Colebrook-Dale Ironworks, a very respectable and opulent company, determined to cover these oak rails with cast iron, not altogether as a necessary expedient of improvement, but in part as a well-digested measure of economy in support of their trade.
>
> From some adventitious circumstance (which I need not take time to relate) the price of pigs became very low, and their works being of great extent, in order to keep the furnaces on, they thought it would be the best means of stocking their pigs, to lay it on the wooden railways, as it would help to pay the interest by reducing the repairs of the rails, and if iron should take any sudden rise, there was nothing to do but take them up, and send them away as pigs.

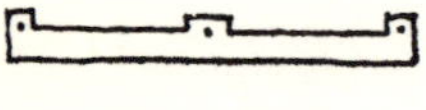

> But these scantlings of iron (as I may call them) were not as those which are now laid in some places [i.e. angled plates], they were

> about 5 feet long, 4 inches broad, and one inch and a quarter thick, with three holes by which they were fastened to the rails, and very complete it was both in design and execution.[27]

Hornblower's tale about the drop in iron prices is quite untrue, for there was no recession in 1767; rather the reverse.[28]

No example of these rails is known to exist today; but all the evidence goes to show that they had projecting ears or lugs, to keep the pins clear of the wheel-treads. According to Hornblower's dimensions, allowing for three lugs 3 in. by 3¾ in., the rail would weigh about 89 lb. John Randall evidently had a rail before him when he wrote: 'These rails were 6 ft. in length (some smaller number being 7 ft.), 3¾ in. broad by 1¼ in. thick, with two and two intermediate snugs for wooden pegs for fixing to the Sleepers.'[29] These would weigh 102½ lb. each. Both weights are lighter than those recorded at Horsehay and the Dale, but there is no reason why the same pattern should have been kept for all the thirty-odd years that these rails were being made, nor why the three iron-works should have used identical patterns.

The only measured drawing we have of this type of rail appears in an unexpected source, and poses some problems. Joseph Ritter von Baader (1763-1835), a mechanically minded native of Munich, came to Britain in 1786 to attend Edinburgh University. In 1791 he turned up at Haigh Iron-works near Wigan, claiming (without a shadow of justification) to have reorganized Coalbrookdale Iron-works and John Wilkinson's foundries. He was at once appointed Manager of Haigh, but was sacked in 1793, leaving a trail of ruin behind him. He returned to Bavaria, where he became Director of Machines and Mines. After another visit to Britain in 1815-16, he published in 1822 a description of his own inventions, a series of somewhat impracticable designs for very narrow-gauge guide-wheel railways. As an introduction, he traces the development of ordinary railways down to his own day. He has much to say about Coalbrookdale, but little of interest – he repeats the gist of Hornblower's tale about the fall in iron prices.[30]

Two of his illustrations, however, (pls. 57-8) are ascribed to the Dale by implication, though not by direct statement. They could very well refer to the Coalbrookdale system, which had small waggons like this as well as large ones (p. 269), or to one of the smaller railways in Shropshire (several of which were plated with cast iron), or even to somewhere else in the Shropshire empire. They bear no resemblance to anything that Tyneside produced. Baader says:

> In a short time, the example of Coalbrookdale was imitated in the rest of the kingdom, and almost everywhere the wooden rails were

> plated with iron bars. Thus began the second phase of this invention [railways], with half wood and half iron tracks, and this was likewise the first step in the introduction of cast iron as a material for track-making. [Pl. 58] shows this arrangement.

His iron rail is only 3 ft. long, 2 in. wide and about 1⅓ in. deep, with two projecting lugs 3 in. by 3¾ in. The weight would be a mere 32 lb. He describes the rails as 'bars of cast iron, on which the wheels run directly; and which are so fastened to the wooden rails by iron nails that the projecting heads of the nails are situated outside the track of the wheels'. He claims that the rolling friction of the waggon wheels was so much reduced by the use of iron rails that the wheel diameter was cut down from 24 to 18 in. At the same time the wheel tread was also narrowed, which might account for the very narrow rails which he draws. The first iron rails were evidently designed to the same width as the wheel tread. Nonetheless, it will be remembered that the Coalbrookdale wooden rails were, at the most, 3½ in. wide, whereas the first iron rails, including the lugs, must have had a maximum width of some 6½ in. So it was not simply a case of applying iron plates to existing rails. Wider rails were essential. Randall records that they measured 4 in. high by 8 in. wide,[31] and Baader's drawing gives 4 by 7½ in. It is significant that 1768 marks the beginning of an intensive spasm of work on the Dale railways, when many of them were relaid with new wooden rails. It is tempting to link this with the advent of the iron plate as a necessary alteration before the iron could be laid. Horsehay alone cast six and a half miles of track in 1768-9.[32]

By about 1780 a large proportion of the system had been plated with iron. Abiah Darby wrote:

> This laying the road with wood begot a scarcity and raised the price of it, so that of late years, the laying of the rails of Cast Iron was substituted; which altho expensive, answers well for wear and duration. We have in the different works, near 20 miles of this road, which costs upwards of £800 a mile.[33]

In 1785, according to Richard Reynolds, there was 'more than twenty miles of iron railways'. A visiting Frenchman, de Givry, wrote in 1784: 'They have in all their establishments a length of thirty miles of railway [*chemin de fer* – the first recorded use of the term], which is estimated to cost upwards of a thousand pounds a mile.' Another Frenchman, P.-C. Lesage, who also was at the Dale in 1784, had this to say:

> All the ways in the area of this foundry, up to two or three miles away, . . . are metalled with iron slag and equipped with iron plates

> three inches wide, on which run the small waggons that carry the iron from the mines and the coal to the forge. . . . Below the [Iron] bridge is the adit of an iron mine . . . whose gallery is six or seven feet wide, equipped with iron plates like the ways in the hills, on which the waggons run.

The latter must have been one of the footrids in the Madeley direction. Iron plates were also used, it seems, on the Caughley railway on the other bank of the Severn.[34]

Elsewhere, in spite of Baader's statement that the Coalbrookdale rails were widely imitated, few examples can be traced. Horsehay sent thirty-six plates to a Bristol agent in 1772,[35] perhaps to try to tempt the South Wales market, where it is conceivable that Sir John Morris used them on his Landore railway – he certainly introduced them underground. Sir Herbert Mackworth applied them on his grandfather's railway at Neath before his death in 1791, inspired, it seems, by the example of Colonel James Fraser at Ravenhead Colliery in Lancashire, which had links with Shropshire. Mackworth was one of the promoters of the British Plate Glass Co. at Ravenhead, which was supplied with coal from Fraser's mine. Both Mackworth and Fraser obtained their rails from Colonel Dalrymple's foundry near Hamilton in Scotland. Though the cost of these Lancashire plates, measuring about 3 in. by 1 in., was, of course, much greater than that of plain wooden track – about 10*s.* 6*d.* per yard run, sleepers and horse-track included – they were far cheaper to maintain, and a wider use of them was advocated – 'the saving in labour in draught cattle would also be astonishingly greater'.

In North Wales the wooden Mancot railway near Hawarden was relaid with iron in 1793 by Beriah Botfield, the proprietor of Old Park furnaces near Coalbrookdale.[36] A few canal railways in the east Midlands, as we shall see, also adopted the cast-iron plate. But by 1800 it could be said of iron-plated Shropshire railways that

> for the want of a proper system in the forming and laying of such roads, they have been found of little or no more service than wood rail-ways, which from the late improvements in iron-roads, are now in disrepute.[37]

The double rails on the Alloa railway in Scotland have already been described. In the early 1780s,

> The next attempt at improvement was the adopting of cast iron, by plating the wooden rails with cast iron in pieces of four feet long, of an inch and a quarter square; each piece of rail having three

> projecting ears with holes, through which a pin was driven to fix the iron rail in its place. This plan proved quite abortive, for the instant the wood gave way or yielded, the cast iron was broken, and it was found so useless and expensive that it was thrown aside altogether. . . . The next step adopted [in 1785] was the double wooden rails before-mentioned, being formed of best Memel or Prussian fir, plated with Swedish malleable iron, 1¾ inches broad and ¾ of an inch thick.

This plan, it was said in 1793, 'is acknowledged to be the most complete in Britain'. The track cost at least 10*s*. per yard run, but for the first twelve years of its life repairs came only to £200 over a distance of more than two miles. Then troubles began. 'This railroad kept long in a good state, but when the wooden rails began to fail, the repairs were never ending and expensive.' About 1810 the wrought-iron-cum-wood was replaced by all-cast-iron rails.[38]

Ralph Allen's railway at Bath was equipped late in its life with similar thin iron plates. Count Frederick Kielmansegge, visiting Allen's quarries in 1761, recorded in his diary that

> The stone is brought in a remarkable way down the hill towards the town. On each side of the road, iron, two or three inches broad, in the shape of carriage-ruts, is laid down; the stones are put into low but strong trolleys, with broad iron wheels, with an outer rim projecting in such a way that they cannot slip out of the iron ruts.[39]

His description is odd, for he clearly saw flanged wheels. It may be that the term 'carriage-ruts' is a translation of the German *Geleise*, which basically means 'rut', but from an early date was applied to ordinary railway track. His reference to the 'outer rim' prompts the thought that the original inside flanges may have proved too flimsy for the job, and been augmented by outside flanges as well. The well-known print of Prior Park in 1750 (pl. 56) certainly shows double flanges on the wheels; but how much trust can be put in the print and in Kielmansegge's description is open to doubt.

The only other known case we find of wrought iron on a Shropshire-inspired railway was at Carron, again in Scotland. The source is late (1868) and not very satisfactory. 'The first line, stretching from Kinnaird Colliery into the interior of the works, was constructed in 1766. The rails in that instance were of wood, covered with a sort of hoop-iron.'[40] The true date of the Kinnaird line is 1760, and no more is known about the iron plates.

The overall costs of building Shropshire railways are almost impossible to work out. The wooden Park Railway near Horsehay seems to have cost £605 for about two miles in 1769-70, materials and labour included;[41] iron rails, if added at 3*s.* 6*d.* per cwt., would account for about another £700 per mile. Abiah Darby's figure of £800 a mile perhaps refers only to the iron, but de Givry's figure of £1,000 might cover the whole cost. The expenses at Alloa, as rebuilt with wrought-iron plates in 1785, were put at £880 per mile.

Turntables are rarely met with. At Steeraway in 1749 a man was paid 'for making ground to lay Rails to save a turn Rail'. A map of Madeley Wood furnaces, dated probably 1757 or 1758, shows that the feeder lines which supplied the furnace mouths with raw materials left the main line by turntable – an arrangement very probably repeated at other furnaces.[42]

The track gauge of Shropshire railways, as far as our evidence goes, was always smaller than that of Tyneside waggonways. The narrowest recorded gauge on Tyneside was 3 ft. 10 in., the widest recorded gauge inspired by Shropshire was 3 ft. 9 in. at Bath. It may be that the Coalbrookdale system, some of whose waggons were large ones possibly copied from the North, also used a wider northern gauge, but there is no evidence for it. An account of 1796 (p. 271) describes a large waggon as being 3 ft. 6 in. wide, presumably at the top of a tapering body, which could thus be accommodated with ease between wheels set at a gauge of 3 ft. 9 in. or less. Baader's drawings of a small waggon confirm this. He gives a gauge of 3 ft. 8 in. to the wooden railway, and 3 ft. 5 in. to the iron-plated one (assuming his scale is on the English standard; if it is on the Bavarian, these figures should be about 2½ in. less). On the early Continental edge railways, most of which seem to have been copied from Shropshire rather than Tyneside, the gauge usually fell between 3 ft. and 3 ft. 6 in., although there were two lines of 4-ft. 2-in. gauge due to William Wilkinson's influence.

So, on the whole, the evidence suggests that the Dale was always conservative in its gauge. Even when angled plate rails began to oust the existing edge rails around 1800, the East Shropshire gauge never rose above 3 ft., and most of the system was 2 ft. 3 in., whereas elsewhere the most common figure was 4 ft. 2 in.

The Tranent and Alloa lines in Scotland were both 3 ft. 3 in.[43] The Hirwaun-Penderyn railway in South Wales probably began life in 1786 with the 3-ft. 2-in. gauge which it had on its relaying with all-iron rail in 1794. Likewise the Landore's gauge of 2 ft. 3 in. in the days of iron was almost certainly continued from its wooden period.[44] On building his railway at Neath, Sir Humphry Mackworth laid his rails 'distant about two Foot from each other'.[45] Did we but know the gauges of the many footrid railways on the Stour and at Broseley and Madeley, we should most probably find them similar to Mackworth's: very narrow, for small waggons capable of working underground.

With so wide a range of gauges, there was naturally a wide range of waggon sizes to fit them. They are listed in Table 8. Little rhyme or reason can be found in Shropshire measures for coal. The stack, originally

Table 8

Shropshire Railway Waggon Capacities
(*For coal unless otherwise stated*)

Neath	1705	14 or 18 cwt.	(misprint in one of two sources)
Madeley (Holland's)	1715	*c.* 20 cwt.	(just over half one usual ton)
Pensnett	1666-75	26+ cwt.	(one stack)
Madeley (Lane pit)	1692	25-40 cwt.?	(probably one stack: coal and ironstone)
where?		22-26 cwt.	(computed from Baader's drawings)
Madeleywood	1756-7	30 cwt.	('half-Ton Waggon': coal and ironstone)
Alloa	1766	30 cwt.	(one Scots chalder)
Tranent	early 19th cent.	30 cwt.	(one Scots chalder)
Rowton	1702	*c.* 40 cwt.?	('usual ton or waggon')
Madeley (Lloyds)	1756	max. 48 cwt.	('ton Wagon')
Coalbrookdale	1751	52½ cwt.	(coal and pig-iron: probably limestone and ironstone too)
Madeleywood	1756-7	60 cwt.	('ton Waggon': coal and ironstone)

it seems equivalent to ten horse-loads of about 2½ cwt. each, is specifically defined on the Pensnett footrail in 1666 as 'to exceede Sixe and Twenty hundred weght';[46] by 1733 at Dawley it had grown to 40 cwt. of 112 lb. each.[47] This is not surprising, the increase being paralleled on Tyneside. The usual or customary ton, by which waggons were often measured, also grew, but in an erratic way. The earliest attempt to define it does not help: a Broseley (Calcutts) lease of 1673 simply lays down the royalty at 12*d.* a ton for best coal, 9*d.* for flint, 'allowing such Tonadge of cole as is usually allowed and sold to Trowemen att Seaverne'.[48] But in 1715, at Holland's footroad in Madeley, the usual ton was 'near forty hundred-weight' and a waggon was computed to hold twenty-one fortieths of a usual ton.[49] At Willey in 1757 a customary ton contained 42 cwt. of 112 lb. each.[50] Smitheman's and the Giffards' lease in 1756 of Lower Lloyds in Madeley ordains 'each ton Wagon not to exceed 48 hundreds in weight',[51] but the very same proprietors, leasing some Madeleywood mines in the very same year, put the ton considerably higher. The lessees were charged royalty of 1*s*. 6*d*. for every ½-ton waggon and 3*s*. for every whole-ton waggon, the weights 'to be the same as sold at the Old Jacobite Footroad in Madeley Wood'; but the load is also defined as 'Thirty Hundred Weight of Coals on each and every Waggon called a half Ton Waggon', thus putting the ton at 60 cwt.[52] It is a relief to find the Coalbrookdale waggons described in 1751 in simple terms: 'Their wagons carry 52½ Hund.'[53]

The waggons shown in illustrations at Broseley in 1788 (pl. 59) and at Landore in 1796 (pl. 55) appear to fall into the 20-30 cwt. category. That this sort of size was indigenous to East Shropshire is suggested by the underground-cum-surface history of the early railways here, and by the apparent fact that horse traction was not known in Shropshire until soon before 1700, and did not become general until the long lines to the iron-works appeared in the 1750s. As late as 1775 John Fletcher, the celebrated Wesleyan Vicar of Madeley, being in some confusion over the terms of a mining lease, took expert advice. His correspondent, R. Chambre, told him:

> By Waggons I certainly mean such as come singly out of the Foot-roads, I never heard of any other nor can associate any other idea to that Term: & surely there is not a Miner who w^{d} act so very dis-ingenuously (to say no more) as to offer you a less sum for this Quantity.[54]

Chambre was evidently no great expert, since he was unaware of the large Coalbrookdale and other waggons, which by no stretch of the imagination could be used underground; but for a footroad mine a waggon normally did mean a comparatively small unit.

It is in the size of some Dale waggons that Tyneside influence is possible. They, and the Madeley waggons of 1756-7, which perhaps imitated them, form the only real break in the pattern of small Shropshire vehicles. Whether their 52½-cwt. capacity deliberately copied the Tyneside 53-cwt. (24-boll) waggons, which were still rare by 1751, we do not know. It may simply be a coincidence, as 52½ cwt. almost exactly equals 2 old stacks, the coal unit used at the Dale in pre-railway days. The waggon in Williams's painting of Coalbrookdale in 1777 (pl. 60), which looks enormous when compared with the man behind, is probably a case of artistic licence, for all the documentary evidence suggests that the capacity of the Dale railways never rose above 53 cwt. The waggon shown in a sketch of 1796 carried about this quantity of coal, with the aid of upward extensions to the sides (fig. 46). The question is complicated further by the fact that four different commodities were regularly carried. So far we have been considering coal waggons. But did the same waggons also carry limestone and ironstone, heavier and therefore more compact than coal? And pig iron, very much heavier still? Or were there special waggons for these materials? Baader's waggon, which would not hold much more than 26 cwt. of coal, could easily accommodate a 53-cwt. load of pig iron. A waggon of very similar size and shape which appears in Loutherbourg's painting of the Dale in 1801 (pl. 61) shows that this small type was to be found at least around the works themselves. But whether both sorts were all-purpose or for specialized traffic only, we just do not know.

The Bath railway was a special line for a special purpose. There, the

FIG 46 Sketch of large Coalbrookdale waggon, 1796.

waggons were loaded 'often with upwards of four Tons', according to Labelye, although John Wood said that blocks rarely weighed 'less than five or six Tun at a time'.[55]

Waggon bodies seem to have been low and narrow in relation to their length. The internal dimensions given by Baader are: length at top, 6 ft. 9 in., at bottom, 6 ft. 2 in.; width at top, 3 ft. 5 in., at bottom, 2 ft. 10 in.; height 2 ft. 3 in. At Bath, where sloping sides were unsuitable for the load, the waggons were 10 ft. 2 in. long, 3 ft. $5\frac{1}{2}$ in. wide and 1 ft. 6 in. high, with detachable sides (pls. 62-3). Elsewhere, the waggon ends were removable for unloading, and even for loading: at Coalbrookdale (pl. 61), Landore (pl. 55), Tranent, Alloa[56] and apparently Broseley (pl. 59).

The constituent parts of a waggon occur rarely in the accounts. The Madeley Lane Pit inventory of 1741 mentions '5 Wagon Sills, 12/6' – the equivalent of the Tyneside soles or main frames – and 'Clift for Wagons 17/6', apparently wood for splitting up to make the vertical ribs. Both recur in the Horsehay accounts in 1768-9, sills at 3*s.* each and cleft at 1*s.* 6*d.* per foot. At the same time, 'waggon staves', perhaps the side planks, appear at 3*s.* for 60, and 'waggon bottoms' at 3*s.* each.[57] The most complete costing of a waggon is that illustrated by the sketch of 1796 (fig. 46), and is worth quoting *in toto.*

Expences of Railway Waggon 10 *ft long* 3 *ft* 6 *i wide*:

	Cwt	Q	lb		£	*s.*	*d.*
2 Ash Sills 10 feet long 8^{i} by 5^{i}							
2 Wrathes 11 feet long 3½ by 4½							
12 Standards 2 ft 6 long – 3 by 4				14½ solid ft			
6 Pieces for Sides before the Clamdogs 2 ft lo – 7 by 2				@ 1/6 p ft	1	1	9
2 Pieces for crosstrees 4 ft 6 lo 3 by 4							
2 Cratches					0	2	0
62 ft of Boards @ 2^{d} ft					0	10	4
Wro^{t} Ironwork	4	0	12	@ 30/ cwt	6	3	0
2 D^{o} Axletrees	1	3	0	@ 30/ cwt	2	12	6
4 Cast Iron Frames	0	2	0	@ 10/ cwt	0	5	0
4 Brass Brasses	0	0	20	@ 1/4 lb	1	6	8
4 Cast Iron Wheels	12	0	8	@ 10/ cwt	6	0	8½
1 Roller	0	0	8	@ 10/ cwt	0	0	8½
Brake Chains & Pole					1	0	0
Making					0	10	0
					£19	12	8

Since the wrathes, the longitudinal members on top of the sides, were 11 ft. long, the 10-ft. length ascribed to the waggon seems to mean the internal length of the body. So too the crosstrees, which apparently capped the waggon ends, were 4 ft. 6 in. long and the internal width was 3 ft. 6 in. The standards were presumably the inclined side members and the clamdogs the iron uprights carrying the cratch or rack-like side extension.[58] Loutherbourg's painting is alone in depicting shafts, an item of equipment not met with elsewhere. It may be due to a misunderstanding on the artist's part of the brake lever, as happened at least once on Tyneside.[59]

The brake levers or jigs (called the jigg-pole at Bath) tended to be straight and fairly short, and bore almost on top of the wheel, whereas on Tyneside they were longer, more curved, and bore more on the front of the wheel. They appear in the Horsehay accounts of 1768 as 'braiks' at 6*d.* each. William Williams's painting of 1777, which includes a waggon descending Jigger Bank (only a small part of which was rope-worked), poses a problem, but his proportions are so astray that his details can hardly be taken as reliable (pl. 60). The waggon has no ordinary brake visible, but the waggon-man is doing something between the rear wheels with a long pole. Over the wheels are lines which look like mudguards, and might be semicircular band brakes that the waggon-man is holding on, using his pole as a lever. Alternatively, he may be sticking his pole into the spokes as a sprag. Spragging was certainly practised at Tranent, and in a sophisticated form at Bath. At Alloa the speed was controlled by drags or beams of wood trailed behind.[60]

The very fine drawings of the Bath waggons (pls. 62 and 63), published in 1734, are the earliest known which depict an English waggon in any detail. The vehicle is quite untypical, however, both in the design of its body and in the elaboration of its brakes. With so heavy a load and so breakneck a gradient, a simple jig or a simple sprag were quite insufficient. In spite of John Padmore's stringent safety devices, it must still have needed considerable nerve to accompany a 4-ton block of stone down a grade of 1 in 10. Each of the rear wheels had an independent 'Jigg Pole', held down by a chain wound round a drum. This drum was rotated, and the chain tightened, by means of a lever in the socket *h*, and held in position by the 'rotchet and click' *ir*. The two central levers *lk* on the back, when dropped from the vertical to the horizontal, each moved a longitudinal shaft under the body, which, by way of a bell-crank, pushed a square sprag, lying parallel to the front axle, out between the spokes of the wheel. One would imagine that the effect of applying these sprags while in motion would simply be to break the spokes, but apparently they worked. Since each wheel had an independent brake, the axle was protected from strains by giving it one square and one round end. One wheel, keyed to the square end, revolved with the axle; the other was loose on the round end, held only by a linchpin.[61]

Wheels were practically always of iron. Of wooden wheels, apart from those discovered in the Broseley area (see p. 106), the only record occurs at Madeley Lane pit in 1741, 'Wooden Bank Wheeles', as opposed to pit wheels for underground use. Yarranton's sketch of his waggons on the River Stour footrails in about 1662, in so far as any trust can be placed in it, appears to show spoked (and therefore iron) wheels (pl. 53). But we cannot be sure until 1729, when Coalbrookdale began to cast wheels. They were small – sometimes very small. One variety weighed 85 lb., but most weighed only about 30 lb., with a minimum of 28 lb. The price, when they were cast in loam, was a moulding charge of 2*s*. per wheel, plus 2*d*. per pound; after 1732 they were usually cast in sand with a charge of 10*d*., plus 2*d*. per pound. Up to 1738, when the Stock Book ends, about half the wheels were bought by John Smitheman, who was working a colliery in Madeley, and most of the large wheels went to William Ashwood at Paddock pit.[62] We can perhaps pick up the trail of these wheels in the inventory of Madeley Coal-works in 1741. At Lane pit, which shared a railway to the Severn with Paddock pit, there were '18 pair Iron Wheels £12' and '15 pair Iron pitt weels £3', or 6*s*. 8*d*. and 2*s*. per wheel, no doubt allowing for depreciation.[63] The 85-lb. wheels were therefore probably for surface use, the 30-lb. ones for underground. Small waggons as well as large were evidently using the Coalbrookdale system in early days, for in 1754 and 1755 the Dale cast wheels weighing only 75 lb. and 94 lb. for the New Railway, at 18*s*. 8*d*. a cwt.[64] The wheels on Ralph Allen's waggons at Bath were 20 in. in diameter in 1734, with six spokes and a fragile-looking flange some 3 in. deep. They weighed, by calculation, about 178 lb. each. It is an interesting coincidence that the wheels on Baader's drawing, though of 24-in. diameter and a different design, are estimated to weigh the same. They were replaced, Baader says, by smaller versions 18 in. in diameter and about 137 lb. in weight. Little more can be said about the smaller type of waggon. At Broseley in 1788 (pl. 59) front and rear wheels were both small, but unequal; at Landore (pl. 55) they were both small and equal; at Alloa in 1793 the diameter was 27½ in. Sandycroft colliery in Flintshire was buying iron wheels from John Wilkinson at Bersham in 1756.[65]

The large Coalbrookdale waggons were given wheels of a size similar to those on Tyneside, and even bigger. In 1755 the Dale began to cast them for the New Railway at 2 cwt. 96 lb. and 3 cwt. 19 lb. each, at a cost of 10*s*. per cwt.; and the 273 wheels it made between 1758 and 1761 averaged 2 cwt. 90 lb. The 1796 waggon had wheels weighing just over 3 cwt.[66] Around Newcastle this was rarely reached, the average wheel weighing perhaps 2½ cwt.

Whereas Tyneside axles were apparently always wrought iron, in Shropshire, as befitted so notable an iron-founding area, they were nearly always of cast iron. On the New Railway in 1754, an axle weighed on average about 100 lb. (some were a mere 84 lb.), but by 1759 had been increased to about 136 lb., which means that the diameter was almost 3½ in., with a length of 5 ft.[67] At an earlier date the Madeley list of 1741 includes '26 Axeltrees at 7s 6d', and Ralph Allen's waggons possessed cast-iron axles of 3 in. diameter. Earlier still, on the Stour Navigation, '20 Martij 1665[/6] S Whyte paid to Crowley for axeltrees for the Waggons on the ffooterayle 5–0–0' – maybe wrought iron at this date. So too, probably, were those on the Charltons' Wombridge line in 1748, when they paid for

'2 cwt. 2 qr. 3 lb. iron to make 6 axel trees £2. 4. 10'. The phraseology (cast-iron axles would be bought ready-made) and the cost of the iron (17*s*. 9*d*. a cwt.) suggest wrought rather than cast iron. The weight of the axles, 47 lb. apiece, also indicates small-sized waggons when compared with the comparable weights on Tyneside (p. 198) and with the 98-lb. wrought-iron axles found on the 1796 waggon.[68] Axle bearings, iron on Tyneside, were always brass here: on the Stour from 1663 (where they were liable to be stolen), at Bath in 1731, at Madeley in 1741 (valued at 1*s*. 3*d*. each), at Horsehay from 1755 (3 lb. each) and at Dale in 1796 (5 lb. each, costing 6*s*. 8*d*.).[69] For lubrication one of several entries in the Steer-away accounts is typical: '(1751) Paid for 16 pound of Grease for the waggons 3/4.'[70]

A new waggon was bought by Coalbrookdale in 1756-7 for £12, and nine more and one 'carriage' were made in the works at 11*s*. each, excluding materials.[71] By 1796 the total cost had risen to £19 12*s*. 8*d*. The waggons at Madeley Lane Pit in 1741 were valued at 9*s*. each, evidently for the body alone: if wheels, brasses and axles are added, the total comes to £2 15*s*. 8*d*. Ralph Allen said of his very elaborate waggons 'that one of them, when compleatly finished, and ready to be used, stands him in upwards of 30 Pounds'; in 1739 he had eleven of them, ten in use and one spare. We know the size of the stock on only two other railways, both small: twelve waggons, with various spare wheels and axles, at Madeley Lane pit in 1741, and eight waggons which Robert Yarranton was to build in 1677 for Ananias Hensey's footrail at Stourbridge.[72] The Coalbrookdale system must have owned a large number of vehicles to carry its heavy traffic, but it is dangerous to guess their number from the output of wheels and axles, a fair proportion of which must have served as replacements.

The amount of traffic carried on the Dale railways is another gap in our knowledge. It must have been very considerable, what with the three ingredients for iron-making converging on all the furnaces, and some of the pig-iron and castings being carried away. The only direct clue we have appears in the accounts of the Horsehay Works, which managed and maintained the New Railway, apparently the two miles from the Coalmoor line to Dawley Bank. Horsehay evidently paid all the expenses of this railway, which it recouped by charging the other constituent works of the Coalbrookdale empire in proportion to the number of their waggons that traversed the line. The entry in question, dated 3 January 1762, reads:

> New Railway. General Charges. Upon settling the acc^t this day it appears to be 3/- a waggon & our Carriage on it since last settling is 1099 wagn @ 3/-. Dale Company D^o 3750 @ 3/-. Ketley Furnace

Co D° 1220 @ 3/-. Dawley Company D° 49 @ 3/-. D° Severn Coal 526 @ 1/6. Lawley Company D° 1520 @ 1/6.

Some of the waggons used only a half of the New Railway before branching off to Lawley or Dawley, and were charged at half-price. The entry gives a total of 8,164 waggons on what was but a small portion of the whole system. But it does not state the period of time involved. A more satisfactory idea can be obtained from the tonnage of pig-iron produced by the Horsehay furnaces. The figures for 1762 do not survive, but in 1761 nearly 10,000 waggon-loads of ore, coal and lime would have been needed to produce the 1,713 tons of pig iron made that year, assuming it all came by rail and that the waggons held $52\frac{1}{2}$ cwt. Over the years 1767-74 the number of waggon-loads needed had risen to an annual average of over 13,000. And these figures are for Horsehay furnaces alone.[73]

When the footrail from Pensnett to Stourbridge was planned in 1661, an annual freight of 15,000 tons (or about 12,000 waggons) was envisaged, but quite certainly never achieved or anything approaching it. Some idea can be gained of the pattern of traffic when the Bath railway was extended to the north bank of the Avon in 1739. The two ferries from Dolemead Wharf across the river each carried two waggons and made four return trips a day. There were:

4 carriages going on the hill constant and over the water,
4 ditto loading on the hill,
2 ditto loading blocks to Dolemeads,
1 ditto spare if any misfortune should happen.[74]

Deplorably little, too, is known of the working of the Shropshire railways. The principle of one horse, one waggon, which was very much the rule on Tyneside, was perhaps the exception here. At Neath, on the surface section of Mackworth's footrid, one horse pulled one waggon. It was the same at Alloa on the first three stages in the railway's development; but on the fourth stage, the very efficient wrought-iron-plated way of 1785, the waggons

> are supposed to weigh altogether about a ton. A waggon carries 30 cwt. of coals and 3 waggons are linked together by chains; so that 1 horse draws $4\frac{1}{2}$ tons of coals at once; and the declivity of the way is so gentle that the same horse draws with ease the 3 empty waggons back to the coal-hill.

The coupling chains were 2 ft. long, to reduce the inertia on starting.[75] The

Neath Abbey Colliery railway was worked in the same way between 1800 and 1804:

> The vehicles in which this fuel is carried, are called dram waggons. Four or five of these heavy machines, linked together, and filled with coal, are drawn upon the rail roads, constructed for the purpose, and only require the assistance of one man for their management.[76]

At Bath the reverse happened, for two horses were needed to pull an empty up the gruelling slope of 1 in 10 and a loaded waggon along the level. Astonishingly, the Coalbrookdale system did likewise. Abiah Darby recorded in 1780, 'One waggon with three horses will bring as much as 20 horses used to bring on horse's backs';[77] and all the evidence bears her out. One would expect that a 53-cwt. waggon could be managed by one horse as on Tyneside, especially once the iron rails were in use; elsewhere they were calculated to improve the efficiency of a horse by 30% by reducing the friction. Much of the trouble no doubt lay in the fact that the loads had to be taken uphill almost as often as downhill, and some of the gradients were fierce. But that is not the whole story. The payments for waggoning were always high – enormously high by Tyneside standards – which can only mean that each waggon-man kept several horses. This holds true even on the railways from Horsehay to Lawley and Dawley, which were comparatively level. Williams's painting of a waggon descending Jigger Bank by gravity (pl. 60) shows four horses trailing behind, which is fair enough – they would be needed to get it back up. But his companion picture of the Dale from the south depicts four or five horses, in line ahead, pulling one waggon *downhill* from the Dale works to the Severn. Why? The gradient, not nearly so steep as Jigger Bank, is still appreciable, although it could hardly need four or five horses to move one waggon back up. The small waggon in Loutherbourg's painting of 1801 is likewise drawn by two horses. One might suspect artist's licence if the other evidence did not agree with Williams and Loutherbourg. Whatever the reason, it was so.

Other forms of motive power were not unknown. Wayleaves at Madeley in 1692 and at Coalbrookdale in 1749 both permit the use of oxen,[78] which were certainly in service at Landore in 1796 (pl. 55). Many of the earlier and smaller Shropshire railways indubitably used manpower, especially those which ran straight out from underground and extended but a short distance on the surface to the Severn. Mackworth's footrid at Neath likewise relied on manpower underground, but on horses on the surface. He also used a cheaper and more ingenious form of power:

> his new Sailing-waggons, for the cheap Carriage of his Coal to the

> Water-side (*whereby one Horse does the Work of ten at all times*; but when any Wind is stirring (which is seldom wanting near the Sea) *one Man and a small Sail does the Work of twenty*) . . . do sufficiently shew what his Genius is capable of in Matters of that nature. And, I believe, he is the first Gentleman, in this part of the World, that hath set up *Sailing-engines on Land, driven by the Wind*, not for any Curiosity, or vain Applause, but for real Profit.[79]

The latter may refer to Mackworth's ventilating devices in his colliery, which are alluded to, along with the sailing waggons, in the verses quoted at the chapter head.

The cost of leading coal along the footrail from Pensnett to Stourbridge, a distance of about two miles, was estimated at 6*d.* per ton in 1662-3, but turned out to be more. The proprietors engaged a special employee in 1666, 'Mr Willm Lewis whoe is ye Clarke at ye footrayle', to record the number of waggons and keep the accounts. On another short footrail at Coalbournbrook nearby, the expenses involved in selling a stack of coal (26 cwt.) in 1675 were: 6*d.* cost, 4*d.* carrying down the footrail, 6*d.* loading and craning, 6*d.* profit – a total of 1*s.* 10*d.*[80] The railway at Bath reduced the cost of transport of stone from 10*s.* to 7*s.* 6*d.* a ton.[81] On the Coalbrookdale system, the coming of the railways brought the transport charges down far more strikingly. Whereas it cost 3*s.* 6*d.* to carry a ton of iron from Horsehay to the Severn by pack-horse, the railway did it for 1*s.* But, as has been seen, the leading charges were still high compared with Tyneside. For example, the price for taking a waggon two and a half miles from Horsehay to the river in 1757 was 2*s.* 6*d.*; ten years later it was 3*s.* 6*d.* to 4*s.* for the return journey, uphill, and 3*s.* for the one and a half miles from the Dale to Horsehay. Occasionally we find men employed, with their teams, by the day: 4*s.* in 1757, 5*s.* in 1768.[82] At this date, a Tyneside waggon-man who owned his horse could not expect more than 10*d.* for a two-and-a-half-mile journey, with little hope of any freight on the return trip. The explanation must lie in the multitude of horses used, although a Dale waggon-man with four horses to keep would still be better off than his northern brother with only one. But he may have had to hand over some of his earnings to his master, for, in the early days at least, waggon-men were supplied by the landowner from whom the railway wayleave was leased. William Ferriday, the Foresters' agent who spied on the Dale and heartily despised the Company and all its works, wrote to his master in 1754: 'I have agreed with Mr Richard Ford to waggon Dubble coals down the Dale Rails.' Again, in 1755: 'I have put another waggoner to push Mr Ford down the Dale. I have likewise set another waggoner on to the old Hall . . . people are now aware of their Sloathfull and Ignorant management for which they are laughed at.'[83]

Maintenance of track and waggons was sometimes let out on contract, as on Tyneside. Andrew Yarranton estimated in 1662-3 for the Pensnett footrail, '3^{d} a Tun allowed for repairing the foote rayle and keeping the wagons and Brasses in ord^{r}', but it was not until the line had been in use for a year that a keeper was found. Lord Windsor wrote to Samuel Baldwin on 26 December 1666:

> The mention hee [Samuel Whyte] makes of the foote raile relates to a proposall made to him at the same time by an other (as hee told mee) who would keepe the Raile in order (wee allowing timber) fr 10^{l} a year, which I Judge agood bargaine for us.

Two days later Whyte said: 'I have had more treaty about keepinge ye foot rail and wagons but have only set ye Man to put it in order now.'[84] In the Coalbrookdale empire the railways were repaired and maintained, and indeed were built, by direct labour, and it is hard to disentangle details from the accounts. Occasionally we find groups of men contracting for piecework, but usually the men are paid individually. Both occur in 1748 in Madeleywood:

Paid John Beddow & Co. Laying the Rails & making Waggons for the Lake Head footrid	£2–15–$8\frac{1}{2}$
p^{d} Jno Rowley for 10 days work at Railway	10–0
pd Cadman for 35 days driving Waggon Way	2– 3–6
pd John Jones for $33\frac{1}{2}$ Days levelling the Rail Way	1–13–6
John Jones 12 days at the Rayls 12/- & 16 days at the Footrid at 1/3 is 20/-	
Geo Cope $2\frac{1}{2}$ d^{s} Work at Lakehead Footrid 2/6 & Fetch^{g} Wood 2/6	5–0
Michael Bayley at Esqr Welds Rayls 2 days[85]	3–0

For maintenance, a gang of men was allocated to a particular railway, and paid 1*s*. to 1*s*. 8*d*. each per day. The General Railway between 1768 and 1774 normally kept three or four labourers busy. Once or twice there are signs that a specialist was employed. On the Bank Railway from 1767 to 1769 a regular entry is 'Francis Perks Keep^{g} them in repair a month £1–16–0' – 1*s*. 6*d*. a day, with the help of a labourer at 1*s*. a day.[86] More specific items appear frequently in the Steeraway lime-kiln accounts, for example:

15 April 1738 Paid severall persons for filling and backing up the Rails laid from the Short Wood to Thomas Jones' pit as by bill on book £30. 18. 09

12 November 1749 Paid Jno Newton for Cleansing the Rails from ye Pitt 5. 0

4 March 1750 [/1] Paid David Jones for getting 21 waggons of Rock in Short wood to mend the Rail Road at 2½ a load 4. 4

6 January 1751 [/2] Paid David Jones for Getting & filling Rock to mend the Rail Road 2. 6

17 December 1752 Paid Chorm for cutting a gutter to take the Watter of Rail Road 1. 0[87]

Some overall figures for keeping and operating a railway are given for Holland's pit at Madeley in 1747, where the cost of coal per waggon included:

Waggoning Bills include waggoning, making Wood for the Waggon and Wind-way, setting Wood, & cleaning the Ways	0 : 0 : 11¼
Finding Wood for the Waggon and Windway & Rails & Sleepers	0 : 0 : 5½
Axtrees, Wheels, Traces, Halters, Shoeing and finding Waggons	0 : 0 : 2
Maintaining 3 Horses, and looking after them	0 : 0 : 6

This makes 2*s*. 0¾*d*. directly attributable to the railway, which, with 2*s*. 7½*d*. for getting the coal and attendant charges and 7*d*. for wharfage, made the total cost 5*s*. 3¼*d*. per waggon. The coal sold at 6*s*. 6*d*.[88]

Our knowledge of the detailed workings of Shropshire railways, it will be clear by now, is regrettably scrappy and slender. But very considerable differences from Tyneside are noticeable, not only in superficial, but even in fundamental matters. It was not a case of adapting Tyneside methods to suit a different type of coalfield, but a basic and early divergence of practice. Economically, materials (especially, of course, iron) tended to be cheaper in Shropshire; ordinary labour cost about the same; but leading charges were very much higher.

15

Canal railways and the coming of the iron rail

And then an aud horse brought a waggon
  A' the way frae the pits to the staith,
But now it appears pretty certain,
  They'll verra suin dee without baith.
For now their fine steam locomotives
  A' other inventions excels,
Aw've only to huik on the waggons,
  And they'll bring a ship-load down their sels.

'The Changes on the Tyne', in J. P. Robson, *Songs of the Bards of the Tyne*, Newcastle (1849).

Canals have already been mentioned frequently, if incidentally. Many inland collieries had railways down to navigable rivers like the Severn, or to partly canalized rivers like the Stour, and with the purely artificial waterways the process was continued. We have seen how the opening of the Sankey Canal in 1757 resulted in a small spate of feeder railways; and once the Canal Age really began in the 1760s, such feeders multiplied. The full effect of the canals in attracting these lines was not felt until the iron rail was established, but there are a fair number of wooden lines to record. They deserve separate treatment because they can rarely be credited with a wholly Shropshire or a wholly Tyneside ancestry. Rather, they form a hybrid breed, with the traits inherited from the two parents emphasized to different degrees. The Caldon Low railway in Staffordshire, for example, is called a Newcastle waggon road, but its track is much more reminiscent of Shropshire. The reason for this is not far to seek. The canal engineers, hailing from many different parts of the country, likewise operated over wide areas by virtue of their calling, disseminating ideas and breaking down local barriers in the technical field as well as the economic. But the first big canals being a basically Midland phenomenon, the railways that came to supply them were inevitably closer to the Shropshire pattern than to Tyneside, at least to begin with. This is noticeable particularly in the wide spread of the words *railway* and *railroad*, compared with the rarer use of *waggonway*. Unfortunately, our knowledge of many of the wooden feeders is limited to their sites and their dates, for minute books give little but the bare bones.

Most canal Acts, early or late, contained clauses enabling feeder railways to be built. The Directors of the Birmingham Canal, for example, in drafting their Bill in 1767, noted that 'Rail or Waggon Ways from the Canal to the respective Coal Mines appears a Matter of great Consequence to this Town'.[1] The Stourbridge Act of 1776, the Dudley of 1776 and the Birmingham and Fazeley of 1783, to name but a few, all empowered the canal company 'to make one or more Railway or Railways' to mines within 1,000 yards of the canal upon agreement with the intervening landowners; if the Company should not, then the mine-owners might.[2] Later Acts increased the distance to three miles or more, or even laid down no limit. With this incentive, wooden feeder railways began to appear. Perhaps the densest conglomeration of them occurred in the Leicestershire, Nottinghamshire and Derbyshire coalfields. In 1778 the Chesterfield Canal Co., who were working a mine between Staveley and Renishaw, built a 'Newcastle Rail Road from the Colliery at Norbriggs to the side Cut there', the engineer being Hugh Henshall, James Brindley's brother-in-law.[3] On the Pinxton branch of the Cromford Canal, opened in 1794, a quarter-mile wooden railway was still in use in 1808 at Pinxton lower pit.[4] The Erewash Canal Committee was told in 1793, 'Earl Stanhope's Tenants to his Iron Works are making a Railway and Canal from Dale Furnace to . . . lock into the Erewash Canal above the Heath Lock', which prompted the building of the Nutbrook Canal to serve the district immediately west of the Erewash. Its opening up to Shipley in 1796 resulted in the closure of a railway, dating perhaps from 1779, which had taken Shipley coal down the steep hill to the Erewash Canal at Newmanleys Mill near Ilkeston.[5] Joseph and Thomas Wilkes were empowered in 1787 to build a 'Rail way' or 'Gangway' from Brinsley Colliery to the canal near Langley Mill, although it was not laid until the late 1790s. *Gangway*, as a synonym for *railway*, was a term confined to Nottinghamshire and Derbyshire. A 'railroad or Gangway' from Beggarlee, and a 'Gangroad' from Newthorpe were likewise authorized in 1789 and 1792. The Erewash Canal Co. built a 'rail'd Road' or 'Gangway' to Ilkeston pits at a cost of £200, and leased it to Messrs. Bourne and Potter in 1783. The same firm laid another wooden railway to the adjoining Nottingham Canal (opened in 1796) from Greasley Colliery.[6]

The limestone quarries at Ticknall and Breedon and the collieries around Coleorton, all on the high ground west of Charnwood Forest in Leicestershire, were the objective of several proposed railways. When a line of canal was first mooted in 1785 from Loughborough westwards across the Forest to Coleorton, it was suggested that the final link to the pits should be completed by a 'Newcastle road'.[7] As it happened, this Forest line of the Leicester Navigation, opened in 1794, did consist of an isolated canal with connecting iron railways at each end, the waggons (they originally hoped) being carried on the boats. William Jessop was

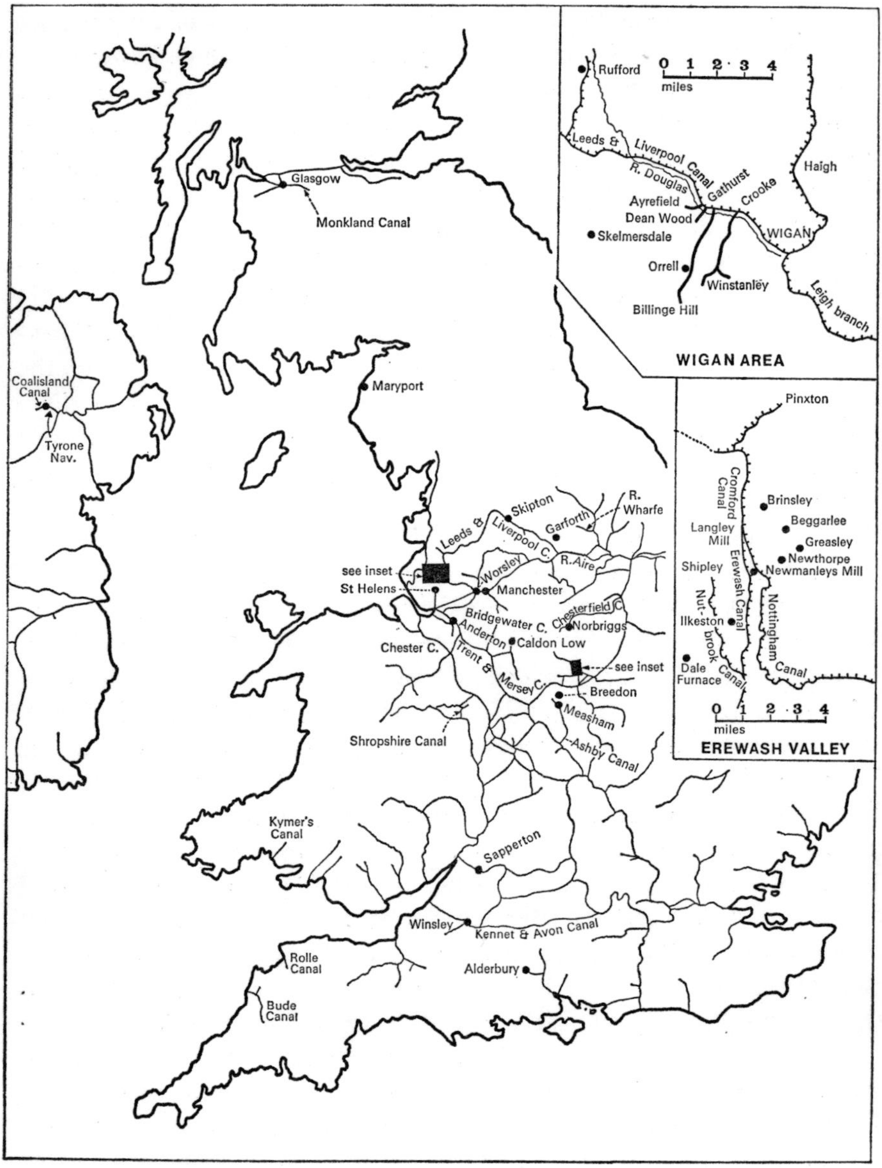

FIG 47 Distribution of wooden railways on canals.

the engineer. In 1787 he had also surveyed a short canal aiming at the same area from the Trent on the west, with a one-mile railway continuing on up to Breedon Quarry at an estimated cost of £1,970. He was not altogether happy with this idea:

> It may on future consideration be found advisable instead of a Rail Way to make a Stone Road, which will be done at less than half the expence of the other: for I am much inclined to think that such a Road properly formed, covered with small beaten Stone, and used only by Carriages with Cylindrical wheels, may answer every purpose of a Railway, & it will be kept in repair at a very small expence.

Nothing came of the scheme, but ten years later it was resurrected by Robert Whitworth, engineer of the Ashby Canal. He anticipated heavy traffic, so the track would have to be strong: 'I think the downward Railway should be altogether made of Iron, or Wood cover'd with Iron which may be nearly the same thing. But the upward one for the empty Waggons may do very well of Wood.' A two-mile railway from Ticknall to Ingleby on the Trent, he estimated, would cost £4,864 (a very considerable expense), on a gradient of 1 in 45. The one-mile Breedon line would be steeper – 1 in 29 – 'which is not near as steep as I observ'd some of them lately in Durham and Northumberland; yet they manage their unwieldy Waggons very well'.[8] Once again, neither railway was built, but both Ticknall and Breedon were linked in 1802 to the head of the Ashby Canal by iron plateways laid by Benjamin Outram.

Further down the Ashby, John Wilkes had a wooden railway at his Measham Colliery. Since it was converted to iron in 1799, and the Canal was only opened in 1798, it may be that it belonged to an earlier, pre-Canal, date.[9]

The longest and most celebrated railway built by a canal company in our period was that of the Trent and Mersey, who from 1769 were supplied with limestone by the owners of the Caldon Low quarries near Leek in Staffordshire. In 1776 they obtained an Act for a branch canal to Froghall, with a three-and-a-half-mile 'Rail-way' thence to the quarries. It was opened probably early in 1779, but it proved unsatisfactory.[10] It had to climb 600 ft., and 'in Frost the waggons slide so much, that it is almost impossible to carry any thing upon it'.[11] 'This line', says John Farey, 'appears to have been set out, before the true principles of this branch of Engineery was well understood, and was very crooked, steep and uneven in its degree of declivity, in different parts.'[12] So in 1783 a new Act was obtaincd, which recited:

> whereas the said Rail-way, so made and completed from Froghall to

> the said Lime-works . . . has been found to be laid and placed in a very inconvenient Course and Direction, and has not answered all the good Ends and Purposes thereby intended; and whereas it appears that a Railway may be made, in a much more easy and convenient Line of Communication,

the Company could rub it out and do it again.[13] The second line was better, but not much, and was replaced by a third, an iron plateway, built by John Rennie in 1803.

Both the first two lines were laid with flat iron plates on wooden rails. John Curr wrote in 1797:

> These roads, which are upon the plan of what is called Newcastle waggon roads, are laid in a firm manner upon wood, (after having been at a great expence of stoneing about ten or twelve inches thick for a foundation;) upon this wood is laid cast iron an inch and a half thick, a part of which weighs in every single yard forward one hundred and forty-one pounds, and other models weigh only eighty-one pounds.

The loaded waggons, he goes on, weigh anything from 3 to 7 tons (surely an exaggeration), 'which burden being laid all upon four feet in length, the above roads, although enormous in the first expence, are nothing too strong'. He advocates his own angled rails for unflanged wheels – which indeed were adopted in 1803 – with trains of smaller waggons.[14]

A 3 ft. rail $1\frac{1}{2}$ in. thick, weighing $70\frac{1}{2}$ lb., averages 5 in. wide; at $40\frac{1}{2}$ lb. it is nearly 3 in. wide. John Farey, writing in 1805, says that the Froghall rails in use long before 1794

> were three feet long, and had two holes, at 18 inches apart, to receive the wooden pins which fastened them down, or rather confined them in their places, on the wooden rails; at one end there was a triangular projection, and at the other a similar notch which fitted into each other . . . and opposite the holes, the rail which was about $1\frac{3}{4}$ inches thick and weighed about 42 lb., was made wider to strengthen that part; yet, with all this precaution, such rails were very liable to break in two at the pin holes, as well as to loose their connecting triangular pieces.

However, Farey gets into a sad muddle, for he goes on to claim that these rails had raised flanges to keep the wheels on the track.[15] It cannot have been so, for Curr would hardly have advocated angled rails if they were already in use. Farey is equating the angled rails of his own day with the flat rails that preceded them. In another book, Farey says that the first line cost £1 per yard, and that on the second line, in 1794, each horse made three journeys a day, four days a week, and two journeys on the other two days. It brought down one waggon at a time, with a load of 66 cwt.[16]

Though Curr calls it a 'Newcastle waggon road', the Caldon line seems to belong more to the Shropshire tradition. The track is pure Shropshire; the engineer was Hugh Henshall, who must have been acquainted with the Lancashire railways; and one of the quarry-owners was John Gilbert, the Duke of Bridgewater's agent, who also owned coal-mines at Donnington Wood on the Coalbrookdale railway system.

There were also some early canal feeder railways in the north. Just above St. Helens and the Sankey Canal lies the Wigan Coalfield on the River Douglas, which had been made navigable, in an erratic way, down to the Ribble. But it was not until the Leeds and Liverpool Canal had bought a controlling interest in the Douglas Navigation, and linked it by canal to Liverpool, that railways first came to the area. On 1 November 1776 an advertisement appeared:

> Messrs WARREN and Co
> Beg leave to acquaint the PUBLIC
>
> THAT they have completed a waggon road from Mr Jackson's Coal Mines in Orrell (near Wigan) down to the Canal, by which means Coal is brought down to their Yard in Liverpool, without being subject to so much breakage, as those from other Mines upon the said Canal.[17]

The 'canal' was still the navigable Douglas, for though the Leeds and Liverpool was planning to extend its artificial cut up to Wigan, parallel to the Douglas, it had not yet begun the work. Indeed, only four days after the advertisement, the Liverpool committee minuted that the Douglas would be closed

> so soon as a Canal shall be completed and a new Navigation made from Wigan to Rufford, and waggon ways continued across the Douglas from the Collieries so as Boats may lie & be loaded in the Canal. Resolved that it is the opinion of this Committee that the Company of Proprietors of the Canal, ought to be at the Expence of making, or continuing, the Waggon Roads from Mr Jackson's colliery, and also one from the Ayref[iel]d Colliery, over the River Douglas to the Canal, in Consideration of the Expence which the Owners, or workers, of these Collieries, have been at in making their present Conveniences for loading; but that the proprietors of the Canal should not afterwards keep them in repair.[18]

The Ayrefield waggon road ran from Roby Mill to the river at Gathurst, a distance of a mile. Jackson's was later known as Hustler's, two miles long from Orrell (and later from Billinge Hill) to Gathurst. Its gauge, when later it was converted to iron rails, was 3 ft. 6 in. Between the Ayrefield and Jackson's lines was a third, owned by the Holme family and running to the river from coal-pits a mile away in Dean Wood. Its date is uncertain, but almost certainly eighteenth-century.[19]

By 1780 the new cut had reached Wigan, and William Berry of Orrell applied for leave to build a waggon road to it from his colliery. His request was granted, provided that

> he make a Strong Rampart Bank ten Yards long against the Bank of the Canal towards the river of equal Height at least with the Bank of the Canal & . . . that the rest of the railed road between the said Canal & the old river be supported upon Pillars & that in the Execution of Such work he do not Stop the Navigation above Seven Days.

The bank, presumably, was to provide a raised level for unloading the waggons into flats. Berry was succeeded by John Clarke, under whose name the railway was better known. By 1797 it was three miles long, from Winstanley to Crooke, with 40-cwt. waggons, each pulled by one horse. Its gauge, when converted to iron rails to carry a Blenkinsop locomotive in 1813, was 4 ft. or thereabouts.[20]

Warren twice attempted, in 1787 and 1790, to build a waggon road to the canal from Skelmersdale, but both times he was thwarted.[21]

On the Yorkshire side, at least two wooden railways were built to the Leeds and Liverpool at Skipton and to the Earl of Thanet's branch canal there. In 1785 the company leased from the Earl his 'Canal, Waggon Road, Lime Rocks &c,' and began immediately to 'prepare proper wood for Sleepers for making and repairing the necessary Waggon Road up to Haw Bank' limestone quarries. Its gauge – an echo of the Middleton line – was 4 ft. 1 in. Next year the canal took over the existing quarries, waggon-way and staith belonging to the Mercer Flats Lime Co. (now Massa Flats), a short distance to the west.[22] A year later 'the Waggons, Rail Road &c belonging Messrs Garforth & Co at Skipton' were valued at £140. This line may have been a third, but was probably one of the other two, being worked for the canal by Garforths, who were a coal- and lime-carrying company.[23]

A canal from the Garforth and Sturton collieries east of Leeds to the Wharfe near Tadcaster was proposed in 1774. Its route included two short lengths of railway to link up isolated sections of waterway. The scheme came to nothing.[24]

On the Bridgewater Canal, in spite of the early proposal for railways on Worsley Brook (p. 253) and the use of waggons underground in Worsley mines (p) 321., we hear of only one surface railway being built. It consisted of a very short incline which climbed up to the top of a lime-kiln on the canal bank near Manchester. On the coal and limestone waggons, which were hauled up by a horse gin, the lower wheels were larger than the upper to compensate for the slope. The unloading arrangements were described in 1780:

> The floor of the Waggon consists of 2 planks, which are hung on hinges on the side planks, and are held together by a rope passing underneath and hooked on the other side, which, when released, allows the load to fall out of its own accord.[25]

Railways were not only built to supply traffic. Occasionally they played a part in the construction of a canal. The Chester Canal minutes record in 1773: 'Ordered that Phil. Norbury do agree with Ellis Meaking and Henry Rigby Carpenters to make a good Sufficient Rail Road in John Griffiths's Garden . . . not to exceed 2/6 per Rood for laying the said Road', and later, 'That the Rail'd Road be lengthen'd at the Company's Expence'. In 1803 the Kennet and Avon Canal placed a contract for 'a wooden Rail Road' on an incline from quarries at Winsley to the canal at Murhill, not far from Bath, to carry stone for building bridges and locks. But it may in fact have been laid with iron rails, for its route is still studded with sleeper blocks. A temporary wooden railway was laid in 1785 in Sapperton tunnel on the Thames and Severn Canal during its construction to remove the spoil. It had but one waggon, with flanged wheels cut from a single block of wood.[26]

A railway is sometimes found as part and parcel of a canal route. There were three reasons why this might be so. Firstly, lack of money. When James Watt surveyed a route in 1769 for the Monkland Canal from the Lanark Coalfield to Glasgow, he decided it would be cheapest if built without locks, ending a mile from Glasgow and continuing into the town and to the Clyde by a 'waggon way or a good Causeway'. The Canal Act, passed in 1770, authorized 'a Road or Waggon-way from the west End of the said Cut or Canal' to the Clyde and to Glasgow. Watt remained in charge of the construction until 1773, when progress was halted. By 1793, when the canal was at last almost complete, 'a waggon road is also finished between the bason and Glasgow'.[27] For a second example, the ill-fated Salisbury and Southampton Canal, struggling up the Avon Valley, ran out of funds when it reached Alderbury, a bare three miles short of its destination, in late 1802. As the cut ended in the middle of nowhere, the company made a last effort to entice some traffic on to the canal by building a wharf and laying a wooden railway to the nearest road. The canal's engineer estimated the cost as:

Beech Rails per Yard	10*d*	
Sleepers per Yard	4½*d*	
Laying down per Yard	4*d*	
Graveling horse Track	3*d*	
629 yards at	1*s* 9½*d*	£56 6*s* 11½*d*
Cutting drain		£5 4*s* 10*d*
		£61 11*s* 11½*d*

Within four years, however, canal and railway were defunct.[28]

At Alderbury goods had to be transhipped from boat to waggon, as they also were on the Monkland Canal, an expensive business. Crates or containers for bulk traffic were not unknown, and several times it was proposed that special boats should carry the actual waggons. When John Smeaton reported on the Earl of Egremont's collieries at Aspatria above Maryport in 1775, he recommended a waggonway pure and simple,

> though an ingenious idea of the ready shipping and unshipping of the waggons, containing their loading, from the canal to the waggon ways, and vice versa, was communicated to me by Mr Johnson, yet it is no ways clear to me, that in so short a navigation any thing will be saved.[29]

William Chapman was a great advocate of the system, suggesting canal barges to hold sixteen full-sized Tyneside waggons.[30] But in the wooden railway period the idea got nowhere – barring the Avon Ferry at Bath (p. 259) and the train-carrying ships at Plymouth (p. 302) – although a few later iron railways connected with canals sent their waggons down by boat.

Rails were also laid, rather than canals cut, where the slope was so long and steep, or water so scarce, that locks were impracticable. Anderton in Cheshire, where the Trent and Mersey Canal passes close to but high above the River Weaver, was an ideal point for transferring the heavy salt traffic from canal boats to river barges, if the steep slope could be overcome. In 1794 John Johnson, the Weaver's Manager, was sent to Coalbrookdale 'for the purpose of examining and informing himself fully of the nature and method of making the different railed roads and other works used there for shipping goods into vessels on the river Severn'. He paid his visit and laid his plans, but it was not until 1799 that he was instructed to make a 'railed way and other conveniences' at Anderton. It was a steep incline. The committee ordered in 1801 that a chain be provided instead of ropes – perhaps there had been accidents. The line remained in operation until the present Anderton vertical lift was installed in 1875.[31]

From time to time the boats themselves were carried up and down a steep slope on railways. We have seen an early example of the canal incline in north Italy (p. 43), and a later instance on Ducart's Coalisland Canal in 1776 (p. 130). It was left to William Reynolds of Ketley Iron-works in Shropshire to build the first successful incline of this sort in the country. He probably knew, via William Jessop, of Ducart's fiasco, and experimented with the idea on his short private Ketley Canal, opened in 1788. The tub-boats, which held 8 tons, were carried on cradles on a double-tracked, self-balanced incline with a fall of about 70 ft. The success of the incline led to its adoption on the longer Shropshire Tub-boat Canal, opened in stages from a junction with the Donnington Wood Canal to the Severn at Coalport between 1789 and 1791 (fig. 44). The most northerly plane, at Wrockwardine Wood, rose 120 ft. in a distance of 350 yards; the middle one, the Windmill plane, fell 126 ft. in 600 yards, and the Hay incline dropped 207 ft. in 300 yards. In 1791 it was decided 'that Men and Horses should be kept at the Inclin'd Planes at the Hay and Wind Mill Farm to draw the Boats up the same', but they were soon replaced by steam engines. Wrockwardine Wood had been steam-worked from the start. The Canal Company charged an extra toll of 3*d.* per ton on the inclines, over and above the ordinary canal tolls, which caused a great deal of bickering and resentment.

The braked winding drum round which the rope was wound was mounted over the sill at the summit of the plane, and could be disengaged to revolve independently of the engine. When a tub-boat, measuring 20 ft. by 6 ft. 2 in. by 3 ft. and holding 5 tons, approached a plane from above, it ran on to a railed cradle submerged in the water (pl. 64). Once the boat was chained down, the engine pulled the cradle up a short reverse slope out of the water, over the sill, and on to the plane proper, which at the Hay and at Windmill was normally worked by counterbalancing. The descending boat pulled up an empty one, which was assisted over the sill at the top by the engine and let down into the upper pound. At the bottom, the cradle ran into the water of the lower pound and the loaded boat floated off. At Wrockwardine Wood, with most of the traffic going uphill, the engine was used much more. The cradles were massive, iron-strapped, wooden platforms, with a raised beam for attaching the rope to. Their wheels were unflanged, 6 in. wide and unequal in diameter – 27 in. at the bottom, 16 in. at the top – to keep the cradle level. An extra pair of 24-in.

wheels, fitted to the outside of the top axle, ran on a specially arranged track on the reverse slope at the incline head, so that even on the contrary gradient the boat stayed reasonably level. Each incline had two cradles in action and one spare. The original haulage ropes were replaced in 1800 by chains. None the less, there were breakaways, with spectacular and expensive results.

The track was exceedingly massive, to carry a load of 8½ tons (cargo, boat and cradle included). The sleepers, measuring 7 ft. by 6 in. by 12 in. were set at 6-ft. intervals, and cost 3*s*. apiece. On them were pegged the main wooden rails, no less than 14 in. square. On top sat the flanged iron plate rails, cast probably at Ketley, on Curr's principle: 6 ft. long, 8 in. wide and 2 in. thick, with a 3-in. high flange. They were pinned down through four holes cast in each rail and mortised at the ends. To make assurance doubly sure, the track was firmly ballasted. James Houghton agreed in 1791

> well and sufficiently to pack the Rails, and pave the Inclin'd plane near the Hay, so that the packing shall project at least Nine Inches on the outside the wood rail, for the whole length, and that the pavement shall be rammed in, and not less in any place than six Inches Deep.

For this he was paid 3*s*. per yard. The gauge was recorded in 1818 as 4 ft. 4¾in., with double track; in 1826-7 as 3 ft. 7 in., with most of the incline laid with three rails, dividing into four for a short passing-place in the middle. At this time the flanges were outside the rails on one track, and inside on the other, so that the centre rail could be shared.

There was also a two-and-three-quarter-mile branch, opened in 1792, from near the Windmill plane to Brierly Hill just above Coalbrookdale itself, and linked to Horsehay by a railway specially built by William Reynolds from the nearest point of the canal. At the Dale end, the traffic was taken from the terminus to the works in an unusual way. Each boat carried four iron crates of 2-ton capacity, but instead of being let down an incline the crates were craned down (and up) vertical shafts. At the bottom they were received on to railway waggons standing in a tunnel ('footroad' as it was called) 120 ft. below, whence they were driven into the open air and to the Dale works. The shafts were modelled on similar ones on the Bridgewater Canal and, nearer home, at Lilleshall on the Donnington Wood Canal, except that neither of these had railways attached. After two years the shafts were replaced by a railway incline (the 'Old Wind') for ordinary waggons, which may have been the first of the new-style plateways in the area. Finally, in 1800, a railway was laid along the towpath from the incline head to the wharf where the Horsehay link railway went off. The Dale thus had a new through line to Horsehay, and the inconvenient Jigger Bank probably fell out of use at this time.[32]

Apart from the Shropshire Canal inclines, there were two more, almost identical, nearby (at Trench on the Shrewsbury Canal and at Lilleshall on the Donnington Wood) and these six were perhaps the most satisfactory examples of their type. The principle had been considered by various engineers several times before and after the Shropshire inclines were built, and a few were tried elsewhere, mostly in south-western England. Edmund Leach was suggesting their use, for example, on the intended canal from Bude across Cornwall to the Tamar in 1774, on just the same principle as Ducart used at Coalisland and Reynolds in Shropshire. John Smeaton was consulted, as he was at Coalisland, but he was evidently never enthusiastic about them. The idea was revived in 1793 by Earl Stanhope, who proposed 'to have iron rail-roads of gradual and easy ascent, on which boats of two tons were to be used, suspended between a pair of wheels of about 6 feet diameter, and to be drawn up or let down the same by horse'.[33] In 1796 the American engineer Robert Fulton, who had been in correspondence with Earl Stanhope, published a treatise on his own scheme for canals and inclines.[34] He advocated, for the sake of cheap construction, small canals with 4-ton tub-boats that were actually fitted with wheels. They could thus move up and down the inclines, which were to replace all locks, without the intervention of cradles. Next year William Chapman wrote a rebuttal of Fulton's treatise, pointing out that any irregularities in the railway would strain the boats' seams,

> to remedy which, I would propose that the carriages, where the boats are large, should move on eight wheels, four on each side, and resting on two axles passing through the middle of parallel bars connecting each pair of wheels; of course, the railway would be less liable to injury, and shock from impediments would be reduced.

This is the first known proposal for bogies.[35] Chapman much preferred his own scheme, the converse of Fulton's plan, that waggons should travel by boat rather than boats travel by railway. In fact, Fulton's design was only used, from 1823, on the Bude Canal, and perhaps in 1825 on the Rolle Canal.[36]

Almost all these canal railways, of one sort or another, lived out their brief lives in the twenty-five years from 1775 to 1800, too early to benefit from the rising public interest in matters industrial and mechanical which has left a mass of information about their successors. The only real exceptions were the canal inclines of Shropshire, which survived and were described. But to all intents and purposes wooden canal railways had disappeared by 1800 and other wooden railways not long after, ousted by the iron rail. With its coming, our terms of reference end; but some brief account must be added to tie up the loose ends and show how the conqueror advanced, and where.

It was on the canal feeders that the all-iron rail first appeared, sweeping across the coalfields in a wave of new railways and converted wooden ones. There were several reasons for this. In the canal mania of the 1790s the sudden increase in the waterways system inevitably meant a sudden increase in the feeder railways. In the two main industrial centres where this happened there were special factors at work. In South Wales, where almost all the canals of the valleys were opened in the 1790s, the rugged country demanded an unusually lengthy system of railways to feed them. Most of these railways were the work of the new iron-masters, who were just then building up the great industrial empires of South Wales, and to whom iron was second nature. In the Midlands, and especially in Nottinghamshire and Derbyshire, it so happened that two of the canal engineers whose services were much in demand were partners in Butterley Iron-works: William Jessop and Benjamin Outram, both of them staunch advocates of railways. It is not surprising that, wherever they went, they left iron railways behind them. It is no accident, too, that the appearance of the all-iron rail in the 1790s and its rapid spread over the first years of the nineteenth century coincided with the Revolutionary and Napoleonic Wars. The price of wood soared, and with the intensification of industry as a result of the wars, the cost of iron fell.[37] All things being equal, iron proved itself far superior in technical and economic terms, for though the initial cost was considerably higher than for wood, the subsequent maintenance and replacement bills were vastly smaller. And so the wooden track gave way to the iron rail and the stone sleeper.

Some of the new canal feeders pioneered the cast-iron edge rail for flanged wheels, a logical development of the wooden, cast-iron-plated track inaugurated at Coalbrookdale. This happened first in South Wales, an area that was always aligned towards Shropshire: some of the great iron-masters hailed from Broseley. The earliest rails they used did away with the wooden rail completely, although the wooden sleepers were retained. The heavy bars, albeit comparatively deep and narrow, were quite clearly based on the Coalbrookdale pattern. William Taitt, a partner in Dowlais Iron-works, wrote in 1791:

> We are now making Rails for our own Waggon Way, which weigh 44 li or 45 li per yard. The Rails are 6 ft long, 3 pin holes in them,

> mitred at the ends, 3 inches broad at Bottom, $2\frac{1}{2}$ Inches Top, and near 2 Inches thick.[38]

This line ran to the head of the Glamorganshire Canal. A similar type of rail was also adopted for the lengthy railways of the Monmouthshire Canal, all laid – another Shropshire touch – to the gauge of 3 ft. 4 in.[39] In the Midlands, William Jessop was long but erroneously credited with the first use of the all-iron edge rail, on the Forest line of the Leicester Navigation in 1789. In fact, the rails were ordered only in 1792 and laid in 1793-4.[40] They had, it seems, a bulbous head for the wheels to run on, and a thin vertical web below, the best disposition of the metal to resist vertical pressure. The material was, of course, still cast iron, and the under-side bulged downwards in a parabolic curve – the 'fishbelly' – for strength; but, in their shape of a vertical beam, these rails were the first ancestors of the modern rail. They did not find much favour in the Midlands, where indeed they remained almost Jessop's monopoly, though he used other types as well.

There emerged at about the same time a quite different design, usually known as the plate rail: L-shaped, for carrying unflanged wheels. We have seen in ch. 4 how this pattern, in wood, went back to the very birth of railways on the Continent, and we shall see in ch. 17 how the same type was applied in British mines, where John Curr finally translated it into iron at Sheffield, originally for underground use. It came to the surface in 1788 at two places: Ketley Canal incline and Wingerworth Furnace near Chesterfield.[41] After a brief pause, it went forth in the mid-1790s to sweep the country, with Benjamin Outram of Butterley Iron-works – his father Joseph had cast underground plates for Curr at his Ripley foundry[42] – as its main protagonist. Outram, more than any other man, determined the shape of industrial railways for a full thirty years. He was responsible, by direct advice or by example, for countless plateways throughout the Midlands. When he was consulted about the Monmouthshire Canal railways in 1799, he recommended a wholesale change-over from edge to plate rail,[43] and practically every railway-owner in South Wales followed his advice. Even after his death in 1805, his disciples carried on the work, building new railways by the score and adapting old ones, until by the 1820s, south of County Durham, the edge rail was very much the exception, the plate rail the rule. There were some pockets of resistance – Devon and Cornwall, for example, North Wales, parts of Lancashire, and the occasional place elsewhere – while Scotland, as ever, was divided in its allegiance. The Coalbrookdale empire had fallen before the conqueror at some indeterminate date, probably around 1800.

The plate rail proved a cul-de-sac in railway development. In spite of all the enthusiastic support it received from expert and amateur, it had serious defects. Its cross-section, with most of the metal thinly disposed in a flat plate and little strengthened by the vertical flange, was deplorably ill-equipped to carry heavy rolling weights. The wide running surface with its retaining flange might almost have been designed to catch and to hold the dirt kicked up by the horses' hooves, to the great increase of the friction on the wheels. The iron edge rail, in contrast, being a vertical beam, was stronger by comparison, and its comparatively narrow and elevated running surface harboured little dirt. Nonetheless, the plate rail did mark an improvement on plain wood, and in most cases one horse could pull a train of several waggons at a time.

It was Tyneside and Cumberland which remained faithful to the edge rail and flanged wheel.* They kept it alive, improved it, and, finally, with the Stockton and Darlington Railway of 1825, dealt the plate rail a blow from which it never recovered. But it was a hard-fought fight. There was strong pressure to build the Stockton and Darlington with plate rail, and only George Stephenson's obstinacy saved the day. Thereafter, although the plate rail is still not quite dead, there was never any real doubt that the edge rail had won. The rail which Tyneside had evolved up to this point was a development of Jessop's fishbelly. It first appeared in the north at Lawson Main on the Heaton waggonway in 1797, and probably at Benwell in the same year. The sleepers were of stone, and the rail was attached to them by the first chairs.[44] Twenty years later the cast-iron edge rail reached its pinnacle in George Stephenson's patent design. But at the same time a new material was slowly creeping on to the scene: wrought iron. It was first used by itself, without a wooden rail below it, in a humble way at Tindale Fell near Brampton in 1808 and at Walbottle about the same time.[45] It became a practical proposition with the patent taken out in 1820 by John Birkinshaw of Bedlington Iron-works, and quickly caught on. Where cast iron is brittle, wrought iron is resilient, so that weight for weight it is twice as strong; and strength for strength, its cost was identical. The Stockton and Darlington, laid partly with cast iron and partly with wrought, not only established the supremacy of the edge rail; it also established the wrought iron rail. With this and the other features of that remarkable line, the locomotive and the passenger traffic, the modern railway had begun.

However, we are leaping ahead in time. Although the cast-iron rail appeared on Tyneside in 1797, it did not immediately sweep the board.

* They suffered only one traitor to the cause, the Wylam waggonway, converted to plate rail in 1808.

Northumberland and Durham were very badly off for iron foundries: the only ones of any size were the Bedlington Iron-works, the Tyne Iron-works at Lemington, and a little later Losh's Walker Foundry. This paucity was no doubt one reason why the wooden wheel had tarried so long, and it certainly explains the delay in adopting the iron rail. If track had to be ordered from other parts of the country, the cost inevitably rose. John Buddle wrote in 1807:

> The principal point in which wooden ways are inferior to iron, is that a horse on the average will do more work by 30 p. cent on an iron way than on a wooden way; but frequently he will do 50 and even 100 p. cent more. Besides, if the way is wanted for a long term of years the expence of upholding the iron is much less than that of wood. The difference in expence in the way depends greatly on the size and Weight of the waggons or Trams to be employed. For a Newcastle Chaldron waggon the Exp^{ce} of Iron will be about double that of wood viz Iron 16/- Wood 8/- p. yard. But iron Rail Ways for Trams or Small Waggons may be made as low as 5/6 or 6/- p. yard.[46]

Eight years later it was estimated that maintenance costs were these:

Main way	Bye way	Traffic per year in tens	Cost per mile per year	Cost per ten per mile
wood	wood	4,000	£432. 5. 4	2/2
wood	iron	4,000	£388. 18. 0	1/11¼
wood	wood	2,200	£310. 8. 2	2/9¾
iron	wood	2,200	£170. 2. 7	1/3[47]

By 1807 these arguments had become irresistible, and many waggonways began to be converted. A visitor to Walker Colliery late that year recorded: 'They are gradually giving up their wood Roads, and introducing Iron, and they make them piece by piece keeping up the old Waggons and running them partly on wood and partly Iron.'[48] The Backworth waggonway was converted in 1810, but the rails had to come from Butterley in Derbyshire.[49] Many other examples could be quoted from around this time. In Tyneside colonies, too, the change-over took place now: at Middleton from 1807; on the Elgin and Halbeath ways about 1810; at Whitehaven from 1811.[50] Five years earlier Curwen of Workington spoke in the Commons against a Bill to impose an Excise duty on iron: 'It had been found desirable to abolish wooden railways for the purpose of substituting iron ones. At a modest calculation the new tax would raise the price of coals by at least a shilling a ton.'[51] At the same time, new waggonways were built of the new material. One of the first, which has already come in for some discussion, was the Kenton and Coxlodge of 1808. Many others followed it.

The wooden waggonway had by no means disappeared by the end of the Napoleonic Wars. The Willington way was converted to iron in 1819, most of the Heaton way in 1821.[52] On the Fordell waggonway, the colliery Manager suggested in 1832 that it was high time the Fordell came up to date; on the iron Halbeath line nearby a horse drew 18 tons a day, on the wooden Fordell only 8½ tons, and its maintenance bills were eight times as high as the Halbeath's. So iron rails were laid between 1833 and 1835.[53] Wooden track could still be found in Cumberland in 1839;[54] the bulk of the great Tanfield waggonway was converted only in the same year.[55] T. J. Taylor could say of Tyneside in 1852 that 'even within the last dozen years the wooden railway was still to be seen in full operation',[56] and another authority, writing in 1841 about wooden rails plated with thin wrought-iron strips, claimed that 'remains of this description may still be seen about Newcastle'.[57] Perhaps the last major line to retain wooden rails was the Garesfield, the last remnant of the Western Way, which kept them until about 1843 and maybe later still.[58] On a small scale, the iron-plated wooden rail survived until 1965 in a timber yard at Sunderland.[59]

The change-over to iron did not greatly alter the operation of the waggonways, except in so far as costs were ultimately reduced and the horses' hauling capacity was increased. What did revolutionize them was the locomotive. A prophecy is attributed to Isaac Wilkinson, father of John and William, who is said to have remarked as an old man in 1779 to a young friend, '*You* will live to see *waggons drawn by steam.* I would have made such a waggon myself if I had had time'.[60] A Newcomen engine on wheels would have been a splendid sight. As it happened, the early development of the locomotive, like the iron rail and the stone sleeper, was due in some degree to the Napoleonic Wars, which had pushed the cost of horses and fodder up to alarming heights. At the same time the steam engine itself had reached a stage, particularly with Richard Trevithick's high pressure boiler, where it was practicable to fit it to a wheeled vehicle. Trevithick himself led the way with experimental engines at Coalbrookdale in 1802 and on the Penydarran tramway at Merthyr Tydfil in 1804. But it was not until 1812 that the locomotive was first used as a commercial proposition, when John Blenkinsop introduced his Matthew Murray-built rack-and-pinion engines to the Middleton line at Leeds. This design was borrowed for several other railways at home and in Germany. This success was followed by experiments in other parts of the country, the initiative being seized in 1813 by Tyneside, which retained it for many years. Christopher Blackett of Wylam had ordered a Trevithick engine in 1805, but did not take delivery, apparently because of fears for the safety of his wooden track.[61] But iron rails, at a long-term cost less than that of wood, could now be made strong enough – just – to withstand an engine's weight and motion, and in 1813 and 1814 the Heaton, Lambton, Wylam, Newbottle and Killingworth waggonways were graced with locomotives built by William Chapman, William Hedley, William Brunton and George Stephenson. John Blenkinsop gave some figures on the comparative costs of horse and locomotive working for the Kenton and Coxlodge way in 1812. He claimed that the capital outlay on locomotives, even including the new track necessary, was 24 per cent. less than on horse traction; and that the operating expenses would be 76 per cent. less for locomotives.[62] This seems unduly optimistic, but then Blenkinsop was trying to promote the sales of his own engines. A more realistic figure for this early period of locomotive development was worked out by John Buddle for a Stephenson engine on the Heaton waggonway in 1821; a saving of 12 per cent. on operating costs if locomotives replaced horses.[63] Once again it was the Stockton and Darlington Railway which, amid a glare of publicity, showed what the steam engine could do, fully confirmed by the astonishing success of the *Rocket* at the Rainhill Trials of 1829.

Almost all of the wooden railways we have described were purely private lines, built by the owners of collieries, iron-works or quarries to transport their own coal, iron and limestone and nothing else. Most canal feeder railways were the same, and it is only in railways which were built by the canal companies as part and parcel of their own transport system that we can see the first stirrings of the public railway on which all comers were entitled to carry goods 'on suitable carriages' (unless the canal company provided the waggons) and on payment of the tolls. This principle was inevitably applied where a canal route was interrupted by a railway, but also held good in theory, if not much in practice, on canal-built railways like the Caldon Low line.

Public railways quite independent of the canals were mooted at an early date. Richard Lovell Edgeworth in the 1760s suggested laying 'rail-ways for baggage waggons on the great roads of England'. But

> one great objection arose from the vast expense of massive rail-ways, and the continued cost of repairs. To obviate this difficulty, it occurred to me to divide the weight that is usually carried on a single waggon into four or five portions and to place them on four or five small carriages; these carriages linked together would be as easily drawn as the same load upon one waggon.

In 1768 he presented models of his waggons to the Society of Arts, who rewarded him with a gold medal. But in his proposals, both for trains of waggons and for public inter-urban railways, he was a long way ahead of his time. Technical and economic difficulties put paid to his schemes for many years. Bailey and Culley remarked in 1797:

> It has been suggested, that a waggon-way might be substituted in lieu of turnpike roads, for the conveyance of heavy goods; but we presume, that the expence of first making, laying out in nearly a level line, or so as to have little or no ascent, and keeping it in repair, would greatly overbalance the additional weight to be drawn by one horse; as the wood rots and wastes nearly as much by a small quantity of carriage passing along it, as a large.

Edgeworth resurrected his plans in 1802, now advocating iron rails and stationary steam engines for cable haulage of traffic:

> What a prodigious saving of expense might thus be effected? If the freedom and facility of intercourse, which has been obtained by good roads and canals, be, as Adam Smith asserts, one of the great causes of our national wealth, how far might the freedom and facility of intercourse be extended by the perfection of the scheme.[64]

This time he was right, and public railways followed, slowly at first, ever more rapidly thereafter. Some too began to cater for regular passenger traffic, which had never been carried (except, no doubt, in a purely informal way) on wooden railways.

Meanwhile wooden railways, although numerous, had always been for private short-distance transport, from colliery to river or quarry to canal. Their total mileage grew something like this:

	1625	1650	1675	1700	1725	1750	1775	1800
Tyneside	—	1	11	37	70	94	123	146
Elsewhere	2	2	5	4	8	39	93	146
Total	2	3	16	41	78	133	216	292

These are but the wildest of estimates, and conservative at that. Even if they are 100 per cent. too small, the total length of wooden railways was minuscule compared with the 22,000-odd route-miles of British public railways at their maximum extent, plus the unrecorded but huge mileage of private railways that served, and still serves, industry throughout the country. But they were important beyond their size, both in what they fulfilled at the time and in what they led up to.

Railways began life simply as part of a colliery's equipment, but in their contribution to the economics of the coal trade they rank alongside the steam engine for pumping mines. One horse and one railway waggon might displace twenty-four packhorses or two or three four-horse wains. If the coal trade had had to rely on these old and inefficient forms of transport, its history would inevitably have been very different; and coal, along with iron, was the foundation of much of Britain's industry. The end product of these early wooden railways was no less important. There is no need to emphasize the economic and social effect of the Victorian railway. The canals had begun the process of linking together the various parts of the country by lines of communication that could carry heavy goods in quantity at small cost. It was not so much this facet of the public railway which was revolutionary as the speed at which it could do it, and especially the speed and ease with which even humble people could make journeys previously quite beyond their imagination. It has always been recognized that, during the nineteenth century, Britain gave the railway to the world. Railways there had been in Europe before, but of a different shape and for a different purpose. The British railway in the form which it finally achieved, both in outline and in detail, was invented and developed in Britain.

Miscellaneous railways

We have seen a portion of this improved rail road, with one of the Carriages suspended, and are really at a loss how to express our admiration of the ingenuity and simplicity with which every part of the apparatus is constructed and arranged.

London Journal of Arts and Sciences, v (1823), 38 on Palmer's Mono-railway.

There still remains a small and miscellaneous collection of wooden railways in Britain which do not fall into any of the three groups so far discussed. A few of them had ordinary track for flanged wheels, the rest were oddities.

Temporary wooden lines were occasionally to be found on constructional works. W. A. Madocks, for example, is said to have used wooden railways in 1800 to assist in the building of his first embankment to reclaim part of Traeth Mawr in North Wales; but there is no certain evidence.[1] Another line was laid in 1836 for one and a half miles along the beach from a stone quarry to the abortive harbour works at Sidmouth in Devon. It included a viaduct and a tunnel a third of a mile long, which proved too small for a locomotive that was delivered. The wooden rails measured $6\frac{1}{2}$ in. by 4 in. by 19 ft. long, with wrought-iron strips 2 in. wide, $\frac{3}{8}$ in. thick and 15 ft. 6 in. long, bolted and spiked on top. The sleepers were of circular section, 6 in. in diameter, except at rail joints, where they were rectangular; and the gauge was about 3 ft. 6 in. The enterprise proved a complete fiasco, no traffic was carried (except locals on joy-rides), and the equipment was sold in 1838, leaving the track to be washed away or buried in sand, where some may still be found.[2] A few temporary tracks laid by contractors in the construction of main-line railways seem also to have had wooden rails, though iron was by far the more common. Pl. 67, originally published in 1848, shows a crude railway of this sort, with a train of three waggons carrying spoil out of a cutting across a rough temporary bridge.

There were, too, a few purely industrial waggonways like this. A strange little line was to be found in the tiny and isolated coalfield at Brora in Sutherland, high up the east coast of Scotland. The pit was linked to the harbour a mile away by a wooden railway, built, it is said, in the early 1770s to a gauge of only 1 ft. 8 in. Early in the nineteenth century the track was relaid with cast-iron fishbelly rails.[3] Again, when the Admiralty awarded a contract to Marc Brunel in about 1810 to improve the timber-working machinery in Chatham Dockyard one of the devices he introduced was a comparatively wide-gauge railway to carry logs to the sawmill. Its track, of wooden rails topped with wrought-iron plates, is said to have inspired his son, Isambard Kingdom Brunel, in designing the permanent way for his G.W.R.[4]

An incline linking the Birk Fell mines with Greenburn Copper Works north-west of Coniston in Lancashire was laid with rails 4 in. square, strapped with wrought iron $1\frac{1}{2}$ in. by $\frac{1}{2}$ in., at a gauge of 2 ft. The date is about 1850.[5] A similar line, of a similar date, carried slate and slate slabs down from a quarry in Nant y Pandy to the Deeside Slate Works just above Glyndyfrdwy in Merioneth, some miles west of Llangollen. The length was one and a quarter miles, the gradient rose in places to 1 in 10, and the gauge was 2 ft. 6 in. The rails, which sat directly on the ballast, measured 4 in. wide and 6 in. deep, and were joined by iron fish-plates and held to gauge by tie-bars every 4 ft. The usual wrought-iron plates, 4 by $\frac{3}{4}$ in., were nailed on top. The locally-made waggons had screw-down brakes acting on all four wheels. They were filled during the day, and in the evening were dispatched down to the works by gravity, not in a train, but singly and at intervals, each with a man aboard to control the brake. Horses took the empties up in the morning. A nasty accident happened in 1907, when the brake failed on the last of four waggons that were going down some distance apart. It rushed down the gradient at the others, one of whose brakesmen jumped off in time; but the man on the third waggon was badly hurt, and the lad on the front one was killed. The line was extended south to another quarry and north to the G.W.R. late last century. These additions were given ordinary iron rails, but the old central section kept its wood right up to its closure in the late 1940s.[6] Much of the track, in a state of advanced decay, is still to be seen (pl. 66).

The last edge railway to describe was equally strange. Known as the Zeal Tor Tramway, it was built in the 1840s to carry peat from Redlake Mire for three and a half miles across Dartmoor and down some horrific gradients to the naphtha works at Shipley Bridge above South Brent. It did not survive for many years. Its peculiarity lay in its wooden rails being bolted down to granite blocks instead of to wooden sleepers.[7] This exceedingly rare, perhaps unique, combination is also attributed, almost certainly falsely, to the Tranent and Cockenzie railway in Scotland, to various railways around Llanelli, and to the Ayleford foundry branch of the otherwise iron-plate-railed Bullo Pill railway in the Forest of Dean.[8] Some of the Zeal Tor blocks are still to be seen.

Not far away is Plymouth. When John Smeaton was building his great masterpiece, the Eddystone Lighthouse, between 1756 and 1759, he laid out a work-yard on the shore of Millbay, where the intricately dovetailed masonry could be cut and fitted before being shipped out to the rock. The blocks weighed up to 3 tons apiece, and to move them around the yard Smeaton used two devices: a sturdy platform mounted on large rollers, and a 'four-wheel carriage that runs along the timber road, commonly called at the *Collieries*, where they are used, a *Rail Road*'. The railroad ran the length of the yard between the working-places, and continued via a 'turn-rail' to the yard entrance, where the stones arrived, and to the wooden jetty, where a pair of sheer legs transferred the block to the ship. Every stone – and there were 1,493 large ones in the Lighthouse – travelled six times on the railway during its stay. The track was most unusual. Instead of flanged wheels running on plain rails, plain wheels ran on flanged rails (fig. 48). At this date in Britain, these rails were only used underground, and the collieries where Smeaton received his inspiration may have been in his native Yorkshire, or more likely in Northumberland, where he was working when summoned to rebuild Eddystone. The rails measured about 8 in. wide by 4½ high, with the flanges rising some 1½ in. on their outside edge. The gauge inside the flanges was about 3 ft. 9 in.[9] It is interesting that William Jessop, whom we have several times met building railways later in the century, began his career here. On the death in 1761 of his father, Josias, foreman shipwright in the dockyard and Smeaton's right-hand man, the sixteen-year-old William was taken under Smeaton's wing and trained as an engineer.

By a strange coincidence – and it may be more than a coincidence – the flanged wooden rail was again used at Plymouth sixty years later. In 1812 John Rennie began the construction of the mile-long breakwater across the entrance to the Sound, using stone quarries at Oreston on the Cattewater. Iron plateways were laid from quarry to quay, where the waggons were run on to special ferry-ships, of which there were ten. When the ship reached the breakwater, the waggons were run up an inclined and projecting ramp at the stern, and emptied over the edge. On board ship rails were of wood, very similar to Smeaton's design. J. U. Rastrick, the engineer, visited the works in 1812 and was not impressed:

> The Method in which they get the Stones and convey them to the Quays: the Construction of the Vessells, the way of loading and discharging is in my oppinion, the *verry verry worst* that could be

thought of by the greatest Bungler there ever was. . . . The Vessells . . . have a lower and upper Deck with two Rail Roads on each Deck paralel with each other and I believe there are seven or Eight Carriages on each Rail Road; these Vessells are about 120 Tons

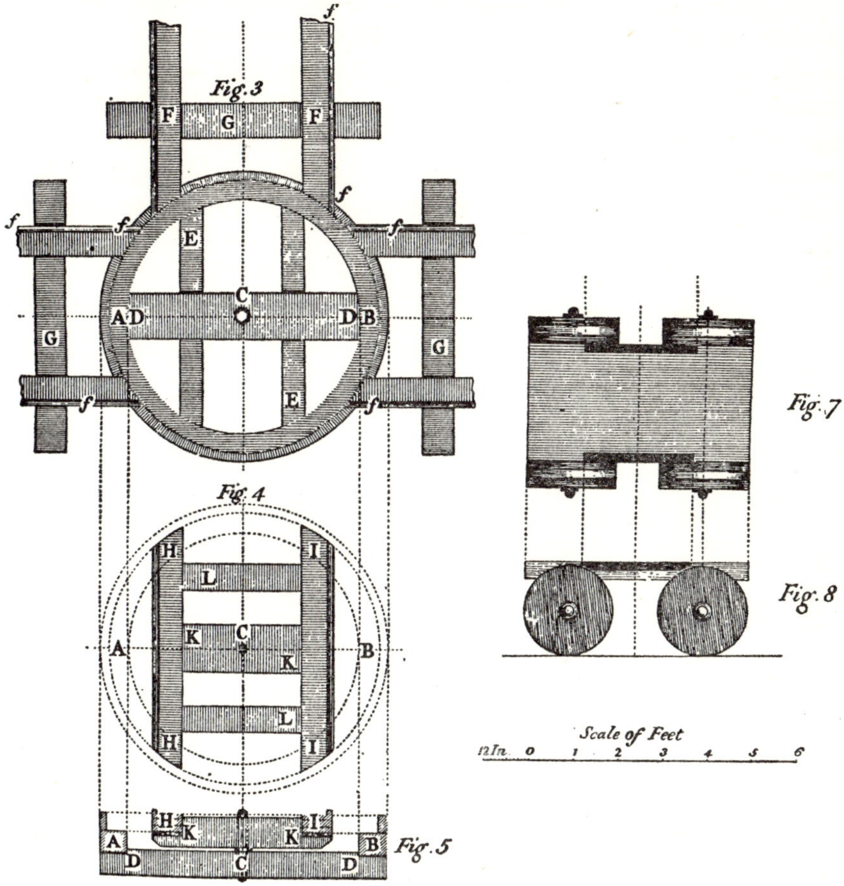

FIG 48 Smeaton's truck and turntable, Plymouth, 1756.

Burden but they do not carry above 60 Tons of Stone . . . and really the length of Time in unloading the Vessell beyond all bounds.

Rennie claimed that the ships were 70-80 tons burden, and that it took from forty to fifty minutes for each boat to moor, discharge and cast-off. On the breakwater itself, as it progressed, quite independent railways were employed for waggons and for cranes to hoist the diving bells. They had wooden edge rails laid at about 3-ft. 9-in. gauge.[10]

A third small and isolated example of the flanged wooden rail was laid about 1890 to a sawmill on the Brightling Park estate in Sussex, about 100 yards long. Its gauge was 2 ft. $6\frac{1}{2}$ in., and it carried a single timber trolley with wooden wheels. The line can still be seen.[11]

On one occasion the guide-wheel principle was applied to a portable and prefabricated railway used in improving a bog in the middle of Ireland for agricultural purposes. Richard Lovell Edgeworth had long been advocating the use of several small waggons in place of one larger one, for spreading the load and economizing on the cost of rails. He put his principles into practice on the uncertain surface of the bogs on his estate on the borders of Longford and Westmeath in 1786. His track (fig. 49) consisted of rails from 8 to 12 ft. long, 5 in. deep and 2 in. broad, strapped with

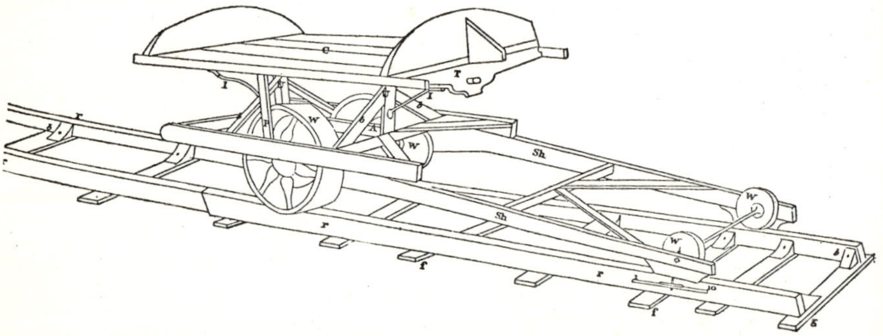

FIG 49 Edgeworth's guide-wheel truck, 1787.

'kieve hoop iron'. They were set 2 ft. 2 in. apart and supported on the sleepers by wooden brackets. The hand-pushed waggons carried half a ton of ash, clay or lime, and were emptied by tipping the body sideways. Their wheels were most peculiar. The principal pair, under the body, were made of iron-shod wood, 2 ft. in diameter and 8 in. broad, with no flange. At the end of the waggon there projected two shafts, armed at the end with two wheels a mere 8 in. in diameter (*ww*) and with two small horizontal guide-wheels bearing against the outside of the rails. To prevent the guide-wheels rising up and derailing the smaller wheels, a narrow ledge (*I*) was nailed to the outside face of the rail. In practice, Edgeworth said, the shafts were kept parallel to the rails by the guide wheels, and so the large wheels stayed on the track. Once a waggon had tipped its load on to the bog, it returned to the filling point by turntables or 'tourniquets' and a parallel track. About twenty waggons were used to every half-mile, and his line was nearly a mile long.[12]

The mono-railway was first patented in 1821 by Henry Robinson Palmer, the chief founder of the Institution of Civil Engineers and engineer to

London Docks. Two lines were built in England on his principle (fig. 50). A mono-railway at work in the Royal Victualling Yard at Deptford as early as 1810 may have been Palmer's own prototype, or a predecessor. At any rate, a full-blown Palmer line was in use at the yard by 1826. A series of cast-iron columns, 10 ft. apart and slotted at the top, held the carrying rail. This consisted of a row of planks each 30 ft. long, 9 in. high and $3\frac{1}{2}$ in. wide, with convex iron strips spiked on top. Each waggon had three pulley-shaped wheels set in line astern and running on the single rail as if on an aerial ropeway. From the axles was suspended the waggon frame, a simple platform slung pannier-wise on each side of the rail and well below it. On the platforms were placed the casks or whatever the load might be. The waggons were moved by manpower. Palmer's other mono-railway, a

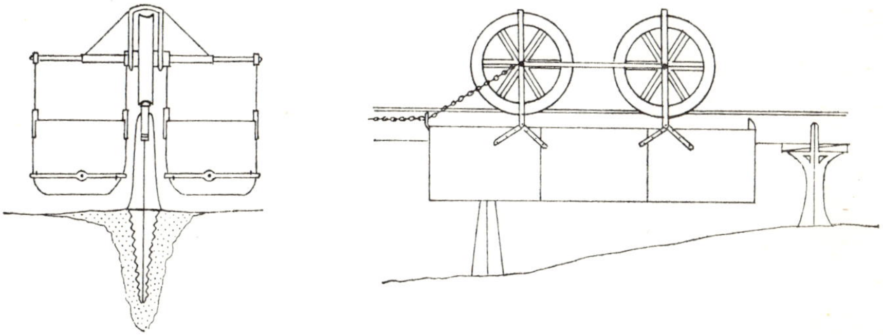

FIG 50 Palmer's mono-railway, 1824.

three-quarter-mile line from brickworks at Cheshunt in Hertfordshire to the River Lee Navigation, was opened in 1825. It was generally similar to the Deptford line, but used wooden posts instead of iron columns, horse traction, and waggons with two large wheels instead of three small. Six or seven waggons formed a train, and one was fitted out as a passenger carriage for occasional use.

The advantages claimed were that the railway took little land and needed no earthworks – on undulating ground the pillars would simply be made higher or lower to keep the rail level. Most important in Palmer's eyes, the friction generated was remarkably small. This was no doubt true when the equipment was new, but as the wheel and rail wore away and fitted each other more closely, so the friction increased. Above a certain height, earthworks would probably have been cheaper than pillars, taking into account the expense of keeping them properly vertical and preventing subsidence. Level crossings and points were cumbrous, being hinged or pivoted lengths of track. Since the two sides of the waggon were suspended from a single point, some care had to be taken to balance the load; and Palmer never explained how the problem of a side wind could be overcome. None the less, the simplicity of the idea caught the public imagination, and one eager supporter even suggested a Palmer mono-railway for the fish (and passenger) traffic from Brighton to London. The trains – fascinating thought – were to be driven by sail.[13]

British railways underground

As me and me marra* was gannin to wark,
We met wi' the devil, it was in the dark;
I up wi' me pick, it bein in the neet,
And knocked off his horns, likewise his club feet.

As me and me marra was putting the tram,
The lowe* it went oot and me marra went wrang;
You would have laughed had you seen the gam,*
And Nick gat me marra, but I gat the tram.

. . .

There is me horse and there is me tram,
Twee horns full of grease will myek her to gan;
There is me hoggers,* likewise me half shoon,
And smash me heart, marra, me puttin's a' done!

'The Collier's Rant', first printed by Joseph Ritson in *The Northumberland Garland*, Newcastle (1793), 67-8.

The earliest coal-mines – distinct, that is, from opencast workings – were bell-pits or shallow shafts opening out at the bottom into small caverns where the coal was dug. As time went on and the most accessible coal was worked out, the pits became deeper and the workings at the bottom became more extensive. The depth of the shaft, being regulated by the depth of the seam, could be as much as 300 or 400 ft. on Tyneside by 1700. But the extent of the underground levels depended on adequate ventilation, not easily provided, and on the economics of transport from the coal-face to the pit bottom. Before the eighteenth century, it was very rare for coal to be worked more than 200 yards from the shaft, for beyond that point it became cheaper to sink a new shaft than to employ extra labour on transport. Thereafter, the deeper the coal was mined, the more expensive it was to sink shafts; and the levels therefore spread further and further.

The most primitive form of transport was a plain basket or leather sack, carried in the hands or on the back, both in the headings and up the shaft on ladders. This was still done by women and girls, in conditions of revolting inhumanity, in some Scottish mines as late as the 1840s. For drawing coal up the shafts, the hand windlass or jack-roll was early introduced, followed by the horse gin and, much later, by the steam engine. In the level, or comparatively level, underground headings the ordinary wheelbarrow was in use by 1356, running, no doubt, on the plain floor at first.[1] Then planks came to be provided as a simple kind of track, the *barrow-way* or *barroway*, whose name survived long after the wheelbarrows were

* *Marra*: mate. *Lowe*: light. *Gam*: game. *Hoggers*: stockings.

replaced by sledges and trams. We first meet this track in the accounts of Raly and Hertkeld mines in County Durham in 1460-1, where the barrow appears as *semivectoria* – 'half-carried', as distinct from the hand-barrow carried by two men like a litter – and barroways as *semivectorie vie*.[2] Barrows survived in some small coal-mines, and in most metalliferous ones, until well into the nineteenth century. In Cornish tin- and copper-mines, for example, the 'barrow-road' consisted of a single plank laid down the narrow level, with a sunken groove along the plank for the wheel to follow. In established mines with hard floors this gave way to a groove cut, and worn, into the floor of the level itself. There are numerous references to 'cutting a barrow-road' in Cornish mine reports, even as late as the 1850s.[3]

The inconvenience of transferring the coal from barrow to basket at the shaft foot led to the introduction of new methods. Most coalfields came to favour the wicker corf or basket – it had different names in different places – which on Tyneside held 4 or 5 cwt. It was filled at the face, taken on some form of transport to the shaft, drawn up by the gin or engine, and emptied 'at bank' – at the pit-head – on to the heap. The underground transport was supplied by three different sorts of vehicle, the sledge, the tram and the rolley. Of these, the sledge sometimes ran on rails, the tram usually, and the rolley always. This multiplicity of types of vehicle was a reflection of the increasing size of underground workings, and all three might be found in the same mine.

Coal was most commonly mined in a parallel grid, the 'bord-and-pillar' method. The bords or passages might be 9 ft. wide, with pillars of coal perhaps 16 ft. square left between them to support the roof. The few large bords which served as the main arteries of transport leading to the shaft were known in Tyneside parlance as the 'outby' bords, and were fed by the smaller 'inby' bords from individual stalls or working-places. In the first place, sledges and trams carried the corves from stall to shaft without a break, but as the workings grew it became usual to restrict small sledges and trams to the smaller inby bords. Where they came to the outby bords, the corves were craned on to the larger sledges or rolleys for transport to the shaft. The transport workers were known by a diversity of names, of which barrower, hurrier and putter were the most common.

By the bord-and-pillar method, by no means all the coal could be extracted – sometimes barely a half was got – but the pillars did to a large extent guard against subsidence of the strata above the coal. The longwall method, invented in Shropshire, allowed all the coal to be dug, but nearly always resulted in subsidence. The working face was pushed forward in a continuous line, and the 'goaf', the area from which the coal had been dug, was filled up with rubbish behind the miners as they advanced. Passages or gates were left through the goaf to carry the coal out. Longwall mines show very much the same distribution as Shropshire-type railways on the surface. Underground, they tended to employ only a single type of transport, not the variety found in bord-and-pillar mines.[4]

The sledge (or sled or slyp) was of the simplest design, a flat platform for the corf with iron or wooden runners. Sometimes it was an integral part of the corf, a rectangular basket with runners.[5] It could be hauled along the floor or thill of the mine – this practice survived in Scotland until the 1840s at least – where grooves might be artificially cut or accidentally worn. Such grooves, 1 ft. 10 in. apart, have been found in a pit of the middle or late eighteenth century at Barwick in Elmet in Yorkshire, along with the iron-shod wooden sleds that used them.[6] It might run along a track of planks or deals, still called the barroway or occasionally the sledgeway. Most of the many references to sledges in the eighteenth century tell us no more. Sometimes, without doubt, the barroway simply formed a flat path to reduce friction, but not to guide the sledge. Occasionally, though, ledges were nailed on to the side of the planks to keep the sledge from straying. This is said to have happened in the late seventeenth century at Whitehaven,[7] and in the eighteenth on Tyneside.[8] In a valuation of Jarrow Colliery in 1809, items like '84 Yards of Penning [rail] in Sledge Way @ 10^d^' suggest, from the price, a fairly elaborate track.[9] When some old workings at Pummer (Darton Main) Colliery near Barnsley were broken into early this century, an antique corf-cum-sledge was found, 2 ft. 4 in. long, 1 ft. 8½ in. wide, and 1 ft. 2½ in. high, as well as 'rails of oak . . . with a flange or protecting rail fastened on the rail for the purpose of keeping the sledge in position as it was hauled along'. This find dated from the end of the eighteenth century at latest.[10] Another Tyneside method of guidance was to lay two rows of planks side by side to form the barroway, with a third laid on top, along the middle of the first pair, for the sledge runners to straddle.[11]

A sledge was moved by two boys – one pushing and one pulling by means of a 'soam' or cord – who occupied a middle position in the underground hierarchy. Boys would start in the pit at the age of six or seven or so as trappers, opening the ventilation doors for traffic to pass through. At eleven or twelve they became foals, then half-marrows, then headsmen, the three subdivisions of the rank of putter. One headsman and one foal, or two half-marrows, worked a sledge (and later a tram), a headsman taking two-thirds, a half-marrow half, and a foal a third of the day's payment. The boy then rose to be a put-and-hewer, spending half the day in getting coal, and half in putting it. Finally, he graduated to full-time hewer.[12] The wages for putting were based on a unit most confusingly called the 'tram'. This was the amount of work which it was agreed one sledge and its barrowmen should do in a day, the quantity of coal put being reduced as the distance increased. A barrowmen's bond at Benwell Colliery in 1703 laid down that when the pits were first opened, four trams were to be worked with a daily total of eighteen score corves. In other words, each sledge with two barrowmen was to carry 90 corves a day from face to shaft. When 792 score had been put (that is, after forty-four days), a fifth tram was added, with a 'bate' or reduction in the quantity to be put to compensate for the increased distance resulting from the advance of the faces. After another 792 score, another tram and another bate were allowed. The regular wage per tram, regardless of the quantity of coal and the length of the put, was to be 1*s.* 10*d.*, or 11*d.* per barrowman if it was equally divided.[13]

The Compleat Collier of 1708, the first detailed account of coal-mining, gives figures which are not dissimilar:

> It is usual to begin with six Trams, which put and hook to the Cable three Score and ten Corves a Piece every Day, which makes up twenty one Scores a Day, and then every thirtieth Day, or so, afterwards, they have a Barrow-Man, or Tram added, still at the same Wages . . . till by chance the Coal will be so far Wrought under Ground, that there must be sixteen Trams; which brings it so, from seventy Corves put, to about twenty six Corves a Tram, so that the more and further a Pit is wrought, you see the dearer she lies in the Charge of Barrow-Men, or putting, for they still keep up the first Price of a Day's Work, were you to add never so many Trams.

This wage, the author says, was usually 1*s.* 8*d.* or 1*s.* 10*d.* with a bonus of 2*d.* if the pull was uphill or the floor was uneven.[14]

Not long after this horses were first introduced underground. They had long been used at bank, to pull corves on sledges from the shaft-head to the heap, where references to 'drawing and sledding' and 'horse sledges' are common. They had probably also made an early appearance in drift mines, approached through horizontal adits. But in mines accessible only by a narrow shaft, the problem of getting the horses down was insoluble until the idea occurred to somebody of driving a special sloping adit for them. In 1716 a contractor for sinking two pits at Bucksnook was to be paid £15 'in consideration of making an open cast to convey horses into the drift to draw or put the coals'.[15] At Whitehaven in 1739 'the coaliers and horses which are made use of under the sea [the mine extended out from the shore] to carry or slype the coals to the bottom of the drawing engine, enter by the side of a hill next the town and go down on a sloping way to lowermost parts of their work'.[16] Horses proved themselves so useful underground, and so few mines could be equipped with this means of access, that before long horses were let down the shafts in slings – perilous proceeding – and by the middle of the century they were in common use, spending most of their lives below ground. William Stukeley, touring the north in 1725, observed that 'horses kept under ground in the coal mines for two or three years, as sometimes they do, have their hair very fine and sleek, and as short almost as that of a mouse'.[17]

Pit horses, or 'galloways' as they were called from their breed, were in fact only ponies. At Stanley (Co. Durham) in 1784 they were valued at £2 5*s*. to £4 13*s*. each, far less than a horse.[18] As they could work only in bords that were 4 ft. 6 in. high or more, hand-putting continued in the lower roads. A bond of about the 1760s for Pontop Pike, Lintz and other pits shows that the underground sledge horse-drivers – only one driver per sledge now – were still paid on the old basis. They were to drive five score and five corves per day as long as the journey was less than 90 yards; when that distance was reached, the stint became four score and four corves up to 130 yards, and so on. Their wage was 10*d*. a day.[19] A Swedish visitor in 1776 wrote:

> Where horses are used, there is a boy, nine to twelve years old, who drives each horse. He generally stands on the sledge behind, uses no reins, and sings continually as long as the horse is in motion. He usually has a light or lamp every 60 yards in the way. . . . Each of these boys is paid a shilling a day.[20]

At Whitehaven a horse travelled nine or ten miles in a nine-hour shift where the roads were nearly level, or six to seven if the gradient was 1 in 9; four corves would fit on to one sledge.[21]

The horse sledge continued in general use in the main bords of Tyneside and Cumberland mines until the second half of the eighteenth century, when it was gradually ousted by the rolley, of which more anon. In the inby bords the hand-worked sledge was gradually replaced by the wheeled tram. We cannot be sure when this first appeared, for the word *tram* did not always imply a wheeled vehicle. Its original meaning was the shaft of a barrow or the frame of a sledge, and many early examples probably mean a sledge or even a barrow plain and simple.* Hett Colliery in Durham, for

* The famous (indeed, notorious) will of Ambrose Middleton in 1555 bequeaths 'To the amendinge of the highewaye or tram, from the weste ende of Bridgegait, in Barnard Castle, 20s'. What was this tram? It can hardly have been a close relative of the colliery vehicle. A timber corduroy road has been suggested, but it may simply be a local name for the street. A wall immediately on the other side of the Tees is still known as the 'tram wall', presumably because it trammels or confines.[22]

instance, had '2 pykes [picks], 2 trammys, et 2 shulys [shovels]' in 1517; Harraton Colliery 'divers Ingines, trammes, shouells' in 1635; Gibside 'Twelve Barrows seven Shovells three Horse sledges [no doubt at bank] & five Trams' in 1692; and at Farnacres near Dunston in 1717 we find 'Repairing Shovel Irons, Tram Shoes &c' – evidently sled runners.[23] Examples could be multiplied. It is not until 1734 that we meet the wheeled tram. The inventory of Felling colliery records '5 Sled Trams, 8.4; 1360 Deals [for the barroway] £36. 14. 10; 31 Wheal Trams £6. 19. 6'. The cost too, gives it away – 4*s.* 6*d.* for a wheal tram, 1*s.* 8*d.* for a sled tram. At Black Close Mine near Blyth there were in 1739 '6 Wheel Trams att 5^{s} p. left under Ground', and at Lands near Derwenthaugh in 1754 '5 Working Trams at 5^{s} 6^{d} p' – as opposed to 'lying' or spare ones.[24] At Plessey Colliery in 1724 the 'corfs are brought to the bottom of the shaft on a little slyp shod with iron', but when Mårten Triewald, the Swedish engineer, estimated the cost of getting coal here in 1740, there were '*hundar* or small trucks, whereon the corves are carried'.[25] Trams also began to oust the sled at the pithead. It was reported in 1759 and 1760 that at the Newcastle Town Moor pits 8*s.* a week was saved 'by Making the Banksmen Sledd out the Coals with a Tram, Instead of the Horse & Sledd'.[26]

Normal underground practice, then, about 1750 was this: If it was possible for a horse and sled to reach the face, they would, to save transferring the corf. But if the roof was too low, a hand-drawn tram (or, in ever-decreasing numbers, a sledge) carried the corf inby to the main bord, where it was craned on to the sledge and pulled by horse outby to the shaft. If the whole mine was too low for horses, then trams alone were used. The boy in front either held the soam over his shoulder, or more often wore it looped round his waist and passing back between his legs. The boy at the back simply pushed with hands and with head, an action which usually resulted in his hair being rubbed off, leaving a bald patch. It was a cruelly hard life, with unremitting labour over long hours. For putters (though not for hewers) the daily stint might be twelve hours or more, so that in winter a boy might never see the daylight.[27] Wages remained at much the same level throughout the eighteenth century, for trams as for sledges, although 'extra allowances' were granted in 1775 by the benign proprietors of the Briton Ferry Colliery in South Wales to the 'drammar' boys.[28] Women were banished from underground on Tyneside quite early in the eighteenth century, but elsewhere they remained below for a long time, either sharing the boys' work in putting or, as at Whitehaven, on lighter jobs, such as hanging corves on to the rope and cleaning the ways. Scotland was quite the worst area for female labour in mines. Many Scottish colliers were serfs – in some respects, indeed, quite literally slaves

– until 1799, and for a full fifty years after that they, and especially their womenfolk, submitted to utterly barbaric toil.[29] Even in the 1840s one gets the impression that a Scottish mine with a railway underground had something to be proud of.

A typical tram is illustrated in fig. 51. It had two curved wooden beams for a frame (they cost about 1*s.* a pair),[30] joined together by cross-bars to form a safe seat for the wicker corf. The wheels were often of wood. At Pitferrane at the head of Lord Elgin's waggonway in Scotland they cost 6*d.* unhooped and 1*s.* hooped in 1796, and some still existed as late as 1818 at Morpeth in Northumberland.[31] Middleton Colliery in Yorkshire began to use cast-iron tram wheels in 1767, weighing 25 lb. each, which was heavy by later standards.[32] Those at Pitferrane in 1796, for example, alongside the wooden ones, and others of 9-in. diameter at Wallsend in 1794,

FIG 51 Tyneside tram and corf, 1773.

all weighed 8 lb. each; and at an unspecified Tyneside colliery in 1805 the wheels were a mere 5½ lb. each. Here the cost of a new tram was 13*s.* 4*d.* against 1 guinea for a horse sledge;[33] elsewhere the valuation cost, allowing for depreciation, was usually in the area of 4*s.* to 6*s.*, with an all-time low of 1*s.* 4*d.* at Jarrow in 1809 for antique wooden-wheeled trams.[34] On Tyneside, a special tram for carrying, not corves, but barroway deals, pit props and the like was known as a 'bogie'; elsewhere terminology differed. At Whitehaven, for instance, the name bogie was occasionally applied to the small inby tram, but more often to the larger horse-drawn outby waggon.[35] In Staffordshire the 'skip' was the local form of tram, a flat platform on which the coal was piled into a pyramid, held in place by bands of iron. An underground truck in Shropshire was a 'dan'; and so on.[36]

The plain, unflanged wheels of the tram occasionally ran on the bare floor, or thill. It was still so in some Scottish collieries in the 1840s, many of which were a good century behind the times. With the ordinary tram wheel, the thill was normally too rough for this to answer even on the lowest economic scale, although some fairly successful experiments were made by John Buddle at Wallsend in the 1790s with extra wide wheels.[37] As with the barrow and sledge, the problems of steering the tram and of reducing the friction over the rough floor were sometimes eased by carving parallel grooves for the wheels to run in. 'Cut ye Thill for a Barrow Way, but it is not yet finished', says an undated report on Long Benton Colliery, probably as late as the 1780s.[38] But this was exceptional, and in respectable mines at least the normal track for a tram was the barroway of planks. Countless references to 'barroway deals' in Tyneside colliery accounts throughout the eighteenth century show its popularity. At Whitehaven the barroway was known as the trailway or corfway, and here too the word *tramway* first appears, at least as early as 1775. The term *tramroad* is first mentioned in 1796 at Pitferrane.[39]

Sometimes, it seems, the barroway was a plain flat surface of wood with no means of guidance: a road completely floored with cross-planks was found in an old mine near Workington which was abandoned before 1781.[40] But writers of the early nineteenth century, when wooden barroways were still within living memory, make it clear that guiding rails were usually provided for the trams, as they sometimes were for sledges, and in the same ways. Either a central row of planks was laid on top of two touching rows, for the wheels to straddle, or more often a strip of wood was nailed on each side of the running planks to form a flange.[41] It was these railways that Smeaton imitated at Plymouth in 1756 (p. 302). In the isolated Somerset Coalfield, a description survives of Mearn's Colliery, High Littleton, in 1792. The coal seams were on a dip, so that inclined planes were used – a *gugg* with double track, a *stipe* with single. At the shaft-foot a windlass was fixed, with a rope

> to the ends of which are hooked two 3-bushell carts on four low wheels which run up and down the guggs, the empty one going down and the full one coming up at the same time . . . A Gugg is a common road about four feet high and wide and 55 yards long, pitches about 9 inches in a yard. In the bottom lies timber framed together down which are two roads of four rutts made for the wheels of the carts to run in so as to pass one by the other going up and down.

Below the gugg was a level twinway (between-way), where the carts were pulled by a man on hands and knees, and beyond this another gugg.[42]

The wooden barroways were not satisfactory, especially since the thin iron tram wheels quickly reduced the flimsy guiding flanges to matchwood. A Tyneside viewer, telling a Parliamentary Commission of the old wooden rails, said, 'the trams were often getting off these boards, and then it was very difficult to get them on again'.[43] The miseries of those days and their solution are best brought out in *The Pitman's Pay*, a long dialect poem written by Thomas Wilson, a Tyneside collier, and first published in 1826-30. Two veterans are yarning in a pub about old times:

When Dicky's corf was fill'd wi' sic,
He let his low*, and stuck't agyend* –
Ax'd Deedy te lay down his pick,
And help him te the heedwis* end.

. . .

The wark now placed, and pit hung on,
The Heedsmen, whether duin or nut,
Mun iv'ry man and mother's son
Lay doon the pick and start te put.

Now then the bitter strife begins,
All pullin', hawlin', pushin', drivin',
'Mang blood and dirt and broken shins,
The waik uns wi' the strang uns strivin'.

Aw mind a tram byeth waik and slaw,
Just streen'd te rags te keep her gannin',
Frae hingin'-on* till howdy-maw,*
Ye hardly knew if gawn or stannin'.

Just pinch'd te death, they're tarn and snarly,
A' yammerin' on frae morn till neet –
Jack off the way, blackgairdin' Charley,
For at the corf nut lying reet.

. . .

Sic, *then*, was the poor putter's fate,
Wi' now and then a stannin' fray,
Frae yokens,* cawd pies,* stoven bait,*
Or cowp'd* corves i' the barrow-way.

. . .

But heavy puttin's now forgotten,
Sic as we had i' former days,
Ower holey thill and dyels a' splettin': –
Trams now a' run on metal ways.

* *Low* = light; *agyend* = again; *heedwis* = headway's; *hingin'-on* = start of work; *howdy-maw* = last corf of day; *yokens* = trams meeting head-on; *cawd pies* = broken axles; *stoven bait* = stolen food; *cowped* = overturned.

God bliss the man wi' peace and plenty,
  That furst invented metal plates!
Draw out his years te five times twenty,
  Then slide him through the heevenly gates.

For if the human frame te spare
  Frae toil and pain ayont* conceivin',
Ha'e ought te de wi' gettin' there,
  Aw think he mun gan strite te heeven.[44]

The man who, one hopes, came to enjoy this bliss in the hereafter was John Curr, a native of Durham and viewer of the Duke of Norfolk's collieries at Sheffield. His scheme for underground railways was to replace the cumbrous Tyneside system of carrying a single basket corf on a tram or sledge inby, and transferring it to a horse-drawn outby sledge or rolley, which held, he said, only two or three corves. He proposed a single type of vehicle for use throughout the mine, the wheel-corf. This had a rectangular wooden body of much the same capacity as a wicker corf, with four unflanged wheels attached. Wheel-corves could be pushed inby by hand one at a time, but marshalled into horse-drawn trains of nine or ten for the outby journey to the shaft, where each corf was raised, wheels and all, to the surface. A horse could pull, Curr claimed, three or four times as heavy a load as with the sledge or rolley. The wheel-corves, in being drawn up the shaft, were guided by vertical rails to save the wear and tear suffered by wicker corves in bumping against each other and against the sides.

His track was light, since the weight of the coal was spread over many wheels. It was simply a conversion of the wooden flanged barroway into iron, not, as has been said, an addition of the flange to existing flat plates for flanged wheels. The rails were L-shaped and usually 6 ft. long, weighed 47 to 50 lb. each, and were nailed to wooden sleepers at a gauge of around 2 ft. The principle was, he pointed out, quite suitable for surface use as well.

Curr described his railways in a book published in 1797, *The Coal Viewer and Engine Builder's Practical Companion*, in which he states at one point: 'The making and use of railroads and corves were the first of my inventions, and were introduced at the Sheffield Colliery about twenty-one years ago.'[45] This sentence has led every railway historian to date these iron rails to 1776. But it must, without a shadow of doubt, be put later, as a few historians of the coal industry have seen.[46] In view of the impact of Curr's rails on underground and on surface transport, it is worth investigating the date of his invention more closely. In the first place, *The Coal Viewer* seems, from internal evidence, to have been written in 1794, so

* *Ayont* = beyond.

that 'twenty-one years ago' takes us to 1773. Curr gives the same sort of date in a report of 1801[47]: 'The First of my Inventions was the Rail Roads which happened 2 or 3 years prior to the Duke of Norfolk's being concerned in Collieries' – 1777, when the Eleventh Duke, then the Earl of Surrey, took over the Sheffield mines on the death of his distant kinsman the Ninth Duke. What, then, were these 'railroads and corves' that Curr introduced between 1773 and 1775? They surely existed, since his statements, even if exaggerated, are so categorical. They cannot have anything to do with the surface line that he built, which he calls a waggonway and which would not use corves; and they cannot refer to his iron rails which, as we shall see, first appeared only in 1787. We simply cannot identify this railway invention of Curr's. When he arrived in Sheffield in 1773, the coal was moved underground in small horse-drawn sledge trams running on the wooden floor, with no means of guidance. The same method was in use in 1787, when the new iron rails began to oust it.[48] It may be that Curr replaced the sledge trams with larger wheel trams running on a flanged wooden barroway, but for some reason had to revert to the original method. But the evidence is simply lacking.

Whatever the answer to this problem, however, there is no doubt that the introduction of Curr's famous iron angled rails is to be dated to 1787. In *The Coal Viewer* he claims that in the iron rail he has 'hit upon a mode superior to anything hitherto practised, as the result of seven years experience informs me.'[49] Seven years from 1794 is 1787. If Curr thought of the idea before, he certainly did not apply it till then. His six reports on the Sheffield collieries from 1773 to 1785 make no mention whatever of iron rails;[50] every report from 1787 on emphasizes their novelty and their usefulness. The only surviving account-book of this period, from 1781 to March 1787, gives no hint of Curr's rails and corves until almost the very end. In February and March 1787 we find '4 Sole Trees 1/8' – which must be for corves or trams, since surface waggon soles cost at least four times as much – and 'to Moddle Wheels . . . 5[sh]', meaning patterns.[51] The clinching evidence is to be found in a report to the Duke of Norfolk on Sheffield Park and Attercliffe collieries by John Buddle, the Newcastle viewer, dated 7 April 1787. He had been asked to comment 'upon the Scheme of Hurrying the Coals which Mr Curr has lately introduced in one of the Pits, and compare the Mode of Hurrying by the Wheel Corf with that of Hurrying with Horses'. He was wholeheartedly in favour. In Park Colliery by the old method with sleds, horses and wicker corves, putting cost 10½*d.* per surface waggon load; by the new, with wheel-corves and manpower, it cost 6¾*d.* Buddle waxed eloquent about the possibility of abolishing the surface waggonway and carrying the wheel-corves in tub-boats on an underground canal. Significantly, it was only three months before that Curr had visited the Duke of Bridgewater's Worsley mines, where such canals were in use. At Attercliffe 'Mr Curr's new Scheme of Hurrying' was not yet applied, but Buddle recommended its instant adoption.[52]

This first use of Curr's iron rails in Sheffield Park Colliery was evidently an experiment limited to one pit, and the corves were moved singly by hand, not in trains by horse, as he later advised. Attercliffe Colliery was not immediately equipped with iron rails throughout, for John Buddle and John Stephenson, taking another view there in April 1789, estimated 'Iron plates wanted to complete the Heading and Gait-Ways of Attercliff Common . . . 2600 yards at 5/6 p'.[53] But, fortified by Buddle's approval, Curr went ahead and completely fitted out Park Colliery with the new rails. His account with Binks, Booth and Hartop, the founders, shows heavy orders for corf wheels and 'road plates' beginning on 20 November 1787.[54] Curr is also said to have been supplied with rails by Joseph Outram, a Ripley ironfounder and Benjamin's Outram's father;[55] and in 1792 he set up his own foundry for railway and engine castings. In 1788 he patented his design for guiding wheel-corves in shafts. He was suddenly dismissed by the Duke of Norfolk in 1801, probably because of his lack of success in undercutting competitors' prices and in dealing with inroads of water in the pits. He protested vigorously, pointing out the benefits his inventions had bought the Duke, but in vain. 'Here', he lamented, 'my Ingenuity has been buried.'[56] He died at Sheffield in 1823.

John Curr's iron rails did not take long to reach other coalfields. The first imitations, strangely enough, were on the surface: in 1788 on the Ketley inclined plane in Shropshire (though here William Reynolds may have developed the flanged rail independently) and at Joseph Butler's Wingerworth furnaces near Chesterfield.[57] The subsequent use on surface railways we have already discussed (p. 293). Underground, the Middleton Colliery at Leeds was quick off the mark. In 1790 it 'Paid John Curr a Gratuity for his Trouble &c respecting the Hurrying & Drawing the Coals agreeable to his Patent £10.0.0', and next year paid 'J. Curr for sundry Castings had from Sheffield £17. 6. 7'.[58] Henceforth cast-iron plates were a regular feature here. The introduction of Curr's rail to Shropshire mines and surface railways has been dated to 1793, when he was called in by the Coalbrookdale Co. to advise on the operation of its railway and shafts at the end of the Shropshire Canal. Although his report makes no mention of his plates, this date may well be correct for underground, where wheeled corves were in use by 1794 and Dale-made Curr rails by 1796; but on the surface the conversion from edge to plate, with the possible exception of the 'Old Wind', probably occurred somewhat later, maybe even after 1800.[59] We first hear of Curr rails in 1796 in South Wales, about 1800 in Cumberland and Scotland, and only in 1802 on Tyneside,[60] where they must have been used some years before – again, the lack of iron foundries probably delayed their widespread adoption.[61] Curr claimed in *The Coal Viewer* (written in 1794?) that his rails 'have been generally imitated, and made use of in most collieries for the last three years, especially in the southern parts of the kingdom'. In his protest of 1801 to the Duke, he wrote:

> I need say no more to recommend Rail Roads, than that about 16 Collieries out of 20 have introduced this mode of conveying Coals in the Countys of York, Lancaster, Salop, Derby, Stafford, Warwick, and great parts of Wales, and is now adopting near London, and in the neighbourhood of Newcastle upon Tyne it has been adopting very much for 2 or 3 years past, and those who live 10 or 15 years will probably see my Rail Roads introduced all over this Kingdom, notwithstanding 12 years passed before they were much imitated.

Curr's corves were not an unmixed blessing. Sometimes they brought back hand-tramming where horses had pulled sleds and rolleys before, and thus led to a wider use of boy labour. But in general it was many years before the wheel-corf, or 'tub', as it came to be known, was adopted. However, there was no doubt about the value of his rails. On Tyneside in particular the outby rolleys remained, and Curr's rails simply replaced the previous wooden ones inby, though still carrying the old type of tram with separate corf. It was here that the real benefit was felt, as *The Pitman's Pay* so graphically describes. The labour force was also much reduced. Whereas with wooden rails there had been about as many putters as hewers, the number of putters could now be cut by a third or even a half.[62]

So much for the flanged rail. The edge rail for flanged wheels developed underground along much the same lines as on the surface. The double tier of rails in the North, the cast-iron armour plating in the Midlands, appeared in the mines as they did on the waggonways. Only the scale was different. In early ways in the Shropshire Coalfield, where horizontal adits or drift mines were common, one and the same railway ran from coal-face to the Severn. We have seen how the flanged wheel may have been invented on this sort of line, how very ancient examples have turned up in the Broseley area, and how *footrid* or *footrail* came almost to presuppose a railway in a mine adit. Such a footrid at a Madeley coal-mine was described by Brokesby in 1711 (p. 232), and another nearby by P.-C. Lesage in 1784: 'Below the [Iron] bridge is the adit of an iron mine . . . whose gallery is six or seven feet wide, equipped with iron plates like the ways in the hills, on which the waggons run.' Perhaps the best-known footrid was Mackworth's at Neath. We have observed it in action on the surface, with horse traction and sometimes sail-power. Its 1200-yard-long underground reaches were worked by hand. Each waggon was 'thrust forth by two Men', the complement of a working gang being three coal-cutters and four waggoners. The waggoners

> have great skill to keep the Waggons upon the Railes through the Tunnels and Windings Under-ground, which are so intricate, and do so greatly vary in every Street of Coal, that a good Waggoner in one Street, will not serve in another; and the Work is so laborious that few Men will undertake it.[63]

This makes it sound as if there was no guidance for the wheels. But Mackworth is simply concerned with extolling the skill and irreplaceability of the colliers whom his rival has abducted. Sir Humphry himself, taking a joy-ride once in a waggon, nearly came to an untimely end. He wrote in his diary on 4 December 1700:

> Ld. make me thankfull for all my deliurances and particularly of lat yt [that] great deliurance i ye Coal-pit wth Sr Ed. Manxell, when I call'd out in hast to Stop, or els I had in all probability Broke my legs; & phaps lost my life; – thre being a waggon in ye way, and my legs hanging our [over] and ye men confident thre was nothing in ye way; I once yielded to them, but providentially Stop't ye very minute wee were running agst it; O wht a Sad thing had it been to haue lost my lif i yt hurry; wthout recollection & repentance.[64]

The vertical shaft, however, was far from unknown in Shropshire and the Shropshire empire, and quite independent edge railways were found in these mines. Waggonways are mentioned in pits around Broseley from 1758, and the inventory of Madeley Lane pit in 1741 allows us to compare the costs of 'pitt Wagons' and their constituent parts with 'Bank wagons' on the surface.[65] Pit waggons were valued at 6*s*. against 9*s*. for bank waggons; pit wheels at 2*s*. each, perhaps weighing 30 lb. (see p. 273), against 6*s*. 8*d*. and perhaps 85 lb.; 'axtrees' 2*s*. against 7*s*. 6*d*.; brasses 1*s*. against 1*s*. 3*d*. In South Wales Sir John Morris, who owned the Pentre pit at Landore near Swansea, wrote in his commonplace book:

> Cast Iron Tram Plates – In Nov'r 1776, I wrote to Messrs Darby & Co. at Coal Brook Dale that I had sent them a pattern in wood about 4 ft. long and 5 in wide, to cast Iron plates for wheeling Coal

> in my Collieries, each plate to weigh ab't 56 lbs: that if they cou'd be supply'd at £8 p ton I shou'd want immediately 100 Tons, and that the introduction of them wou'd occasion a vast consumption of metal *never before* us'd for such purpose. This pattern was made in the summer of 1776.

And later: 'In 1788 there was about 240 tons of Cast Iron Tram Plates underground in Landore Coll'y', or three and a half miles of track.[66] It is possible that Morris anticipated John Curr in using iron flanged rails, but much more likely he was ordering flat plates (nearly 1 in. thick) for flanged wheels, on the pattern that Coalbrookdale had pioneered. Whether in this case his claim of novelty was justified we do not know. If he meant flat plates in a coal-mine, it is possible.

Another special case was the Duke of Bridgewater's mines at Worsley, near Manchester. Here, in the words of Arthur Young, who inspected the mine in 1768, coal was carried to boats on the underground canal

> in little four-wheeled waggons, which contain ten *cwt.* of coals, and are pushed along by a man setting his head and hands against them. . . . The roads all lead to the [canal] tunnel. When the man with the waggon comes over a well (of which there are several) that is sunk from the road through the arch of the tunnel, and under which the boats are fixed, he stops on a frame work of wood, which turns on pivots, and is so contrived, that upon drawing up a part of one end of the waggon, some of the coals drop out, and then the waggon is tilted up, and all the rest follow them, falling into the boat beneath.

Wherever two lines crossed each other, turntables were provided.[67] This unloading arrangement was evidently copied from Shropshire practice (p. 259), and is met with in the same form on the Continent (p. 334).[68]

In the Tyneside empire, the wooden edge railway for flanged wheels appeared underground in the first half of the eighteenth century. John Buddle thought the date was about 1775,[69] but this is much too late. Matthias Dunn is more likely right when he says, talking of the early eighteenth century: 'Horses were as yet scarcely introduced underground; but when they were, the roads were constructed in the same manner as those above ground, the rollies carrying two or three corves each.'[70] To begin with, horses usually pulled sledges in the outby bords, but as the century wore on, these sledges (and even more primitive forms of transport) were gradually replaced by the rolley, which was always horse drawn. The first indication we get is at Whitehaven in 1727, eight years before the first surface waggonway appeared, when the accounts record payments for waggons, waggoners, '4 Doz pins 2^{s} for Wag^{n} Way', 'makeing up & leading a parcel of Wheels', and so forth. Soon afterwards the Sheffield Colliery accounts record, in 1730, 'By mending the Railed Way the Wood & Workmanship £123. 11. 0'. From its position in the accounts, it must mean an underground line, in contrast to the surface waggonway. Several more indeterminate entries mention the railed way during the next few years, but the matter is clinched by an entry in 1736: 'Robert Greaves for laying the rail in the pitt 4/6.'[71] Both underground and surface lines had vanished by the time John Curr arrived in 1773.

The horse waggonway was established in Tyneside before 1750, when Michael Meinzies patented his self-acting incline for bords that sloped at more than 1 in 15 and were impracticable for horses. He was a Scot by birth and domicile, but from his vocabulary was certainly best acquainted with Tyneside mines. He proposed to lay a 'double waggon way' in the sloping bords, each track 'two feet wide only and about half a foot distance from one another'. Round large pulley wheels at top and bottom ran two endless ropes with downward-projecting hooks, to which wheeled 'boxes' were attached, the downward loaded ones pulling up the empties. If the dip went the other way – if the coal had to be taken up the incline – it could be balanced by containers of water going down. The boxes should be 2 ft. wide and 4 ft. long and high, with iron-shod wheels 2 ft. in diameter and 2 in. thick. They were to be supplied by 'corves laid on common four wheel trams', although he advised using the boxes inby if possible, connecting with the incline in the main bord by turntables. The rails might be made of stone instead of timber, or even cut into the floor if it was suitable. Meinzies anticipated John Curr by suggesting that the boxes be drawn complete up the shafts, and emptied on the heap by similar inclines built up on portable trestles. He also anticipated the Duke of Bridgewater by proposing underground canals to remove coal from the mine if conditions were suitable.[72]

In 1754 William Brown of Throckley was in Scotland:

> At Burroughstoness [Bo'ness] I was assisting to Contrive a Waggon Way underground on which will have to bring the Coales 600 yards at first starting tho at present they are Convey'd from the face of the workings to the Pitts Bottoms and in some places up the Shafts upon womans Backs which poor Creatures carries some of them 13 Stones weight of Coales at a time in there Creals.

Three months later John Burrel of Bo'ness told Brown: 'This six Weeks past I have had three horses imployd below Ground which answers to my wish and design to have 3 more at work in 2 Weeks time.'[73]

Gabriel Jars visited several mines in 1765. Near Newcastle he found horses underground which

> led the coal from the most distant parts of the mine to below the gin shafts. For this purpose, wooden ways are laid like those used on the surface of the ground, on which run four-wheeled waggons to carry the corves full of coal that are drawn to day by the gins. In the places where such roads are not made, young boys use small sledges on which they set the corves and draw them to the shaft, or to the horse way.

He found similar waggonways at the Carron Company's mines near Falkirk (although the waggons had wooden bodies for the coal, not a flat platform for corves), and at Whitehaven, with double track and an incline on Meinzies' principle, balanced by water-containers.[74]

The Tyneside underground waggon and waggonway, or rolley and rolley-way, as they came to be called, were a miniature version of the surface variety, except that the rolleys were flat to hold the corves. Curr, belittling the 'Newcastle-road' to emphasize his own plans, claimed that a rolley held three corves at most, but John Buddle records that at St. Anthony's Colliery, Byker, in 1776 'the Waggons carried six 20 peck Corves each – the Corves stood two abreast on the Waggons which were drawn by a Single Horse'.[75] By 1810 rolleys held up to nine corves.[76] Wooden wheels survived until 1796 at least in Pitferrane Colliery in Scotland, where 'Rolley wheels shod' are valued at 1*s*., 'unshod' at 8*d*. More usually the wheels were cast iron – 14 in. is one diameter given – and weighed around 75 lb. each. A new rolley in 1805 cost £3 1*s*. 11*d*.[77]

The rolleyways were sometimes single-track with loops, but more often double-track. They were built, as above ground, by waggonway wrights. A Wallsend contract of 1792 reads:

> Let Thomas Hall, Waggon Way Wright, the laying of the Waggon Way along the East Mothergait in the C Pit, viz: The Main and Bye way both to lie in the same Bord. The Main Way to be double, that is a slit Rail on a whole one at 5^{d} p. yard. The Bye Way at 4^{d} p do. False Rails at 2^{d} p. do. Sleepers to be distant 3 feet center to center. The Sleepers to be sunk in the Foul Coal. The Wright to get in his stuff from the Shaft. Width of Way to be 2 ft. 6 Ins.

Subsequent agreements were made to gutter and crease the way – i.e. make ditches and keep the flangeways clear[78]. Sleepers were often of fir, costing from 3*d*. to 4½*d*. each; rails were usually of oak at about 8*d*. per yard.[79] Typical prices for laying appear in the contract above. By the 1780s wrought-iron plates were being fixed to the rolleyways. An estimate of the cost of developing Heaton Colliery in 1786 mentions 'Smith work . . . sharping, sledge shoes, Trams, Waggs & Wagg^{n} Way Plates, Mauls, Wedges &c', and an agreement of Harraton, Lambton and Bourn Moor collieries in 1784 lays down: 'General Lambton [the owner] to make and keep all Waggonways, Barrowways, Rolleys, Sledges, in good and sufficient Repair and to plate the Rails of the said Waggonway with Iron.'[80] At Wylam from 1807 the rolleyways were likewise covered with 'Roll'd Plates' weighing about 24 lb. each – probably 12 ft. at 6 lb. per yard.[81] The wages for the waggoners at Lambton in 1784 were 8*d.* per score of corves on a distance of 300 yards, rising 1*d.* for every 50 yards up to 1*s.* 2*d.* for 600 yards. Again as on the surface, rails made wholly of cast iron were introduced to the rolleyways in the early 1800s, both on Tyneside and in Cumberland.

The pattern of transport, then, typical to a Tyneside colliery around the turn of the century was this: corves were taken singly from the face to the crane on inby trams running on the flanged barroway. If the barroway was still wooden, a tram needed two boys; if Curr's plates had come in, only one boy. Six or eight corves were then craned on to an outby rolley, which had flanged wheels running on wooden edge rails, and was drawn by horse to the shaft foot. When the corves had been drawn up to bank, they were either emptied via the screen into the surface waggons or set on trams or sledges to be tipped on the waste heap.

One other underground waggonway on Tyneside which does not fit into this pattern deserves description. East Kenton Colliery, drowned in 1715, was recovered by the viewer Christopher Bedlington, who drove a three-mile drainage drift which debouched into the Tyne at Scotswood and became known as Kitty's Drift. The wayleaves were obtained in 1791, but work did not begin until early in 1796. It served not only as a drain, but also as the colliery's only exit for coal, with a single-track wooden waggonway. It was a regular sight for the tourist, one of whose descriptions, published in 1807, is worth quoting at length.[82]

> Having previously obtained permission of a viewer, or some other person concerned in the colliery, who will give you any directions that may be necessary, we would recommend you to provide yourself with a small hand lantern (as more convenient to be carried than a candle, stuck on a piece of clay, which is usually practised) a light being necessary for each person. It would also be adviseable to take with you a change of dress, as, from the dirt you are likely to contract in the mine, you may feel uncomfortable without removing, at least,

your upper cloathes. Strong boots, to keep the feet dry, and an old hat, are also necessary. Being thus prepared, proceed to the staith, which is by the river side, about four miles above Newcastle, a pleasant excursion by water. There, some of the men, who know of your coming, will assist in seating you on a set of small empty coal waggons, capable of containing two persons each, seven of which are drawn along a railway by one horse. As soon as you are placed, with your candles lighted, you set off at full speed, with a boy in the first waggon, for your charioteer, into a tunnel, or subterraneous passage six feet high, about the same breadth, and three miles in length. You will find it an advantage to have one of the men for a guide, to point out anything, that may appear striking on your passage to the pit. It is particularly necessary to guard against putting your hands suddenly out of the waggon, as the tunnel, in most places, is only wide enough to admit the waggon and horses, and you are of course by doing so in danger of receiving an injury; but if you sit undisturbed, you ascend very smoothly and pleasantly up the tunnel, on an inclined plane, till you arrive at the place where the men are working the coals. At your first entrance into the tunnel you are struck with the noise of the waggons, which, being fastened with chains to each other, and going at the rate of ten miles an hour, make a reverberatory noise resembling thunder. The passage is in general hewn out of solid rock, composed of what miners call a metal stone, a sort of schistus. Where there is no rock, it is arched with brick or stone.

The water from the pit runs down by the side of the rail-way to the river. The waggons are deep and square; wider at the top, than at the bottom, and are fast on all sides. The bottom has hinges, and can be let down to discharge the coals, of which these waggons contain about three bolls each. At intervals there are double railways; and where you come to one of these, the boy stops his horse, and a dead silence ensues, forming a striking contrast to the noise you have just heard. After calling aloud, he listens to hear if any loaden waggons are coming down, that they may there pass each other. The candle of the boy coming down appears like a star at a distance through the gloom, and has a very pleasing effect, as it gradually approaches.

When he is past, your driver renews his speed, until he reaches the next interval, when he repeats his call, and should no answer be heard in return, he proceeds. If, by the negligence of the boys, the waggons should meet, where there is no double railway, the boy with the empty waggons unlooses his horse, which is taught to turn round, and force the waggons back with its breast, until they reach the double part. The full ones having passed you, you set forward again. . . . The air up the tunnel is cold, but perfectly pure. Your approach to the workings is announced by a sensible warmness. You alight from your waggon, to view the different operations to which your guide will conduct you . . . East Kenton pit is so much more easy of access than others, from its being the only one, which you can enter by a tunnel or railway, that it has been visited by many ladies. It is usual to recompence the waggon boy, and any of the men to whom you give any trouble, with a few shillings each.

The waggonway was largely superseded in 1808 by the Kenton and Coxlodge line to Wallsend, and any traffic that still used the drift was stopped by a roof-fall in 1810. Thereafter it was kept solely as a drain.[83]

One final category of underground railways remains: those in metalliferous mines. In spite of the widespread use of the Hund on the Continent and indeed in the Mines Royal in the Lakes, most British ore-mines managed without railways of any sort until the nineteenth century. There seems to have been practically no application, for example, of the wooden railway in Cornish mines, where lodes were deep and near-vertical, allowing little scope for the levels or drifts. The only wooden track here known to the writer was in Nangiles Mine near St. Day, with rails about 3 in. square set at a gauge of some 20 in. Their date is probably early last century.[84]

The waggonway was not entirely unknown, too, in Welsh metalliferous mines. Lewis Morris reported in 1754 on the Company of Mine-Adventurers' lead-mine at Grogwynion above Aberystwyth – not far from Talybont, where a Hund may have existed in 1669 (p. 17) – that

The great level was not cut broad enough to make a waggon way so that they are obliged to wheel the ore and deads in five narrow wheel barrows at a very great expence. . . . In order to make the work compleat, the great level ought to be widened so as to make a waggon way as far as the vein at least.[85]

Another waggonway was to be found in the Llangynog lead-mine in Montgomeryshire in the 1780s.[86] A copper-mine at Llanberis, opened in 1791, boasted a wooden railway when the Rev. W. Bingley visited it in 1801:

Whilst I, and three gentlemen . . . were watching the women break the ore, a loaded waggon was brought out of the level. This gave me the opportunity of returning in an empty one to examine the interior of the work. We therefore leaped into the waggon that was to convey

> us, and with one miner before to drag, and two others behind to push us along, we entered into the narrow cavern. This was about two hundred yards in length, and in the whole way it was seldom more than six feet wide, and seven or eight feet high. In some places the rugged arch of the roof was so low, that we were under the necessity of stooping down in order to pass it. The day was one of those excessively hot ones that we frequently have about the middle of August; and the chilling damp which immediately struck us on entering, added not a little to the terrors of the place. The level lying in a direct line, we took no lights along with us. I sat with my back towards the entrance, and the perfect darkness of the place, the confused noises (from the arched and low roof) of the talking of our guides, and the rumbling of the wheels, reechoed in deep and imperfect sounds along the cavern, added to the frequent jolting of our vehicle, from the badness of the wooden rail-road, inspired me with some ideas not perhaps altogether agreeable. The length of the journey seemed more than double what it really was.[87]

A similar description in more emphatic terms of romantic horror is given of Ecton copper-mines in Staffordshire in 1769. Here an adit ran into the hill from the River Dove for about 400 yards. The first 60 yards were 4 ft. 6 in. high, the rest 6 ft. In the central chamber of the mine, the ore

> is put into four-wheeled waggons that will hold about a ton and a half each. These waggons have cast brass wheels, and are run in grooves thro' the Adit, by boys from 12 to 14 years of age, with great facility. . . . Indeed, such a horrid gloom, such rattling of waggons, noise of workmen boring of rocks under your feet, such explosions of blasting, and such a dreadful gulph to descend, present a scene of terror, that few people, who are not versed in mining, care to pass through.

The brass wheels sound unlikely, and the grooves may well be another misapprehension of the point of flanged wheels (see p. 136). It is, however, possible that we have a case of Continental influence here, for the mine was worked soon after the Restoration 'by some *German* miners, sent for over by *Prince Rupert*'.[88]

A long-abandoned wooden railway still exists in the North Bradda lead- and copper-mines near Port Erin in the Isle of Man, dating perhaps from the 1790s.[89]

Wooden railways belonging to metalliferous mines were perhaps most common in the northern Pennines, especially in the lead-mines of Allendale and Alston Moor, where Tyneside influence was strong. Late in the seventeenth century shaft-working began to be replaced by long, horizontal levels, and eventually the two largest firms, the Beaumont Co. and the London Lead Co., introduced horse railways into these levels. Barneycraig Horse Level in West Allendale, begun in 1760, was one of the earliest.[90] Gabriel Jars found his way to the nearby Coalcleugh Mine, which also belonged to the Beaumont Co., in 1765. He reported that there was in the upper level a railway just like the surface lines around Newcastle, except that the gauge was 2 instead of 4 ft., and that the waggons, two of which were pulled by one horse, were small and low. Since the London Lead Co., which was working the adjoining mine of Rampgill, 'does not have the same facilities for the transport of the waste, it pays the other company for passage at a fixed sum per waggon'.[91] John Wallis wrote of Coalcleugh in 1769: 'It is said to be the deepest mine in England, 100 fathoms, a subterraneous waggon-way, of a mile in length, leading to its *ostium*.'[92] A typical Allendale waggon measured 6 ft. by 1 ft. 6 in. at the top, narrowing to 4 ft. by 10 in. at the bottom; the body was 2 ft. 6 in. deep, the total height being 4 ft.[93]

Specimens of old wooden track are still occasionally discovered underground. Oak rails from Beaumont mines at Scraith Hole in Allendale and at Groverake, near Blanchland in the Derwent Valley (pl. 65), measure some 4 in. square and 6 ft. long, with a gauge of about 1 ft. 9 in.[94] Larch rails have been found in Calvert Mine at Tynehead only 2¼ in. by 2½ in., set at around 1-ft. 6-in. gauge, and, much further south, at Merryfield sett in Nidderdale. Here the gauge was 1 ft. 8½ in., and the rails 2¼ in. high by 3¼ in. wide by 6 ft. long. They were held down by wooden pins to extremely crude sleepers 6 or 7 in. wide, but only 1½ in. thick, set at 3-ft. centres.[95] The Scraith Hole rails may come from the eighteenth century, the rest probably from the nineteenth: the Merryfield ones belonged to a level whose arch was dated 1851. The Groverake rails had a wrought-iron-clad point frog, now lost.

Flat cast-iron plates came early to the northern Pennines. The London Lead Co., a Quaker firm, is said to have introduced them into their mines about 1769, which is not unlikely, since they had close connections with the Coalbrookdale Co. over steam engines. The first certain instance we have, however, is in 1776, when John Smeaton began to drive the famous Nent Force level for the Greenwich Hospital Commissioners. The level, which ran from near Alston eastwards to Nent Head for almost five miles, was intended both for drainage and transport. A cast-iron railway, almost certainly in the regular form of flat plates on wood, was used to remove the rock excavated in driving the level, whose minimum section was 9 ft. square. Once completed, it served in its lower half as an underground canal for barges.[96] Thomas Sopwith wrote: 'Nentforce level set the example of cast iron railways, which, in the London Lead Company's works and numerous other mines, were speedily adopted in the place of wooden rails.' To make sure the way was clear, the waggonmen practised an early form of signalling:

> When a signal is to be made to some distance, it is done by beating on the rails or posts, five beats, the first two slow, the other three quick, and this is repeated several times. The same signal is used in the Newcastle coal-mines, where it is denominated 'jowling'.[97]

In fine, the underground railways of the larger metalliferous mines partook of characteristics of both Shropshire footrids and North Country rolley-ways, although in the close confines of mine galleries there was no room for any wide variety of design, either in waggon or in track.

18

English railways on the Continent

C'est une heureuse imitation du mode de roulage qui est usité avec le plus grand succès en Angleterre.

A. M. Héron de Villefosse, *De la Richesse Minérale*, Paris, ii (1819), 535, on the Königsgrube railway in Silesia.

It was not until the last quarter of the eighteenth century that the English waggonway was first exported to the continent of Europe. In contrast to the anonymity of most previous railway developments on the Continent, enough evidence survives for us to identify the major figures responsible for bringing the idea over from England, and how and where they applied it.

Those who introduced the English railway to Hanover, Prussia and France formed a complicated group of friends and relations. The railway came to Prussia as an indirect result of the establishment by Frederick the Great of a Department of Mines and Ironworks, to make the best use of the resources of the lands that Prussia was acquiring in her rapid expansion. Friedrich Anton von Heinitz, who was promoted head of this department in 1777, proceeded to set Prussian mining on its feet. A man of considerable experience himself – he had been Captain in the mines of the Upper Harz, had travelled widely (England included) and had founded the Freiberg Mining Academy – he also appointed exceptionally able men to take charge of the various local offices, notably Reden in Silesia and Stein in the Ruhr.[1]

Friedrich Wilhelm Graf von Reden (1752-1814), a nephew of Heinitz and a Hanoverian by birth, was trained in the Hanoverian mines of the Upper Harz by another uncle, Claus Friedrich Freiherr von Reden, who was Mine Captain and Engineer at Clausthal from 1769 to 1791. After a long tour of mining areas of the Continent and of England, the young Reden joined the Prussian service, and was made head of the Silesian office in 1779. Here he completely reformed the industry. He imported miners from the Harz and Mansfeld, reopened Tarnowitz lead-mine for the first time since the sixteenth century, and here and at Waldenburg introduced English methods of timbering, pumping and transport – the English waggonway. In 1786 he was visited by Samuel Homfray of Penydarran Ironworks in Glamorgan, from whom he acquired a steam pumping engine for Tarnowitz. Three years later he invited over another iron-master, William Wilkinson, who had already built the first two English waggonways in France. Wilkinson (1744?-1808), the brother and overseas agent of the more famous John of Willey, Bradley and Bersham Works, was always overshadowed, and often shabbily treated, by his domineering brother, who was renowned for his personality, ambition and lack of scruple.[2] The purpose of William's visit to Silesia was to introduce the smelting of lead with coke at Tarnowitz, but he struck up a friendship with Reden, and the two returned to England together in 1789, along with Fritz, the son of Reden's Clausthal uncle. One concrete result of this trip for Reden was that he acquired another steam engine, built by John Wilkinson at Bersham in defiance of Watt's patents. On his return to Silesia, Reden began to open up a new colliery at Königsgrube, near Gleiwitz. It was in full swing in 1795, and used English railways. He also persuaded one John Baildon, an iron expert from the Carron Works in Scotland, to come out and build a new furnace for him at Gleiwitz, which was blown in in 1796, and another at Königshütte, where the first blast was begun in 1802. An English waggonway linked this furnace with Königsgrube coal-mines. Reden succeeded his uncle Heinitz as head of the Department of Mines in 1802, and died in 1814.[3]

Heinrich Friedrich Karl Freiherr vom und zum Stein (1757-1831), a brilliant mining expert and administrator, and a great personal friend of Reden besides, entered the Prussian mining service in 1780 after the usual training and travels. Only four years later he was posted by Heinitz to the County of Mark, to superintend the Ruhr collieries. His junior colleague, F. A. A. Eversmann, was sent to England in 1784-5, and returned with plans (approved by Stein and Heinitz and soon fulfilled) to build the first English railway in the Ruhr. The years 1786 and 1787 saw Stein himself in England, with two companions. One was a Reden – almost certainly the Clausthal one; the other was Friedrich (his Christian names are not recorded), a young engineer in Stein's own service and the son of a Clausthal Mine Captain and magistrate, Johann Konrad Friedrich. One of the two Friedrichs is credited with building the first English railway on the Continent, at Clausthal in the 1770s. On his return to the Ruhr, Stein was probably visited in 1788 by Samuel Homfray. Stein transformed the organization of the Ruhr collieries, but, although he revolutionized their transport by road, he seems to have sponsored only one English railway, and he introduced little new mechanization: the first steam engine there did not arrive until 1798. He was promoted in 1796 to the Presidency of the Westphalian Chambers, and thereafter rose to become a Minister of State. He was largely responsible for laying the foundations of Prussia's subsequent greatness in both commercial and military spheres.[4]

Heinitz, Stein and Graf von Reden all saw that England provided the best available model, and the best machines, for industrializing Prussia; Claus von Reden doubtless had the same objects for Hanoverian Harz, which was, of course, under the same crown as England. The reason for their frequent visits to Britain, and for their inviting British experts to the Continent, was largely the desire to learn all that they could of our mining and iron-making techniques, and particularly to penetrate the jealously guarded secrets of Watt's steam engines. The list, no doubt incomplete, of the journeys they and their agents made to this country is impressive: Graf von Reden in 1776 and perhaps 1779; Heinitz in 1779; Claus von Reden in 1782; Eversmann in 1784-5; Stein, the younger Friedrich and probably Claus von Reden in 1786-7; Friedrich perhaps in 1788; Graf and Fritz von Reden in 1789. One gets the impression that Boulton and Watt's Soho Works were under almost constant surveillance by stealthy foreigners (as indeed they were by stealthy Englishmen). It was industrial espionage pure and simple, for the Germans did not scruple to bribe engine-men, or incapacitate them with drink, in order to catch a glimpse of their machinery; they enticed skilled workmen to Germany with the promise of better pay; they tried to insinuate themselves into the Soho Works under false names and derisory disguises. Boulton wrote to Watt in 1779, 'I doubt not but you will be properly cautious with these Prussian Pimps'; but the Pimps in question did meet with some success. Friedrich, for example, returned to Stein in 1788 with plans of a fire engine and the Birmingham Canal; Baildon was inveigled to Silesia in 1794 to build furnaces for Reden. But on the whole the British, masters and workmen alike, resisted their offers and blandishments; except, that is, the Wilkinsons, who had a reputation for unpatriotic conduct in supplying foreign powers with information, machinery and even armaments. Reden got from them a good knowledge of smelting techniques and a pirated Watt engine,[5] as well as a similar engine from Homfray. It was partly these visits to England, and partly the visits by Englishmen to Germany, that resulted in the introduction of the English railway. Whatever the Germans discovered (or failed to) about the steam engine, the waggonways of England were no secret, and lay there openly for the copying.

The first such railway in Germany is sometimes supposed to have been a German invention. John Francis tells a story, from an unspecified

German source,[6] of the late engineer, Mr. Friederichs [*sic*], son of a Harz miner, who toured the mines of Europe on a grant from the Hanoverian Government. What he saw prompted him to invent a new means of conveyance for ore, consisting of a locomotive, a waggon, and iron rails, an idea ultimately taken to England, where it was widely advocated by Thomas Gray. Friedrich Wilhelm von Reden, grandson of Claus von Reden, gives a shorter but similar account,[7] supported, he says, by the diaries of his grandfather and the Silesian von Reden. According to him, Engineer Friedrich, as a young man in the 1770s, invented and built an iron railway with waggons from the Dorothea Mine at Clausthal to the stamps, and that this type of waggon was exported to England in 1810. What Francis' source says about the invention of the locomotive, however, is demonstrably false; and Thomas Gray, the author of *Observations on a General Iron Rail-way* (1820), which urged the nation-wide use of railways and locomotives, was very well aware of the origin of both. But a few basic truths remain. It does seem that the first railway of this type on the Continent was at Clausthal, and was built by Friedrich. But track and waggons cry their English ancestry aloud, and we may be sure that the inspiration came, directly or indirectly, from this country. Whether the builder was the elder or the younger Friedrich is not clear. The father, having been magistrate since 1755, could perhaps hardly be called 'young' in the 1770s; on the other hand, the son, when in England with Stein in 1787, could still be described as a

> young man . . . an excellent engineer & Modeller . . . the son of a very worthy comptrolling Captain. . . . He has been brought up to the practical parts of mining from his infancy; & has gone thro' all its different branches from the Stamps up to the function of a Carpenter, Engine Man, Captain & Surveyor.[8]

No source dates the Clausthal railway more closely than the 1770s until Ržiha, writing in 1867, acknowledges its English origin and dates it to 1775.[9] It may be so, although Ržiha is not remarkable for his historical accuracy. One might guess that Friedrich von Reden, on his return from England to the Harz in 1776, described the railways he had just seen to Uncle Claus, and to his colleague, Johann Friedrich, or his son, who thereupon set about copying them to replace the existing pack-horse transport. Alternatively, Claus von Reden, or Friedrich, may have undertaken a journey themselves to compare the mining techniques of their sister-kingdom with their own.

It may well be that this railway was at first a purely wooden one, for the Upper Harz Mines Archives record that difficulties arose in 1806 in the course of experiments to strengthen the wooden rails with iron bars. At any event, success had been achieved by 1807. The first detailed description and drawings of this Clausthal railway were published in 1819[10] (fig. 52). It was a cast-iron *voie angloise*, carrying silver ore from the Dorothea Mine adit to the washery and stamps about 700 yards away. The waggon, built of wood, held about 1 cu. yard and ran on four cast-iron flanged wheels each some 1 ft. 8 in. in diameter. The wheels were slightly staggered, for each had its own tapering axle, an idea designed to reduce friction and ease the passage of curves. The waggon ran down a gentle gradient by gravity, with a waggon-man perched on the small platform at the back with the brake, bearing on one front wheel, easy to hand. On reaching the

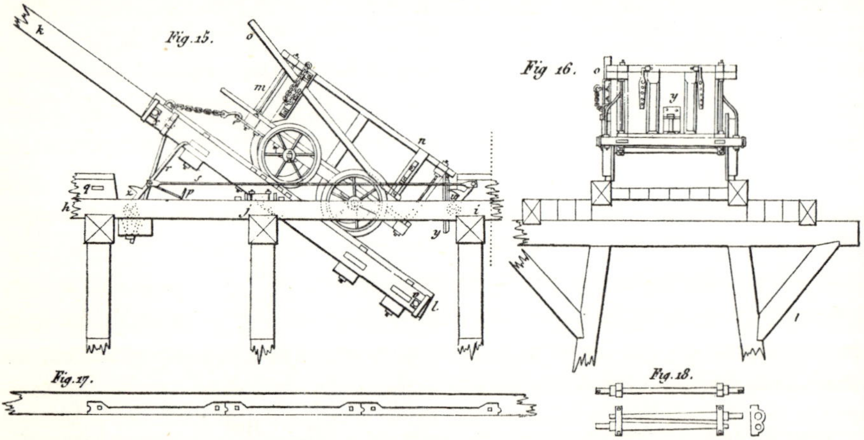

FIG 52 English waggon, tippler and cast-iron rail, Clausthal, 1819.

washery, the waggon was chained to a tipping section of track, which was released by a lever and a pair of catches, tilting the waggon forwards so that the contents shot out through its hinged front. The waggoner then brought the platform back to the level by pulling down a counter-balance beam at the back, and pushed his waggon back up the slope. The track consisted of wooden rails about 7 in. wide and 9 in. high, set at a gauge of some 3 ft. 4 in.; on them were fixed cast-iron bars 4 ft. long and about $2\frac{1}{2}$ in. by $1\frac{1}{2}$ in. in section. These rails were held down by pegs through holes cast in side-swellings at each end, and were further prevented from shifting by simple morticed joints. The whole railway, like almost all the early English ones on the Continent, is redolent of Shropshire practice rather than Tyneside.

This Clausthal line is also distinguished for the first royal train journey on record. When the Harz was briefly under French domination as part of the kingdom of Westphalia, its puppet King, Jérôme Bonaparte,

Napoleon's youngest brother, and his Queen Katharina visited the Dorothea Mine on 5 August 1811, along with twenty-one courtiers and ministers and a great many more of lesser degree. The order of ceremony read:

> Travel on the iron way to Dorothea ore-washery. Inspect the ore-washery at work. The retinue then proceeds *via* the Elisabeth crushers back to the iron way. Their Majesties travel on the Hunds back to the Mine Office.[11]

Francis' anonymous German source describes the same occasion:

> The director of mines caused a carriage of the invention of Mr Friederichs to be fitted up for an excursion; the king, attended by ten gentlemen of his court, mounted the first vehicle, and the queen, attended by ladies of her suite, stepped into the second carriage, which closely followed the first. They started at the same time, and the distance which usually occupied forty-five minutes took only five.[12]

Three-quarters of an hour seems an unduly pessimistic estimate, even for a royal retinue, to cover 700 yards.

Similar railways soon appeared elsewhere in the Harz, nearly always described as 'English'. Another Clausthal mine, Alter Segen, had a line two-thirds of a mile long, with similar waggons to Dorothea's, except that here the front wheels were twice the diameter of the rear ones, and were set nearly under the centre of gravity. They were tipped by lifting the back and pivoting the body on the front axle, and were loaded by automatic hoppers.[13] A truck of this sort, with manual loading, is illustrated in pl. 68. There was another line at Zellerfeld, 1,500 yards long,[14] and yet another from 200 yards inside the Adolph adit at Wildenmann, running down at 1 in 100 for a total of 1,290 yards to the stamps. It was very like that at Clausthal: the sleepers were 7 ft. long, with oak or pine rails 7 to 8 in. square at 3-ft. 6-in. gauge, and cast-iron bars varying in length from 3 to 6 ft. – a typical 4-ft. one weighed 45 lb. The track was expensive, about 3½ reichsthaler, or 17*s.* 6*d.*, per yard. The waggons held about 22 cwt., ran down by gravity, and were pushed up by two men, who worked eight trucks in an eight-hour shift.[15] This type of track, though gradually ousted by all-iron rails without the longitudinal baulks, survived in places in the Harz well into the 1860s.[16]

As soon as Stein was appointed to the Mining Office in the Ruhr, he put his finger on one of the weakest points of the colliery management there. He wrote to Heinitz in June 1784 bemoaning the haphazard supply of coal to the shipping-places on the river, and suggesting that several coal stores be built, which could hold a good stock and so balance out the vagaries of supply.

Nothing could more assist the situation than the improvement of the

> barrow-ways [*Schiebewege*]. The supply of coal to the stock-piles suffers from all their bad qualities, just like the whole economy of the mines, which are never worked regularly, but sometimes exploited with zeal, sometimes entirely abandoned. To prevent the unhappy consequences of this arrangement, our only solution is to improve the roads. . . .[17]

Heinitz's answer was to send Stein's junior colleague, F. A. A. Eversmann, on a protracted eighteen-month tour of Britain, to acquaint himself, not only with transport methods, but other mining techniques as well, including, of course, Watt's recent improvements to the steam engine.[18] On his return at the end of 1785, Eversmann waxed enthusiastic about that boon and blessing to any colliery, the English waggonway. So he and Stein wrote a report, sent to Heinitz in February 1786, recommending that an English coal-way (*englischer Kohlenweg*) be built from the Dahlhausen Colliery to the Ruhr.[19] Heinitz gave the go-ahead, and by June Stein could tell him:

> Eversmann's projected coal-way is coming along like a house on fire – they almost are ready for laying the iron, though it can't be done yet because of shortage of water in the Ruhr, which is holding up its transport.[20]

The railway was no doubt finished when the water in the Ruhr allowed traffic to move again. Its original iron rails would certainly have been either flat cast-iron bars as at Coalbrookdale, or thin wrought-iron strips as used elsewhere in Britain. The first detailed account of the line dates from 1823, when it was 1,700 yards long.[21] The track was supported over the soggy ground by a continuous stone wall, 20 in. wide, under the sleepers. Both sleepers and rails were oak and 6 in. square, with the rails sunk $1\frac{1}{2}$ in. into the sleepers. The gauge was 3 ft., and the space between the rails was filled with stone and gravel to form a horse-path. On top of the wooden rails, whose life was some ten or twelve years, were nailed the straps – by now, if not before, made of wrought iron – 2 in. wide by $\frac{3}{8}$ to $\frac{1}{2}$ in. deep and weighing $13\frac{1}{2}$ lb. per yard.

The early date of this line disposes of the claim[22] that the first railway in the Ruhr was that at Hattingen, which was apparently built by Heintzmann soon after 1802.[23] A description of 1830[24] says that it was some 2,000 yards long, of very similar construction to the Dahlhausen one, although the gauge was only 1 ft. 10 in. and the materials were correspondingly lighter – the wrought-iron plates measured 1 in. by $\frac{5}{12}$ in. The waggons (still called Hunds) were small, with bodies 1 ft. 6 in. by 1 ft. 7 in. by 1 ft. 8 in. They weighed 416 lb., carried a load of about 900 lb., and ran on English flanged wheels 1 ft. 3 in. in diameter. One horse could pull twenty loaded waggons downhill to the river in twenty-nine minutes, and the same number of empties back to the mine in twenty-four minutes. The gauges in the Ruhr tended to be narrow, and the waggons small, because, as in Shropshire, the railways often began deep underground, continuing straight out of the adits and down to the river without interruption. The Schwarz Adler Colliery had such a line in 1835, 600 yards of which were underground and 1,100 yards on the surface; a putter was

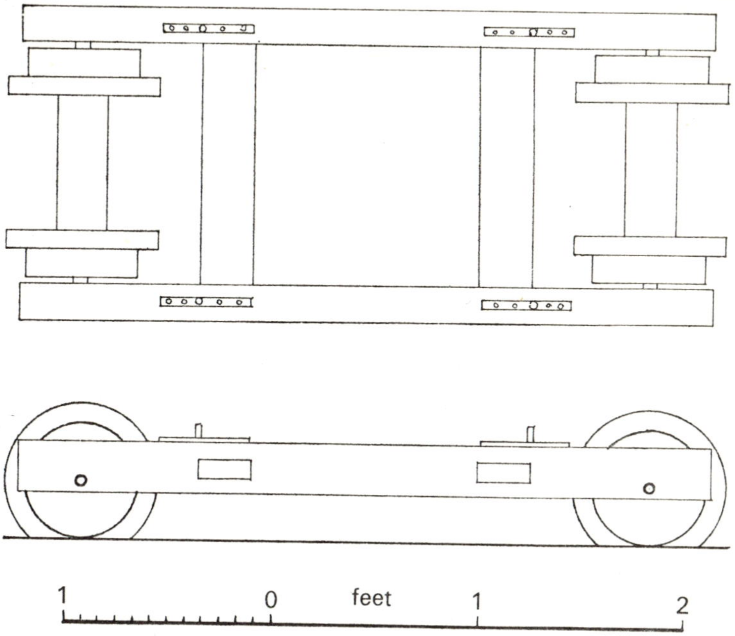

FIG 53 English waggon used underground in the Ruhr, 1823.

expected to push seven ½-ton waggons, which meant covering nearly fourteen miles, per shift.[25] On purely underground lines, apart from the flanged rails already noticed (p. 57), English rails were also used. The trucks often had detachable bodies, which were hoisted up the shafts, and their wheels were of varying degrees of sophistication: of wood, turned in a single piece or in two pieces, with iron tyres; or with wooden tread and wrought-iron flange; or all of cast iron (fig. 53). The underground track usually had no iron laid on it, but merely an extra strip of beech, as in the English double way; its size varied according to the gauge and load. Sleepers ranged in section from 3½ by 1½ in. to 6 in. square, rails from 2 to 6 in. square, with top strips 1½ in. thick. The gauge might often be less than 1 ft., but was occasionally as large as 2 ft. 8½ in.[26] Horses had sometimes replaced manpower underground as early as 1790,[27] and usually did on the surface, except on the shortest routes. Hand-pushed 'Hunds' survived on the old Louisenbahn, near Dortmund, for example, as late as 1847, when it was in a decidedly decrepit state. It was a tortuous line of 1-ft. 8-in. gauge, 900 yards long and with several passing loops.[28]

In the 1820s and very early 1830s, nearly twenty railways came into existence from various collieries to the Ruhr or to main roads (fig. 54), the

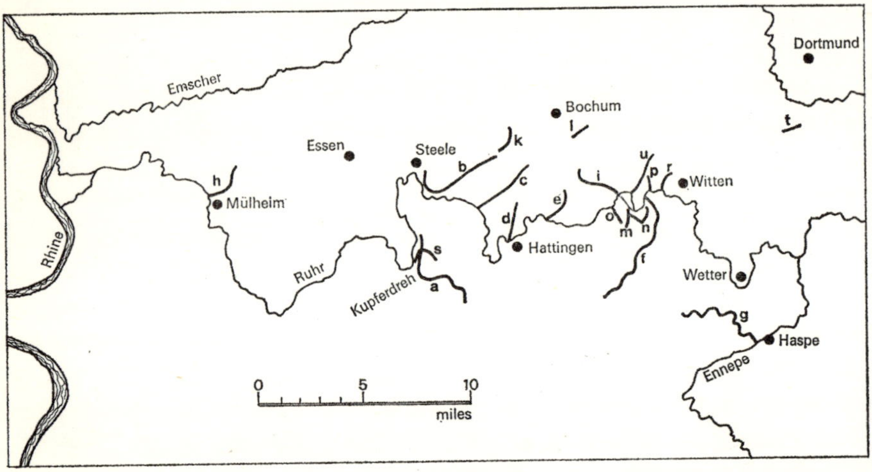

FIG 54 Early wooden railways in the Ruhr. *a*, Prinz Wilhelm (Deilbacher); *b*, Ruhr-Wilhelms; *c*, Dahlhauser; *d*, Heintzmann'sche (Hattingen); *e*, Brockhauser; *f*, Hardensteiner (Muttental); *g*, Schlebuscher; *h*, Sellerbecker; *i*, Ruhr-Neuemissgunst; *k*, Eppendorfer; *l*, Altenbochumer; *m*, Orion; *n*, St. Johannes; *o*, Helena; *p*, Hammerbank; *r*, Neusteinkuhl; *s*, Hinsbecher; *t*, Louisenbahn; *u*, Urbanus.

longest being the Deilbach Railway at Kupferdreh, with five miles on the surface and two and a half underground. It was built by the first German railway company (promoted by Friedrich Harkort in 1826), was begun in 1830, and, after a visit by the Prince in 1831, was renamed the Prinz Wilhelm Railway. A detailed description was published just before it was converted to an ordinary standard-gauge line in 1846-7. Although locomotives were originally envisaged, it was horse-worked. The gauge was only 1 ft. 11 in., but the trackwork was heavy. The rails were 3 in. wide and 5 in. thick, sunk $\frac{1}{2}$ in. into the cross-sleepers, and fastened with wooden pins. The sleepers, at 2-ft. 9-in. intervals, were 5 in. thick, 4 ft. 6 in. long and 7 in. wide under the joints of the rails (5 in. wide elsewhere), and rested at each end on stone blocks. Wrought-iron running rails, 1 in. by $\frac{1}{2}$ in., were fixed to the wooden ones with countersunk screws. The horse-path was filled with gravel to $1\frac{1}{2}$ in. above the top of the wooden rails, but two rows of stone slabs, set on edge, kept the gravel clear of the flangeways. The waggons, with strapped wooden bodies 6 ft. long and 2 ft. 8 in. high, weighed 4 cwt. and held about 10 cwt. They ran on 15-in. iron wheels with a 2-ft. wheelbase. The waggons were made up into trains – one horse could pull 6 tons along the level – coupled together by three-link chains. Every mile and a half or so was a passing loop, where the horse was changed, each keeping to its own section. As an extra luxury, the outer rails were superelevated on curves. Experiments had been made with cast-iron fishbelly rails, which, although satisfactory, proved too expensive.[29]

Harkort had built several short railways during the 1820s, and in 1826 experimented with a mono-railway on the Palmer system (see p. 305) at Wetter.[30] Another line, likewise built by him between 1829 and 1832, was the Schlebusch Railway, running nearly five miles down to the Ennepe (a tributary of the Ruhr) at Haspe. The gauge and track were identical to the Prinz Wilhelm Railway's, but the engineering was much heavier, including a 30-ft.-high embankment and bridges 400 ft. long and up to 60 ft. high. The waggons came in two sizes, 6 cwt. and 12 cwt., and were sometimes made of plate iron. The cast-iron wheels were usually under the body, with a small wheelbase for curves, and had six spokes each and a flange 1 in. deep. There were dumb buffers and chain couplings. Trains ran with twenty-five small waggons, or the equivalent in large ones, headed by two horses. Only two trains ran each way per day, leaving the mine at 4 or 5 a.m. and at midday, but the rate of progress was slow – it took some five hours to complete the round trip.[31]

Until well into the 1830s, new colliery railways in the Ruhr were built with the typical track of wooden rails protected with thin wrought-iron strips, and only gradually did all-iron rails begin to oust them. Although the 'German' flanged rail was not uncommon underground, it was only once applied here to a surface line.[32]

In Silesia Reden's improvements did not, as far as we know, include railways until late in his career there. Before 1800 he equipped a blast furnace at Falkenberg with an inclined plane to feed it with coke and ore. A water-wheel provided the power, the gradient was about 1 in 3, and the wooden track was double, with a common centre rail and a gauge of roughly 1 ft. 8 in. The long, flat waggons, on cast-iron flanged wheels, were piled with baskets of fuel.[33] In 1803, as we have seen, the colliery of Waldenburg was given hand-pushed, container-carrying trucks to work on flanged rails between the face and the underground canal. They displaced, or partly displaced, an earlier system of horse-operated waggons running on a *chemin de fer*, probably similar to the Harz, Ruhr and Königsgrube lines.[34]

One feels that the changeover was not to the best advantage of the mine. At Tarnowitz lead-mine, the English railway did not appear until after Reden's time. It was only in 1810 that the wheelbarrows in the main levels and the Mansfeld Schlepphunds in the minor ones were replaced by English waggons on wooden rails and by Hungarian Hunds respectively.[35]

Reden's most famous railways, however, were at the collieries and furnaces near Beuthen. The mine of Königsgrube was in production by 1795, and very probably used underground lines from the start: they were certainly there in 1802, when horses were first introduced into the galleries.[36] Our fullest description of the workings is given by Héron de Villefosse, who had had charge of the Ruhr collieries during the French occupation, and who published his inestimable *De la Richesse Minérale* in 1819.[37] His illustrations of Königsgrube, however, are based on drawings made fifteen years or more earlier.[38] The underground waggons, he says, were of two basic sizes. One had two axles, held three iron-strapped wooden containers each of $1\frac{1}{2}$-bushel capacity, and was either pushed by two men or pulled by a horse. The other, more usual, type had four axles (as at Clausthal), held three containers each of $4\frac{1}{2}$ bushels or two of $7\frac{1}{2}$, and was always horse-drawn (pls. 69 and 70). One horse worked the coal from three separate faces to the shaft, each distance being about 200 yards, and each return journey taking twelve and a half minutes. It would lead an empty from the shaft to the first face, exchange it for a loaded waggon (leaving the empty in a short spur to be loaded), take the full waggon to the shaft, wait while the containers were replaced by empties, lead the waggon to the second face, and so on in rotation. One horse would manage forty-three round trips in nine hours, an unusually cheap and efficient organization. The track was made up of half-log pine sleepers every 3 ft., fir rails some 4 in. square pegged down by oak nails boiled in oil, and sometimes (though by no means always) with cast-iron bars on top. The gauge was about 3 ft. 4 in. Villefosse also gives us one of the very few drawings of pointwork on a wooden railway (pl. 71), with iron plating on the short pivoted switch-rails, on one stock-rail, and on the crossing. He also indicates the advantages of the curved point over the more primitive straight one, in that it allows the departing line to curve away from the main line so much more quickly.

A container took twelve minutes to wind up the shaft and to empty. On reaching the top, it was set on a small tip-waggon frame (pl. 69), held upright by chains, and pushed by hand along one of four lines on a staging, until it was over a hole in the floor. The chain was released, the container tipped over on its trunnions, and emptied itself into the waggon standing below. These large waggons also ran on four axles, each with its own wheel of about 2-ft. 2-in. diameter (pl. 72). When one was full – they held 50 cwt., and only fourteen were normally filled in nine hours – it was pulled by a horse down the gentle gradient of 1 in 180 to the coking ovens of Königshütte, some 830 yards away, where it was emptied through a door in its side. The track was similar to that underground, only larger. The sleepers were at 4-ft. bearings; the rail was 5 in. by 6 in.; the cast-iron bars were 4 ft. 2 in. long by about 3 in. square, with mortised ends and ears for nailing down; and the gauge was 4 ft. 2 in. The points, except that they were fully iron-clad, were similar to those underground.

Whether or not the underground railways had appeared before the end of the eighteenth century, this first surface line, again copied directly from English practice, was opened in 1802, the same year that the Königshütte furnaces, which it fed, were blown in.[39] The shaft, however, which Villefosse describes (the Einsiedelschacht) was but one of several, each of which had its own similar underground system. Some of them also had separate surface railways, which varied in gauge and in waggon-size: it is recorded of one, for example, that the gauge was 3 ft. 7 in., and that one horse in ten hours could transport enough coal to keep three furnaces going for twenty-four hours. There was yet another railway below, to carry the iron from the furnaces to the fineries, which had all-cast-iron rails and a 3-ft. gauge.[40] It was probably this line that was the destination of a Blenkinsop-type rack locomotive built in 1816 by the Royal Iron Foundry in Berlin. The engine was satisfactorily tested in June on a 3-ft. gauge track at the Foundry, and was sent to Königshütte in thirteen packing cases, arriving in October. Many were the red faces when it was put together and found to be of the wrong gauge.[41] This lower line at Königshütte alone had the solid cast-iron rails that were necessary to carry the rack, and Berlin is hardly likely to have blundered over the gauge to the extent of 7 or 14 in. The trouble may well have been that Königshütte asked for an engine for the 3-ft. gauge, forgetting to say that their feet were Silesian, while Berlin assumed that they meant Prussian feet. Unfortunately, 3 ft. Silesian equalled some 3 ft. 3½ in. Prussian.

The first English railway in France* was built at Indret on the mouth of

* A drawing which purports to show a pleasure railway of 1714, built by command of Louis XIV in the park of the royal palace at Marly-le-Roi, is of doubtful authenticity.[42] In the picture, an elaborately baroque carriage (whose wheels are not visible), filled with elegant passengers, is being pushed by three flunkeys. The track, emerging from an arched pavilion on to a turntable, has, apparently, narrow rails set vertically in chairs. The drawing as we have it is of nineteenth-century date, and can be traced no further back than 1894; the track looks suspiciously modern; and no confirmation can be found in written sources. The picture may be a nineteenth-century rehash, with added details, of a genuine eighteenth-century engraving, but the evidence as it stands demands a verdict of not-proven.

the Loire as a part of a brand-new ordnance factory commissioned by the French Government. The builder, whom we have already met in Silesia, was William Wilkinson, the most obvious man to choose for designing and equipping such a works, since he was the foreign agent for his brother John, and owned jointly with him the foundry at Bersham where John's recently invented cannon-boring machines were at work. The invitation came from the Minister of Marine in March 1777. Exactly a year later, France declared war on England. John Wilkinson was often accused, whether justly or not, of supplying the French in time of war with cannon disguised as cast-iron pipes for the Paris waterworks. This reputation can hardly have been improved by the undoubted fact that his brother William was not in the least put off by the declaration of war, but proceeded to provide the enemy with a fully equipped armaments factory on the English model.

Dr. P. J. Wexler has recently assembled a most useful collection of contemporary sources which shed light on the previously obscure Indret Railway.[43] A memoir of either 1778 or 1779 says 'the *chemin ferré** which leads from the foundry to the boring-mill can be completed in a fortnight at the most'. An inventory of 15 September 1780 makes it clear that a railway is meant: 'the *chemin ferré*, for which the 7,294 ft. of cast-iron bars make 1,215 fathoms 4 ft., weighing 134 lb. per fathom, total 162,900 lb.' Three years later the cost of materials, workmanship and laying the wood and iron worked out at 53 francs per fathom. By 1800 the railway was gone, or at least rebuilt; an estimate of April in that year says:

> The old *chemins ferrés* were of a defective construction; the cast-iron bars were placed upon the oak sleepers and held by nothing but wooden pegs, being neither let into the sleepers nor supported inside or outside, with the result that the waggons pushed many bars out of place every day. These *chemins* have been rebuilt in accordance with the plan.

This makes it sound as if the rails were solely of iron, with no wood longitudinals at all. The design of Wilkinson's slightly later lines at Le Creusot, however, and other references to wood railways at Indret, go to suggest that the track was, as usual at this date, of basically wooden construction.

On the completion of the Indret Works, Wilkinson moved on in 1782 to begin the famous Le Creusot Foundry (Saône-et-Loire), where once

* Although here the term *chemin ferré* means *chemin de fer*, it is not always so. From the Middle Ages a *chemin ferré* or *voie ferrée* has also meant a metalled road, with no railway connotation at all.

again Dr. Wexler has assembled the contemporary evidence.[44] As at Indret, railways carried the cannon from foundry to boring-mill, but here there was another line, to bring coal some 800 yards from a nearby pit at Montcenis to the furnaces. The boring-mill line was probably made in 1782, the colliery line perhaps not till the furnaces were blown in in 1785. Both railways seem to have been built of wood alone until the furnaces they served had cast the flat iron rails to plate them with. Le Métherie gives this description in 1787:

> The furnace and forges lie 3 or 400 fathoms from the coal shafts. The fuel is carried from the mine to the platform [of the coking ovens] in waggons which run on *chemins de fer*. . . . In their construction, the ground is first prepared by giving it a slope of $\frac{1}{3}$ or $\frac{1}{2}$ in. per fathom. . . . On the ground thus prepared are laid sleepers at three foot intervals, to which rails are nailed with wooden pegs. These rails are laid at a distance apart equal to the gauge of the waggon. On the wooden rails are laid cast-iron bars, as the blast furnaces supply them, which reduce the friction and strengthen the track. Most of these bars have already been made with cast iron supplied by the furnaces. The waggon consists of a platform on which is fixed a body for holding the coal. The wheels are of cast iron, and each wheel has its own axle. Since the waggon runs on a perfectly regular and inclined surface, one horse, which on ordinary ground can hardly pull 10 or 12 cwt., is able to haul 50 cwt. The result is a saving of four-fifths of the cost of transport.

Several sources confirm the English origin of these railways. Ignace de Wendel (who had accompanied De Givry on his visit to Coalbrookdale in 1784) wrote of Le Creusot in 1787:

> The *chemins de fer* built at Montcenis in imitation of those in England, where some are as long as 5 or 6 leagues, seem at first glance, like fire engines, to be very expensive; but when one sees a single horse pulling five horses' load along the railways, one no longer feels any qualms about the point of such expense.

J. J. Ferber visited Le Creusot in 1788, and left a German account of the railway and some rather crude drawings[45] (pl. 73). When the coal reached the surface, he said, it was put in baskets into large 'Hunds' on a *Rollweg* called an English coal-way (*englischer Kohlenweg*), and pulled by horse to the coking plant. The waggons had cast-iron flanged wheels fixed to the axles, running on cast-iron bars (*aaa*), 'which are fastened together on wooden beams by wooden pins at *bbb*'. There were occasional passing-loops, and a weigh-house. Here the wheel flanges ran onto a platform (*B*) between the rails, transferring the weight of the waggon to the point *e*. The weigh-bar, pivoted at *g*, was balanced by weights at *h*. This machine, it was claimed, was accurate to ½ lb.

In the Revolution, Le Creusot fell into incompetent hands and went rapidly downhill. A visitor in May 1796, who had previously seen the works in all their glory, lamented: 'I saw that they were almost completely ruined, those beautiful railways whose construction I had once admired, and that the iron bars which had adorned them were either displaced or removed.' The colliery line probably disappeared before 1800, but the boring-mill railway survived. An inventory made in 1813, when the works were sold, merely says: 'The cannon are taken to the boring-mill on trucks called English waggons, which run on a *chemin ferré* starting within reach of the foundry cranes, and are set, by means of turntables, under the boring-mill cranes.' The works were re-equipped with more up-to-date iron railways after 1828.

Wilkinson's source of inspiration for the railways he built at Indret and Le Creusot may have been his own Bersham Works, or his brother's Willey Furnace, though we do not know if the railways at either works were yet fitted out with cast-iron strips. But, even if they were not, he would quite certainly have known all about the railways at Coalbrookdale, the *fons et origo* of the type. The gauge of the Indret line is not known at all; that of Le Creusot is not known for sure. Lunier, writing in 1805 of railways in general and Le Creusot in particular,[46] says that in building railways 'a track 4 ft. 2 in. wide is left between the parallel bars'. This is in fact the gauge recommended by Benjamin Outram, but Lunier may be referring to the Le Creusot line. If so, it is interesting to note that the waggons here bore a distinct resemblance to those on the Königsgrube-Königshütte Railway of 1802, in that both had four axles, a 50-cwt. capacity, and apparently the same gauge. That Wilkinson had considerable influence in Silesia is certain, and this similarity in the railways suggests that some credit for the design of the Königshütte waggons is due to him as well as to Reden. But any close comparison between waggons in different parts of the Continent and between Continental and English ones is always bedevilled by the different standards of measurement in use. The Königshütte gauge of 4 ft. 2 in. is on the Silesian standard, equalling about 3 ft. 11 in. English; Lunier's gauge of 4 ft. 2 in., assuming it refers to Le Creusot and assuming it is on the old French standard (already being superseded by the metric system), equals about 4 ft. 1½ in. English. In the same way, it may be significant that both the Clausthal and the underground Königsgrube lines had gauges of 3 ft. 4 in. or slightly more, measured on the

Brunswick and Silesian scales respectively. Both would equal roughly 3 ft. $1\frac{1}{2}$ in. English. We do not, unfortunately, know the gauge of the Coalbrookdale system with any certainty (see p. 267), but the most likely figure is 3 ft. 5 in. To make bad worse, this 3 ft. 5 in., though probably English, might be Bavarian (about 3 ft. $2\frac{1}{2}$ in. English).

The first French public railway, the St. Etienne and Andrézieux, was begun in 1824 with cast-iron fishbelly rails, and its continuation, the St. Etienne and Loire, in 1828 with wrought iron.[47] A few later mining and constructional lines, however, were given wooden rails plated with wrought iron. From 1832 Faymoreau Colliery (Vendée), besides a wooden flanged-rail track (see p. 61), also used oak edge rails underground, measuring $2\frac{1}{2}$ in. by $3\frac{1}{4}$ in., set at 2-ft. $4\frac{3}{8}$-in. gauge. Thin iron plates $1\frac{1}{2}$ in. by $\frac{1}{8}$ in. were held on top by countersunk nails. Two men pushed a waggon holding 9 cwt. of coal along a 350-yard line to the adit mouth, where they emptied it through an end door.[48] St. Etienne (Loire) Colliery used a similar form of track before 1829.[49] So did a constructional line built beside the Burgundy Canal before 1833.[50] Perhaps the best-known French railway of this type, however, was laid in 1830 to carry blocks of stone for the building of Cherbourg Harbour[51] (fig. 55). The rails were of the now familiar pattern, 8 in. high and 6 in. wide, resting on sleepers set 2 yards apart, and wrought-iron bars 2 in. by $1\frac{1}{4}$ in. were nailed on. The cost, it was said, was half that of all-iron rails in chairs on stone blocks. There were, however, several peculiar features at Cherbourg. Angle-blocks were fixed to either side of the rails where they lay on the sleepers, as an extra safeguard to maintain the gauge (4 ft. 11 in. outside the rails). Then, since the whole railway, 650 yards long and of double track, was on a continuous curve, the inner wheels were made nearly $\frac{1}{4}$ in. smaller in diameter than the outer. Lastly, the flanges were on the outside of the wheels. Why, it will be asked, was the flange so regularly kept inside on the English railway? The reason for outside flanges at Cherbourg was to keep the flangeways clear of the horse-track between the rails, an excellent idea in theory, but in practice, as M. Virla, the engineer, admitted, it was very different. With inside bearings, the load in a waggon tends to bend the ends of the axle up; but whereas the pressure of inside flanges against the rail on curves counteracts this tendency, with outside flanges it aggravates it. Then it is usually the flange on the outer wheel which keeps the waggon on the rail on curves, weighed down by the weight of the waggon; but if the flanges are outside, the inner wheel takes the strain, although the weight is off it. Again, if a flange rides on to the rail, it will usually run back again, other things being equal, because it is of bigger diameter than the tread; with an outside flange the opposite happens. Lastly, pressure of an outside flange on the rail tends to loosen the wheel, though with inside flanges the wheel is only pushed more tightly on to its axle. In fact, examples of outside-flanged wheels are very few and far between: on two early underground lines in Saxony, soon to be discussed; an experimental use of them on the Baltimore and Ohio in America, which soon gave them up; and a cement works railway of 1825-6 at Swanscombe in Kent. Double-flanged wheels

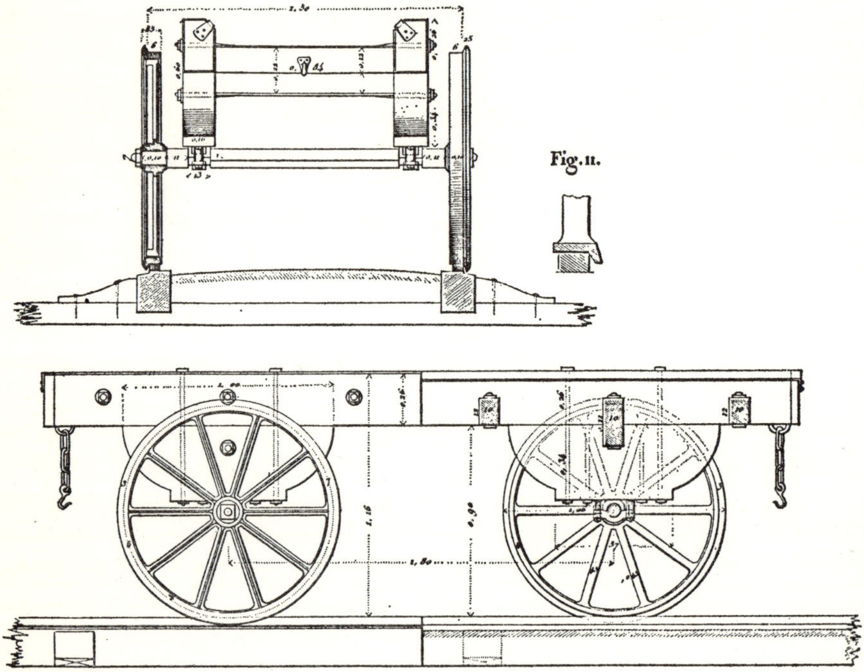

FIG 55 Outside-flanged waggon at Cherbourg, 1830.

were not so rare, once the all-iron rail became established, but they are a different kettle of fish.

The English waggonway came to Belgium in almost exactly the same way that it reached France – it was installed in a cannon foundry at Liège in 1804, though we have no details.[52] In Russia its origin was again the same, and here we know a little more about it. Catherine the Great began building the Alexandrovsk cannon and armament works at Petrozavodsk on Lake Onega in 1773-4, with A. S. Yartzov as Manager. In 1788 a short railway about 180 yards long was laid to carry the castings, as at Indret and Le Creusot and Liège, from foundry to boring-mill. The rails were of wood, possibly plated with iron.[53] Although the credit is given by Russian sources to Yartzov, it is pretty certainly due to Charles Gascoigne, the former managing partner of the Carron Ironworks, and Charles Bird or Byrd, a Carron moulding expert, who had come to Alexandrovsk in the summer of 1786 with machinery and patterns for the cannon foundry. Gascoigne had played a big part in the building up of Carron's reputation, and had been largely responsible for the design and development of the carronade (originally called the 'gasconade') on which that reputation was firmly based. He left Scotland, not in disgrace, but as a bankrupt and pursued by the hostility of some of his partners and of Matthew Boulton. A promise by the Empress of a salary of £2,000 a year was no doubt an extra enticement, and he and Bird served her so well in equipping her foundry that they ended their days as Russian citizens, Gascoigne at least enjoying considerable wealth and power.[54] The Alexandrovsk line was apparently rebuilt in the first quarter of the nineteenth century with English cast-iron flanged rails,[55] as a result of correspondence in 1804 between William Vaughan, one of the engineers of London Docks, where such plates were in use, and the Russian Minister of Marine.[56] Thereafter English edge rails apparently found only one home in Russia, on Pyotr Kuz'mich Frolov's Zmeinogorsk Railway of 1806 in the Altai, until the Russian railway system proper began with Tserepanov's short line at Nizhne-Tagilsk in 1835.[57] Both the Zmeinogorsk and Nizhne-Tagilsk railways had all-iron rails.

English influence was certainly responsible for the introduction of the flanged wheel to one more country in the eighteenth century. This was Sweden. In 1797 a new colliery was opened at Höganäs, near Helsingborg, owned by a Swedish company, but managed by a native of Newcastle, John Stafford. The following year he built a railway, just under a mile long, from the several pits to the Harbour. It began life with wooden rails and a gauge of 3 ft., although cast-iron rails of unspecified type were in use by 1818. The waggons were small – much smaller than any on Tyneside – and were pulled in trains, up to ten at a time, by horsepower. John Watson, the celebrated Newcastle viewer, was mining adviser to the colliery for a time. The line was converted in 1873 to 2-ft. 6-in. gauge for locomotives.[58] The great copper-mine of Falun received its first English-type waggons in 1809, but they ran on solid iron rails,[59] as they did on a number of other lines of slightly later date in the iron-mines of eastern Värmland.

The flanged wheel met with a mixed reception in the Saxon Erzgebirge. One underground version, in use by 1791, was manifestly inefficient[60] (pl. 15). It had a very small, coffin-like body, which held a paltry 5,000 cu. in. The wheels, 1 ft. in diameter and sometimes of wood, sometimes of cast iron, had tapering outside flanges that fitted the chamfered rails so tightly that friction was unnecessarily high. The gauge was 2 ft. 1 in.,

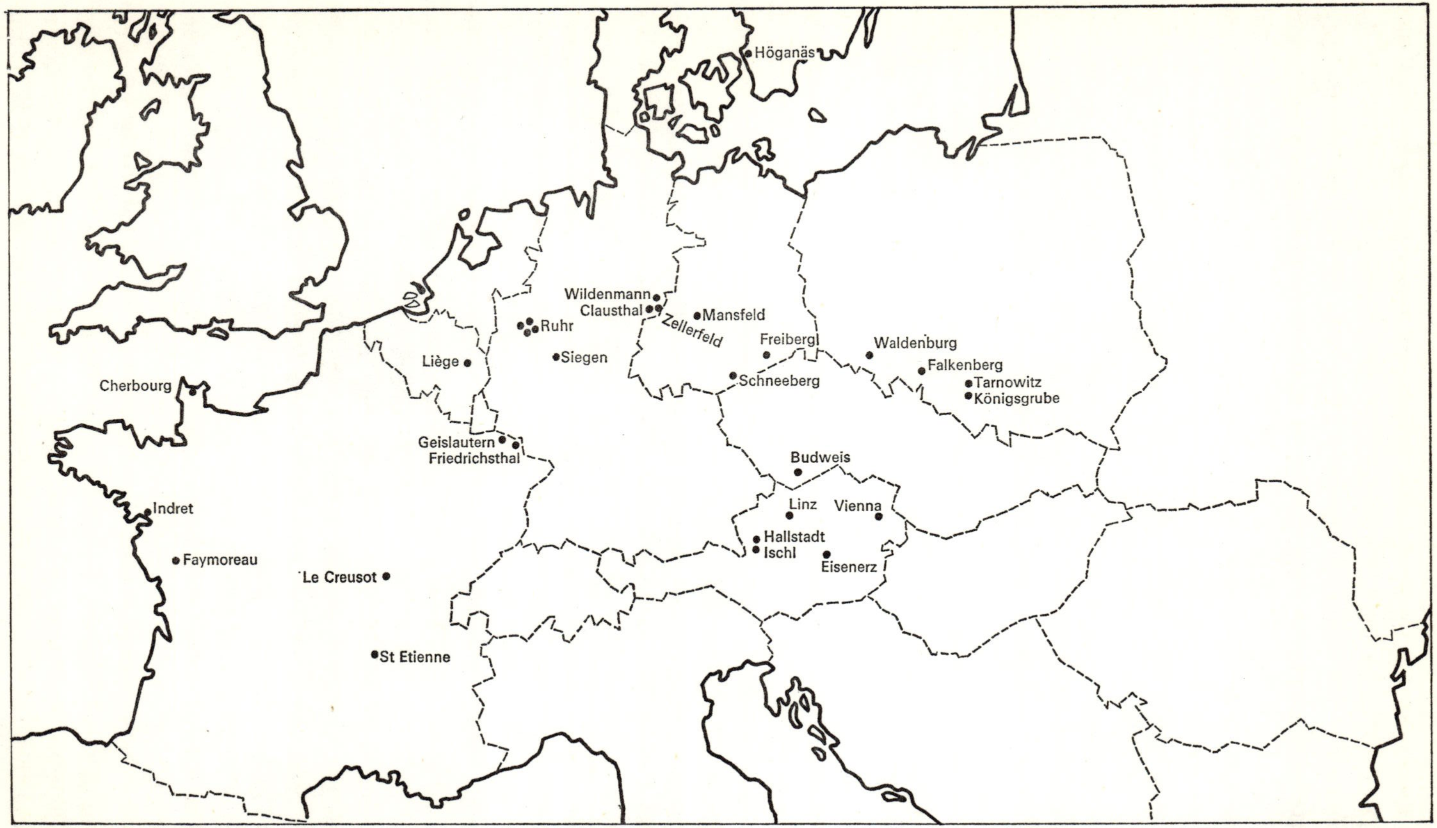

FIG 56 Distribution of English wooden railways on the Continent (add Alexandrovsk on Lake Onega).

measured outside the rails. Like Baldauf's guide-wheel Hund at Schneeberg (p. 73), the front wheels were not far in advance of the centre of gravity, and the rear axle was free of the body, being attached only by a forked arm to pivot on the front axle. The back of the body could thus be lifted to shoot the contents out of a door in the front, while all four wheels remained on the track. In spite of its light load, a boy could manage only ten or twelve trips with it in eight hours over a 60-fathom run, a performance that was bad by Freiberg standards, and quite shocking compared with Schemnitz (see Appendix).

A cobalt-mine at Schneeberg, however – the selfsame one where Herr Baldauf experimented with his guide-wheel Hunds – was early in trying out the flanged wheel too[61] (pl. 74). Once again Baldauf was responsible, along with his fellow-engineer, Mende, whom we have also met introducing cable haulage of Hunds at Johanngeorgenstadt (p. 30). They succeeded in making a far more useful machine than Baldauf's guide-wheel Hund and the flanged-wheel example just described. The only similarity, indeed, between that version and theirs lay in the outside flanges. In all else they were as different as chalk from cheese. Baldauf's waggon was large – it measured 3 ft. 7 in. long by 1 ft. 10 in. wide by 2 ft. 4 in. high inside, which is 26,488 cu. in. The hefty construction of 2-in.-thick walls, dovetailed and strapped together, made it heavy as well – 6 cwt. empty. Of the two wrought-iron axles, the rear one was bolted inside the body, but the front one, mounted on a block outside the body, was pivoted in the centre to allow a limited play on curves. The wheel-naves, running loose on the axles, had steel bearings inside, and were held on by split pins. Eight wrought-iron spokes were let into the nave and bolted at the other end to the combined flange and tread, again of wrought iron – the earliest recorded use of this material for the purpose. The tread was 4 in. wide and 18 in. in diameter; the flange, slightly tapering, was $\frac{1}{8}$ in. thick and 4 in. deep. The whole of the bottom of the waggon could be dropped to empty the load: it was held by hinges on one side, and on the other by two linked catches, released by a lever above. The waggons cost no less than 50 thaler, or some £12. The rails, of softwood 4 in. wide, 5 in. high and 12 ft. long, were nailed to cross-sleepers, and were latterly protected on top by strips of beech 3 in. by 1 in., which reduced both friction and wear on the rail proper. Curves were made either with curved rails or with short, straight sections. The putter walked along two rows of planks laid on the sleepers between the rails. The gauge (outside) was 3 ft. 1 in., which allowed a little play for the flanges. This was a surface line, built in 1785 to replace wheelbarrows on the 100-yard run from the horse-gin at the shafthead to the tips. At the tip end, where the track was mounted on a staging, a gap was left between the rails for the waste to fall through; the staging was of course extended from time to time. When the line was first built, Baldauf found that the wheels did not stand up well to the impact of the waste being tipped from a height at the shafthead, so, ever ingenious, he devised a remedy. He set a wooden platform between the rails, one end of which was hinged, the other connected to a lever with a long arm. When this lever was pressed, the platform rose up under the waggon's floor and lifted it bodily 1 in. or so off the rails. Thus elevated, the wheels were immune to the shock of loading. A similar waggon appeared in 1793 in another Schneeberg mine. This time the wheels were wood: less sensitive, one supposes, and certainly lighter and cheaper – the cost of this waggon was only two-fifths of Baldauf's. In spite of these early specimens, however, the flanged wheel was slow in catching on in the Erzgebirge. The first iron rails in the mines appeared only in 1838, in the Kurprinz at Freiberg,[62] and even then they were flat wrought-iron plates on wooden longitudinals. By 1851, however, vertical bar-rails of solid iron were in general use, with double-flanged wheels.[63]

The appearance of the English railway in the Austrian Empire was largely due to the efforts of Anton von Gerstner, who is chiefly remembered as the builder of the first public railway in Austria, the Moldau and Danube (or Budweis and Linz). This line was proposed in 1808 and again in 1813, the primary object being to carry salt away from the Salzkammergut. Gerstner visited a number of English railways in 1822 to collect ideas, and, after laying an experimental line in the Prater at Vienna, began construction from Budweis on the Moldau, southwards towards Mauthausen on the Danube. The first thirty-nine-mile section was opened in September 1827, and thus yielded priority as the earliest public railway on the Continent to the St. Etienne and Andrézieux, which first carried traffic in May 1827, though it was not officially opened until October 1828. By 1832 the whole length to Linz was complete. It was horse-worked.[64]

Gerstner had, however, had experience of building railways before the Moldau and Danube scheme was realized. The iron-mines of Eisenerz in Styria lay in the mountains high above the furnaces, which perforce occupied the river valley, and transport of ore from one to the other was for centuries a headache. The medieval method was horse and cart, an expensive process and wickedly destructive to both. A new device was introduced in 1564, a light two-wheeled barrow to carry a leather sack full of ore. Every worker, on coming off shift, trundled one down the steep and tortuous mountain path, and likewise took an empty back up when going on shift.[65] In 1567 Hans Gasteiger, a local architect and engineer, proposed to overcome these transport difficulties by means of an inclined railway. He obtained a patent from Karl II which gave him the right to

> carry the ore from the Erzberg at Eisenerz on rails and in large trucks [*Truhen*] from the mine down to the smelting and blast furnaces, and to bring back up, in the trucks returning empty, every sort of necessity for the mine, which . . . machine is of a peculiar sort that has never hitherto been seen.[66]

We cannot say what type of rail he envisaged, but the project was one of some novelty for the time, since overland railways of any length were then limited to Lower Hungary, where they had but recently been introduced. The movements of Styrian experts on forestry Riesen (see p. 69) to and from Hungary may have inspired Gasteiger's proposal. However, a proposal it remained, and for nearly 250 years the sack-and-barrow system survived. Then in 1810 Anton von Gerstner (or possibly his father) fulfilled Gasteiger's project by building a railway from the mines to the furnaces, dropping 400 m. in 4 km. Part of this line was worked by man-power at first, with horses later: the rest was presumably cable-operated. The rails were wooden, with flat iron bands on top.[67] This example was followed in 1814 when Sybold, Director of the mines, began to replace the underground Leitnagel Hunds by English waggons running on wrought-iron rails.[68]

On the Moldau and Danube, the track was of the usual type. The rails, of pine or fir, 6 in. wide, 7 in. deep and up to 36 ft. long, rested on cross-sleepers every 6 ft. The original wrought-iron strips, supplied by Bohemian forges, measured 18 ft. long by 2 in. wide by $\frac{1}{3}$ in. deep, held down by countersunk nails or screws every 2 ft. Of the £94,400 cost of the first section, £8,303 was spent on these iron strips. The gauge was 3 ft. 6 in. The road bed was normally 8 ft. wide, the track being ballasted with gravel; but on embankments the sleepers rested on a continuous dry-stone wall to prevent subsidence. In a few places, it seems, the wrought iron was replaced by cast-iron rails 3 ft. long, $1\frac{3}{4}$ in. wide and 1 in. deep, held down by a nail through each end.

The wrought-iron-plated rail appeared on other Austrian mining railways only at Hallstadt and Ischl in the Salzkammergut. English waggons were used in 1855 alongside the older Leitnagel Hunds: at Ischl could be seen the remarkable sight of Hund track laid between the English rails.[69]

Three more mining areas in Germany proper can be quickly disposed of. In the Mansfeld copper-mines, English waggons began to displace the Schlepphund (see p. 40) in 1815. The wooden rails here were 3 in. by

4 in., with a life of five or six years; the wrought iron was 3 in. by $\frac{3}{8}$ in. weighing 10 lb. per yard, with a life of ten years. The cost was 2 reichsthaler 7 groschen per yard – roughly 12*s.* The hand-pushed waggons, with side-doors, came in two sizes: 4 ft. 10 in. by 1 ft. 9 in. by 1 ft. 6 in., with 1-ft. 6-in. wheels on two axles, and 4 ft. 10 in. by 2 ft. by 1 ft. 8 in., with 2-ft. wheels on four axles. Although they cost 53 reichsthaler, or approaching £13, their superiority over the old Schlepphund was obvious (see Appendix).[70] The mines of Siegen in Nassau and Friedrichsthal in the Saar had very similar waggons and track in 1846.[71] This last line was something of an anomaly. All the other railways in the area seem to have been either plateways with flanged rails, or else, from very soon after the Saar became Prussian in 1815, to have had all-iron rails. One example of the latter was the cast-iron railway built in 1818 from Geislautern to the Saar, for which the Royal Iron Foundry at Berlin built another Blenkinsop-type locomotive. As with their engine for Königshütte, this attempt proved a complete fiasco too, for the steam pressure was insufficient to move the engine, and the boiler leaked.[72]

Joseph Ritter von Baader, a convinced advocate of railways who had spent eight years in Britain, suggested a horse railway from Nürnberg to Fürth in 1814. It did not come to pass for twenty-one years, when it was opened at last, with iron rails and locomotives, the first fully-fledged public railway in Germany.[73] Indeed, apart from the Moldau and Danube, few other public railways on the Continent ever used the iron-plated wooden rail – the Magdeburg-Leipzig line was an example[74] – and it never achieved wide favour as it did in America.

The railway first appeared in the New World in 1795 when, it is said, a short wooden line was laid on Beacon Hill at Boston to carry building materials for State House. It was followed by a few similar railways, all in the eastern states, for serving brickworks, powder mills, furnaces and the like.[75] Canada received its first railway, seven miles long, in 1818 at the Albion coal-mines near Pictou in Nova Scotia.[76] By the middle 1820s railway promotion on a larger scale was under way. Colonel John Stevens was responsible for the first locomotive in America, which ran on a circular track on his lawn, in 1825; and it is interesting that when he found the curve too tight for the flanged wheels, he replaced them with horizontal guide-wheels.[77] In the same year the Pennsylvania Society for the Promotion of Internal Improvements sent William Strickland to England to inspect the railways here,[78] a visit which resulted in the opening in 1832 of the iron-railed Portage Railway from Philadelphia.[79] The Quincy Railway, to carry granite to the Neponset near Boston, dates from 1826. Its massive track was originally formed of granite cross-sleepers at

8-ft. bearings, supporting pine rails 12 in. high by 6 in. wide, which were capped by oak strips 3 in. by 2 in.; the whole assembly was crowned by wrought-iron plates $2\frac{1}{2}$ in. wide and $\frac{3}{8}$ in. thick. The first waggon, which boasted wheels 6 ft. in diameter, carried the granite blocks slung under the axles. Before long, however, the wooden rails were replaced by granite ones, again plated with iron. Another early mineral line was the coal railway of 1827 at Mauch Chunk (Lehigh Valley) in Pennsylvania, worked by gravity and horse-power: a very normal line, with sleepers every 4 ft., 4-in. by 6-in. rails, and the usual wrought-iron straps, as they became known in America, on top.[80]

Towards the end of the 1820s larger lines, some of them for passenger traffic, began to spring up. Three examples must suffice. The Delaware and Hudson Railroad (which is, incidentally, the oldest company still operating a public railway anywhere in the world) sent one of its engineers, Horatio Allen, to England in 1828 to order locomotives. The railway, which had sixteen miles of 4-ft. 3-in. gauge track, was laid with light straps and spruce stringers or rails on sleepers set at 8-ft. 6-in. intervals. It was opened in 1829, and although the famous *Stourbridge Lion*, ordered by Allen from J. U. Rastrick, was tried in steam, none of the four engines was in fact put to regular use, as the track was thought to be too light for them.[81] The Camden and Amboy Railroad likewise used wooden track from its opening in 1830 until the prototype flat-bottom rail, rolled at Dowlais to the design of Robert Stevens (son of Colonel John Stevens), arrived the next year, with a Stephenson locomotive.[82] With the Baltimore and Ohio, whose first section was also opened in 1830, the early period of American railroads may be said to have been over. Once again wooden rails and iron straps formed the track, though the line was unusual in experimenting with outside flanges[83] (see p. 346) and with some granite rails.[84] On edge railways, stone rails are met with, it seems, only here, on the Quincy Tramway, and on some sidings of the Plymouth and Dartmoor Railway. Examples of flanged stone rails are two colliery lines in southern France (see p. 61), and the Haytor and Conisbrough tramways in England.

There is not room in this book to follow the subsequent course of American wooden railways: they deserve a volume to themselves. Suffice it to say that, long after 1830, any number of lines had to be content with iron-strapped wooden rails. Timber was plentiful and cheap, the expense of importing rails from England was huge, and iron foundries were few. Those that did exist could cope with nothing more elaborate than thin iron straps until 1844, when the first all-iron rails were rolled in America. There were problems, of course, with strap rail, especially when they tore loose from their nails and impaled passenger carriages.[85]

On the whole, however, during the first period of railroad expansion, the wooden stringer and iron strap served well enough, but as the foundries grew wood was relegated more and more to branch lines. In the second big period of expansion, during the Civil War and after it, the foundries were again unable to meet the demand, and some railways, even in the 1870s, were built of wood without any strapping at all. The cost of construction, and even of relaying in wood as the need arose, was small compared with the interest on the capital that would otherwise be tied up in iron rail. Wooden rails even now survive on some logging lines in the U.S.A., as they do on similar routes in such countries as Canada and Australia.

The origin of these Continental and American railways which used the flanged wheel and the wooden rail (whether iron-plated or not) is in many cases quite clearly visible. Either a Briton was responsible for building them, as at Indret, Le Creusot, Höganäs, Alexandrovsk, and perhaps Silesia, or a foreign engineer brought the idea back from Britain, as in the Ruhr, Austria, probably the Harz, America, and doubtless other cases. By 1800 the trickle of Continental engineers visiting this country had become almost a flood; but the principle of the English railway was probably well enough known in Europe by now to need no direct influences. Of areas which used the flanged wheel before 1800, there remain but two where the origins are unexplained. One, the Saxon Erzgebirge, where flanged-wheel trucks appeared in 1785, most probably did have an English source, even though we cannot pinpoint it. Heinitz himself, who had close connections with Saxony, and who visited England in 1779, is a possible intermediary. The other area, Transylvania and the Banat, has already been discussed (ch. 6). The flanged wheel, or, rather, flanged roller, is first recorded here in 1794, and though its origin may have been English, it was perhaps more probably an independent invention, in view of the distance and isolation of those parts. The frequency with which the flanged-wheel was termed the 'English railway' – and never 'German' or 'Hungarian' – leaves no doubt at all that the Continent was indebted for it to England, and, with the sole possible exception of Transylvania and the Banat, played no part in its early development.

Some indication has been given above of when the basically wooden railways described were replaced by all-iron ones. Both in these cases, and in mines where the all-iron English railway directly superseded the Hund without the intermediate stage of wooden English track, the pattern is similar. A very few all cast-iron lines appeared on the Continent in the 1800s and 1810s; in the 1820s rather more were built, with the emphasis now on wrought iron; from the 1830s onwards there came a positive flood. By 1860 there can have been very few Continental mines of any size that could not boast a railway with flanged wheels and all-iron rails, although, naturally, the old tended to survive alongside the new, whether Leitnagel or Hungarian Hund or old-fashioned English rail.

The reasons for the final adoption of the English waggon are not far to seek. On the surface, its superiority over any cart, road waggon, or pack-horse was quite obvious. Overland or underground, it easily overcame the guide-wheel truck, which in any case was firmly established in only two places, because in its best form it produced less friction and had fewer wheels to come to grief. As in Britain, the engineers of the Continent often succumbed to the ostensible charms of the flanged rail, but, again as in Britain, the all-iron edge railway finally won the day by virtue of its stronger section and of the smaller surface on which dirt could lodge and so impede the wheels. The Leitnagel Hund long resisted change, but was eventually vanquished by its own small capacity and the high friction engendered by its small wheels and guiding-pin.

The struggle between Hungarian Hund and English waggon, even more protracted, was in effect the struggle between man and machine. In the end, as throughout the history of railways, mining and indeed of all human activity, the machine inevitably won. Becker, the advocate of the Hungarian Hund, waxes eloquent on the subject.[86] Where there is any form of mechanical guidance, he says, the truck is the master of the man; there is no room for skill, and the putter's strength is wasted in overcoming the friction. 'Why should the waggon pick its own way, and the putter perform the duty of a horse?' Is it not obvious, he argues, that the putter is the most economical way of steering the truck? Every truck must have a putter, and a good putter is quite capable of steering. There is thus no friction from mechanical guidance, and the putter can therefore work faster. Artificial guidance involves the cost of double guidance, with very much reduced efficiency. Becker counts the guide-wheel truck one grade worse than the Leitnagel Hund, and the English waggon one grade worse still. While this last stricture, though over-severe, is partly true of the wooden rail, the iron rail transformed the value of the English waggon. His eulogy of the Hungarian Hund is fully justified in theory, and even in practice, given perfect putters. But there lay the rub. Lower Hungary and a few mines elsewhere did somehow manage to train putters almost to perfection, but elsewhere this problem was insoluble, and even in Lower Hungary the mines were eventually forced, by the difficulty of providing these putters, to turn to Becker's abhorred flanged-wheel trucks. With a Hungarian Hund, a bad putter was worse than useless. So the English railway won the day.

Appendix

Comparative Efficiency of Continental Railway Vehicles

		Shift in hours	*Run in fathoms*	*Cu. ft. per Hund*	*Cwt. per Hund*	*Hunds per shift per putter*	*Cwt. per shift per putter*	*Power needed in fathom cwt./hour*	
Wheelbarrow:									
Marienberg[1]	1777	8	139	2	3	18	54	936	—
Leitnagel Hund:									
Marienberg[1]	1777	8	400	3	4·25	7	30	1,467	—
Freiberg[2]	1791	6	297	5·8	5·4	7	38	1,871	—
„[2]	1791	6	113	4·34	4	27	108	2,201	—
„[2]	1791	6	188	4·65	4·33	10	43	1,357	—
Johanngeorgenstadt[2]	1791	6	120	4·65	4·33	12	52	1,040	—
Schneeberg[2]	1791	6	110	—	6	20	120	1,100	Two-man
„[2]	1791	6	206	—	6	14	84	1,442	Two-man
Altenberg[3]	1823	8	240	—	8	10	80	2,400	—
Mansfeld[4]	1825	8	80	5·2	3·7	10	37	370	Coal
Braubach[5]	1790	8	170	11·5	12·5	12	150	3,188	Pivot-axle 1 in 50
Freiburg-i.-B.[6]	1846	9	50	3·7	4·3	42	181	1,033	—
„[6]	1846	9	100	3·7	4·3	26	112	1,242	—
„[6]	1846	9	500	3·7	4·3	6	26	1,290	—
Roche-la-Molière[7]	1836	8	300	—	8	13	104	3,900	Pivot-axle coal: 1 in
Hall (Tyrol)[8]	1854	5¼	757	2·6	2·25	6	14	1,900	Salt: 1 in
Idria[9]	1819	8	100	—	2	20	40	500	—
Felsöbanya[8]	1854	6	507	—	2·46	8	20	1,665	1 in 58 dc
Zalathna[8]	1854	8	525	3·2	3·1	6	19	1,221	1 in 24 dc
Bochnia[8]	1856	7½	217	—	5·6	21	118	2,268	Salt: one c men per H
Hungarian Hund:									
Schemnitz[10]	1805	6	25	3·1	6	80	480	2,000	—
„[10]	1805	6	50	3·1	6	50	300	2,500	—
„[10]	1805	6	100	3·1	6	35	210	3,500	—
„[10]	1805	6	200	3·1	6	20	120	4,000	—
„[10]	1805	6	500	3·1	6	10	60	5,000	—
„[10]	1805	6	1,000	3·1	6	5	30	5,000	—

arative Efficiency of Continental Railway Vehicles (cont.)

		Shift in hours	*Run in fathoms*	*Cu. ft. per Hund*	*Cwt. per Hund*	*Hunds per shift per putter*	*Cwt. per shift per putter*	*Power needed in fathom cwt./hour*	
arian Hund (cont.):									
nberg[1]	1777	8	139	3	4·25	29	123	2,137	—
rg[6]	1846	5	50	3·3	4·25	24	102	1,020	—
	1846	5	500	3·3	4·25	6	26	2,550	—
hal[6]	1846	8	50	5·35	6	32	192	1,920	—
[6]	1846	8	500	5·35	6	8	48	3,000	—
urg-i.-B.[6]	1846	9	50	2·5	4·3	42	181	1,003	—
,,[6]	1846	9	500	2·5	4·3	7	30	1,436	—
ühel[8]	1854	7	375	2·8	2·63	10	26	1,409	1 in 48 down
erötz[8]	1854	8	258	5·8	7·05	13	92	2,954	1 in 36 down
witz[11]	1853	9	50	3·3	4·25	55	234	1,300	—
,[11]	1853	9	200	3·3	4·25	15	64	1,422	—
phund:									
feld[4]	1825	8	110	5	7·25	12	87	1,196	—
[4]	1825	8	380	5	7·25	5	36	1,722	—
-wheel Hund:									
eberg[12]	1791	6	390	8·4	7	5	35	2,276	1 in 178 up
,,[12]	1794	6	390	6	5	7	35	2,276	1 in 178 up
d-wheel Riesen:									
g[10]	1805	8	800	—	10	3	30	3,000	Horse
h waggon:									
birge[2]	1791	8	60	3	c. 4	12	c. 48	360	—
feld[4]	1825	8	186	16	23	10	230	5,347	—
[4]	1825	8	380	16	23	7	161	7,647	—
[6]	1846	8	275	—	13	13	169	5,792	—
rg[6]	1846	8	253	—	10	20	200	6,325	—
nmann[6]	1846	8	750	—	22	8	176	16,500	Gravity
sgrube[6]	1846	8	330	—	13·3	14	186	7,681	Horse: coal
ichsthal[6]	1846	8	50	—	10	19	190	1,180	Coal
,,[6]	1846	8	900	—	10	5	50	5,625	Coal

The capacities given here may differ from those in the text. Here, the capacity is the um, piled; there, it is usually the cubic content of the body, level.

Glossary

Bracketed letters indicate that the use of a word was restricted:

(T) Tyneside and colonies

(S) Shropshire and colonies

(M) metalliferous mines

(C) the Continent

Unless otherwise indicated, all words enjoyed a reasonably wide circulation. Only the meanings used in this book are given.

adit (originally M) – drift, entrance to a drift.

ashing (T) – strewing the rails with ashes to control a waggon's speed in rain.

backing and filling packing ballast under the track.

ballast broken stone, cinders etc. forming the foundation of railway track.

bank embankment; steep gradient; of mines, pit-head; (adj.) surface, opposite to 'pit'; **banksman** man in charge of drawing to bank.

bar-rail wrought-iron rail of a basically square, rectangular or circular section.

barrower, barrowman man engaged in underground transport, whether with wheelbarrow, sledge or tram; **barroway, barrow-way** planked path for wheelbarrows; later railway for underground sledges and trams.

bate (T) – reduction in stint of a barrowman.

battery (T) – embankment.

bellpit an early type of mine, small workings swelling out at the foot of a shallow pit.

Bergtruhe see *Truhe.*

Bergwerksmarke (Harz) – token paid to putter for every Hund put.

binding baulks (T) – transverse strengthening ties on a wooden bridge.

blow in of a blast furnace, to bring into production.

bogie (T) – underground truck for carrying props, deals, etc.; (Cumberland) outby vehicle = Tyneside *rolley.*

boll measure of coal, usually about 8 pecks (see p. 184); (verb) to measure a waggon.

bord underground passage in a coal-mine; **bord and pillar** system of mining coal (see p. 308).

bottom board (T) – waggon bottom.

brace (T) – side support for a wooden bridge.

branch subsidiary line, corresponding both to the modern branch and to the modern siding.

brass (S) – brass bearing for waggon axle.

breast (T) – bar of wood nailed to convoy as a brake-block; **breast** or **land breast** (T) – abutment of a bridge.

bridge (C) – wooden staging outside adit for tipping from; (T) – elevated approach to a staith.

bulwark (T) – embankment.

bush to mount a wheel on its axle.

bye-stand (T) – a passing-loop.

bye way, occasionally **side way** (T) – track for empty waggons.

bye-work (T) – a return load carried from staith to colliery.

casseyway (causeway) (Scottish) – railway.

cast to dig.

chaldron measure of coal, up to 53 cwt. on Tyneside, 30 cwt. in Scotland, about 26 cwt. in London; later = a waggon.

check rail or **false rail** (T) – subsidiary rail to ensure that the wheels stay on the running rails.

clamdog (S) – iron upright on waggon supporting the cratch.

cleadings, cleathings (T) – planks forming waggon sides.

cleft (S) – probably vertical rib of a waggon = Tyneside *sheth.*

cold pie (T) – broken axle.

convoy, rarely **tiller, rudder** (T) – brake lever.

coop (T) – waggon body.

corf wicker basket for coal; **corfway** (Whitehaven) barroway.

cratch (S) – upward extension of the waggon side.

crease, creese (T) – wheel flange; **crease plate, nook plate** iron strip nailed to inside of flange; **creaser** workman who cleaned the flangeways of a waggonway.

creel (Scotland) – basket for carrying coal on the back.

crosstree (S) – transverse timber forming the top of a waggon's end (=Tyneside *howtree*).

crowntree (T) – transverse beam at the top of a wooden bridge.

cut (T) – cutting.

dan (S) – underground truck.

deads (M) – waste rock.

deals (T) – planks; underground rails.

dip steepest slope of coal strata.

double way (T) – track with two layers of wooden rails; only means two tracks side by side when **double railway** (S) or **waggonway** (T).

dozen measure of coal, in Yorkshire frequently a *chaldron.*

drag (S) – a beam pulled behind a waggon as a brake.

dram (Wales) – tram.

draw to wind up a mine-shaft.

drift a mine driven in from a hillside, not approached by a shaft.

edge rail rail for flanged wheels.

English waggon, waggonway; Englischer Schienenweg, voie anglaise etc. (C) – flanged-wheel waggon and its track.

false rail see *check rail.*

fitter (T) – coal-master's commercial agent.

flank wall (T) – retaining wall.

flat (Lancashire) – barge.

flume (C) – artificial water channel for floating logs in forestry.

foal (T) – a junior putter.

footgang (T) – path for waggon-man beside track.

footrid, footroad, footridge or **footrail** (S) – originally an opencast mine, then a drift mine; sometimes a railway.

fother (T) – a wain-load.

frame general term for rails and sleepers together.

gaillelot, galhiot (Liège) – large underground truck.

galloway (T) – a pit pony.

gangroad, gangway (E. Midlands) – railway.

gate, gait (T) – a journey with a waggon.

gate, gateway underground passage.

geers (T) – uprights of wooden bridges or staiths.

German Hund (C) – originally a self-steering Hund; later a Hund whose centre of gravity fell between the axles (see p. 56).

German Rail (C) – flanged rail, wooden or iron.

Gestänge (C) – usual German term for underground rails.

gin horse-driven winding engine.

goaf worked-out area in mine.

going (T) – of waggons and track, working, useable (opposite of *lying*).

groove rail (so called by misapprehension).

gugg (Somerset) – underground incline with double track.

guide-wheel subsidiary wheel mounted horizontally under a waggon to keep it on the rails.

gutter ditch.

half marrow (T) – middle grade of putter, between foal and headsman.

haypoke (T) – bag for hay carried on a waggon.

head, heading, headway passage driven into the coal in a mine.

headsman (T) – senior putter.

heap, heapstead coal or waste tip.

hewer miner who actually digs ore or coal.

hind-iron, long-iron (T) – vertical bar on waggon side to guide convoy.

howtree (T) – transverse beam forming the top of a waggon's front or back.

Hund (C) – generic German name for a mining truck (see pp. 7-8); **Hundstösser** or **Hundläufer** its putter.

HUND = HOUND = DOG
GROWLS • SOUND IN ACTION

Hungarian Hund (C) – Hund with no means of guidance (see ch. 3); later, any Hund whose centre of gravity fell over, or nearly over, the rear wheels.

hurry (Cumberland) – a spout; (Ireland) – an incline; (Yorkshire) – to transport coal underground; **hurrier** a putter.

inby (T) – the area of a mine between face and transfer-point of corves from tram to rolley.

incline, inclined plane gradient down which waggons are let by rope; **balanced** or **self-acting incline** one where the loaded waggon going down pulls up an empty.

inset (S) – a drift mine or footrid; occasionally a railway.

jig (S) – brake lever; **jigger** brakesman.

jowling (TM) – underground signalling.

keel (Yorkshire) – river and canal sailing vessel; (T) – coal lighter; **keelmen, keelroom** its crew, mooring place.

keep to maintain track or waggons; **keeper** or **undertaker** the man responsible.

kibble (M) – bucket for drawing ore up shaft; also (S) used for coal.

lead to transport coal by railway.

ledges (T) – extra boards fitted to a waggon to increase its capacity.

Leitnagel, Spurnagel (C) – guide-pin that steered many Hunds.

level a level or near-level passage in a mine.

linchpin pin to keep a wheel on its axle.

lode (M) – vein of ore.

loiter-pin (T) – retaining pin at foot of convoy.

long-iron (T) – see *hind-iron*.

longwall (S) – a system of mining coal (see p. 308).

lying (T) – of waggons, out of use, spare (opposite of *going*).

main rail (S) – main line.

main way or occasionally **waggonway** (T) – track for loaded waggons.

margin (T) – wheel flange.

marrow (T) – pull a waggon with two horses; a second horse or man.

metal often rubble or earth.

mount embankment.

Newcastle Road (T) – waggonway.

nook plate see *crease plate.*

offgate (T) – track leading off a staith.

ongate (T) – track leading on to a staith.

outby (T) – the area of a coal-mine between shaft and transfer-point of corves from tram to rolley.

overing (T), **overend** (Whitehaven) – longitudinal forming the top of a waggon's side.

pan rubbish (T) – refuse from salt-pans used as ballast.

parting (T) – junction.

pass-bye (S) – passing-loop.

penning (T) – underground rail.

plane see *incline.*

plate a flat wrought-iron or cast-iron strip nailed on wooden rail; **plate rail** from 1787, angled iron rail; **plateway** a railway formed of plate rails.

putter (orig. T) – man who pushed (*put*) trams underground; **put-and-hewer** grade of miner between putter and hewer.

Reibeisen (C) – iron strip attached to rails to reduce wear.

Riesen and compounds, **Risswerk** (C) – originally a forestry slide; later a surface railway in Slovakia and Transylvania; a waggon for such a railway, with either guide-wheels or, later, flanged wheels; occasionally a road cart; **Riesenhund** a large Hund.

ring-iron (T) – ring holding lower end of convoy.

Rinnwerk (C) – wooden chute for ore.

Risswerk (C) – see *Riesen.*

rolley (T) – underground flanged-wheel waggon used outby; **rolleyway** its track.

rook (Midlands) – old coal measure, about 30 cwt.

Rücken (C) – single beam forming the main frame of a Hund.

rudder see *convoy.*

run (T) – a gradient where waggons could run by gravity; **run amain** of a waggon, to get out of control.

saddle (C) – an extra container placed on top of a Hund.

Schlepphund (C) – a Hund for low galleries (see p. 40).

sclon (Mons) – an underground sledge or tram.

scotches (T) – wedges to hold convoy in ring-iron; **scotch-box** box on waggon to hold them.

screen (T) – combined chute and sieve for loading waggons at the pit-head.

sett (M) – area of a mining lease.

sheth, sheath (T) – vertical rib of a waggon (*side* or *corner sheth*); transverse beam of main frame (*bottom sheth*).

shod (T) – of wheels, with iron treads.

shoe sledge runner.

shoveller (C) – man who filled the Hund in a mine (trans. German *Säuberer*).

side rail (S) – a branch.

side way (T) – see *bye way*.

siding (T) – normally a passing loop; rarely a branch.

sill (S) – longitudinal beam of waggon main frame (=Tyneside *sole*); (T) – longitudinal timber of a wooden bridge forming base of track.

single way (T) – track with a single layer of rails (see *double way*).

skip (Staffs) – tram.

sledge, sled, slyp underground vehicle, usually with runners, occasionally with wheels.

soam (T) – cord for pulling sledge or tram.

sole (T) – longitudinal beam forming main frame of waggon or tram; transverse foundation beam of a wooden bridge.

sough drain.

sough metal, sow metal cast iron.

sployon (Liège) – small underground truck.

spout (T) – chute on a staith.

sprag (S) – a bar inserted into the spokes of a wheel as a brake.

Spur (C) – the gap between two rails in which the Leitnagel fitted; **Spurnagel** see *Leitnagel*.

stack (S) – measure of coal, from 26 to 40 cwt.

staith wharf or staging where waggons are unloaded into boats; **staith-disk** (T) – turntable.

stall (T) – individual working face in a coal-mine.

stamps (M) – machines for crushing ore.

standard (S) – upright member of waggon body (=Tyneside *sheth*).

staves (S) – probably planks for waggon sides.

steer (Cumberland) – a staith.

Stege or **Unterlage** (C) – sleepers.

stipe (Somerset) – underground incline with single track.

stock rail in points, the fixed outer rail against which the switch rail bears.

strap iron band to strengthen a waggon; (America) – iron strip laid on a wooden rail.

Strassbäume (C) – rails.

stretcher (T) – longitudinal strengthening tie on a wooden bridge.

stringer (America) – a wooden rail.

switch, switch rail in points, the shifting rail; a set of points.

ten (T) – measure of coal, varying from about 25 to 53 tons (see p. 188). **tentale** payment according to number of tens of coal carried.

thill (T) – floor of a coal-mine.

tie (esp. America) – a sleeper; **tiebar** an iron rod to keep rails to gauge;

tiller see *convoy*.

ton not necessarily 20 cwt.: in Cumberland up to 25 cwt., in Shropshire up to 60 cwt. (see pp. 189, 268); also a measure for wood, producing 120 yards of rail.

tourniquet (Ireland) – turntable.

Tragwerk (C) – wooden flooring to a mine-level, on which rails were laid.

trailway (Whitehaven) – barroway.

tram originally an underground sledge or wheelbarrow; later, a wheeled truck; (T) – unit of work done by one sledge or tram in a day; later still a surface waggon; **tramroad, tramway** (T) – underground railway for trams; (orig. S. Wales) – surface railway.

transporter incline one where an ordinary waggon is carried on, and held level by, a host waggon.

trapper (T) – boy who opened and shut ventilation doors underground.

trod tread of a wheel; **trod-plate** (T) – strip of iron nailed to the tread.

trow (Severn) – sailing vessel.

Truhe, Bergtruhe (C) – usually a synonym for Hund (see p. 8); occasionally a non-wheeled container for ore; a road cart; **Truhenläufer** its putter.

trunk (T) – a staith.

tub-boats floating containers towed in trains along small canals.

turnrail turntable.

twinway (Somerset) – level track between underground inclines.

undertaker see *keeper*.

Unterlage see *Stege*.

viewer (T) – colliery engineer.

Wagen (C) – a road waggon; a mine vehicle to carry a Truhe or box; later any railway vehicle with equal-sized wheels.

waggon road (esp. Yorks and Lancs) – railway.

waggonway normal Tyneside word for *railway*, also frequent elsewhere; see also *main way*.

wain road cart for coal.

wall coal face.

washery (M) – the place where crushed ore and rock are separated.

wayleave permission to lay a railway across another's property; the cost thereof; occasionally (T) = *waggonway*.

wheel-corf underground truck with wheels.

windway, wind (S) – ventilation passage in a mine; sometimes apparently an inclined adit.

winze (M) – subsidiary shaft between levels.

worm screw.

wrathe (S) – longitudinal forming the top of a waggon's side (= Tyneside *overing*).

Abbreviations

AfB	*Archiv für Bergbau und Hüttenwesen*, Berlin.
AfM	*Archiv für Mineralogie, Geognosie, Bergbau und Hüttenkunde*, Berlin.
AdesM	*Annales des Mines*, Paris (continuation of *JdesM*).
BM	British Museum.
Boehmer	Karl Friedrich von Boehmer, *Ueber die Grubenförderung*, i, Freiberg und Annaberg (1791) – only one volume published.
BTHR	British Transport Historical Records.
CRO	Cumberland, Westmorland and Carlisle Record Office.
CW	*Transactions of the Cumberland and Westmorland Antiquarian and Archaeological Society*, Kendal.
Dendy Marshall	C. F. Dendy Marshall, *A History of British Railways down to the year 1830*, London (1938).
DRO	Durham County Record Office.
GPL	Gateshead Public Library.
Hughes	Edward Hughes, *North Country Life in the Eighteenth Century* [vol. i], *The North-East 1700-1750*, Oxford (1952); vol ii, *Cumberland and Westmorland 1700-1830*, London (1965).
Jars	Gabriel Jars, *Voyages Métallurgiques*, Lyon: i (1774), ii (1780), iii (1781).
JdesM	*Journal des Mines*, Paris (continued as *AdesM*).
LCL	Leeds Central Libraries, Archives Dept.
NEIMME	Library of the North of England Institute of Mining and Mechanical Engineers, Newcastle.

NRO	Northumberland Record Office.
NUL	Newcastle University Library.
PRO	Public Record Office.
PSAN	*Proceedings of the Society of Antiquaries of Newcastle upon Tyne.*
SBL	Shrewsbury Borough Library.
SCL	Sheffield City Library.
Schmidt	Franz Anton Schmidt, *Chronologisch-systematische Sammlung der Berggesetze der Königreiche: Ungarn, Kroatien, Dalmatien, Slavonien und des Grossfürstenthumes Siebenbürgen*, Wien: 17 Vols., 1834-36.
SRO	Shropshire Record Office.
TNS	*Transactions of the Newcomen Society*, London.
ZBHSPS	*Zeitschrift für das Berg- Hütten- und Salinenwesen in dem Preussischen Staate*, Berlin.

Notes

Chapter 1

1 Lee, Charles E., *The Evolution of Railways*, 2nd ed., London (1943), 2ff.

2 See, for example, H. Godwin in *Proceedings of the Prehistoric Society*, xxvi (1960), 1-36.

3 Gracie, Inst. Capt. H. S., in *Antiquity*, 110 (June 1954), 91-8.

4 N. M. Verdelis in *Illustrated London News*, 19 October 1957, 644-51.

5 *Heronis Alexandrini Opera quae supersunt omnia*, ed. W. Schmidt, Leipzig (1889), i.342.

6 Staunton, Sir G. L., *An Authentic Account of an Embassy from the King of Great Britain to the Emperor of China*, London (1797), pl. 34.

7 This summary of medieval mining is drawn largely from Professor J. U. Nef's excellent article, 'Mining and Metallurgy in Medieval Civilization', in the *Cambridge Economic History of Europe*, Cambridge (1952), ii.429ff.

8 Agricola, Georgius, *De Re Metallica libri XII*, Basel (1556), 113, 380, 412; trans. by H. C. and L. H. Hoover, New York (1950), 156, 180.

9 An explanation given in several Czech dictionaries, e.g. Václav Machek, *Etymologícky Slovník*, Praha (1957), 150.

10 Körner, Georg, *Eine philologisch-historische Abhandlung von dem Alterthume des böhmischen Bergwerks*, Schneeberg (1758), 12. Sternberg, Graf Kaspar, *Urkundenbuch zur Geschichte der böhmischen Bergwerke* (the second part of *Umrisse der Geschichte des Bergbaues und der Berggesetzgebung des Königreichs Böhmen*, Band 2), Prag (1838), 212.

11 Gritzner, Max Joseph, *Commentar der Ferdinandeischen Bergordnung vom Jahre 1553*, Wien (1842), 58.

12 Schönberg, Abraham von, *Ausführliche Berg-Information*, Leipzig (1693), 20.

13 Neuberg, C., *Goslars Bergbau bis 1552*, Hannover (1892), 206.

14 Vlachovič, Jozef, *Slovenská Med v 16 a 17 Storočí*, Bratislava (1964), 52.

15 Schmidt, i.321.

16 Ferber, J. J., *Abhandlungen über die Gebirge und Bergwerke in Ungarn*, Berlin (1780), 287.

17 *Amtlicher Führer durch die Sammlungen*, Deutches Museum, München (1925).
18 Winkelmann, Heinrich, *Der Bergbau in der Kunst*, Essen (1958), 70.
19 *Zeitschrift des deutschen und österreichischen Alpenvereins*, xxii (1891), 228.
20 Worms, Stephen, *Schwazer Bergbau im fünfzehnten Jahrhundert*, Wien (1904), 101.
21 Keyssler, J. G., *Neueste Reise durch Teutschland, Böhmen, Ungarn, die Schweitz, Italien und Lothringen*, Hannover (1740): English translation, 2nd ed., London (1757), iv.71.
22 Sternberg, *op. cit.*, 212.

Chapter 2

1 Prodan, D., and Goldenberg, S., in *Anuarul Institutului de istorie din Cluj*, vii (1964), 140, 143.
2 Schmidt, v.519.
3 Reproduced in Ratkoš, Peter, *Povstanie Baníkov na Slovensku roku 1525-1526*, Bratislava (1963), 52.
4 Vlachovič, Jozef, *Slovenská Med v 16 a 17 Storočí*, Bratislava (1964), 31, 150.
5 Peithner, J. T. A., *Versuch über die natürliche und politische Geschichte der böhmischen und mährischen Bergwerke*, Wien (1780), 426.
6 Kořan, Jan, in *Sborník pro dějiny přírodnich věd a techniky*, ii, Prague (1955), 19.
7 *Beiträge zur Geschichte der Technik und Industrie*, xii, Berlin (1922), 180ff.
8 For his biography, see introduction to Herbert C. and Lou Henry Hoover's translation of *De Re Metallica*, London (1912).
9 Mathesius, Johann, *Bergpostilla oder Sarepta*, Nürnberg (1571), 23*a*, 126*a*, 137*b*.
10 Bornhardt, Wilhelm, *Geschichte des Rammelsberger Bergbaues von seiner Aufnahme bis zur Neuzeit*, Berlin (1931), 195.
11 *Zeitschrift des Harz-Vereins für Geschichte und Altertumskunde*, xliv, Wernigerode (1911), 46.
12 Wagner, Thomas, *Corpus Juris metallici recentissimi et antiquioris*, Leipzig (1791), 594.
13 *Revue Lorraine Illustrée*, iii, Nancy (1908), 156. Now in Bibliothèque Municipale de Saint-Dié.
14 *Ibid.*, 153ff. Winkelmann, Heinrich, *Bergbuch des Lebertals*, Dortmund (1962). Now in Bibliothèque de l'Ecole supérieure des Beaux Arts, Paris.
15 His, E., in *Jahrbuch der königlich Preussischen Kunstsammlungen*, xv (1894), 207ff. Now in BM.
16 *Zeitschrift für die Geschichte des Oberrheins*, NF ii (1887), 436.
17 Wolfstrigl-Wolfskron, Max Reichsritter von, *Tiroler Erzbergbaue 1301-1665*, Innsbruck (1903), 324.
18 *Berg- und Hüttenmännisches Jahrbuch der k.k. montanistischen Hochschule zu Leoben und Přibram*, lii, Wien (1904), 438, 445.
19 Three at Innsbruck, one each at Vienna, Munich, Bochum, Leoben, and Wertheim/Main. The Bochum MS. was published privately by the Gewerkschaft Eisenhütte Westfalia Lünen in 1956. A bad version of some of the

text was printed by Franz Friese in *Berg- und Hüttenmännisches Jahrbuch der k.k. Bergakademien Schemnitz und Leoben und der k.k. Montan-Lehranstalt Přibram*, xiv, Wien (1865), 125-172. Text and illustrations used in this book are from the best MS., Tiroler Landesmuseum Ferdinandeum, Innsbruck, FB 4312.

20 Winkelmann, Heinrich, *Der Bergbau in der Kunst*, Essen (1958), fig. 51.

21 Ržiha, Franz, *Lehrbuch der gesammten Tunnelbaukunst*, Berlin (1867), i.255. Haarman, A., *Das Eisenbahn-Geleise*, Leipzig (1891), i.10. Tomlinson, W. W., *The North Eastern Railway, Its Rise and Development*, Newcastle (1914), 4. All claimed or suggested that Germans brought the railway to Cumberland, but without giving any evidence. This is to be found in Collingwood, W. G., *Elizabethan Keswick* (vol. viii of Cumberland and Westmorland Antiquarian and Archaeological Society Tract Series), Kendal (1912).

22 See also Donald, M. B., *Elizabethan Copper*, London (1955).

23 Collingwood, *op. cit.*, 43, 50, 52, 72, 80, 89, 90. The original accounts are in the Augsburg Stadtarchiv.

24 BM Sloane MS. 2487. Lebeter, F., in *Der Anschnitt*, xvi (1964), 23, saw that these rowle wagons might be railway vehicles.

25 Lewis, W. J., in *Ceredigion*, i, Aberystwyth (1950-51), 177ff. Gough, J. W., *The Superlative Prodigall, a Life of Thomas Bushell*, Bristol (1932), 36ff.

26 Pettus, Sir John, *Fodinae Regales*, London (1670), 38.

27 Gumilevsky, L. I., ЖЕЛЕЗНАЯ ДОРОГА. ТРЕТЬЕ ПЕРЕРАБОТАННОЕ И ДОПОЛНЕННОЕ ИЗДАНИЕ, Moscow (1950), 41. Vil'gel'm de Gennin, ОПИСАНИЕ УРАЛЬСКИХ И СИБИРСКИХ ЗАВОДОВ, Moscow (1937), 546-7

28 Hermann, B. F. J., *Mineralogische Reisen in Sibirien vom Jahre 1783 bis 1796*, St. Petersburg (1797), i, 311. Gumilevsky, *op. cit.*, 41. Virginsky, V. S., and Savelev, N. Y., СТРОИТЕЛЬСТВО ВОДОДОЙСТВУЮЩИХ УСТРОЙСТВ ИА АЛТАЕ В ХУIII ВЕКЕ, Moscow (1955), 13.

29 Beaunier and Gallois in *JdesM*, xcii (Floréal An XII), 109. d'Aubuisson, J. F., in *JdesM*, cxxi (January 1807), 37-8. Laurent de Villedeuil, P. C., *Bibliographie des Chemins de Fer*, Paris (1903), 85. Sée, H., in *Annales de Bretagne*, xxxvii (1925-6), 402ff.

30 Combes, C., in *AdesM*, 3rd series, viii (1835), 446-7.

31 Böbert, K. F., in *AfM*, v (1832), 260.

32 Ržiha, *op. cit.*, i.532.

33 Gaun, Franciscus, S.J., *Iter per Salinas Tyrolenses*, Innsbruck (1707), 11. Huyssen in *ZBHSPS*, ii (1855), 34-5. Le Noble, J., in *Abhandlungen einer Privatgesellschaft in Böhmen zur Aufnahme der Mathematik, der vaterländischen Geschichte, und der Naturgeschichte*, iv, Prag (1779), 322.

34 de la Goupillière, Haton, *Cours d'exploitation des mines*, 4th ed., Paris (1931), ii.816.

35 *Südost-Forschungen*, xviii, München (1959), 79; xxii (1963), 207-8.

36 *JdesM*, clxx (1811), 126.

37 Agricola, Georgius, *De Re Metallica libri XII*, Basel (1556), 85, 113.

38 Browne, Edward, *A brief Account of some Travels in Hungaria, Thessaly, Austria, Styria, Carinthia, Carniola and Friuli*, London (1673), 100, 108.

39 *Neue Berg-Ordnung dess Königreichs Ungarn . . . Erleutterungen Zweyer Alten Berg-Ordnungen der Sieben königl. Freyen Berg-Städte*, Wien (1703), 1.

40 Delius, Christoph Traugott, *Anleitung zu der Bergbaukunde*, Wien (1773), 258, 285.

41 d'Aubuisson, J. F., in *JdesM*, cxxi (January 1807), 37-8.

42 Daub in *AfM*, xx (1846), 606.

43 d'Aubuisson, J. F., *Des Mines de Freiberg en Saxe et de leur Exploitation*, Leipsic (1802), iii.26.

44 Calvoer, Henning, *Acta Historico-Chronologico-Mechanica circa Metallurgiam in Hercynia Superiori oder . . . Maschinenwesens und der Hülfsmittel bey dem Bergbau auf dem Oberharze*, Braunschweig (1763), ii.29.

45 Héron de Villefosse, A. M., *De la Richesse Minérale*, Paris (1819), ii, pl. 16 fig. 12-14.

46 Stockicht in *Bergbaukunde*, ii, Leipzig (1790), 49ff.

47 Böbert, K. F., in *AfM*, v (1832), 251.

48 d'Aubuisson, *op. cit.*, iii.77.

49 Stockicht in *Bergbaukunde*, ii, Leipzig (1790), 49ff.

50 Gervoy, M., in *AdesM*, 3rd series, x (1836), 447.

51 e.g. Schulz in *AfB*, iv (1821), 75.

52 e.g. Schemnitz: Delius, C. T., *Anleitung zu der Bergbaukunde*, Wien (1773), 258.

53 e.g. Dürrenberg in the Salzkammergut, *c.* 1720: Kirnbauer, F., in *Blätter für Technikgeschichte*, xxiv, Wien (1962), 103.

54 Cancrin, Franz Ludwig, *Beschreibung der vorzüglichsten Bergwerke in Hessen, in dem Waldekkischen, an dem Haarz, in dem Mansfeldischen, in Chursachsen, und in dem Saalfeldischen*, Frankfurt aM (1767), 415.

55 Boehmer, 85. Beyer, August, *Beytraege zur Bergbaukunde*, Dresden (1794), 282.

56 *Jahrbuch für den Berg- und Hütten Mann*, Dresden (1841), 119.

57 Boehmer, 81. d'Aubuisson, *op. cit.*, iii.276.

58 Cancrin, *op. cit.*, 415.

59 Brard, C. P., *Grundriss der Bergbaukunde*, Berlin (1830), 224.

60 Stockicht in *Bergbaukunde*, ii, Leipzig (1790), 49ff.

61 Combes, C., *Traité de l'exploitation des mines* (1845), iii.12.

62 Bornhardt, Wilhelm, *Geschichte des Rammelsberger Bergbaues von seiner Aufnahme bis zur Neuzeit*, Berlin (1931), 195.

63 *Magazin der Bergbaukunde*, vi, Dresden (1789), 119. Böbert, K. F., in *AfM*, v (1832), 254.

64 *Kalendar für den Sächsischen Berg- und Hüttenmann*, Freiberg (1835), 39.

65 Rössler, Balthasar, *Speculum Metallurgiae Politissimum oder Hell polierter Berg-Bau-Spiegel*, Dresden (1700), 53.

66 d'Aubuisson, *op. cit.*, iii.26. Trebra, F. W. H. von, *Erfahrungen vom Innern der Gebirge*, Dessau und Leipzig (1785), 227.

67 Boehmer, 77.

68 Combes, C., *Traité de l'exploitation des mines* (1845), iii.9.

69 Wolfstrigl-Wolfskron, Max Reichsritter von, *Tiroler Erzbergbaue 1301-1665*, Innsbruck (1903), 401.

70 Scheuermauer, L., *Die Fugger als Montanindustrielle in Tirol und Kärnten*, München und Leipzig (1929), 440.

71 e.g. at Idria: Kircher, Athanasius, S.J., *Mundus Subterraneus*, Amsterdam (1664), ii.153.

72 Böbert, K. F., in *AfM*, v (1832), 260.

73 Boehmer, 82.

74 *Jahrbücher der Berg- und Hüttenkunde*, v, Salzburg (1801), 158-9.

75 e.g. Calvoer, Henning, *Acta Historico-Chronologico-Mechanica* . . . Braunschweig (1763), ii.29.

76 Berward, Christian, *Interpres Phraseologiae Metallurgicae*, Frankfurt-a-M. (1684), 36. Richter, C. F., *Neuestes Berg- und Hütten Lexicon*, Leipzig (1806); s.v. Hund, Hundskette.

77 Karsten, C. J. B., *Metallurgische Reise durch einen Theil von Baiern, und durch die süddeutschen Provinzen Oesterreichs*, Halle (1821), 322.

78 Boehmer, 89.

79 Héron de Villefosse, A. M., *De la Richesse Minérale*, Paris (1819), ii.303.

80 Vlachovič, Jozef, *Slovenská Med v 16 a 17 Storočí*, Bratislava (1964), 150 (Neusohl, 1604). Schmidt, xii.413, 481 (Nagybanya and Transylvania, 1766 and 1767).

81 Schmidt, xi.5; xiv.302.

82 Wolfstrigl-Wolfskron, Max Reichsritter von, *Tiroler Erzbergbaue 1301-1665*, Innsbruck (1903), 314.

83 Wutke, Konrad, *Codex Diplomaticus Silesiae*, 21: *Schlesiens Bergbau und Hüttenwesen, Urkunden und Akten (1529-1740)*, Breslau (1901), 206.

84 *Berg- und Hüttenmännisches Jahrbuch der k.k. montanistischen Hochschulen zu Leoben und Přibram*, lii, Wien (1904), 438, 445. Scheuermauer, L., *op. cit.*, 440.

85 Vlachovič, *op. cit.*, 150, 259.

86 Worms, Stephen, *Schwazer Bergbau im fünfzehnten Jahrhundert*, Wien (1904), 101. Schmidt, xi.5, xii.413.

87 Böbert, K. F., in *AfM*, v (1832), 253, 260.

88 Heyse, G., in *Zeitschrift des Harz-Vereins für Geschichte und Alterthumskunde*, iii, Wernigerode (1871), 669ff., 997, iv.419, v.213, xv.200. Ernst, C. von, in *Oesterreichischer Zeitschrift für Berg- und Hüttenwesen*, xxxiii, Wien (1885), 140.

89 Callon, J., *Lectures on Mining* (1881), ii.80.

Chapter 3

1 Vlachovič, J., in *Historické Štúdie*, vii, Bratislava (1961), 8.

2 *Ibid.*, 226.

3 Marsili, Aloysius Ferdinandus Comes, *Danubius Pannonico-Mysicus*, The Hague and Amsterdam (1726), iii, tab. 10.

4 Schmidt, xi.5.

5 Becker, Wilhelm Gottlob Ernst, *Journal einer Bergmännischen Reise durch Ungarn und Siebenbürgen*, Freiberg (1815), i.120-58.

6 Schmidt, xi. 349 (Kremnitz and Felsöbanya, 1760).

7 Grimm, J., in *Jahrbuch der k.k. Geologischen Reichsanstalt*, viii, Wien (1857), 719. Serlo, Albert, *Beitrag zur Geschichte des Schlesischen Bergbaues in dem letzten hundert Jahren*, Breslau und Berlin (1879), 90.

8 Trebra, Friedrich Wilhelm Heinrich von, *Erfahrungen vom Innern der Gebirge*, Dessau und Leipzig (1785), 215.

9 Boehmer, 72, 92.

10 *Fragment einer Bergmännischen Reise nach Freiberg im Erzgebirge*, Leipzig und Flensburg (1785), 17.

11 *Magazin der Bergbaukunde*, vi, Dresden (1789), 111. Boehmer, 72.

12 Gatterer, Christoph Wilhelm Jakob, *Anleitung den Harz, und anderer Bergwerke*, Göttingen (1785), i.46-7. Böbert, K. F., in *AfM*, v (1832), 289ff.

13 Dietrich, Baron de, *Description des Gîtes de Minérai, Forges, Salines, Verreries, Tréfileries, Fabriques de fer-blanc, Porcelaine, Faience, etc., de la Haute et Basse-Alsace*, Paris (1789), 258.

14 Daub in *AfM*, xx (1846), 602-20.

15 Ržiha, Franz, *Lehrbuch der gesammten Tunnelbaukunst*, Berlin (1867), i.532.

16 Becker, *op. cit.*, i, 131-2.

17 d'Aubuisson, J. F., *Des mines de Freiberg en Saxe et de leur Exploitation*, Leipsic (1802), i.198.

18 Böbert, K. F., in *AfM*, v (1832), 255, 290.

19 Serlo, Albert, *Leitfaden zur Bergbaukunde*, 3rd ed., Berlin (1878), ii.10.

20 Delius, Christoph Traugott, *Anleitung zu der Bergbaukunde*, Wien (1773), 258.

21 Hartmann, Carl, *Der innern Gebirgswelt Schätze und Werkstätten, oder Gemeinfassliche Darstellung der Bergbaukunde*, Stuttgart (1838), 188.

22 *Kalendar für den sächsischen Berg- und Hüttenmann*, Freiberg (1835), 38-9.

23 Schmidt, xvi.328.

24 Trebra, *op. cit.*, 227. Böbert, K. F., in *AfM*, v (1832), 254.

25 Information from Mr. M. Badik of Spišská Nová Ves.

26 Kiesslingen, J. G., *Gegründete Nachricht von dem Bergbau und Schmelzwesen in der Grafschaft Mansfeld*, Leipzig (1747), 6.

27 Böbert, K. F., in *AfM*, v (1832), 262ff.

28 Serlo, *op. cit.*, ii.13.

29 Treptow, E., *Grundzüge der Bergbaukunde und Aufbereitung*, Wien und Leipzig (1891), 216.

30 Cancrin, Franz Ludwig, *Beschreibung der vorzüglichsten Bergwerke in Hessen, in dem Waldekkischen, an dem Haarz, in dem Mansfeldischen, in Chursachsen, und in dem Saalfeldischen*, Frankfurt-a-M. (1767), 8, 71, 83, 263, 271 (Frankenberg, Riegelsdorf, Wisenau, Rothensburg, Wettin). Héron de Villefosse, A. M., *De la Richesse Minérale*, Paris (1819), ii.500 (Schaumburg). Seyfert in *ZBHSPS*, iv (1857), 186 (Riestadt). *JdesM*, xliv (Floréal An VI), 612-13 (Meisenheim and Landsberg).

31 Delius, *op. cit.*, para. 379. d'Aubuisson, J. F., in *JdesM*, ci (Pluviose An XIII), 341. Du Bosc in *AdesM*, iii (1818), 23.

Chapter 4

1 Bayerische Staatsbibliothek, München, MS. Lat. 197, fol. 35v, 36r. For the Anonymus, see also Gille, B., *Les ingénieurs de la Renaissance*, Paris (1964), 60-1.

2 Ramelli, Agostino, *Le Diverse et Artificiose Machine*, Paris (1588), 214.

3 Lorini, Buonaiuto, *Delle Fortificationi*, Venetia (1596), 194-6.

4 *Hans Georg Ernstingers Raisbuch*, Stuttgart (1877), 39.

5 Zonca, Vittorio, *Nuovo Teatro di Machine et Edificii*, Padova (1607), 58-60. See also Keller, A. G., *A Theatre of Machines*, London (1964), 64.

6 For example, de Servière, Grollier, *Recueil d'Ouvrages curieux*, Lyon (1719), 61.

7 Reisch, Gregor, *Margarita Philosophica*, Freiberg (1503), Liber ix.

8 Petrarca, Franciscus, *Von der Artzney bayder Glück, des güten und widerwertigen . . . Teütsch gezogen*, Augsburg (1532), book i, p. lxx verso.

9 Agricola, Georgius, *De Re Metallica libri XII*, Basel (1556), 72, 74. Albinus, Petrus, *Meissnische Bergk Chronica: Darinnen fürnemlich von der Bergkwercken des Landes zu Meissen gehandelt wird . . .*, Dresden (1590), title-page and facing p. 1. For four-wheeled barrows cf. *Eyn wolgeordent vnd nützlich büchlin wie man Bergwerck suchen vn finden sol*, Worms (1518), title-page; Müller, Johann Christoph, *Mappa Geographica Regni Bohemiae*, Augsburg (1720), vignette.

10 Published by Lee, C. E., in *TNS*, xxxiii (1960-1), 3 and pl. iii (*a*). See also Dollfus, C., *Histoire de la Locomotion Terrestre*, Paris (1935), i.36.

11 Published by Treptow, E., in *Beiträge zur Geschichte der Technik und Industrie*, xii, Berlin (1922), 180ff.; and Winkelmann, Heinrich, *Der Bergbau in der Kunst*, Essen (1958), 74 and fig. 40.

12 Winkelmann, *op. cit.*, fig. 133.

13 In the Musées Royaux des Beaux Arts, Brussels.

14 Schatzkammer der Residenz, München, cat. no. 43. See also Winkelmann, *op. cit.*, 151.

15 Munsterus, Sebastianus, *Cosmographiae Universalis Lib. VI*, Basileae (1550), 431.

16 *Cosmographia, Bschreibung aller Lender durch Sebastianum Munsterum*, Basel (1545), 385. *Revue Lorraine Illustrée*, iii, Nancy (1908), 157. Burmeister, Karl Heinrich, *Sebastian Munster*, Basel (1963), 128.

17 In Tekniska Museet, Stockholm. See also Johnson, W. A., *Christopher Polhem*, Hartford, Conn. (1963), 173.

18 Hülphers, A., *Dagbok öfwer en resa igenom de under Stora Kopparbergs höfdingedöme lydande lähn och Dalarne år 1757*, Wästeräs (1762), 610.

19 Leupold, Jacob, *Theatrum Machinarum Hydrotechnicarum*, Leipzig (1724), ii.105 and pl. xxii.

20 Semyonov, Yuri, *The Conquest of Siberia*, London (1944), 270.

21 Danilevksy, V. V., РУССКАЯ ТЕХНИКА, Leningrad (1948), 148. Virginsky, V. S., ВОЗНИКНОВЕНИЕ ЖЕЛЕЗНЫХ ДОРОГ В РОССИИ ДО НАЧАЛА 40х ГОДОВ XIX ВЕКА, Moscow (1949), 54.

22 Virginsky, V. S., ТВОРЦЫ НОВОЙ ТЕХНИКИ В КРЕПОСТНОЙ РОССИИ, Moscow (1957), 117.

23 Winkelmann, *op. cit.*, fig. 14.

24 Le Noble, J., in *Abhandlungen einer Privatgesellschaft in Böhmen zur Aufnahme der Mathematik, der vaterländischen Geschichte, und der Naturgeschichte*, iv, Prag (1779), 320ff.

25 Huyssen in *ZBHSPS*, ii (1855), 34-5.

26 Köhler, G., *Lehrbuch der Bergbaukunde*, Leipzig (1884), 291.

27 *Blätter für Geschichte der Technik*, i, Wien (1932), 144.

28 Phillips, E., in *AdesM*, 4th series, viii (1845), 239ff.

29 For this classification, see von Oeynhausen and von Dechen on Ruhr colliery transport in *AfB*, vii (1823), 104.

30 *Ibid.*, 121ff.

31 Reden, Freiherr Friedrich Wilhelm von, *Die Eisenbahnen Deutschlands*, Band I, Theil 2, ii, Supplement ii, Berlin, Posen and Bromberg (1846), 405f.

32 Ponson, A. T., *Traité de l'Exploitation des Mines de Houille*, Liège (1853), iii.3-4.

33 *AfB*, vii (1823), 108-15.

34 Serlo, Albert, *Leitfaden zur Bergbaukunde*, 3rd ed., Berlin (1878), ii.11.

35 Hartmann, Carl, *Der innern Gebirgswelt Schätze und Werkstätten, oder gemeinfassliche Darstellung der Bergbaukunde*, Stuttgart (1838), 188.

36 Daub in *AfM*, xx (1846), 622.

37 Héron de Villefosse, A. M., *De la Richesse Minérale*, Paris (1819), ii.522ff.

38 Ponson, *op. cit.*, 3-4.

39 *Beiträge zur Geschichte der Technik und Industrie*, i, Berlin (1909), 27f.

40 Dendy Marshall, C. F., *A History of Railway Locomotives down to the End of the Year 1831*, London (1953), 234. Reden, Freiherr F. W. von, *Die Eisenbahnen Deutschlands*, Band I, Theil 2. ii, Berlin, etc. (1844), 888. Henderson, W. O., *The State and the Industrial Revolution in Prussia 1740-1870*, Liverpool (1958), 68, 151.

41 Pinno in *ZBHSPS*, xii (1864), 155.

42 Treptow, E., *Grundzüge der Bergbaukunde und Aufbereitung*, Wien und Leipzig (1891), 216f.

43 Decamps, Gonzalès, on the Mons collieries in *Mémoires et Publications de la Société des Sciences, des Arts et des Lettres du Hainaut*, Mons, Année 1879, 181-2; Années 1888-9, 140-1.

44 Morand, *L'Art d'exploiter les mines de charbon de terre*, Paris (1773), ii.224, 233. Baillet in *JdesM*, lix (Thermidor An IX), 829. Decamps, *op. cit.*, 1888-9, 140.

45 Fournel, Henri, *Etude des Gîtes houillers et métallifères du bocage Vendéen, faite en 1834 et 1835*, Paris (1836), 129-30.

46 Wexler, Peter J., *La Formation du Vocabulaire des Chemins de Fer en France (1778-1842)*, Genève et Lille (1955), 33. Gervoy, M., in *AdesM*, 3rd series, x (1836), 448. *Bulletin d'industrie de la Loire*, ii (1824), 257ff.

Chapter 5

1 Browne, Edward, *A Brief Account of some travels in Hungaria, Thessaly, Austria, Styria, Carinthia, Carniola, and Friuli. As also some observations on the Gold, Copper, Quick-silver, Baths and Mineral Waters in those parts*, London (1673), 101.

2 Elton, Sir Arthur, in *Proceedings of the Somersetshire Archaeological and Natural History Society*, cvii, Taunton (1963), 35. Lee, C. E., in *TNS*, xxxiii (1960-1), 7.

3 *Historické Štúdie*, vii, Bratislava (1961), 230.

4 Jars, ii, 147.

5 Delius, Christoph Traugott, *Anleitung zu der Bergbaukunde*, Wien (1773), 285-6.

6 *Neue Berg-Ordnung dess Königreichs Ungarn . . . Erleütterungen Zweyer Alten Berg-Ordnungen der Sieben Königl. Freyen Berg-Städte . . .*, Wien (1703), 64.

7 Marsili, Aloysius Ferdinandus Comes, *Danubius Pannonico-Mysicus*, The Hague and Amsterdam (1726), iii, tab. 11.

8 Schmidt, i.413.

9 *Ibid.*, ii.371-2.

10 *Ibid.*, ii.379, 380.

11 *Ibid.*, iii.414.

12 *Ibid.*, iv.53.

13 *Ibid.*, iv.671, ix.485.

14 *Ibid.*, iii.61.

15 *Ibid.*, iii.138, 156-7.

16 An engraving 'Das Berg-werck', Schemnitz (1685): reproduced by H. Winkelmann, *Der Bergbau in der Kunst*, Essen (1958), fig. 144.

17 Schmidt, iv.74.

18 *Ibid.*, viii.195.

19 Delius, *op. cit.*, 285.

20 Schmidt, xiii.172.

21 *Ibid.*, xiv.232.

22 The English waggon had arrived by 1846: *AdesM*, 4th series, x (1846), 655-6.

23 Schmidt, xi.216, 224, 244, 362, 441, xii.413.

24 Gatterer, Christoph Wilhelm Jakob, *Anleitung den Harz, und andern Bergwerke*, Göttingen (1785), i.94. Scheuchenstuel, Carl von, *Idioticon der Oesterreichischen Berg- und Hüttensprache*, Wien (1856).

25 For early Austrian slides, see *Oesterreichische Weisthümer* (ed. Siegel, H., Tomaschek, K., and others), Wien, 9 vols. (1870-1909), *passim*. Srbik, Heinrich Ritter von, *Studien zur Geschichte des Oesterreichischen Salzwesens*, Innsbruck (1917), 224. Wagner, Thomas, *Corpus Iuris Metallici Recentissimi et Antiquioris*, Leipzig (1791), 41, 49, 51.

26 Kaindl, R. F., *Geschichte der Deutschen in den Karpathenländern*, Gotha (1907), i.317, ii.101.

27 Deffontaines, Pierre, *La Vie Forestière en Slovaquie*, Paris (1932), 48.

28 Schmidt, i.350.

29 e.g. *Ibid.*, i.376, 403-8, iv.21, 164, xi.394.

30 Deffontaines, *op. cit.*, 50.

31 For more modern slides, see Gayer, Karl, *Die Forstbenutzung*, Aschaffenburg (1863), i.392ff. Förster, G. R., *Das forstliche Transportwesen*, Wien (1885), *passim*. Bryant, R. C., *Logging*, New York (1923), 262ff.

32 Winter, G., *Niederösterreichische Weisthümer*, Wien (1886), i.328.

33 Agricola, Georgius, *De Re Metallica Libri xii*, Basileae (1556), 127.

34 Lindroth, Sven, *Gruvbrytning och Kopparhantering vid Stora Kopparberget intill 1800-talets början*, Uppsala (1955), i.627-9.

35 In Rinmansarkivet, Tekniska Museet, Stockholm.

36 Bergius, P. J., *Anteckningar från en resa till Dalarna äv 1751* (1951), 77.

37 Beckmann, Johann, *Schwedische Reise in den Jahren 1765-1766: Tagebuch*, Uppsala (1911), 64. Schreber, Daniel Gottfried, *Neue Sammlung verschiedener in die Cameralwissenschaften*, Bützow und Wismar (1762), i.86 (article by Johann Christian Daniel Schreber).

38 Uppsala Universitetsbibliotek MS. S. 29b, Ritning i anonym resebeskrivning.

39 Linnaeus, C., *Dalaresa, Utlandsresan och Bergslagsresan*, Stockholm (1953), 397.

40 Outhier, [R], *Journal d'un voyage au Nord en 1736 & 1737*, Paris (1744), 178.

41 Jars, iii.40, ii.86.

42 Calvoer, Henning, *Acta Historico-Chronologico-Mechanica circa Metallurgiam in Hercynia Superiori oder . . . Maschinenwesens und der Hülfsmittel bey dem Bergbau auf dem Oberharze*, Braunschweig (1763), ii.77-8.

43 Johnson, W. A., *Christopher Polhem*, Hartford, Conn. (1963), 38-9.

44 Boehmer, 84.

45 Beyer, August, *Beytraege zur Bergbaukunde*, Dresden (1794), 282-95.

46 *Bergmännisches Journal*, Jahrg. v Bd. i, Freyberg & Annaberg (1792), 483-91.

47 d'Aubuisson, J. F., *Des Mines de Freiberg en Saxe et de leur Exploitation*, Leipsic (1802), i.202, 230.

48 Lempe, C. T., in *Magazin der Bergbaukunde*, ix, Dresden (1792), 225ff.

49 Baader, Joseph von, *Neues System der fortschaffenden Mechanik*, München (1822), *passim*.

50 See Lee, C. E., *The Evolution of Railways*, 2nd ed., London (1943), 92-3.

Chapter 6

1 Feldhaus, F. M., *Ruhmesblätter der Technik*, Leipzig (1910), 515.

2 See, e.g., Fürst, Artur, *Die Welt auf Schienen*, München (1918), 82, and Lee, Charles E., *The Evolution of Railways*, 2nd ed., London (1943), 11.

3 Daicoviciu, Constantin, *Din Istoria Transilvaniei*, Bucureşti (1961), 180.

4 Tollius, Jacobus, *Epistolae Itinerariae*, Amsterdam (1700), 166.

5 Müller, Konrad, *Siebenbürgische Wirtschaftspolitik unter Maria Theresia*, München (1961), 22-4.

6 Böhm, L., *Geschichte des Temeser Banats*, Leipzig (1861), ii.91, 97.

7 Kaindl, R. F., *Geschichte der Deutschen in den Karpathenländern*, Gotha (1907), ii.101.

8 Griselini, Franz, *Geschichte des Temeswarer Banats*, Wien (1780), ii.65-7, 75.

9 Esmark, Jens, *Kurze Beschreibung einer mineralogischen Reise durch Ungarn, Siebenbürgen und das Bannat*, Freyberg (1798), 64.

10 Schmidt, xii.29.

11 *Ibid.*, 481.

12 Born, Ignatz Edler von, *Briefe über mineralogische Gegenstände, auf seiner Reise durch das Temeswarer Bannat, Siebenbürgen, Ober- und Nieder-Hungarn, an den Herausgeber derselben, Johann Jacob Ferber*, Frankfurt and Leipzig (1774), 101.

13 Stütz, Andreas, *Physikalisch-Mineralogische Beschreibung des Gold- und Silber Bergwerkes zu Szekerembe bey Nagyag in Siebenbürgen*, Wien (1803), 48-9.

14 Becker, Wilhelm Gottlob Ernst, *Journal einer Bergmännischen Reise durch Ungarn und Siebenbürgen*, Freyberg (1816), ii.192.

15 *Studii:Revistă de istorie*, x.2, Bucureşti (1957), Planşa iv. Franzenau, F., *Systematische Beschreibung des Bergwerks Szekeremb oder Nagyag*, MS. in Arhivă Stat Cluj, Arhivă Zlatna, No. 1528, pp. 97-9, kindly communicated by Drs. S. Goldenberg and A. Neamţu.

16 *Jahrbuch der k.k. Geologischen Reichsanstalt*, viii, Wien (1857), 135-6.

17 *Verhandlung und Mittheilungen des siebenbürgischen Vereins für Naturwissenschaften zu Hermannstadt*, ii, Hermannstadt (1851), 88.

18 *Jahrbuch der k.k. Geologischen Reichsanstalt*, viii, Wien (1857), 719-20.

19 Esmark, *op. cit.*, 89, 105.

20 Fridvaldszky, Joannes, *Minero-Logia Magni Principatus Transilvaniae*, Claudiopoli (1767), 78.

21 *Berg- und Hüttenmännisches Jahrbuch der k.k. montanistischen Hochschulen zu Leoben und Přibram*, liii, Wien (1905), 147.

22 I am indebted for this information to Bergassessor a.D H.Conrad of the Bergbau-Museum, Bochum.

23 Kaindl, *op. cit.*, 101.

24 Schmidt, vi.312.

Chapter 7

1 Nef, J. U., *The Rise of the British Coal Industry*, London (1966), i.19-20. For readers wishing to pursue the economic side of the coal industry, this book is much to be recommended, along with Ashton, T. S., and Sykes, J., *The Coal Industry of the Eighteenth Century*, Manchester (1964): both are reprints of earlier-published works. On the more technical side, Galloway, R. L., *Annals of Coal Mining and the Coal Trade*, London (1898), has still not been superseded, although Atkinson, Frank, *The Great Northern Coalfield 1700-1900*, Barnard Castle (1966), covers a part of the subject.

2 *Transactions of the Royal Irish Academy*, Dublin, v (1793), 271-2; Mott, R. A., in *TNS*, xxxvii (1964-5), 2.

3 Taylor, T. J., 'The Archaeology of the Coal Trade', in *Memoirs Chiefly Illustrative of the History and Antiquities of Northumberland . . .* (Archaeological Institute of Great Britain), London, i (1858), 169-70; Nef, *op. cit.*, ii.376; *Extracts from the Records of the Company of Hostmen* (Publications of the Surtees Society, vol. cv), Durham (1901), 38; Proceedings of Committee of House of Commons, 1738-9, in NEIMME, Buddle Coll., 14/288; carts are illustrated on the map in Gardiner, Ralph, *England's Grievance discovered in Relation to the Coal Trade*, Newcastle (1655).

4 *Acts of the Privy Council 1595-96*, London (1901), 32.

5 Edington, R., *A Treatise of the Coal Trade*, London (1817), 133, 116-18.

6 Paris, Matthew, *Historia Major*, London (1640), 661.

7 Surtees, R., *The History and Antiquities of the County Palatine of Durham*, London (1820), ii.209.

8 PRO, Durham 4/i/107, 226.

9 *Acts of the Privy Council 1623-25*, London (1933), 362, 487.

10 'England's First Rails, a Reconsideration', in *Renaissance and Modern Studies*, iv, Nottingham (1960), 119-34.

11 For his career, see Smith, Richard S., 'Huntingdon Beaumont, Adventurer in Coalmines', in *Renaissance and Modern Studies*, i (1958), 115-53.

12 *Ibid.*, 120-1.

13 Historical Manuscripts Commission, *Report on the MSS. of Lord Middleton*, London (1911), 177.

14 PRO, C2 James I, S 27/4.

15 Gray, W., *Chorographia, or a Survey of Newcastle upon Tine*, Newcastle (1649), 24-5.

16 In Gateshead Public Library.

17 *Renaissance and Modern Studies*, iv, Nottingham (1960), 129.

18 PRO, LR 1/203/198.

19 BM Stowe MS. 326, 7ff, quoted in full by Nef, *op. cit.*, ii.417-19.

20 *Transactions of the North of England Institute of Mining Engineers*, Newcastle, xv (1865-6), 188. Lee, C. E., 'The World's Oldest Railway', in *TNS*, xxv (1945-7), 142.

21 *Memoirs Chiefly Illustrative of the History and Antiquities of Northumberland* . . ., London (1858), i.180.

22 In DRO.

23 Newcastle Record Office, Act Book of the Common Council, 1656-1722, 59.

24 Newcastle Record Office, Journal of the Common Council, 1650-9, 364.

25 PRO, St. Ch. 8, 310/16.

26 SRO, 1224/163.

27 PRO, St. Ch. 8, 109/8.

28 Administrations, Somerset House.

29 All these documents are in SRO, Tyrwhitt-Jones Coll., Box 43, except the map, SRO, 1224/1/32.

30 SRO, 1224/163.

31 *Ibid.*, Tyrwhitt-Jones Coll., Box 43.

32 *Ibid.*

33 Pepper, J. H., *The Play book of Metals*, London (1861), 23.

34 Nottingham University Library, Middleton MSS. 177/37-38; information kindly supplied by Dr R. S. Smith. The site of Lime Pit Fields is established by the Tithe map of Kinlet, Tithe Redemption Office 29/181, the site of Bower Hill by the 6-in. O.S.

35 *Register of the Privy Council of Scotland*, 1st series (1885), vii.278-9.

36 *The Acts of Parliament of Scotland*, vi, part 1 (1870), 606.

37 *Transactions of the North of England Institute of Mining Engineers*, Newcastle xv (1865-6), 177.

38 *Renaissance and Modern Studies*, iv, Nottingham (1960), 120.

39 Bailey, F. A., *Transactions of the Historical Society of Lancashire and Cheshire*, xc, Liverpool (1947), 1ff. Harris, J. R., in *Journal of Transport History*, ii.3, Leicester (1956), 175.

40 I am grateful to Mr. J. Langton for telling me that in his researches into the history of the South Lancashire Coalfield he has found no tittle of evidence for waggonways at Prescot or elsewhere in the seventeenth century.

41 PRO, E 178/5276.

42 Historical Manuscripts Commission, *Report on the MSS. of Lord Middleton*, London (1911), 175.

43 For a more detailed discussion of this question, see Nef, *op. cit.*, i.240-5 and ii.446-51.

44 SRO, 1224/163.

45 Randall, John, *Broseley and Its Surroundings*, Madeley (1879), 96.

46 *Edgar Allen News*, Sheffield, 44 (September 1965), 205. The wheel is now (1966) in the possession of Mr. B. C. Roberts of Broseley.

47 Historical Manuscripts Commission, *Report on the MSS. of Lord Middleton*, London (1911), 494-7.

48 Nef, *op. cit.*, i.65.

49 *Transactions of the Shropshire Archaeological and Natural History Society*, xi (1888), 10ff.

50 *Ibid.*, 426.

51 Broseley Parish Registers.

52 *Acts of the Privy Council 1601-4*, London (1907), 214.

Chapter 8

1 Sopwith, Thomas, *Guide to Newcastle* (1838), 103.

2 NUL, Misc. MSS. 10, f.67 (James Clavering's letterbook).

3 NEIMME, 23/10.

4 Gibson, John, *Plan of the Collieries on the Rivers Tyne and Wear, also Blyth, Bedlington and Hartley . . .*, 1787 (first issue). This map is the source wherever the date of 1787 occurs in this chapter; but it is not completely reliable either for topography or for fact.

5 PRO, FEC 1/W.37; *History of Northumberland*, Newcastle (1909), ix.231ff.

6 NRO, 2DE 7/8.

7 Lee, C. E., in *Railway Magazine*, Mar-Apl 1944, 111; *History of Northumberland* (1907), viii.20-1.

8 NEIMME, 24/37. Charles E. Lee's article, 'Tyneside Tramroads of Northumberland', in *TNS*, xxvi (1947-9), 199-229, though useful, is not always completely accurate.

9 *History of Northumberland* (1907), viii.20, 412; NEIMME, 49/4/199, 21/1, Buddle Coll. 23/106ff.

10 NRO, ZWI 1 (Willington lease-book); NEIMME, Watson Coll. 10/54/46, 67.

11 NRO, ZWI 1; ZAN M 17/197a 62; ZWI 8. NEIMME, 22/6-7; Johnson Coll. 4/39-44, 66-107. Newcastle Record Office, Journal of Common Council 1766-85, 190-3, 457-8. Richardson, William, *History of the Parish of Wallsend*, Newcastle (1923), 213-16, 475.

12 Casson, W., *Plan shewing Collieries and Waggonways on the Rivers Tyne and Wear* (1801). NEIMME, 5/10/32a, Johnson Coll. 12/8.

13 *History of Northumberland*, Newcastle (1930), xiii.47; NEIMME, Watson Coll. 11/7/50.

14 Newcastle Record Office, Journal of Common Council 1743-66, 188-90, 216.

15 GPL, Cotesworth AC 2/3. Newcastle Record Office, Journal of Common

Council 1718-43, 8, 52. GPL, Cotesworth CB 1/1-11. NEIMME, 18/1/24, 32.

16 Newcastle Record Office, Journal of Common Council 1718-43, 317, 326; 1743-66, 13, 26, 79. NEIMME, Watson Coll. 10/54/1, 4; 20*a*/9; 18/1/74, 98. *History of Northumberland*, Newcastle (1930), xiii.437.

17 Newcastle Record Office, Journal of Common Council 1718-43, 461. *History of Northumberland*, Newcastle (1930), xiii.264. NEIMME, 49/4/86, 149.

18 NEIMME, Johnson Coll. 3/13. Dendy Marshall, fig. 71. NEIMME, Johnson Coll. 12/10.

19 *Edgar Allen News*, Sheffield (1925), 635.

20 NRO, ZDT Box 12; Newcastle Record Office, Journal of Common Council 1718-43, 15. NRO, ZAN M 17/197a 45. NEIMME, 25/2, 8.

21 *PSAN*, 3rd series, iv (1910), 46. Hughes, i.170.

22 Welford, Richard, *A History of the Parish of Gosforth*, Newcastle (1886), 80. *Archaeologia Aeliana*, 3rd series, v, Newcastle (1909), 110. NEIMME, 49/4/1.

23 Casson, W., *Plan shewing Collieries and Waggonways on the Rivers Tyne and Wear*, 1801.

24 NRO, ZAN M 17/197a 24, 48. NEIMME, 16/1/35.

25 Hedley, O. D., *Who invented the Locomotive Engine?*, London (1858), 1. Baxter, B., *Stone Blocks and Iron Rails*, Newton Abbot (1966), 150.

26 PRO, Durham 4/ii/193.

27 *History of Northumberland*, Newcastle (1926), xii.164-5. SCL, Bagshawe Coll. 3305-6.

28 NRO, Armstrong Coll., Plan of Crawcrook Waggonway 1789. Newcastle Central Library, Gowland MSS., vol. 1.

29 Taylor, T. J., in *Memoirs chiefly illustrative of the History and Antiquities of Northumberland* (Archaeological Institute of Great Britain), London (1858), i.180. NRO, ZCO viii/1/4 (counterpart in DRO, LO/F/136).

30 *PSAN*, 3rd series, viii.153. SCL, Wharncliffe MS. 140/10.

31 NRO, ZCO viii/1/13; ZCE 10/2; Armstrong Coll., A Map of y[e] Estates of John Clavering Esq, 1721. GPL, Cotesworth BS 9/1, 4; BS 3/10. GPL, Ellison A 35.

32 Bourn, William, *History of the Parish of Ryton, including the Parishes of Winlaton, Stella and Greenside*, Carlisle (1896), 87.

33 *Ibid.*, 184. NRO, ZDT Box 12. PRO, FEC 1/W 26/1/10. GPL, Cotesworth BS 3/7.

34 *TNS*, xxv (1945-7), 141-62.

35 Surtees, R., *The History and Antiquities of the County Palatine of Durham*, London (1820), ii.209n. Bailey, John, *General View of the Agriculture of the County of Durham*, London (1810), 35.

36 SCL, Wentworth Woodhouse Muniments, BR 72/2, 6, 8, 10, 19.

37 DRO, Strathmore Papers (lease and leading lists). BM Add. MS 40747, 160.

38 SCL, Wharncliffe MSS. D 652; 115/3, 7, 8.

39 GPL, Ellison A 35; Cotesworth BV 1/8. *Archaeologia Aeliana*, 3rd series, v, Newcastle (1909), 138.

40 NUL, Misc. MS. 10, 4ff. Hughes, i.16f, 196. GPL, Cotesworth, BV 1/9.

41 GPL, Cotesworth BT 3/6.

42 DRO, Strathmore Papers. BM Add. MS. 40747, 184.

43 GPL, Cotesworth CL 1, CB 1, CA 13. *PSAN*, 3rd series, viii, 153. NUL, Misc. MS. 10, 178.

44 Hughes, i.235.

45 GPL, Cotesworth CA 13. Hughes, i.235. NEIMME, 18/1/4.

46 GPL, Cotesworth BT 4/24.

47 NEIMME, 18/1/51, 74. *PSAN*, 3rd series, viii.153.

48 NUL, Misc. MS. 10, 178, 182-3. NEIMME, 18/1/2.

49 NEIMME, 18/1/6, 10, 11, 44; Johnson Coll., 1/198.

50 NEIMME, Buddle Coll. 14/275, 281; Johnson Coll. 1/27. DRO, D/X 35/8.

51 DRO, D/XP 14.

52 NEIMME, 46/19. Casson, W., *Plan shewing Collieries and Waggonways on the Rivers Tyne and Wear*, 1801. *Transactions of the North of England Institute of Mining Engineers*, xv, Newcastle (1865), 215. 'Map of the Railways in the Newcastle on Tyne Coal Field in 1812' in Galloway, R. L., *The Steam Engine and its Inventors*, London (1881).

53 GPL, Cotesworth AC 4/10. Newcastle Record Office, Journal Book of Common Council 1699-1718, 19*a*.

54 NRO, ZAN M 17/197c, 1, 15, 16. GPL, Cotesworth AC 4/10.

55 NEIMME, Buddle Coll. 18/93. Newcastle Record Office, 0020/5/10.

56 Hughes, i.7.

57 Brand, John, *The History and Antiquities of . . . Newcastle upon Tyne*, London (1789), ii.125. NRO, ZAN M 17/197c 54, 43, 48. Sykes, John, *Local Records*, Newcastle (1833), i.350.

58 NUL, Misc. MS. 10, 59. NEIMME, 33/18, a map of 1776 which does not mark the Beamish extension, usually said to date from 1763. NRO, ZAN M 17/197c 43.

59 Sykes, John, *op. cit.*, i.317. NRO, ZAN M 17/197c 43.

60 NEIMME, Buddle Coll. 45/1/44. Burleigh and Thompson, *A Plan of the River Wear from Newbridge to Sunderland Barr*, 1737, a most useful map for the river end of waggonways at that date.

61 Dendy Marshall, 22. Railway and Canal Historical Society and North Eastern Railway Association Tour, 19 June 1966: *Historical Notes of Waggonways of the River Wear*, by C. H. A. Townley.

62 DRO, LO/F/338. *Archaeologia Aeliana*, new series ii, Newcastle (1858), 112-13.

63 Hopkinson, G. G., in *Transactions of the Hunter Archaeological Society*, Sheffield, viii (1963), 131, giving no source. SCL, Arundel Castle MSS. S 192-195B.

64 Wood, John, *Essay towards a Description of Bath*, London (1749), ii.424-5.

65 CRO, D/Lons: John Spedding's notebook on management of Whitehaven collieries, *c.* 1699-1703. Moore, R. W., in *Victoria County History*, Cumberland, ii (1905), 360.

66 CRO, D/Lons: Howgill and Whingill Colliery Abstract of Accounts 1728-1751, and John Spedding's letter-book 1734-6 (16 November 1735), and 1737-9 (24 March 1737/8). *Cumberland Pacquet*, 10 August 1802.

67 CRO, D/Lons: John Spedding's letter-book 1737-9 (11 January 1737/8). NEIMME, 16/1/260. CRO, D/Lons: Plan of Enterance into Town, Wagon Road and Steaths (1754?). Moore, R. W., in *Transactions of the Federated Institution of Mining Engineers*, vii, Newcastle (1893-4), 625.

68 Fletcher, Isaac, in *CW*, iii (1876-7), 296, 298. CRO, D/Cu/7/1; D/Cu/4/218; D/Lons: plan of Workington Harbour and ground adjoining; D/Sen, Shipping 10 (plan of Harbour of Workington 1777).

69 NEIMME, 16/1/25. CRO, D/Lons: papers relating to Seaton Colliery, 1719-41; D/Sen, Shipping 10. Fletcher, Isaac, in *CW*, iii (1876-7), 298.

70 CRO, D/Cu/4/216. *Term Reports*, i, 560 (Senhouse v Christian, January 1787). Hughes, ii.142-3.

71 Moore, R. W., in *Victoria County History*, Cumberland (1905), ii.380. Baxter, B., in *Journal of the Railway and Canal Historical Society*, 8.4 (1962), 57.

72 *Journal of the House of Commons of the Kingdom of Ireland*, iv, 1731-40, appendix clxxxii-cxcvi. Patterson, E. M., *The Ballycastle Railway*, Dawlish (1965), 12-13.

73 *A Letter to a Commissioner of the Inland Navigation concerning the Tyrone Collieries*, Dublin (1752), 9. McNeill, D. B., *Ulster Tramways and Light Railways*, Belfast (1956), 8.

74 McCutcheon, W. A., *The Canals of the North of Ireland*, Dawlish (1965), 69-74. *Reports of the late John Smeaton, F.R.S.*, London (1812), ii.279 (report of 1774).

75 NEIMME, Watson Coll. 5/8/13. McNeill, *op. cit.*, 8. Chapman, William, *Observations on the Various Systems of Canal Navigation*, London (1797), 7. *Edinburgh Encyclopaedia* (1830), sub. Navigation, Inland.

76 Rees, Abraham, *The Cyclopaedia, or Universal Dictionary of Arts, Sciences and Literature*, vi, London (1819), sub Canal. Parker, J., *Illustrated Rambles from Hipperholme to Tong*, Bradford (1904), 34.

77 *Bradford Antiquary*, n.s. iii, Bradford (1911), 315. *Industrial Archaeology*, vi, Newton Abbot (1969), 77.

78 Cusworth Hall Museum, Doncaster, miscellaneous coal-mining MSS., *Inquisition ad quod damnum* of 1755. I am most grateful to Mr. J. F. Goodchild for telling me of this source.

79 31 Geo II, cap. 22 (9 June 1758).

80 *Leedes Intelligencer*, 26 September 1758.

81 NEIMME, Watson Coll. 8/8/532.

82 For the Middleton Railway in its wooden phase, see (Garnett, D.), *The Middleton Colliery Railway, Leeds*, 2nd ed., Leeds (1960), and Rimmer, G., in *Yorkshire Bulletin of Economic and Social Research*, vii, No. 1, Hull (1955), 41-57.

83 BTHR, York, ACN/1/5.

84 *Ibid.*, ACN/1/5. LCL, DB 212/10/14.

85 BTHR, York, ACN/1/7; 4/122.

86 *Ibid.*, ACN/1/7.

87 *York Courant*, 22 October 1745.

88 Roberts, George, *Topography and Natural History of Lofthouse and Its Neighbourhood*, London and Leeds (1882), i.75-6.

89 BTHR, York, ACN/1/5; 4/122; 4/136.

90 Yorkshire Archaeological Society, Wakefield Manor Court Rolls, 1783, entry 50.

91 BTHR, York, ACN/1/7. *Journal of the Railway and Canal Historical Society*, v, No. 3 (1959), 51.

92 Galloway, R. L., *Annals of Coal Mining*, London (1898), i.321.

93 SCL, Arundel Castle MSS. S218, 215, 217. *The Universal Magazine of Knowledge and Pleasure*, lv, London (December 1774), 330.

94 Leader, R. E., *Sheffield in the Eighteenth Century*, Sheffield (1901), 341. SCL, Arundel Castle MSS. S198.

95 *Transactions of the Hunter Archaeological Society*, Sheffield, viii (1963), 131. *Gentleman's Magazine*, London, xxxiii (1763), 533. West Riding Record Office, Pontefract Quarter Sessions Rolls, April 1766.

96 14 Geo III, cap. 93.

97 *York Courant*, 12 September 1769.

98 NRO, ZAN M 17/197b 64.

99 Fleming, J. A., *Scottish and Jacobite Glass*, Glasgow (1938), 155. Dott, George, *Early Scottish Colliery Waggonways*, London (1947), 31-2.

100 *Ibid.*, 2. Sinclair, Sir John, *Statistical Account of Scotland*, Edinburgh (1792), ii.270. *Sporting Magazine*, xxix (1807), 248.

101 NEIMME, 16/1/261-3

102 Pococke, Richard, *Tours in Scotland*, Edinburgh (1887), 276. Sinclair, *op. cit.* (1795), xvi.519. NEIMME, Buddle Coll. 6/230.

103 Inglis, J. C. and F., *The Fordell Railway*, Larbert (1946), 6-7. Sinclair, *op. cit.* (1795), xv.270. *New Statistical Account of Scotland*, Edinburgh (1845), ix.186.

104 NEIMME, Johnson Coll. 2/59-73

105 *Ibid.*, Johnson Coll. 2/113. Earl of Elgin's papers, Broomhall, Fife. Pennant, Thomas, *A Tour in Scotland, 1772*, London (1776), ii.217. Sinclair, *op. cit.* (1794), xiii.467.

106 DRO, LO/F/338.

107 Clerk, Sir John: Scottish Record Office, GD/18, 2115 and 2106, quoted in *CW*, 2nd series, lxv (1965), 307, and *Transactions of the Architectural and Archaeological Society of Durham and Northumberland*, xi, Gateshead (1965), 426.

108 *Journal of the House of Commons*, xxxiv (1774), 604.

109 Chapman, William, *Observations on the Various Systems of Canal Navigation*, London (1797), 42.

110 *Term Reports*, vii (1798), 599.

111 Stukeley, W., *Itinerarium Curiosum*, centuria ii, London (1776), 68. Pococke, Richard, *Northern Journey, 1760* (Publications of Surtees Society, vol. cxxiv), Durham (1915), 241.

112 Desaguliers, J. T., *A Course of Experimental Philosophy*, London (1734), i.275. Evelyn, Helen, *The History of the Evelyn Family*, London (1915), 179. Shaw, Rev. S., *Tour to the West of England in 1788* (1789), reprinted in J. Pinkerton, *General Collection of Voyages and Travels*, London (1808), ii.252.

113 *Westminster Review*, January 1835, 135.

114 Shaw, *op. cit.* Defoe, Daniel, *A Tour thro' the Whole Island of Great Britain*,

2nd ed., London (1738), ii.246; Kielmansegge, Count Frederick, *Diary of a Journey to England in the Years 1761-1762*, London (1902), 132. Owen, *Observations on the Earth, Rocks, Stone and Minerals for Some Miles about Bristol* (1754), 240.

115 Loveday, John, *Diary of a Tour in 1732*, Edinburgh (1890), 172. *Gentleman's Magazine*, February 1769, 59. *TNS*, xxxii (1959-60), 21.

Chapter 9

1 Spearman, Gilbert, *An Enquiry into the Ancient and Present State of the County Palatine of Durham* (1729), 61, 73, 87.

2 Hughes, i.7, 154. Tomlinson, W. W., *The North Eastern Railway, Its Rise and Development*, Newcastle (1914), 6-7.

3 GPL, Cotesworth CH 1/1-4.

4 NRO, Armstrong Coll. St. Mary's Gateshead Churchwardens' Accounts 1650-1807, 215.

5 GPL, Cotesworth CA 13.

6 *Term Reports*, i.560 (Senhouse *v.* Christian, January 1787).

7 Tomlinson, *op. cit.*, 7. Barnardiston, Thomas, *Reports of Cases determined in the Court of King's Bench*, London (1744), i.318-19. NEIMME, Johnson Coll. 1/281.

8 NRO, 2DE 7/1.

9 North, Roger, *The Life of the Right Honourable Francis North, Baron of Guilford*, London (1742), 136.

10 *Extracts from the Records of the Company of Hostmen* (Publications of the Surtees Society, vol. cv), Durham (1901), 151.

11 e.g. 1714: GPL, Ellison A 35; 1739: *Journal of the House of Commons*, 28 February 1738/9, NEIMME, Buddle Coll. 14/251ff.

12 Hughes, i.154.

13 Historical Manuscripts Commission, *Report on the MSS. of the Duke of Portland*, London (1901), vi.104.

14 DRO, Strathmore Papers.

15 Newcastle Record Office, Journal of Common Council 1766-85, 190ff.

16 13 Geo. II, cap. 14.

17 20 Geo. II, caps. 9, 12 and 13 (Newcastle and Belford, Durham and Tyne Bridge, and Sunderland and Durham Road Acts of 1746).

18 31 Geo. II, cap. 22.

19 NEIMME, Watson Coll. 5/8/13.

20 GPL, Cotesworth AC 2/30.

21 On the Dunston way: *Archaeologia Aeliana*, 3rd series, v, Newcastle (1909), 136.

22 e.g. Bucksnook, GPL, Cotesworth BV 1/8.

23 Historical Manuscripts Commission, *Report on the MSS. of the Duke of Portland*, London (1901), vi.104.

24 SCL, She. 160. NEIMME, Johnson Coll. 2/85. Stephen, William, *Story of Rosyth and Inverkeithing*, Edinburgh (1938), 102.

25 NEIMME, 46/19.

26 *Ibid.*, 18/1/4, 54.

27 Wood, Nicholas, *A Practical Treatise on Rail-Roads*, London (1825), 85. Dunn, Matthias, *An Historical, Geological and Descriptive View of the Coal Trade of the North of England*, Newcastle (1844), 39.

28 GPL, Cotesworth CA 13.

29 NEIMME, Buddle Coll. 24/45.

30 GPL, Ellison A 35.

31 *Encyclopaedia Britannica*, 2nd ed., Edinburgh (1778), iii.2043.

32 NEIMME, 16/1/257-8.

33 Inglis, J. C. and F., *The Fordell Railway*, Larbert (1946), 10. Earl of Elgin's papers, Broomhall, Fife.

34 Patent Office Library, Specification 653 of 1750.

35 LCL, MC 50, 202.

36 de Saint-Fond, B. Faujas, *Travels in England and Scotland, 1784* (1907), 139f.

37 *Leedes Intelligencer*, 11 June 1798.

38 Curwen MSS., Workington; quoted in Hughes ii.143.

39 Sanderson, P., *The Antiquities of the Abbey or Cathedral Church of Durham, also A particular Description of the County Palatine of Durham*, Newcastle (1767), part 2, 27.

40 Goodchild, J., in *Journal of the Railway and Canal Historical Society*, 5.3 (1959), 51.

41 BM Add. MS. 40747, 184. GPL, Cotesworth CL 1, CB 1.

42 Hutchinson, W., *History and Antiquities of the County Palatine of Durham*, Carlisle (1794), iii.498.

43 NUL, Misc. MS. 10, 182-3.

44 NEIMME, Buddle Coll. 14/352-3.

45 Stukeley, William, *Itinerarium Curiosum*, centuria ii, London (1776), 69. Loveday, John, *Diary of a Tour in 1732*, Edinburgh (1890), 172.

46 NEIMME, 23a/27-31.

47 *Ibid.*, Buddle Coll. 18/123-4.

48 *Ibid.*, Watson Coll. 10/54/55.

49 *Ibid.*, 49/4/30-1.

50 Jars, i.200. Andersson, B. Q., in *Kongl. Swenska Vetenskaps Academiens Handlingar*, xxxvii, Stockholm (1776), 308.

51 NEIMME, 38/1/7.

52 GPL, Cotesworth CL 1, CB 1.

53 NEIMME, 18/1/4. Loveday, *op. cit.*, 172.

54 Hutchinson, *op. cit.*, Newcastle (1778), ii.423.

55 Brand, John, *The History and Antiquities of . . . Newcastle upon Tyne*, London (1789), ii.306n. It is sometimes said that the first attempt was in wood, but Brand implies, and Stukeley confirms, that it was in stone.

56 *Kongl. Swenska Vetenskaps Academiens Handlingar*, xxxvii, Stockholm (1776), 308.

57 GPL, Cotesworth BT 4/24.

58 DRO, Strathmore Papers. NEIMME, 18/1/9.

59 GPL, Cotesworth CB 1.

60 Honeyman, H. L., in *PSAN*, 4th series, iii (1929), 111-15.

61 NRO, ZRI 35/15.

62 CRO, D/Lons, Plan of Enterance into Town, Wagon Road and Steaths. *Reports of the late John Smeaton, F.R.S.*, London (1812), iii.396f. Moore, R. W., in *Transactions of the Federated Institution of Mining Engineers*, vii, Newcastle (1893-4), 625.

63 *Term Reports*, i.560 (Senhouse *v.* Christian, January 1787). *CW*, iii (1876-7), 304.

64 McCutcheon, W. A., *The Canals of the North of Ireland*, Dawlish (1965), pl. xii.

65 GPL, Ellison C xx.

66 NUL, Misc. MS. 10, 157. GPL, Ellison A 35.

67 NUL, Misc. MS. 10, 9.

68 GPL, Ellison C xx. DRO, Strathmore Papers.

69 NEIMME, Watson Coll. 8/10/31-4.

70 NRO, ZAN M 17/197c 64, 76.

71 *Ibid.*, ZCO viii/1/4.

72 For a general description of staiths and spouts, see Jars, i.201.

73 Newcastle Record Office, Foreshore Papers, 137.

74 GPL, Cotesworth BS 3/24.

75 Andersson, B. Q., in *Kongl. Swenska Vetenskaps Academiens Handlingar*, xxxvii, Stockholm (1776), 309.

76 Fisher, Joseph, in *Transactions of the Royal Irish Academy*, v, Dublin (1793), 272.

77 Pennant, Thomas, *A Tour in Scotland and Voyage to the Hebrides, MDCCLXXII*, Chester (1774), 48.

78 Dunn, Matthias, *A Treatise on the Winning and Working of Collieries*, Newcastle (1848), 144.

Chapter 10

1 Jars, i.200.

2 Wood, Nicholas, *A Practical Treatise on Rail-Roads*, London (1825), 40.

3 NEIMME, Buddle Coll. 15/201.

4 *History of Northumberland*, viii, Newcastle (1907), 21.

5 GPL, Cotesworth BS 9/4. SCL, Arundel Castle MS. S219.

6 NEIMME, 16/1/258; Watson Coll. 8/10/31-4.

7 e.g., *Ibid.*, Watson Coll. 5/9/20 (in 1805).

8 *Ibid.*, 16/1/23; Earl of Elgin's Papers, Broomhall, Fife. *Journal of the House of Commons of the Kingdom of Ireland*, iv, 1731-48, Appendix cxci.

9 CRO, D/Lons: drawing of iron waggon wheel, 1735. Hughes, ii.200-3. Dendy Marshall, 28.

10 Wood, *op. cit.*, 39. NEIMME, Watson Coll. 11/15.

11 CRO, D/Cu/3/65.

12 de Saint-Fond, B. Faujas, *Travels in England and Scotland, 1784*, (1907), 139. Mackenzie, E., *A Historical and Descriptive View of the County of Northumberland*, Newcastle (1811), i.194.

13 NEIMME, Buddle Coll. 18/17.
14 Wood, *op. cit.*, 3rd ed. (1838), 20.
15 NEIMME, Buddle Coll. 18/201.
16 *Ibid.*, Buddle Coll. 18/93.
17 *Ibid.*, 49/4/199.
18 Wood, *op. cit.*, 1st ed. (1825), 42; 3rd ed. (1838), 22.
19 *Penny Cyclopaedia*, London (1841), xix.245.
20 DRO, Strathmore Papers.
21 31 Geo. II, cap. 22, p. 4. LCL, MC 1, MC 2, MC 50.
22 West Riding Record Office, Pontefract Quarter Sessions Rolls, April 1766. Guest, John, *Relics and Records of Men and Manufactures at Rotherham*, Rotherham (1866), 55.
23 Jars, i.201.
24 NEIMME, Watson Coll. 8/7/57ff.
25 *Ibid.*, 46/19.
26 *Ibid.*, Buddle Coll. 3/46. Smiles, Samuel, *Life of George Stephenson*, 2nd ed., London (1857), 195. NEIMME, Watson Coll. 8/15, 9/13.
27 Details of the Sunderland line kindly supplied by Mr. Frank Atkinson. Ayr: Butt, John, *Industrial Archaeology of Scotland*, Newton Abbot (1967), 218.
28 Dendy Marshall, 28. Tomlinson, W. W., *The North Eastern Railway, Its Rise and Development*, Newcastle (1914), 13.
29 Cumming, T. G., *Illustrations of the Origin and Progress of Rail and Tram Roads*, Denbigh (1824), 10.
30 *Prize-Essays and Transactions of the Highland Society of Scotland*, vi, Edinburgh (1824), 7.
31 Rees, Abraham, *The Cyclopaedia, or Universal Dictionary of Arts, Sciences and Literature*, vi, London (1819), sub Canal.
32 SCL, Arundel Castle MS. S194.
33 NRO, ZCO iv/47/18a.
34 Ed. Armstrong, Sir W. G., *The Industrial Resources of the Tyne, Wear and Tees*, Newcastle (1864), 280.
35 NRO, 2DE 7/8, 8/2/3.
36 *Trans. Hunter Archaeological Society*, Sheffield, viii (1963), 132. *Letters of William Scott*, Newcastle (1848), *passim.*
37 GPL, Cotesworth BS 9/4. PRO, FEC 1/W 26/1/7. NEIMME, Johnson Coll. 4/66-107.
38 Hughes, ii.200-3.
39 LCL, MC 1, 2, 64 etc.
40 *Journal of the House of Commons of the Kingdom of Ireland*, iv, 1731-48, Appendix cxci.
41 NUL, Misc. MS. 10, 59.
42 NEIMME, 16/1/258.
43 GPL, Ellison C xx.
44 NRO, ZAN M 17/197a 24, 197c 76.
45 Andersson, B. Q., in *Kongl. Swenska Vetenskaps Academiens Handlingar*, xxxvii, Stockholm (1776), 307-8.

46 NEIMME, 16/1/257.
47 *Ibid.*, Buddle Coll. 24/47.
48 *Ibid.*, 46/19.
49 Earl of Elgin's papers, Broomhall, Fife.
50 NEIMME, Watson Coll. 11/7/50.
51 *Ibid.*, Watson Coll. 8/7/57ff. NEIMME, 46/19.
52 *Ibid.*, Watson Coll. 9/13.
53 *Ibid.*, 29/52. *Transactions of the North of England Institute of Mining Engineers*, xvii, Newcastle (1867-8), Appendix 1, 163.
54 Tomlinson, *op. cit.*, 407ff. Dendy Marshall, fig. 62.
55 NEIMME, 16/1/264.
56 Hutchinson, William, *The History of the County of Cumberland*, Carlisle (1794), ii, map facing p. 41. Moore, R. W., in *Transactions of the Federated Institution of Mining Engineers*, vii, Newcastle (1893-4), 625.
57 NRO, ZCE 10/2 (8 June 1714).
58 Dixon, Joshua, *The Literary Life of Wm. Brownrigg, with an account of the Coal-mines near Whitehaven*, London (1801), 110n.
59 GPL, Ellison C xx (Ravensworth). DRO, N.C.B. deposited papers (Tanfield). NRO, Armstrong Coll., Report on Garesfield Colliery.
60 GPL, Cotesworth BV 1/11. *Journal of the House of Commons of the Kingdom of Ireland*, iv, 1731-48, Appendix cxci. NEIMME, 49/4/199; Watson Coll. 9/13.
61 Jars, i.200. *Encyclopaedia Britannica*, 2nd ed., Edinburgh (1778), iii.2043.
62 NEIMME, 16/1/258; Buddle Coll. 45/1/6.
63 *Ibid.*, 18/1/28, 34. SCL, Wharncliffe MS. 115/7.
64 *Prize-Essays and Transactions of the Highland Society of Scotland*, vi, Edinburgh (1824), 22. Dendy Marshall, fig. 62.
65 Strickland, William, *Report on Canals, Railways, Roads and Other subjects*, Philadelphia (1826), fig. 42. Oeynhausen, C. von, and Dechen, H. von: see *TNS*, xxix (1953-5), 7.
66 Dendy Marshall, fig. 71.
67 Tomlinson, *op. cit.*, 83. Inglis, J. C. and F., *The Fordell Railway*, Larbert (1946), 10. *Transactions of the Architectural and Archaeological Society of Durham and Northumberland*, xi, Gateshead (1965), 426, quoting Scottish Record Office, GD/18, 2106.
68 Document of 1835 quoted in Warren, J. G. H., *A Century of Locomotive Building*, Newcastle (1923), 331.
69 NEIMME, Watson Coll. 9/13.
70 Earl of Elgin's papers, Broomhall, Fife: Chas Landale's commonplace book.
71 CRO, D/Lons: drawing of Iron Wagon Wheel, 1735.
72 Hedley, O. D., *Who Invented the Locomotive Engine?*, London (1858), 33. Tomlinson, *op. cit.*, 83.
73 Lee, Charles E., *The Evolution of Railways*, 2nd ed., London (1943), 2.

Chapter 11

1 GPL, Cotesworth CG 3/11, A 5/18.
2 30 Car. II, cap. 8. E.g. NEIMME, Watson Coll. 10/54/4.

3 DRO, Strathmore papers.

4 NEIMME, 18/1/52, 58.

5 Newcastle Record Office, Journal of Common Council 1743-66, 188, 216.

6 NEIMME, 49/4/17. DRO, Strathmore papers.

7 Newcastle Record Office, Journal of Common Council 1766-85, 190-3. NRO, ZWI 1.

8 GPL, Cotesworth BT 4/1, AC 4/7, 10, 18.

9 6-7 Will. III, cap. 10. 11 Geo. II, cap. 15.

10 DRO, N.C.B. deposited papers.

11 Nef, J. U., *The Rise of the British Coal Industry*, London (1966), ii.368-70. 30 Car. II, cap. 8. 6-7 Will. III, cap. 10.

12 SCL, Wentworth Woodhouse Muniments BR 72/6.

13 NRO, ZCO viii/1.

14 *Reports of the late John Smeaton, F.R.S.*, iii, London (1812), 412. BTHR, York, ACN/1/5,7.

15 NEIMME, 16/1/258. Hutchinson, William, *The History of the County of Cumberland*, Carlisle (1794), ii.66-7. *Mariner's Mirror*, liv (1968), 72.

16 CRO, D/Lons: John Spedding's notebook on management of Whitehaven collieries, *c.* 1699-1703; Papers relating to Seaton Colliery 1719-41; letter-book of John Spedding 1737-9 (19 February 1737/8, 28 May 1738). Hutchinson, *op. cit.*, ii.140.

17 Hughes, ii.139, 146-7, 177. NEIMME, 49/4/302.

18 NEIMME, Johnson Coll. 2/85. Dendy Marshall, 119. Sinclair, Sir John, *Statistical Account of Scotland*, Edinburgh (1795), xv.270. *New Statistical Account of Scotland*, Edinburgh (1845), ix.186. *Sporting Magazine*, xxix (1807), 248.

19 Scottish Record Office, GD/18, 2106, quoted in *Transactions of the Architectural and Archaeological Society of Durham and Northumberland*, xi, Gateshead (1965), 427. NEIMME, Watson Coll. 20a/9.

20 *The London Magazine, or Gentleman's Monthly Intelligencer*, xxxiii, London (1764), 144-5. *The General Magazine of Arts and Sciences*, London (1764), 285.

21 Morand, *L'Art d'exploiter les mines de charbon de terre*, ii, section 4, Paris (1776), 865.

22 GPL, Ellison C xx.

23 *Kongl. Swenska Vetenskaps Academiens Handlingar*, xxxvii, Stockholm (1776), 306.

24 1679, Stella Grand Lease: NRO, ZCO iv/47/8a. 1765, Willington: NEIMME, Watson Coll. 10/54/67.

25 SCL, Arundel Castle MS. S193. GPL, Ellison C xx. LCL, MC 1.

26 North, Roger, *The Life of the Right Honourable Francis North, Baron of Guilford*, London (1742), 136-7.

27 PRO, FEC 1/W 26/1/7.

28 GPL, Ellison C xx.

29 Dendy Marshall, 138 – the authors in question are Brand, Hutchinson, and Faujas de Saint-Fond.

30 GPL, Cotesworth CA 7/5. Wood, Nicholas, *A Practical Treatise on Rail-Roads*, London (1825), 77.

31 GPL, Ellison C xx.

32 NRO, 2DE 7/8. NEIMME, Buddle Coll. 46/13.

33 *General Magazine* (1764), 285f. Jars, i.202. LCL, MC 1.

34 Fawcett, Rev. R. J., *History of the Parish of Dipton* (1911), quoted in *TNS*, xxv (1945-7), 144. NEIMME, 46/19. Bourn, William, *Whickham Parish: Its History, Antiquities and Industries*, Carlisle (1893), 29. NRO, ZAN M 17/197a 48.

35 Patent Office Library, Specification 529 of 1731.

36 Loveday, John, *Diary of a Tour in 1732*, Edinburgh (1890), 172.

37 GPL, Ellison C xx. NEIMME, Johnson Coll. 2/63.

38 Dunn, Matthias, *An Historical, Geological and Descriptive View of the Coal Trade of the North of England*, Newcastle (1844), 39.

39 SCL, Arundel Castle MS. S194.

40 CRO, D/Lons: drawing of iron waggon wheel, 1735; drawing of Whitehaven hurry; Stock Account Book, Whitehaven, 1775-9. Scottish Record Office, GD/18, 2115, quoted in *CW*, 2nd series*i*, lxv (1965), 307. Dixon, Joshua, *The Literary Life of Wm. Brownrigg, with an account of the Coal Mines near Whitehaven*, London (1801), 112. CRO, D/Lons: papers relating to Seaton Colliery, 1719-41.

41 LCL, MC 2.

42 NEIMME, 16/1/258.

43 NRO, 2DE 7/8.

44 NEIMME, Buddle Coll. 18/42. Parker, James, *Illustrated Rambles from Hipperholme to Tong*, Bradford (1904), 295.

45 NEIMME, Buddle Coll. 14/52.

46 SCL, Arundel Castle MS. S198. GPL, Cotesworth CA 7/6.

47 NRO, 2DE 7/8.

48 SCL, Wharncliffe MS. 115/7. LCL, MC 1. NEIMME, Watson Coll. 10/54/67.

49 SCL, Arundel Castle MS. S194, S192. CRO, D/Lons: Stock Account Book, Whitehaven, 1775-9.

50 Andersson, B. Q., in *Kongl. Swenska Vetenskaps Academiens Handlingar*, xxxvii, Stockholm (1776), 307.

51 Brand, John, *The History and Antiquities of . . . Newcastle upon Tyne*, London (1789), ii.687.

52 NRO, ZCO iv/47/8a.

53 Jars, i.203.

54 NEIMME, 16/1/258.

55 *Ibid.*, 16/1/260.

56 Rees, Abraham, *The Cyclopaedia, or Universal Dictionary of Arts, Sciences and Literature*, vi, London (1819): sub Canal.

57 Hedley, O. D., *A Descriptive Account of the Means used on the Tyne and Wear for effecting the safe Transit of Railway Carriages in ascending and descending Steep Planes, together with a historical Sketch of the Brake as applied to railway Carriages*, Newcastle (1834), 11-17.

58 Inglis, J. C. and F., *The Fordell Railway*, Larbert (1946), 13. Dott, George, *Early Scottish Colliery Waggonways*, London (1947), 14.

59 Dendy Marshall, 119. GPL, Ellison C xx. NEIMME, Watson Coll. 10/54/67.

60 NEIMME, 18/70; Buddle Coll. 24/1, 15/201.

61 NRO, ZCO iv/47/8a.

62 NEIMME, Buddle Coll. 18/17.

63 SCL, Wharncliffe MS. 115/7. LCL, MC 203, 204. CRO, D/Cu/7/1.

64 *Journal of the House of Commons of the Kingdom of Ireland*, iv, 1731-48, Appendix cxcii.

65 In order: NEIMME, 18/70. NRO, 2DE 7/8. NRO, ZCO, iv/47/11. NEIMME, 18/69. Hughes, i.170. PRO, FEC 1/P 29, FEC 1/W 26/1/8. Hughes, i.170. NEIMME, 18/1/6.

66 NEIMME, 12/65/1-21. Inglis, J. C. and F., *The Fordell Railway*, Larbert (1946), 14. NEIMME, Johnson Coll. 2/85.

67 NRO, 2DE 7/8.

Chapter 12

1 *Transactions of the Institution of Mining Engineers*, lxxxi, London (1931), 293.

2 Hughes, i.254.

3 NEIMME, 18/1/10.

4 Sinclair, Sir John, *Statistical Account of Scotland*, Edinburgh (1794), x.507. *New Statistical Account of Scotland*, Edinburgh (1845), x.186.

5 NRO, Blackett (Wylam) papers.

6 *Ibid.*, Armstrong Coll.

7 *Ibid.*, Blackett (Wylam) papers. See also Washington Bond of 1793, NEIMME, Buddle Coll. 18/89.

8 NEIMME, 18/1/58. Hughes, i.141.

9 *Commercial and Agricultural Magazine*, iii, London (1800), 243. SCL, Wharncliffe MS. 140.

10 Quoted in Hughes, i.254.

11 LCL, MC 1, MC 65.

12 GPL, Cotesworth BT 3/6. Andersson, B. Q., in *Kongl. Swenska Vetenskaps Academiens Handlingar*, xxxvii, Stockholm (1776), 309.

13 NEIMME, 18/1/6. Hughes, i.255.

14 NRO, Blackett (Wylam) papers. LCL, MC 64. *The London Magazine, or Gentleman's Monthly Intelligencer*, xxxiii, London (1764), 145.

15 NEIMME, Buddle Coll. 13/1. NRO, Armstrong Coll.

16 SCL, Arundel Castle MS. S196. LCL, MC 204 and *passim*.

17 NEIMME, Buddle Coll. 6/3/12.

18 Swann, E. W., in *TNS*, xxiv (1943-5), 111-112. NRO, ZWI 1 (1774 Willington lease).

19 Quoted in Hughes, i.255.

20 Hughes, i.251.

21 NRO, ZCO iv/47/8a.

22 NUL, Misc. MS. 10, 94, 99, 100. *Correspondence of James Clavering*, Publications of Surtees Society, Vol. 178, Gateshead (1967), 68, 102, 226.

Ashton, T. S., and Sykes, Joseph, *The Coal Industry in the 18th Century*, Manchester (1964), 89-91, 116-19.

23 NRO, Armstrong Coll., abstract of farms of John Clavering.

24 NEIMME, Buddle Coll. 18/17. NRO, Blackett (Wylam) papers. NEIMME, Buddle Coll. 3/28-30.

25 Jars, i.203.

26 *The County Palatine of Durham, survey'd by Capt. Armstrong* . . . (1768), vignette; here the artist apparently misinterprets the brake lever as a shaft.

27 Ayton, R., *Voyage around Great Britain* (1814), 150f.

28 NEIMME, 16/1/258, 264.

29 GPL, Ellison A 35. NEIMME, Buddle Coll. 14/326.

30 GPL, Cotesworth CL 1, CB 1.

31 Wood, Nicholas, *A Practical Treatise on Rail-Roads*, London (1825), 87-8.

32 *London Magazine*, xxxiii (1764), 145. *General Magazine* (1764), 285ff.

33 Bourn, William, *History of the Parish of Ryton, including the Parishes of Winlaton, Stella and Greenside*, Carlisle (1896), 23, 24, 184.

34 NRO, ZCE 10/2.

35 SCL, Arundel Castle MS. S194.

36 NUL, Misc. MS. 10, 93, 42.

37 *CW*, xv (1899), 392ff.

38 NRO, ZCO iv/47/5-18 *passim*.

39 *Archaeologia Aeliana*, 3rd series, v, Newcastle (1909), 136.

40 GPL, Ellison A 35.

41 NEIMME, 18/1/4.

42 *Ibid.*, 18/1/11; Johnson Coll. 1/84, CRO, D/Lons: Papers relating to Seaton Colliery 1719-41.

43 NEIMME, Buddle Coll. 18/17; Johnson Coll. 2/70.

44 NEIMME, 49/4/302.

45 GPL, Ellison C xx; Cotesworth CA 7/4.

46 NRO, ZCO iv/47/8a.

47 GPL, Ellison C xx. NEIMME, Buddle Coll. 18/93.

48 Brand, John, *The History and Antiquities of . . . Newcastle upon Tyne*, London (1789), ii.687n.

49 NEIMME, Buddle Coll. 18/17. (Baillie, Rev. John), *An Impartial History of the Town and County of Newcastle upon Tyne and Its Vicinity*, Newcastle (1801), 484.

50 NEIMME, Johnson Coll. 12/10. Inglis, J. C. and F., *The Fordell Railway*, Larbert (1946), 13. Bailey, J., and Culley, G., *General View of the Agriculture of the County of Northumberland*, 2nd ed., Newcastle (1797), 10. SCL, Arundel Castle MS. S215/16.

51 NEIMME, 9/13.

52 PRO, FEC 1/W 37.

53 SCL, Arundel Castle MS. S193, S194.

54 GPL, Cotesworth BS 9/1*a-b*, 4.

55 GPL, Ellison C xx.

56 NRO, ZCO iv/47/8a.

57 NEIMME, 18/1/54.
58 GPL, Cotesworth BS 3/7.
59 NUL, Misc. MS. 10, 76, 153, 173. NRO, ZCE 10/2.
60 NRO, ZAN M 17/197c 76.
61 GPL, Cotesworth CB 1/10. NEIMME, 18/1/54. DRO, Strathmore papers.
62 GPL, Ellison C xx.
63 *Ibid.*, Cotesworth BS 9/7.
64 NEIMME, Buddle Coll. 18/93. GPL, Ellison C xx. PRO, FEC 1/W 26/1/7.
65 GPL, Cotesworth A 5/18.
66 *Ibid.*, BS 9/4.
67 Brand, *op. cit.*, 687n.
68 Sykes, John, *Local Records, or, Historical Register of Remarkable Events*, Newcastle (1833), i.304, 317. *Leedes Intelligencer*, 26 September 1758.
69 BM Add. MS. 40747, 110.
70 Quoted in Hughes, i.17. See also Hughes, i.196.
71 GPL, Cotesworth BV 1/9.
72 NUL, Misc. MS. 10, 4, 9, 11.
73 St. Mary's Gateshead Churchwardens' Accounts 1680-1807, 215.
74 GPL, Cotesworth CG 3/14.
75 NRO, ZRI 35/15.
76 *The Universal Magazine of Knowledge and Pleasure*, lv, London (1774), 320.
77 Curr, John, *Railway Locomotion and Steam Navigation*, London (1847), 179.
78 *Archaeologia Aeliana*, New Series, ii, Newcastle (1858), 111-14.
79 DRO, Durham Quarter Sessions Order Book 1765-75, 84; 1774-83, 11.
80 GPL, Cotesworth BS 9/4.
81 NUL, Misc. MS. 10, 186.

Chapter 13

1 *The Itinerary of John Leland* (ed. Thomas Hearne), Oxford (1744), vi.93.
2 *A History of Wednesbury*, Wolverhampton (1854), 99.
3 Dudley, Dud, *Mettallum Martis, or Iron made with Pit-coale, Sea-coale, &c.*, London (1665), 35.
4 Plot, Robert, *The Natural History of Stafford-shire*, Oxford (1686), 129.
5 SRO, Tyrwhitt-Jones Coll., Box 43.
6 *Ibid.*
7 SRO, 1681 (Ferriday Family), Box 183/15.
8 *Ibid.*, Tyrwhitt-Jones Coll., Box 43.
9 Ele, Martin, in *Philosophical Transactions*, xix, No. 228, London (1697), 544.
10 Randall, John, *Broseley and Its Surroundings*, Madeley (1879), 59, 57.
11 SBL, Map and Admeasurements of Broseley Hall Estate.
12 Randall, *op. cit.*, 74, 97.
13 SRO, 1224/143. Savage, R. F., and Smith, L. D. W., *The Waggon-Ways and Plate-ways of East Shropshire*, Birmingham School of Architecture thesis

(1965), 134-6. I am much indebted to this work for the basis of the Shropshire railway system.

14 *Ibid.*

15 SRO, Madeley Vestry papers (on temporary loan).

16 SBL, MS. 331, 124-38 *passim*. The site of Lakehead is established by the Tithe map of Madeley, Tithe Redemption Office 29/225, and by BM Map K xxxvi. 16. 1. Raistrick, Arthur, *Dynasty of Ironfounders*, London (1953), 70, erroneously puts Lakehead at the top of the Coalmoor railway from Coalbrookdale; Mott, Dr. R. A., in *TNS*, xxxvii (1964-5), 10, puts Lakehead at Willey. His two papers, 'English Waggonways of the Eighteenth Century', deal largely with Coalbrookdale and its railways, but some of their conclusions have to be treated with caution.

17 SBL, MSS. 329, 331.

18 SRO, 1681 (Ferriday Family), Box 183/14.

19 Library of Society of Friends, Norris MS. viii 66-69.

20 Wexler, P. J., *La Formation du Vocabulaire des Chemins de Fer en France (1778-1842)*, Genève et Lille (1955), 19.

21 SRO, 1681 (Ferriday Family), Box 183/1; 676 (Charlton of Apley Estate Accounts) 26 September 1747.

22 Savage and Smith, *op. cit.*, 53, 112-13, quoting leases in SRO.

23 SRO, 1224/299. Savage and Smith, *op. cit.*, 59, 70, 124.

24 For its general history, see Raistrick, *op. cit.*

25 *Ibid.*, 174.

26 *Victoria County History*, Shropshire (1908), i.463 (J. Randall).

27 Library of Society of Friends, Journal of Abiah Darby.

28 SRO, 1681 (Dawley Deeds), Box 1.

29 This section is based on SBL, MS. 333, with details culled from Raistrick's *Dynasty of Ironfounders*, Mott in *TNS*, xxxvii (1964-5), and Messrs. Savage and Smith's thesis.

30 William Salt Library, Stafford, D 1788 61/5. Almost all the rest of the evidence for the Stour Navigation footrails is contained here and in D 1788 37/8, 59/3 and 61/7.

31 Historical Manuscripts Commission, 5th Report, part i, Appendix 160 (Duke of Sutherland's papers); 15th Report, part x, 65 (MSS. of Shrewsbury Corporation). *Reasons for making Navigable the Rivers of Stower and Salwerp . . .* (1661?). *An Answer as well to a Paper, Entituled Reasons wherefore the making Navigable of the Rivers of Stower and Salwerp . . . will be of great advantage to the County of Salop* (1661?).

32 14 Car. II (copy in William Salt Library, Stafford, D 1788 45/14).

33 Hadfield, Charles, *The Canals of the East Midlands*, Newton Abbot (1966), 27, 15. Information kindly supplied by Philip Weaver.

34 Randall, *op. cit.*, 97.

35 All the evidence for Mackworth's railway can be found in the following broadsides and pamphlets: *The Case of Sir Humphry Mackworth, and of the Mine-Adventurers, with respect to the Irregular Proceedings of severall Justices of the Peace for the County of Glamorgan, and of their Agents and Dependents* (pp. 1-16; pagination continues in) *Affidavits, Certificates and Presentments, Proving the facts mentioned in the Case of Sir Humphry Mackworth . . .*,

London (1705), 5, 10-12, 28, 36-7, 42, 52. *The Case of Sir Humphry Mackworth in the Report of his Complaints for Breach of Privilege*, [London, 1706], 4-6. *Case . . . with Respect to the Extraordinary Proceedings of the Agents . . . of the Rt. Hon. Sir Thomas Mansell, Bar.*, London (1707), 6. *The Case of John Morgan and others*, London (1705). *A View of the Advantages arising to the Copartners or Company of the Mineral Manufacturers at Neath . . .*, London (1720), frontispiece. Mackworth, Sir Humphry, *The Second Part of the Book of Vouchers to Prove the Case and Defence of the Deputy Governor and Directors of the Company of Mine-Adventurers*, London (1711), 8. Waller, William, *The Mine Adventurers laid open . . .*, London (1710), x. *Journal of the House of Commons*, 18 December 1705.

36 Earl of Elgin's papers, Broomhall, Fife.

37 *Blue Books*, 1810, iv.170.

38 Phillips, D. Rhys, *History of the Vale of Neath*, Swansea (1925), 290.

39 Jones, W. H., *The Port of Swansea*, Carmarthen (1922), 118, 260. *Blue Books*, 1810, iv.152.

40 Swansea University Library, Morris MSS. Information kindly supplied by Mr. J. Hollingworth. *TNS*, xxxii (1959-60), 22.

41 Ed. Lloyd, Sir John E., *A History of Carmarthenshire*, Cardiff (1939), ii.342. Hadfield, Charles, *The Canals of South Wales and the Border*, London (1960), 31.

42 Morris, W. H., in Cambrian Archaeological Association, *Annual Meeting at Carmarthen, Programme* (1967), 11.

43 Dendy Marshall, 85.

44 Howell, Jenkin, in *Cambrian Notes and Queries*, 1902. Hadfield, *Canals of South Wales*, 118f. BTHR, Aberdare Canal Minute Book, 12 July 1799.

45 For an excellent account of this railway, see Sir Arthur Elton's Presidential Address in *Proceedings of the Somersetshire Archaeological and Natural History Society*, cvii, Taunton (1963). Desaguliers, J. T., *A Course of Experimental Philosophy*, London (1734), i.274-9.

46 Evelyn, Helen, *The History of the Evelyn Family*, London (1915), 179. Defoe, Daniel, *A Tour thro' the Whole Island of Great Britain*, 2nd ed., London (1738), ii.246.

47 Shaw, Rev. S., *Tour to the West of England in 1788* (1789), reprinted in Pinkerton, J., *General Collection of Voyages and Travels*, London (1808), ii.252.

48 Raistrick, *op. cit.*, 58. Palmer, A. N., in *Transactions of the Honourable Society of Cymmrodorion*, London (1897), 31. *Journal of the National Library of Wales*, viii, Aberystwyth (1953-4), 341.

49 National Library of Wales, Hawarden Deeds 1327, 1704, 1852. Rawson, C. Rees, in *Archaeologia Cambrensis*, xciv, Cardiff (1941), 130f. [Willet, Richard], *A Memoir of Hawarden Parish, Flintshire*, Chester (1822), 99, 101-5.

50 Clegg, H., in *Transactions of the Lancashire and Cheshire Antiquarian Society*, lxv (1955), 92.

51 *Transactions of the Historical Society of Lancashire and Cheshire*, lxxxii (1930), 163ff.; xc (1938), 91ff.

52 *Ibid.* c (1948), 55ff.

53 Barker, T. C., and Harris, J. R., *A Merseyside Town in the Industrial Revolution: St Helens, 1750-1900*, London (1959), 35, 145.

54 *Williamson's Liverpool Advertiser*, 16 December 1757; *Liverpool General Advertiser*, 19 June 1778. I am grateful to Mr. J. Langton for these and other Lancashire references.

55 PRO, PL 6 85/35. *Liverpool General Advertiser*, 14 July 1769.

56 John Ryland's Library, Legh Muniments Box 43.

57 Harris, J. R., in *Journal of Transport History*, ii.3, Leicester (1956), 175-6. Barker and Harris, *op. cit.*, 86.

58 Murray, David, *The York Buildings Company*, Glasgow (1883), 65.

59 Chambers, Robert, *Domestic Annals of Scotland*, Edinburgh (1861), iii.472.

60 Dendy Marshall, 113.

61 Brown, Kenneth, in *Railway Magazine*, January 1938, 1-4, and references therein.

62 Hogg, James, *The Jacobite Relics of Scotland*, Edinburgh (1821), ii.113.

63 Hamilton, Henry, *An Economic History of Scotland in the Eighteenth Century*, Oxford (1963), 209. Earl of Elgin's papers, Broomhall, Fife (Chalmers correspondence, 4 February 1769).

64 Bald, Robert, *General View of the Coal Trade of Scotland*, Edinburgh (1812), 10. Sinclair, Sir John, *Statistical Account of Scotland*, Edinburgh (1793), viii.617. *New Statistical Account of Scotland*, Edinburgh (1845), viii.30-1.

65 SRO, Tyrwhitt-Jones Coll., Box 43.

66 William Salt Library, Stafford, D 1788 59/3.

67 SRO, Tyrwhitt-Jones Coll., Box 43.

68 Randall, John, *Broseley and Its Surroundings*, Madeley (1879), 74.

69 BTHR, Brecon and Abergavenny Canal minute book 17 October 1798.

Chapter 14

1 SRO, Tyrwhitt-Jones Coll., Box 43.

2 Randall, John, *Broseley and Its Surroundings*, Madeley (1879), 74. Savage, R. F., and Smith, L. D. W., *The Waggon-Ways and Plateways of East Shropshire*, Birmingham School of Architecture thesis (1965), 135, 112.

3 National Library of Wales, Hawarden Deeds, No. 1327. Earl of Elgin's Papers, Broomhall, Fife.

4 Peach, R. E. M., *The Life and Times of Ralph Allen*, London (1895), 78-81.

5 SRO, 1224/299.

6 Morris, W. H., in Cambrian Archaeological Association, *Annual Meeting at Carmarthen, Programme* (1967), 11.

7 SBL, MS. 333/185-6, 193, 205.

8 Elton, Sir Arthur, in *Proceedings of the Somersetshire Archaeological and Natural History Society*, cvii, Taunton (1963), 48. Savage and Smith, *op. cit.*, 53.

9 Library of Society of Friends, Norris MS., xiii.11.

10 Randall, John, *History of Madeley*, Madeley (1880), 291.

11 *Annals of Agriculture*, iv, London (1785), 166-7.

12 Gregory, W., *Ralph Allen and Prior Park*, Bath (1886), 37.

13 SRO, Tyrwhitt-Jones Coll., Box 43.

14 Gregory, *loc. cit.*

15 Savage and Smith, *op. cit.*, 134. Wexler, P. J., *La Formation du vocabulaire des Chemins de Fer en France (1778-1842)*, Genève et Lille (1955), 19.

16 Sinclair, Sir John, *Statistical Account of Scotland*, Edinburgh (1793), viii.617. John Ryland's Library, Legh muniments, Box 43.

17 SBL, MS. 333/323. Joseph von Baader, *Neues System der fortschaffenden Mechanik*, München (1822), Taf. 1. Sinclair, *op. cit.*, 617.

18 SBL, MS. 333/143. Phillips, D. Rhys, *History of the Vale of Neath*, Swansea (1925), 238.

19 SBL, MS. 333/123. *Affidavits, Certificates and Presentments proving the Facts mentioned in the Case of Sir Humphry Mackworth*, London (1705), 52. Desaguliers, J. T., *A Course of Experimental Philosophy*, London (1734), i, pl. 22.

20 SRO, Madeley Vestry papers (on temporary loan).

21 SBL, MS. 332/217.

22 *New Statistical Account of Scotland*, Edinburgh (1845), viii.30-1.

23 Rathbone, H. M., *Letters of Richard Reynolds with a Memoir of His Life*, London (1852), 27.

24 Raistrick, Arthur, *Dynasty of Ironfounders*, London (1953), 178.

25 SBL, MS. 333/277, 148.

26 Mott, R. A., in *TNS*, xxxvii (1964-5), 29-30.

27 *Blue Books*, 1809, iii.578-80.

28 Mott, R. A., in *TNS*, xxxvii (1964-5), 23-4.

29 *Victoria County History*, Shropshire (1908), i.465 (J. Randall).

30 Hauck, Ehrhard, *Joseph Ritter von Baader*, München (1933). Birch, A., in *Bulletin of John Ryland's Library*, 35 (1952-3), 324. Baader, Joseph von, *Neues System der fortschaffenden Mechanik*, München (1822), 25-6.

31 Randall, John, *History of Madeley*, Madeley (1880), 291.

32 Mott, R. A., in *TNS*, xxxvii (1964-5), 26-9.

33 Raistrick, *op. cit.*, 174.

34 Wexler, *op. cit.*, 19. Randall, John, *Broseley and Its Surroundings*, Madeley (1879), 77.

35 SBL, MS. 333/407.

36 Earl of Elgin's papers, Broomhall, Fife: letterbook of Rose, factor, 1792. Barker, T. C., and Harris, J. R., *A Merseyside Town in the Industrial Revolution: St. Helens, 1750-1900*, London (1959), 35. Holt, John, in *Communications to the Board of Agriculture*, i, London (1797), 192. [Willett, Richard], *A Memoir of Hawarden Parish, Flintshire*, Chester (1822), 104.

37 Wilkes, John, in *Communications to the Board of Agriculture*, ii, London (1800), 477.

38 Sinclair, *op. cit.*, viii.617. *New Statistical Account*, viii.30-1. Erskine of Mar, J. F., in *Communications to the Board of Agriculture*, i, London (1797), 203.

39 Kielmansegge, Count Frederick, *Diary of a Journey to England in the years 1761-1762*, London (1902), 132.

40 Gillespie, Robert, *Round About Falkirk* (1868), 154.

41 Mott, R. A., in *TNS*, xxxvii (1964-5), 30.

42 SRO, 1224/299. BM map-room K 36. 16. 1.

43 Scott, A., in *Prize-Essays and Transactions of the Highland Society of*

Scotland, vi, Edinburgh (1824), 23. Dott, George, *Early Scottish Colliery Waggonways*, London (1947), 29.

44 Hadfield, Charles, *The Canals of South Wales and the Border*, London (1960), 119. Oeynhausen, H. von, and Dechen, C. von, in *AfB*, xix (1829).

45 *The Case of John Morgan and Others* (1705).

46 William Salt Library, Stafford, D 1788 61/5.

47 SRO, 1681 (Ferriday Family), Box 183/1.

48 *Ibid.*, Tyrwhitt-Jones Coll., Box 43.

49 *Ibid.*, Madeley Vestry papers (on temporary loan).

50 *Ibid.*, 1224/143.

51 Library of Society of Friends, Norris MS. viii.66-9.

52 SRO, 1681 (Ferriday Family), Box 183/14.

53 Library of Society of Friends, Norris MS., xiii.11.

54 SRO, Madeley Vestry papers (on temporary loan).

55 Desaguliers, *op. cit.*, 276. Elton, Sir Arthur, in *Proc. Soms. Arch. Nat. Hist. Soc.*, cvii.44.

56 Dott, *op. cit.*, 20, 30.

57 SRO, Madeley Vestry papers (on temporary loan). SBL, MS. 333/122-3, 199-201.

58 SRO, Labouchere Coll., Particulars of the Expence in making a Railway Waggon, 1796.

59 *The County Palatine of Durham, survey'd by Capt. Armstrong* . . . (1768), vignette.

60 Dendy Marshall, 113. Dott, *op. cit.*, 30.

61 Desaguliers, *op. cit.*, 275-6.

62 SBL, MS. 329/194, 211; 331/64, etc. Coalbrookdale Company Ltd. MS. 1.

63 SRO, Madeley Vestry papers (on temporary loan).

64 SBL, MS. 332/3, 6.

65 Sinclair, *op. cit.*, viii.617. Rawson, R. Rees, in *Archaeologia Cambrensis*, xciv, Cardiff (1941), 131.

66 SBL, MS. 332/18, 89, 177, 196. SRO, Labouchere Coll.

67 e.g. SBL, MS. 332/43, 177.

68 William Salt Library, Stafford, D 1788 61/7. SRO, 676 (Charlton of Apley Estate Accounts), October 17 1748.

69 SBL, MS. 332/127.

70 SRO, 1224/299.

71 SBL, MS. 332/54.

72 William Salt Library, Stafford, D 1788 37/8.

73 SBL, MS. 332/246. Horsehay iron output figures from *TNS*, xxxi (1957-9), 280.

74 William Salt Library, Stafford, D 1788 61/5. Gregory, W., *Ralph Allen and Prior Park*, Bath (1886), 37.

75 Sinclair, *op. cit.*, viii.617.

76 Donovan, E., *Descriptive Excursions through South Wales and Monmouthshire in the Year 1804 and the four preceding Summers*, London (1805), i.xxii.

77 Raistrick, *op. cit.*, 174.

78 SRO, Madeley Vestry papers (on temporary loan). *Victoria County History, Shropshire* (1908), i.463. (J. Randall).

79 Waller, William, *An Essay on the Value of the Mines, late of Sir Carbery Price*, London (1698), Preface.

80 William Salt Library, Stafford, D 1788 59/3, 61/7.

81 Elton, Sir Arthur, in *Proc. Soms. Arch. Nat. Hist. Soc.*, cvii.44.

82 Mott, R. A., in *TNS*, xxxvii (1964-5), 13, 14, 24.

83 Savage and Smith, *op. cit.*, 128f.

84 William Salt Library, Stafford, D 1788, 59/3, 61/7.

85 SBL, MS. 331/125, 129, 132.

86 *Ibid.*, MS. 333/28.

87 SRO, 1224/299.

88 *Ibid.*, Madeley Vestry papers (on temporary loan).

Chapter 15

1 BTHR, BCN 1/1 (25 September 1767).

2 16 Geo. III, cap. 28 and 66; 23 Geo. III, cap. 92.

3 BTHR, CHC 1/1 (30 November 1778).

4 Farey, John, *General View of the Agriculture of Derbyshire*, London (1817), iii.340.

5 BTHR, EWC 1/1 (19 January 1793). Stevenson, Peter, in *Stanton and Staveley News*, iii (1966), 71. Farey, *op. cit.*, 288, 364.

6 BTHR, EWC 1/1 (12 January 1787, 1 May 1787, 5 May 1789, 28 December 1792, 5 May 1778, 6 May 1783), Farey, *op. cit.*, 288.

7 Patterson, A. Temple, in *Transactions of the Leicestershire Archaeological and Historical Society*, xxvii (1951), 74.

8 BTHR, ASHC 2/4.

9 Farey, *op. cit.*, 288.

10 Hadfield, Charles, *The Canals of the West Midlands*, Newton Abbot (1966), 33-4. 16 Geo. III, cap. 32. Hollick, J. R., in *Railway Magazine*, June 1937, 435ff.

11 William Salt Library, Stafford, Sneyd Papers, HM 37/9.

12 Farey, *op. cit.*, 453f.

13 23 Geo. III, cap. 33.

14 Curr, John, *The Coal Viewer and Engine Builder's Practical Companion*, Sheffield (1797), 13.

15 In Rees, Abraham, *The Cyclopaedia, or Universal Dictionary of Arts, Sciences and Literature*, vi, London (1819), sub Canal.

16 Farey, *op. cit.*, 453.

17 *Liverpool General Advertiser.*

18 BTHR, LLC 1/41.

19 Clegg, H. E., in *Transactions of the Institution of Mining Engineers*, cxvii (1957-8), 784ff. Information from Mr. D. Anderson of Parbold.

20 BTHR, LLC 1/41 (21 October 1780). Holt, J., in *Communications to the Board of Agriculture*, i, London (1797), 192. Information from Mr. D. Anderson.

21 BTHR, LLC 1/3 (30 March 1787); LLC 1/43 (19 December 1790).

22 *Ibid.*, LLC 1/3 (27 January 1786).

23 *Ibid.*, LLC 1/3 (30 March 1787).

24 Hudson, G. S., in *National Register of Archives, West Riding, Northern Section, Annual Report and Bulletin*, x (1967), 21.

25 Hogrewe, J. L., *Beschreibung der in England seit 1759 angelegten und jetzt gröstentheils vollendeten schiffbaren Kanäle*, Hannover (1780), 146.

26 BTHR, CC 1/2 (30 April 1773, 12 November 1773). Elton, Sir Arthur, in *Proceedings of the Somersetshire Archaeological and Natural History Society*, cvii (1963), 56-7. Household, Humphrey, *The Thames & Severn Canal*, Newton Abbot (1969), 58.

27 *Scottish Historical Review*, Edinburgh, xxix (1950), 121ff. 10 Geo. III, cap. 105. Sinclair, Sir John, *Statistical Account of Scotland*, Edinburgh (1793), vii.383.

28 Welch, Edwin, *The Bankrupt Canal: Southampton and Salisbury 1795-1808*, Southampton (1966), 17-18.

29 *Reports of the late John Smeaton, F.R.S.*, London (1812), iii.396f.

30 Chapman, William, *Observations on the Various Systems of Canal Navigation*, London (1797), 42ff.

31 Willan, T. S., *The Navigation of the River Weaver in the Eighteenth Century*, Manchester (1951), 42ff.

32 This account of the East Shropshire Canals in general, and their inclines in particular, is drawn from Hadfield, Charles, *The Canals of the West Midlands*, Newton Abbot (1966), 150-65; Williams, W. Howard, in *Journal of Industrial Archaeology*, Newton Abbot, 2.3 (1965), 91-105; BTHR, SHPC 1/1, especially under 24 February, 22 October, 10 December 1790; 29 August, 6 October 1791; 4 October 1792; 8 April 1793; 26 September, 2 October 1800; 15 April 1801; Plymley, J., *General View of the Agriculture of Shropshire* (1803), 292ff.; Dutens, J. H., *Mémoires sur les Travaux Publics de l'Angleterre*, Paris (1819), 42; Oeynhausen, C. von, and Dechen, H. von, in *AfB*, xix (1829).

33 Hadfield, Charles, *The Canals of Southern England*, London (1955), 57-9.

34 Fulton, Robert, *A Treatise on the Improvement of Canal Navigation*, London (1796).

35 Chapman, William, *Observations on the Various Systems of Canal Navigation*, London (1797), 42.

36 Hadfield, Charles, *The Canals of Southern England*, London (1955), 242-4, 257, 370.

37 See, for example, Ashton, T. S., *Iron and Steel in the Industrial Revolution*, Manchester (1963), 142ff.

38 Elsas, Madeleine, *Iron in the Making: Dowlais Iron Company Letters 1782-1860*, Cardiff (1960), 171.

39 Coxe, Rev. William, *An Historical Tour in Monmouthshire*, London (1801), ii.230. *Blue Books*, 1850, xxxi.192.

40 Abbott, R., in *Transactions of the Leicestershire Archaeological and Historical Society*, xxxi (1955), 51ff.

41 Farey, *op. cit.*, 288.

42 Leader, R. E., *Sheffield in the Eighteenth Century*, Sheffield (1901), 341.

43 National Library of Wales, Maybery papers, Bundle 14.

44 Dendy Marshall, 148. Nemnich, P. A., *Neueste Reise durch England, Schottland und Ireland*, Tübingen (1807), 64.

45 Thompson, James, in *Newcastle Courant*, 18 December 1824. Wood, Nicholas, *A Practical Treatise on Rail-Roads*, London (1825), 61.

46 NEIMME, Buddle Coll., 15/202.

47 *Ibid.*, Johnson Coll., 12/1ff.

48 Earl of Elgin's papers, Broomhall, Fife: Chas. Landale's commonplace book, 1 December 1807.

49 NEIMME, Buddle Coll., 23/106-115.

50 *Ibid.*, Watson Coll., 8/8/532. Fernie, John, *A History of the Town and Parish of Dunfermline*, Dunfermline (1815), 116. *New Statistical Account of Scotland*, Edinburgh (1845), ix.245. Moore, R. W., in *Transactions of the Federated Institution of Mining Engineers*, Newcastle, vii (1893-4), 632. Lysons, D. and S., *Magna Britannia*, London (1816), iv.cxxii.

51 *Parliamentary Debates*, vii.75ff.

52 NEIMME, Johnson Coll., 12/8. Lee, C. E., in *TNS*, xxvi (1947-9), 210-11.

53 Dott, George, *Early Scottish Colliery Waggonways*, London (1947), 6.

54 Tomlinson, W. W., *The North-Eastern Railway, Its Rise and Development*, Newcastle (1914), 13.

55 Lee, C. E., in *TNS*, xxv (1945-7), 155.

56 Taylor, T. J., in *Memoirs chiefly illustrative of the History and Antiquities of Northumberland* (Archaeological Institute of Great Britain), London (1858), i.181.

57 *History and Description of Fossil Fuel, the Collieries, and the Coal Trade* (1841), 353.

58 *Reid's Handbook to Newcastle-upon-Tyne* (1863), 193.

59 Atkinson, Frank, *The Great Northern Coalfield 1700-1900*, Barnard Castle (1966), 49.

60 Smith, W. M. Hawkes, *Birmingham and South Staffordshire, or Illustrations of the History, Geology and Industrial Operations of a Mining District* (1838), 13.

61 Dendy Marshall, C. F., *A History of Railway Locomotives down to the End of the Year 1831*, London (1953), 21.

62 NEIMME, Watson Coll., 9/13.

63 *Ibid.*, Buddle Coll., 6/3/12.

64 Edgeworth, R. L., in *Nicholson's Journal of Natural Philosophy, Chemistry and the Arts*, new series, i, London (1802), 221f. Bailey, J., and Culley, G., *General View of the Agriculture of the County of Northumberland*, 2nd ed., Newcastle (1797), 10.

Chapter 16

1 *Byegones relating to Wales and the Border Counties*, Oswestry (1900), 356. Lewis, M. J. T., *How Ffestiniog got Its Railway*, Caterham (1965), 2.

2 *Railway Magazine*, July 1967, 412. Further information kindly supplied by Mr. A. B. Jeffery and Mr. M. J. Messenger.

3 Frew, I. D. O., in *Railway Magazine*, January 1960, 55ff.

4 Rolt, L. T. C., *Isambard Kingdom Brunel*, London (1957), 14.

5 Information by courtesy of Mr. J. Kenneth Major.

6 Boyd, J. I. C., in *Railway World*, March 1954, 68. *Blue Books*, 1908, xix.339 (reports of Inspectors of Mines). O.S. 1-in., 1837; 6-in., 1875. Measurements from surviving remains.

7 Crossing, William, *A Hundred Years on Dartmoor*, 5th ed., Plymouth (1902), 38; *Idem*, *Guide to Dartmoor*, part v, Plymouth (1914), 66. Information from Mr. A. B. Jeffery.

8 Dendy Marshall, 113. Innes, John, *Old Llanelly* (1902), 74. Paar, H. W., *The Great Western Railway in Dean*, Dawlish (1965), 65.

9 Smeaton, John, *A Narrative of the Building and a Description of the Construction of the Edystone Lighthouse with Stone*, London (1791), 94, 196.

10 *TNS*, iv (1923), 53. Rennie, Sir John, *An Account of the Breakwater in Plymouth Sound*, London (1848), 19, 21. Dupin, Charles, *Voyages dans la Grande Bretagne: Force Navale*, Paris (1820), pl. v.

11 Information kindly supplied by Mr. J. Kenneth Major.

12 Edgeworth, R. L., in *Blue Books*, 1810-11, vi.784-5; 1813-14, vi, part ii, 104 (reports to Commissioners for improving bogs of Ireland). Bland, F., in *Edgar Allen News*, Sheffield, October 1928, 240.

13 *TNS*, xviii (1937-8), 16. Palmer, H. R., *Description of a Railway on a New Principle*, London (1824). Dendy Marshall, 171. *Repertory of Patent Inventions*, August 1825, 129ff. *Mechanics Magazine*, August 1825, 287. *Register of Arts and Sciences*, 1824, 113ff.

Chapter 17

1 *Victoria County History*, Durham (1907), ii.322 (Professor H. Louis).

2 *Ibid.*, 324.

3 Information kindly supplied by Mr. D. B. Barton.

4 For mining methods, see Nef, J. U., *The Rise of the British Coal Industry*, London (1966), i.363ff.

5 A specimen from Allerton Colliery, Yorkshire, is in the Science Museum.

6 *Blue Books*, 1842, xvi.892 (Reports to Commissioners on Employment of Children). Barwick sled in Wakefield Museum. Information supplied by Mr. W. N. Slatcher.

7 *Victoria County History*, Cumberland (1905), ii.353f. (Moore, R. W.).

8 Heslop, R. O., in *PSAN*, 3rd series, vi (1915), 25.

9 NEIMME, Watson Coll., 8/15.

10 Ball, Thomas, in *PSAN*, 3rd series, ix (1919), 120. See also, for South Gloucestershire, *Blue Books*, 1842, xvii.40.

11 Dunn, Matthias, *A History of the Coal Viewers*, MS. in NEIMME, 15/1.

12 Ashton, T. S., and Sykes, Joseph, *The Coal Industry of the Eighteenth Century*, Manchester (1964), 20.

13 Louis, Professor H., in *Transactions of the Institution of Mining Engineers*, lxxix, London (1929-30), 292-3.

14 J. C., *The Compleat Collier*, reprint by Richardson, M. A., Newcastle (1845), 36, 39.

15 Hughes, i.253, quoting from the Cotesworth MSS.

16 Clerk, Sir John: Scottish Record Office GD/18, 2115, quoted in *CW*, 2nd series, lxv (1965), 307f.

17 Stukeley, William, *Itinerarium Curiosum*, centuria ii, London (1776), 68-9.
18 Swann, E. W., in *TNS*, xxiv (1943-5), 109.
19 Louis, Professor H., in *Transactions of the Institution of Mining Engineers*, lxxix, London (1929-30), 294.
20 Andersson, B. Q., in *Kongl. Swenska Vetenskaps Academiens Handlingar*, xxxvii, Stockholm (1776), 170.
21 Dixon, Joshua, *The Literary Life of Wm. Brownrigg, with an account of the Coal mines near Whitehaven*, London (1801), 101-2.
22 *Wills and Inventories*, Vol. ii (Publications of the Surtees Society vol. 38), Durham (1860), 37n. *Oxford English Dictionary*.
23 *Abbey of Durham Account Rolls*, Vol. ii (Publications of Surtees Society, Vol. 100), Durham (1899), 293. PRO, Durham 4/i/212. DRO, Strathmore papers. GPL, Cotesworth BT 2/32.
24 NEIMME, 18/69; 49/4/42, 164.
25 Clerk, Sir John: Scottish Record Office GD/18, 2106; quoted in *Trans. of the Architectural and Archaeological Society of Durham and Northumberland*, xi, Gateshead (1965), 427. *Kongl. Swenska Vetenskaps Academiens Handlingar*, i, Stockholm (1740), 321.
26 NEIMME, 36/38/16.
27 Ashton and Sykes, *op. cit.*, ch. 4.
28 John, A. H., *The Industrial Development of South Wales*, Cardiff (1950), 145.
29 See, for example, Pike, E. Royston, *Human Documents of the Industrial Revolution in Britain*, London (1966), 245ff.
30 *TNS*, xxiv (1943-5), 109.
31 Earl of Elgin's papers, Broomhall, Fife: inventory of stock on Pitferranc, Urquhart and Balmule collieries, 1796. NEIMME, 5/10/7.
32 LCL, MC 2.
33 NEIMME, Buddle Coll., 18/42, 24/23.
34 NEIMME, 8/15.
35 Wood, O., in *Economica*, New Series, xxi, London (1954), 58ff. Fletcher, Isaac, in *CW*, iii (1876-7), 308.
36 e.g. *Blue Books*, 1842, xvi.32. Dunn, Matthias, *A Treatise on the Winning and Working of Collieries*, Newcastle (1848), 26.
37 NEIMME, Buddle Coll., 15/204. Galloway, R. L., *Annals of Coal Mining and the Coal Trade*, London (1898), i.315.
38 NEIMME, Johnson Coll., 3/71.
39 CRO, D/Lons: Howgill and Whingill colliery accounts, 1727-32; stock account book, Whitehaven, 1775-9. Earl of Elgin's papers, Broomhall.
40 Fletcher, Isaac, in *CW*, iii (1876-7), 308.
41 *Ibid*. Wood, Nicholas, in *Transactions of the North of England Institute of Mining Engineers*, iii, Newcastle (1854-5), 246f. Dunn, Matthias, *A Treatise on the Winning and Working of Collieries*, Newcastle (1848), 26. *Idem*, *An Historical, Geological and descriptive View of the Coal Trade of the North of England*, Newcastle (1844), 41.
42 William Smith's notes: MS. in Dept. of Geology and Mineralogy, University of Oxford, quoted in *Proceedings of the Somersetshire Archaeological and Natural History Society*, xcvii, Taunton (1952), 46ff.
43 *Blue Books*, 1842, xvi.665.

44 Wilson, Thomas, *The Pitman's Pay*, Gateshead (1843), 27-31.

45 Curr, John, *The Coal Viewer, and Engine Builder's Practical Companion*, Sheffield (1797), 5-6.

46 Galloway, *op. cit.*, i.322. Ashton and Sykes, *op. cit.*, 64.

47 SCL, Arundel Castle MS. S214.

48 *Ibid.*, S215, 223.

49 Curr, *op. cit.*, 9.

50 SCL, Arundel Castle MSS. S215, 217, 218, 219, 221, 222.

51 *Ibid.*, S198.

52 *Ibid.*, S223.

53 NEIMME, Buddle Coll., 46/14.

54 SCL, Arundel Castle MS. S201.

55 Leader, R. E., *Sheffield in the Eighteenth Century*, Sheffield (1901), 341.

56 SCL, Arundel Castle MS. S214.

57 Farey, John, *A General View of the Agriculture of Derbyshire*, London (1817), iii.288.

58 LCL, MC 51, MC 203.

59 *TNS*, xi (1930-1), 127. SRO, Labouchere Coll., particulars of the expence of a corve, 1794. *The Hatchett Diary*, ed. Raistrick, A., Truro (1967), 60.

60 Ashton and Sykes, *op. cit.*, 108. Hughes, ii.203. Paynen, P. L., in *Business History*, iii, Liverpool (1961), 75ff. NRO, Blackett (Wylam) papers, colliery ledger. NEIMME, Buddle Coll., 15/7.

61 *Blue Books*, 1842, xvi.574f.

62 Ashton and Sykes, *op. cit.*, 68.

63 *The Case of Sir Humphry Mackworth, and of the Mine-Adventurers, with respect to the Irregular Proceedings of severall Justices of the Peace . . .*, London [1705], 5, 36.

64 Phillips, D. Rhys, *History of the Vale of Neath*, Swansea (1925), 263.

65 SBL, Map and Admeasurements of Broseley Hall Estate, 6, 10, 22. SRO, Madeley Vestry papers (on temporary loan).

66 Morris MSS., Swansea University Library.

67 Young, Arthur, *A Six Months' Tour through the North of England*, 2nd ed., London (1771), iii.226.

68 See also Hogrewe, J. L., *Beschreibung der in England seit 1759 angelegten und jetzt gröstentheils vollendeten schiffbaren Kanäle*, Hannover (1780), 151. Malet, Hugh, *The Canal Duke*, Dawlish (1961), 107.

69 NEIMME, Buddle Coll., 15/201; 30/94.

70 Dunn, Matthias, *An Historical, Geological and descriptive View of the Coal Trade of the North of England*, Newcastle (1844), 41.

71 CRO, D/Lons: Howgill and Whingill colliery accounts, 1727-32. SCL, Arundel Castle MSS. S192, 195a.

72 Patent Office Library, Specification 653 of 1750.

73 NEIMME, 16/1/261-3.

74 Jars, i.192, 268, 242.

75 NEIMME, Buddle Coll., 30/94.

76 *Ibid.*, Buddle Coll., 14/152.

77 Earl of Elgin's papers, Broomhall, Fife. NEIMME, Buddle Coll., 14/152, 18/42, 24/2.

78 *Ibid.*, Buddle Coll., 18/19.

79 e.g. NEIMME, Watson Coll., 8/15.

80 NEIMME, Johnson Coll., 3/13; Watson Coll., 11/7/15.

81 NRO, Blackett (Wylam) papers, colliery ledger.

82 It was not built about 1770, as often claimed. NEIMME, Johnson Coll., 2/19. *History of Northumberland*, Newcastle (1930), xiii.37. *The Picture of Newcastle upon Tyne . . .*, Newcastle (1807), 181-6.

83 Mackenzie, E., *A Historical and Descriptive Sketch of the County of Northumberland*, Newcastle (1811), i.126.

84 I am indebted to Mr. M. J. Messenger for telling me about this line and providing details.

85 BM Add. MS. 15024, f. 17.

86 Lewis, W. J., *Lead Mining in Wales*, Cardiff (1967), 330.

87 Bingley, Rev. W., *North Wales, including Its Scenery, Antiquities, Customs, etc.*, London (1804), i.230ff.

88 Effard, W., in *Gentleman's Magazine*, London (February 1769), 59. Plot, Robert, *The Natural History of Stafford-shire*, Oxford (1686), 165. Watson, R., *Chemical Essays*, Cambridge (1781), i.342-3.

89 Boyd, J. I. C., *The Isle of Man Railway*, Lingfield (1962), 200.

90 Raistrick, A., and Jennings, B., *A History of Lead Mining in the Pennines*, London (1965), 151.

91 Jars, ii.542.

92 Wallis, John, *Natural History and Antiquities of Northumberland*, London (1769), i.120.

93 *Archaeologia Aeliana*, 4th series, xxxvi (1958), 247.

94 Raistrick and Jennings, *op. cit.*, 151. Groverake rails in York Railway Museum – details by courtesy of Mr. R. J. Hunter.

95 Calvert rails measured by courtesy of Mr. Charles E. Lee. Merryfield rails in author's collection.

96 Raistrick and Jennings, *op. cit.*, 151, 201.

97 Sopwith, T., *An account of the Mining Districts of Alston Moor, Weardale and Teesdale*, Alnwick (1833), 30ff.

Chapter 18

1 For Heinitz' career, see Henderson, W. O., *The State and the Industrial Revolution in Prussia 1740-1870*, Liverpool (1958), 9f. Dennert, Herbert, *Kleine Chronik der Oberharzer Bergstädte und ihres Erzbergbaus*, Clausthal-Zellerfeld (1954), 144.

2 For William Wilkinson, see *The Engineer*, 163, London (1937), 61. Henderson, W. O., *Britain and Industrial Europe 1750-1870*, Liverpool (1954), 40-3, 56, 152. Wexler, P. J., *La Formation du vocabulaire des Chemins de Fer en France (1778-1842)*, Genève et Lille (1955), 20ff.

3 For the Redens, see Henderson, *The State and the Industrial Revolution*, 11-18. Ford, G. S., *On and Off the Campus*, Minneapolis (1938), 178-9. Schwemann, A., in *Beiträge zur Geschichte der Technik und Industrie*, xiv,

Berlin (1924), 22ff. Dennert, *op. cit.*, 144. *ZBHSPS*, i (1854), 201ff. Pounds, N. G., *The Upper Silesian Industrial Region*, Bloomington, Ind. (1958), 28ff.

4 For Stein's career, see Henderson, *Britain and Industrial Europe*, 150-2; *The State and the Industrial Revolution*, 32-3. Ford, *op. cit.*, 169. Srbik, H. Ritter von, in *Historische Zeitschrift*, cxlvi, München (1932), 482ff. *Freiherr von Stein, Briefwechsel, Denkschriften und Aufzeichnungen* (ed. Botzenhart, Erich), Berlin (1931), i, passim. On the Friedrichs, Botzenhart, *op. cit.*, 156, 161. Ford, *op. cit.*, 169. Dennert, *op. cit.*, 144.

5 Ashton, T. S., *Iron and Steel in the Industrial Revolution*, Manchester (1963), 80.

6 Francis, John, *A History of the English Railway*, London (1851), i.89-90.

7 Reden, Freiherr Friedrich Wilhelm von, *Deutsches Eisenbahn-Buch*, Berlin (1845), ii.

8 Birmingham Assay Office Library, Matthew Boulton Coll.: Raspe to Boulton, 26 January 1787.

9 Ržiha, Franz, *Lehrbuch der gesammten Tunnelbaukunst*, Berlin (1867), i.256.

10 Information kindly supplied by Oberbergrat Herbert Dennert. *Das Heimatmuseum mit seiner Bergwerksanlage in Zellerfeld*, Clausthal/Zellerfeld (no date), 23. Héron de Villefosse, A. M., *De la Richesse Minérale*, Paris (1819), ii.280-1.

11 *Zeitschrift des Harz-Vereins für Geschichte und Alterthumskunde*, l, Wernigerode (1917), 163.

12 Francis, *op. cit.*, 90.

13 Hartmann, Carl, *Der innern Gebirgswelt Schätze und Werkstätten, oder gemeinfassliche Darstellung der Bergbaukunde*, Stuttgart (1838), 226-8. Dennert, *op. cit.*, fig. 8.

14 Hartmann, *op. cit.*, 226.

15 Böbert, K. F., in *AfM*, v (1832), 294-5. Daub in *AfM* xx (1846), 622.

16 Groddeck, A. von, in *ZBHSPS*, xiv (1866), 292.

17 Botzenhart, *op. cit.*, 95.

18 [Heinitz, F. A. von], *Mémoire sur les Produits du Règne Minéral de la Monarchie Prussienne*, Berlin (1786), 34. Beck, L., *Die Geschichte des Eisens in technischer und kulturgeschichtlicher Beziehung*, Braunschweig (1898), iii.928.

19 Beck, *op. cit.*, 960.

20 Botzenhart, *op. cit.*, 147.

21 Oeynhausen, H. von, and Dechen, C. von, in *AfB*, vii (1823), 140.

22 Beck, *op. cit.*, 757.

23 Svedenstjerna, Erich Thomas, *Reise durch einen Theil von England und Schottland in den Jahren 1802 und 1803*, Marburg and Cassel (1811), 43.

24 Dietlein in *Journal für die Baukunst*, ii, Berlin (1830), 288f.

25 Combes, C., in *AdesM*, 3rd series, viii (1835), 458.

26 Oeynhausen, H. von, and Dechen, C. von, in *AfB*, vii (1823), 138-9, 148.

27 Botzenhart, *op. cit.*, 169.

28 Reden, Freiherr Friedrich Wilhelm von, *Die Eisenbahnen Deutschlands*, Band I, Theil 2.ii, supplement ii, Berlin, etc. (1846), 405ff.

29 Reden, *op. cit.*, Band I, Theil 2.ii (1844), 884f.; *ibid.*, supplement ii (1846), 406f. Köllman, Wolfgang, *Friedrich Harkort*, Düsseldorf (1964), i.95f., 101. Henderson, *The State and the Industrial Revolution*, 153.

30 Köllman, *op. cit.*, 93. *Beiträge zur Geschichte der Technik und Industrie*, x, Berlin (1920), 28.

31 Reden, *op. cit.*, I, 2.ii, 884-8; *ibid.*, supplement ii, 407f. Köllman, *op. cit.*, 99.

32 Reden, *op. cit.*, I, 2.ii, supplement ii, 405.

33 *Beiträge zur Geschichte der Technik und Industrie*, i, Berlin (1909), 27.

34 d'Aubuisson, J. F., in *JdesM*, lxxxvi (Brumaire An XII), 100.

35 Serlo, Albert, *Beitrag zur Geschichte des schlesischen Bergbaues in den letzten hundert Jahren*, Breslau and Berlin (1879), 90.

36 Serlo, *op. cit.*, 71.

37 Villefosse, *op. cit.*, ii, 535-55.

38 *Beiträge zur Geschichte der Technik und Industrie*, i, Berlin (1909), 12-15.

39 *Zeitschrift des Vereins für Geschichte Schlesiens*, lxxii, Breslau (1938), 295f.

40 Werner, E., in *Beiträge zur Geschichte der Technik und Industrie*, xii, Berlin (1922), 21ff.

41 Dendy Marshall, C. F., *A History of Railway Locomotives down to the end of the Year 1831*, London (1953), 233.

42 *The Picture Magazine*, iii, London (1894), 88: copy in *Railway Magazine*, lxxxvi, London (September 1940), 520.

43 Wexler, *op. cit.*, 20-2.

44 *Ibid.*, 22-5.

45 Ferber, J. J., *Mineralogische und metallurgische Bemerkungen in Neuchatel, Franche Comté und Bourgogne im Jahr 1788*, Berlin (1789), 56-7.

46 Lunier, *Dictionnaire des Sciences et des Arts*, Paris (1805), i.289.

47 Coste and Perdonnet, in *AdesM*, 2nd series, vi (1829), 167ff.

48 Fournel, Henri, *Etude des Gîtes houillers et métallifères du bocage Vendéen, faite en 1834 et 1835*, Paris (1836), 121-7.

49 Coste and Perdonnet in *AdesM*, 2nd series, vi (1829), 169.

50 Minard in *Journal für die Baukunst*, ix.2, Berlin (1836), 103-4.

51 Virla, M., in *Annales des Ponts et Chaussées*, Paris, 1835, 211ff.

52 Clapham, J. H., *Economic Development of France and Germany, 1815-1914*, Cambridge (1936), 58.

53 Virginsky, V. S., ВОЗНИКНОВЕНИЕ ЖЕЛЕЗНЫХ ДОРОГ В РОССИИ ДО НАЧАЛА 40х ГОДОВ XIX ВЕКА, Moscow (1949), 57, 247. Urodkov, S. A., ПЕТЕРБУРГО-МОСКОВСКАЯ ЖЕЛЕЗНАЯ ДОРОГА, Leningrad (1951), 23.

54 Campbell, R. H., *Carron Company*, Edinburgh and London (1961), 144-53.

55 Virginsky, V. S., ТВОРЦЫ НОВОЙ ТЕХНИКИ В КРЕПОСТНОЙ РОССИИ, Moscow (1957), 172, 279.

56 АРХИВ ИСТОРИИ НАУКИ И ТЕХНИКИ ЛЕНИНГРАД, iv (1954), 400-8.

57 Gumilevsky, L. I., ЖЕЛЕЗНАЯ ДОРОГА. ТРЕТЬЕ ПЕРЕРАБОТАННОЕ И ДОПОЛНЕННОЕ ИЗДАНИЕ, Moscow (1950), 43ff.

58 Thompson, Thomas, *Travels in Sweden during the Autumn of 1812*, London (1813), 305. NEIMME, Watson Coll., 31/35, 35a/23, 5/9/43. Allen, P., and Whitehouse, P. B., *Narrow Gauge Railways of Europe*, London (1961), 97-9.

59 Lindroth, Sven, *Gruvbrytning och Kopparhantering vid Stora Kopparberget intill 1800-talets början*, Uppsala (1955), i.630. Stora Kopparbergets Bergslagets arkiv, Falun: *Clas Wallmanns band ritningar ar konstar*, 1800.

60 Boehmer, 82.

61 Beyer, August, *Beytraege zur Bergbaukunde*, Dresden (1794), 271-80.

62 Hoffmann, Walter, *Bergakademie Freiberg*, Frankfurt-a-M. (1959), 23.

63 Heuchler, Eduard, *Album für Freunde des Bergbaus*, Freiberg (1851-2), pls. 4, 7, 11.

64 On Moldau and Danube in general, see Gerstner, A. von, in *Journal du Génie Civil*, iv, Paris (1829), 483ff.; v (1829), 37. *AdesM*, 2nd series, vi (1829), 168ff. *Journal für die Baukunst*, ii, Berlin (1830), 383ff. Enderes, B., in *Beiträge zur Geschichte der Technik und Industrie*, xvi, Berlin (1926), 13ff. Reden, Freiherr F. W. von, *Die Eisenbahnen Deutschlands*, Band I, Theil 2.i, Berlin, etc. (1844), 183ff.

65 Bittner, L., in *Archiv für österreichische Geschichte*, lxxxix, Wien (1901), 483-4. Jars, i.33.

66 Mayer, F. M., *Geschichte der Steirmark*, Graz (1898), 296.

67 Klein, A. A., in *Zeitschrift des historischen Vereines für Steiermark*, lv, Graz (1964), 8.

68 Gerstner, A. von, in *Journal du Génie Civil*, iv, Paris (1829), 486.

69 Huyssen in *ZBHSPS*, ii (1855), 34-5.

70 Böbert, K. F., in *AfM*, v (1832), 259-70.

71 Daub in *AfM*, xx (1846), 622.

72 Dendy Marshall, C. F., *A History of Railway Locomotives down to the End of the Year 1831*, London (1953), 234.

73 Henderson, *Britain and Industrial Europe*, 158.

74 Serlo, Albert, *Leitfaden zur Bergbaukunde*, 3rd ed., Berlin (1878), ii.15.

75 Day, J. R., *More Unusual Railways*, London (1960), 22-3.

76 Lockie, G. R., in *Railway Magazine*, lxxviii, London (February 1936), 112.

77 Dendy Marshall, C. F., *A History of Railway Locomotives down to the End of the Year 1831*, London (1953), 242.

78 Strickland, William, *Reports on Canals, Railways, Roads, and Other Subjects*, Philadelphia (1826).

79 Vowles, H. P., in *Railway Magazine*, lxxii, London (May 1933), 337.

80 Day, *op. cit.*, 21, 25.

81 Loree, L. F., in *TNS*, iv (1923), 64ff.

82 Raidabaugh, G. P., in *Journal of the Iron and Steel Institute*, xcv, London (1917), 241.

83 Earle, T., in *Journal of the Franklin Institute*, ix, Philadelphia (1830), 168. Bruton, J. F., in *TNS*, xxx (1955-7), 86.

84 *Journal of the Iron and Steel Institute*, xcv, London (1917), 241.

85 *Ibid.*, 242.

86 Becker, W. G. E., *Journal einer bergmännischen Reise durch Ungarn und Siebenbürgen*, Freyberg (1815), i.150, 136, 139.

Appendix

1 Trebra, F. W. H. von, *Erfahrungen vom Innern der Gebirge*, Dessau and Leipzig (1785), 224, 227.

2 Boehmer, 94ff., 100ff., 114ff., 82.

3 *AdesM*, viii (1823), 876.
4 Böbert, K. F., in *AfM*, v (1832), 260, 262, 259.
5 Stockicht in *Bergbaukunde*, ii, Leipzig (1790), 49ff.
6 Daub in *AfM*, xx (1846), 617ff., 622.
7 Gervoy, M., in *AdesM*, 3rd series, x (1836), 447.
8 Ržiha, F., *Lehrbuch der gesammten Tunnelbaukunst*, Berlin (1867), i.532, 534.
9 Héron de Villefosse, A. M., *De la Richesse Minérale*, Paris (1819), ii.347.
10 Becker, W. G. E., *Journal einer bergmännischen Reise durch Ungarn und Siebenbürgen*, i, Freiberg (1815), 120; ii (1816), 192.
11 Carnall, R. von, in *ZBHSPS*, i (1853), 31.
12 Beyer, August, *Beytraege zur Bergbaukunde*, Dresden (1794), 282, 286.

Index

Poetry z songs